Textile-led Design for the Active Ageing Population

The Textile Institute and Woodhead Publishing

The Textile Institute is a unique organisation in textiles, clothing and footwear. Incorporated in England by a Royal Charter granted in 1925, the Institute has individual and corporate members in over 90 countries. The aim of the Institute is to facilitate learning, recognise achievement, reward excellence and disseminate information within the global textiles, clothing and footwear industries.

Historically, The Textile Institute has published books of interest to its members and the textile industry. To maintain this policy, the Institute has entered into partnership with Woodhead Publishing Limited to ensure that Institute members and the textile industry continue to have access to high calibre titles on textile science and technology.

Most Woodhead titles on textiles are now published in collaboration with The Textile Institute. Through this arrangement, the Institute provides an Editorial Board which advises Woodhead on appropriate titles for future publication and suggests possible editors and authors for these books. Each book published under this arrangement carries the Institute's logo.

Woodhead books published in collaboration with The Textile Institute are offered to Textile Institute members at a substantial discount. These books, together with those published by The Textile Institute that are still in print, are offered on the Elsevier website at: http://store.elsevier.com/. Textile Institute books still in print are also available directly from the Institute's website at: www.textileinstitutebooks.com.

A list of Woodhead books on textile science and technology, most of which have been published in collaboration with The Textile Institute, can be found towards the end of the contents pages.

Woodhead Publishing Series in Textiles:
Number 142

Textile-led Design for the Active Ageing Population

Edited by

Jane McCann and David Bryson

The Textile Institute

AMSTERDAM • BOSTON • CAMBRIDGE • HEIDELBERG
LONDON • NEW YORK • OXFORD • PARIS • SAN DIEGO
SAN FRANCISCO • SINGAPORE • SYDNEY • TOKYO
Woodhead Publishing is an imprint of Elsevier

ELSEVIER

WP WOODHEAD PUBLISHING

Published by Woodhead Publishing Limited in association with The Textile Institute
Woodhead Publishing is an imprint of Elsevier
80 High Street, Sawston, Cambridge, CB22 3HJ, UK
225 Wyman Street, Waltham, MA 02451, USA
Langford Lane, Kidlington, OX5 1GB, UK

British Library Cataloguing-in-Publication Data
A catalogue record for this book is available from the British Library

Library of Congress Control Number: 2014940723

ISBN 978-0-85709-538-1 (print)
ISBN 978-0-85709-878-8 (online)

For information on all Woodhead Publishing publications visit our website at http://store.elsevier.com/

Typeset by TNQ Books and Journals
www.tnq.co.in
Printed and bound in the United Kingdom

Contents

List of contributors

S. Benton University of Westminster, London, UK

P. Borcherding University of Cincinnati, Cincinnati, OH, USA

J. Bougourd University of Wales, Newport, UK

D. Bryson University of Derby, Derby, UK

J. Bubonia Texas Christian University, Fort Worth, TX, USA

W. Burns University of Ulster, Londonderry, UK

J. Cassim Royal College of Art, London, UK

M. Fuller University of Leeds, Leeds, UK

J. Greengrass Greengrass Design Studio Ltd, Glossop, UK

V. Haffenden University of Brighton, Brighton, UK

M.J. Head Inov-8 Ltd, Staveley, UK

R. Hibbert Line Consultants, London, UK

S. Hinder RaFT Research, Clitheroe, UK

C.J. Hussey Consultant, UK

C. Johnston Royal College of Art, London, UK

D.C. Jones Fibretronic Limited, Kowloon, Hong Kong, China

V.L. Knowles Pebble International, Manchester, UK

C.M. Lewis University of South Wales, Newport, UK

W. Lu China Women's University, Beijing, China

B.R.M. Manning University of Westminster, London, UK

J. McCann University of South Wales, Newport, UK

C.S. Porter Loughborough University, Loughborough, UK

M.C. Price Wearable Futures Design Group, Bristol, UK

M. Sinfield Older People's Reference Group (OPRG), Sheffield, UK

J. Smith University of Brighton, Brighton, UK

K. Stevens University of Innsbruck, Dornbirn, Austria

D. Taylor University of South Wales, Newport, UK

M.W. Timmins Consultant, UK

J. Twigg University of Kent, Canterbury, UK

T. Williamson University of Salford, Salford, UK

Woodhead Publishing Series in Textiles

1 **Watson's textile design and colour Seventh edition**
Edited by Z. Grosicki
2 **Watson's advanced textile design**
Edited by Z. Grosicki
3 **Weaving Second edition**
P. R. Lord and M. H. Mohamed
4 **Handbook of textile fibres Volume 1: Natural fibres**
J. Gordon Cook
5 **Handbook of textile fibres Volume 2: Man-made fibres**
J. Gordon Cook
6 **Recycling textile and plastic waste**
Edited by A. R. Horrocks
7 **New fibers Second edition**
T. Hongu and G. O. Phillips
8 **Atlas of fibre fracture and damage to textiles Second edition**
J. W. S. Hearle, B. Lomas and W. D. Cooke
9 **Ecotextile '98**
Edited by A. R. Horrocks
10 **Physical testing of textiles**
B. P. Saville
11 **Geometric symmetry in patterns and tilings**
C. E. Horne
12 **Handbook of technical textiles**
Edited by A. R. Horrocks and S. C. Anand
13 **Textiles in automotive engineering**
W. Fung and J. M. Hardcastle
14 **Handbook of textile design**
J. Wilson
15 **High-performance fibres**
Edited by J. W. S. Hearle
16 **Knitting technology Third edition**
D. J. Spencer
17 **Medical textiles**
Edited by S. C. Anand
18 **Regenerated cellulose fibres**
Edited by C. Woodings
19 **Silk, mohair, cashmere and other luxury fibres**
Edited by R. R. Franck
20 **Smart fibres, fabrics and clothing**
Edited by X. M. Tao
21 **Yarn texturing technology**
J. W. S. Hearle, L. Hollick and D. K. Wilson
22 **Encyclopedia of textile finishing**
H-K. Rouette

23 **Coated and laminated textiles**
W. Fung
24 **Fancy yarns**
R. H. Gong and R. M. Wright
25 **Wool: Science and technology**
Edited by W. S. Simpson and G. Crawshaw
26 **Dictionary of textile finishing**
H-K. Rouette
27 **Environmental impact of textiles**
K. Slater
28 **Handbook of yarn production**
P. R. Lord
29 **Textile processing with enzymes**
Edited by A. Cavaco-Paulo and G. Gübitz
30 **The China and Hong Kong denim industry**
Y. Li, L. Yao and K. W. Yeung
31 **The World Trade Organization and international denim trading**
Y. Li, Y. Shen, L. Yao and E. Newton
32 **Chemical finishing of textiles**
W. D. Schindler and P. J. Hauser
33 **Clothing appearance and fit**
J. Fan, W. Yu and L. Hunter
34 **Handbook of fibre rope technology**
H. A. McKenna, J. W. S. Hearle and N. O'Hear
35 **Structure and mechanics of woven fabrics**
J. Hu
36 **Synthetic fibres: Nylon, polyester, acrylic, polyolefin**
Edited by J. E. McIntyre
37 **Woollen and worsted woven fabric design**
E. G. Gilligan
38 **Analytical electrochemistry in textiles**
P. Westbroek, G. Priniotakis and P. Kiekens
39 **Bast and other plant fibres**
R. R. Franck
40 **Chemical testing of textiles**
Edited by Q. Fan
41 **Design and manufacture of textile composites**
Edited by A. C. Long
42 **Effect of mechanical and physical properties on fabric hand**
Edited by H. M. Behery
43 **New millennium fibers**
T. Hongu, M. Takigami and G. O. Phillips
44 **Textiles for protection**
Edited by R. A. Scott
45 **Textiles in sport**
Edited by R. Shishoo
46 **Wearable electronics and photonics**
Edited by X. M. Tao
47 **Biodegradable and sustainable fibres**
Edited by R. S. Blackburn
48 **Medical textiles and biomaterials for healthcare**
Edited by S. C. Anand, M. Miraftab, S. Rajendran and J. F. Kennedy
49 **Total colour management in textiles**
Edited by J. Xin

50 **Recycling in textiles**
Edited by Y. Wang
51 **Clothing biosensory engineering**
Y. Li and A. S. W. Wong
52 **Biomechanical engineering of textiles and clothing**
Edited by Y. Li and D. X-Q. Dai
53 **Digital printing of textiles**
Edited by H. Ujiie
54 **Intelligent textiles and clothing**
Edited by H. R. Mattila
55 **Innovation and technology of women's intimate apparel**
W. Yu, J. Fan, S. C. Harlock and S. P. Ng
56 **Thermal and moisture transport in fibrous materials**
Edited by N. Pan and P. Gibson
57 **Geosynthetics in civil engineering**
Edited by R. W. Sarsby
58 **Handbook of nonwovens**
Edited by S. Russell
59 **Cotton: Science and technology**
Edited by S. Gordon and Y-L. Hsieh
60 **Ecotextiles**
Edited by M. Miraftab and A. R. Horrocks
61 **Composite forming technologies**
Edited by A. C. Long
62 **Plasma technology for textiles**
Edited by R. Shishoo
63 **Smart textiles for medicine and healthcare**
Edited by L. Van Langenhove
64 **Sizing in clothing**
Edited by S. Ashdown
65 **Shape memory polymers and textiles**
J. Hu
66 **Environmental aspects of textile dyeing**
Edited by R. Christie
67 **Nanofibers and nanotechnology in textiles**
Edited by P. Brown and K. Stevens
68 **Physical properties of textile fibres Fourth edition**
W. E. Morton and J. W. S. Hearle
69 **Advances in apparel production**
Edited by C. Fairhurst
70 **Advances in fire retardant materials**
Edited by A. R. Horrocks and D. Price
71 **Polyesters and polyamides**
Edited by B. L. Deopura, R. Alagirusamy, M. Joshi and B. S. Gupta
72 **Advances in wool technology**
Edited by N. A. G. Johnson and I. Russell
73 **Military textiles**
Edited by E. Wilusz
74 **3D fibrous assemblies: Properties, applications and modelling of three-dimensional textile structures**
J. Hu
75 **Medical and healthcare textiles**
Edited by S. C. Anand, J. F. Kennedy, M. Miraftab and S. Rajendran
76 **Fabric testing**
Edited by J. Hu

77 **Biologically inspired textiles**
Edited by A. Abbott and M. Ellison
78 **Friction in textile materials**
Edited by B. S. Gupta
79 **Textile advances in the automotive industry**
Edited by R. Shishoo
80 **Structure and mechanics of textile fibre assemblies**
Edited by P. Schwartz
81 **Engineering textiles: Integrating the design and manufacture of textile products**
Edited by Y. E. El-Mogahzy
82 **Polyolefin fibres: Industrial and medical applications**
Edited by S. C. O. Ugbolue
83 **Smart clothes and wearable technology**
Edited by J. McCann and D. Bryson
84 **Identification of textile fibres**
Edited by M. Houck
85 **Advanced textiles for wound care**
Edited by S. Rajendran
86 **Fatigue failure of textile fibres**
Edited by M. Miraftab
87 **Advances in carpet technology**
Edited by K. Goswami
88 **Handbook of textile fibre structure Volume 1 and Volume 2**
Edited by S. J. Eichhorn, J. W. S. Hearle, M. Jaffe and T. Kikutani
89 **Advances in knitting technology**
Edited by K-F. Au
90 **Smart textile coatings and laminates**
Edited by W. C. Smith
91 **Handbook of tensile properties of textile and technical fibres**
Edited by A. R. Bunsell
92 **Interior textiles: Design and developments**
Edited by T. Rowe
93 **Textiles for cold weather apparel**
Edited by J. T. Williams
94 **Modelling and predicting textile behaviour**
Edited by X. Chen
95 **Textiles, polymers and composites for buildings**
Edited by G. Pohl
96 **Engineering apparel fabrics and garments**
J. Fan and L. Hunter
97 **Surface modification of textiles**
Edited by Q. Wei
98 **Sustainable textiles**
Edited by R. S. Blackburn
99 **Advances in yarn spinning technology**
Edited by C. A. Lawrence
100 **Handbook of medical textiles**
Edited by V. T. Bartels
101 **Technical textile yarns**
Edited by R. Alagirusamy and A. Das
102 **Applications of nonwovens in technical textiles**
Edited by R. A. Chapman
103 **Colour measurement: Principles, advances and industrial applications**
Edited by M. L. Gulrajani

104 **Fibrous and composite materials for civil engineering applications**
Edited by R. Fangueiro
105 **New product development in textiles: Innovation and production**
Edited by L.Horne
106 **Improving comfort in clothing**
Edited by G. Song
107 **Advances in textile biotechnology**
Edited by V. A. Nierstrasz and A. Cavaco-Paulo
108 **Textiles for hygiene and infection control**
Edited by B. McCarthy
109 **Nanofunctional textiles**
Edited by Y. Li
110 **Joining textiles: Principles and applications**
Edited by I. Jones and G. Stylios
111 **Soft computing in textile engineering**
Edited by A. Majumdar
112 **Textile design**
Edited by A. Briggs-Goode and K. Townsend
113 **Biotextiles as medical implants**
Edited by M. W. King, B. S. Gupta and R. Guidoin
114 **Textile thermal bioengineering**
Edited by Y. Li
115 **Woven textile structure**
B. K. Behera and P. K. Hari
116 **Handbook of textile and industrial dyeing. Volume 1: Principles, processes and types of dyes**
Edited by M. Clark
117 **Handbook of textile and industrial dyeing. Volume 2: Applications of dyes**
Edited by M. Clark
118 **Handbook of natural fibres. Volume 1: Types, properties and factors affecting breeding and cultivation**
Edited by R. Kozłowski
119 **Handbook of natural fibres. Volume 2: Processing and applications**
Edited by R. Kozłowski
120 **Functional textiles for improved performance, protection and health**
Edited by N. Pan and G. Sun
121 **Computer technology for textiles and apparel**
Edited by J. Hu
122 **Advances in military textiles and personal equipment**
Edited by E. Sparks
123 **Specialist yarn and fabric structures**
Edited by R. H. Gong
124 **Handbook of sustainable textile production**
M. I. Tobler-Rohr
125 **Woven textiles: Principles, developments and applications**
Edited by K. Gandhi
126 **Textiles and fashion: Materials design and technology**
Edited by R. Sinclair
127 **Industrial cutting of textile materials**
I. Viļumsone-Nemes
128 **Colour design: Theories and applications**
Edited by J. Best
129 **False twist textured yarns**
C. Atkinson

130 **Modelling, simulation and control of the dyeing process**
R. Shamey and X. Zhao
131 **Process control in textile manufacturing**
Edited by A. Majumdar, A. Das, R. Alagirusamy and V. K. Kothari
132 **Understanding and improving the durability of textiles**
Edited by P. A. Annis
133 **Smart textiles for protection**
Edited by R. A. Chapman
134 **Functional nanofibers and applications**
Edited by Q. Wei
135 **The global textile and clothing industry:** *Technological advances and future challenges*
Edited by R. Shishoo
136 **Simulation in textile technology: Theory and applications**
Edited by D. Veit
137 **Pattern cutting for clothing using CAD: How to use Lectra Modaris pattern cutting software**
M. Stott
138 **Advances in the dyeing and finishing of technical textiles**
M. L. Gulrajani
139 **Multidisciplinary know-how for smart textiles developers**
Edited by T. Kirstein
140 **Handbook of fire resistant textiles**
Edited by F. Selcen Kilinc
141 **Handbook of footwear design and manufacture**
Edited by A. Luximon
142 **Textile-led design for the active ageing population**
Edited by J. McCann and D. Bryson
143 **Optimizing decision making in the apparel supply chain using artificial intelligence (AI): From production to retail**
Edited by W. K. Wong, Z. X. Guo and S. Y. S. Leung
144 **Mechanisms of flat weaving technology**
V. V. Choogin, P. Bandara and E. V. Chepelyuk
145 **Innovative jacquard textile design using digital technologies**
F. Ng and J. Zhou
146 **Advances in shape memory polymers**
J. Hu
147 **Design of clothing manufacturing processes: A systematic approach to planning, scheduling and control**
J. Gersak
148 **Anthropometry, apparel sizing and design**
D. Gupta and N. Zakaria
149 **Silk: Processing, properties and applications**
Edited by K. Murugesh Babu
150 **Advances in filament yarn spinning of textiles and polymers**
Edited by D. Zhang
151 **Designing apparel for consumers: The impact of body shape and size**
Edited by M.-E. Faust and S. Carrier
152 **Fashion supply chain management using radio frequency identification (RFID) technologies**
Edited by W. K. Wong and Z. X. Guo
153 **High performance textiles and their applications**
Edited by C. A. Lawrence
154 **Protective clothing: Managing thermal stress**
Edited by F. Wang and C. Gao
155 **Composite nonwoven materials**
Edited by D. Das and B. Pourdeyhimi

156 **Functional finishes for textiles: Improving comfort, performance and protection**
Edited by R. Paul
157 **Assessing the environmental impact of textiles and the clothing supply chain**
S. S. Muthu
158 **Braiding technology for textiles**
Y. Kyosev
159 **Principles of colour appearance and measurement**
A. K. R. Choudhury

Part One

Understanding the active ageing population

Technological culture and the active ageing: a lifetime of technological advances

1

D. Bryson
University of Derby, Derby, UK

1.1 Introduction

In an age of rapid textile and technological innovation, we are now at a point when electronics are becoming embedded into textiles and garments, not just as novelties but tried and tested and ready for mass production. It is important, therefore, for designers to understand the challenges and limitations, in terms of the usability of the technology, from the perspective of the rapidly growing ageing community (50+).

This chapter aims to provide an overview of the technological changes that have taken place during the lifetime of the active ageing from the perspective of a personal journey highlighting the changes that have taken place but also looking at the effect of these changes on working practices and living patterns at the time, now and in the future. My experiences, in the workplace and in education, have led me to contribute to cross-disciplinary research and education through enhancing design students' understanding of human anatomy and physiology, technology, the development of interactive learning materials, program development, and the co-editor's earlier book *Smart Clothing and Wearable Technology*.

There have been many authors who have looked at the changes and identified the types of people inhabiting this new world as (1) digital natives, those who have grown up only knowing this new digital world of the past 20 years; and (2) digital immigrants (Prensky, 2001), those who have lived through the changes of the past 50 or so years of rapid change. These changes are not just about availability of technological devices but how working practices have changed and the impact of these on how we go about our daily lives.

As a personal journey, the transitions I have gone through in my career will be different from others growing up at the same time. Hopefully the reflection and thinking about the impact of the changes will be similar to others even if readers who are digital natives find many of the ways we used to work incomprehensible or, as students have expressed it to me, "Just plain weird."

1.2 Learning and teaching

What we experience in school, college, university, and after in work greatly influences personal directions for both good and ill. The transition for me came full circle as I

Textile-led Design for the Active Ageing Population. http://dx.doi.org/10.1016/B978-0-85709-538-1.00001-8

moved from work as a medical photographer back into education, so learning and teaching and what has occurred in terms of changes is equally part of my working life as well as my formative years.

1.2.1 Transitions in technologies

The transition in technologies is most noticeable in retrospect from using logarithm tables and a slide rule at school for any calculations. We only got to use a manual calculating machine for oneday, so the main choice was to use the slide rule (Slipstick, Figure 1.1). King's School Worcester did build an extension onto the gymnasium to house a computer donated by Metal Box Co. Ltd while I was doing my "A" Levels, with the whole of the computer needing a massive air-conditioned space.

At university, one student had a new calculator in the first year, 1973. In 1976 for an anthropometry project, measuring the head shapes and stature of medical students, the initial calculations were done using Fortran. This calculating machine had a limit on the number of steps, and first you had to load all the steps and then the data you wanted to analyze for regression and correlation. Similarly, the first programmable calculator I bought in 1979 (Figure 1.1) to make calculations easier still required programing before inputting data (Toth, 2012).

A requirement for university was definitely a good set of colored crayons, alongside pens and pencils, whether initially for the geological diagrams and then for embryology, with stages in growth and development illustrated by Professor David Sinclair once I had changed to anatomy. I copied from blackboard or greenboard to paper, and often rewrote so I could read what I had written.

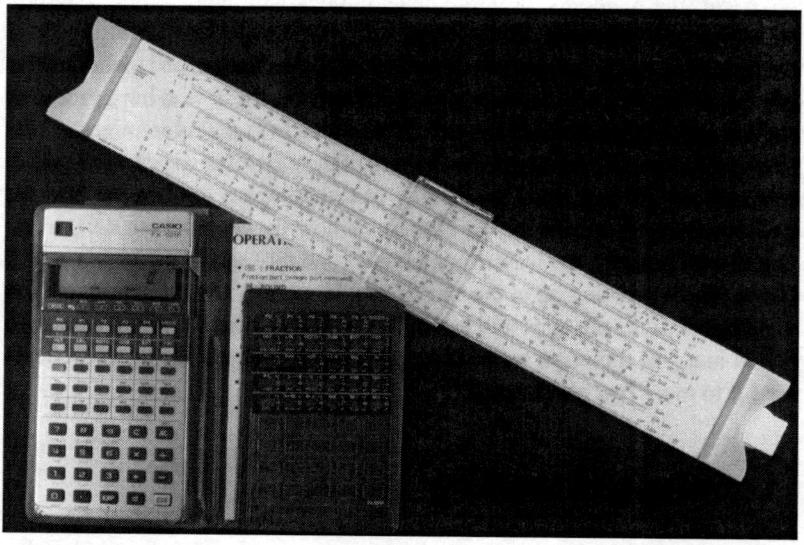

Figure 1.1 Top at angle, Thornton Slide Ruler. Bottom, Casio Programing Calculator FX-501P.
Photographer: David Bryson.

There were a range of audiovisual technologies, even if they had been around for a long time: the overhead projector, epidiascope projecting diagrams and radiographs from anatomical texts onto the screen in the anatomy dissection lecture theater with its steeply raked seats at Marischal College, and sets of stereo photographs of dissections that could be viewed with a 3D Stereo Viewmaster (Whiting, 2009). For teaching practice, it was more using the overhead projector and acetates. For creating handouts, it was a Banda machine rather than a photocopier, with the smell of alcohol coming off the sheets from either handwritten or occasionally typed originals for learning activities and handouts.

Training as a clinical photographer at the University Hospital of Wales, Cardiff it was very much conventional technologies for the student trainees with the PET computer and other devices restricted to the Head of Department, Professor Ralph Marshall. Indeed while working as a medical and medicolegal photographer, a computer was very far from a necessity, as manual equipment still worked and was more affordable than computers (see Section 1.3 for changes in Photographic practice).

Before desktop computers, there was a distinct development in typewriters, as technology moved from manual to electric, with work for copying in photography being typed with an IBM golf-ball typewriter with Baryta paper and Helvetica font, and then moving on to electronic computers, or really just an electric computer with a small amount of memory and a one line digital display. There was then a transition to typewriters as word processors before computers with printers attached took over.

1.2.2 The impact of inclusive design and technology on learning and teaching

"Scientists have discovered that digital natives' lifelong exposure to technology means that their brains are developing differently. Educators are beginning to understand that reaching them requires a new style of education that accommodates the ways in which these students learn. Blended and hybrid learning programs offer powerful benefits in reaching and challenging digital natives" (Rudi, 2011).

Initially back to teaching, first the part-time HNC in Medical and Scientific Photography at Berkshire College of Art & Design and then the BSc(Hons) Biological Imaging at the University of Derby, traditional technologies were used, with an overhead projector with acetates. A computer soon became a key tool to support communication as well as a tool for developing learning materials that with time could overcome one of the barriers to learning—that of needing to have the lecturer in the classroom to be able to listen to and view what was to be learnt.

The aim of inclusive design for learning has become more feasible with technology allowing multiple ways to access the same information. For example, PowerPoint slides as a ppt file or slides saved as a pdf for viewing before, during, or after a lecture or session; videos of lectures as screencasts that can also be listened to as podcasts; handouts as notes, and a series of lecture notes specific to modules brought together into a pdf file or even a course book or booklet.

The difficulty is that all of these are possible, but it is unlikely given time constraints that every possible alternative would be available, so a subset of these is more likely, e.g., screencasts, slide handouts, and links to further resources including quizzes and learning tools or interactive presentations, screencasts, and collated lecture notes. The flexibility of being able to make different materials can leave the lecturer unsure which is the best combination, and many developments can only take place over time as materials are built on and developed further year after year.

The onus on digital immigrants to cope with the requirements of digital natives is for many a step too far and for others a challenge to use the new technology to enhance learning.

Developing active learning materials requires lecturers to use models for learning that are more intensive and absorbing for the student. This was always the best way to learn, but the accessibility to information to back-up and support this style of learning was not as accessible as it is now. Flipping classes where lectures and materials are online and classes are either for practical learning, debates, cases, problem-based learning, or discussions, is going back to experiential learning, and any questions can be asked of the lecturer supervising or looked up online while working in the class.

The move from walking the shelves, to hardwired, to mobile shows the change in accessibility of research material and resources that has reinvigorated experiential learning. Table 1.1 shows the changes in finding a reference to support learning before and after the Internet and e-journals.

The change from one per desk to one or two per pocket or bag, depending on the size of laptop, mobile, or tablet, has freed the Internet from the length of a cable to anywhere with a WiFi, 3G, or 4G signal.

1.2.3 Social networks and communication modalities

The rise of social media from early explorations like Computer Mediated Communications in Higher Education (Steeples, Unsworth, Bryson, et al., 1996) to the massive

Table 1.1 Changes in finding references and other resources

Library-based exploration	Online exploration
Walk to the library Locate *Index Medicus* Search for your subject Make a list of references that might be suitable Check to see which journals are held in stock Write reference citation details on a card Walk to where journals are kept Bring back journals for photocopying Photocopy, trim, and staple **1 h minimum**	Open laptop, mobile phone, or tablet Username/password to access Internet via WiFi if needed Open web browser Go to Google Scholar or University's search pages Search for papers Click on paper; if have access to journal, open in web browser, and read **1–4 min depending on complexity**

growth in modern social networks has changed the way we communicate with people in the same organization and other organizations. The change has been from letters, derogatively termed snail mail, to e-mail, to Skype and video built into mobile phones, to business-orientated social media like LinkedIn, to researchers having their own social networks (e.g., academia.edu and ResearchGate) allowing for easy access to papers and Q&A from those who have written papers or are engaged in research.

This has allowed for better subject networks, which is especially useful for smaller disciplines and specializations where there are insufficient numbers to allow collaboration at a local level. The possibility of sending out a request and getting support from anywhere in the world has become a regular occurrence rather than a long-term goal.

This development—from hard-wired and slow to WiFi and fast broadband access—has freed students and lecturers alike. Comments like these from Steeple's 1996 paper about Computer Mediated Communication (CMC)—"Disciplined patterns of networked activity need to be encouraged, particularly to counterbalance the periods of the day when use of the Internet is busy. Queuing to gain access to PCs can be demotivating too"—are no longer seen in the UK.

The development of virtual learning environments (VLEs) followed by social media tools has led to a split in how these tools are seen with a structured environment of BlackBoard or Moodle for storing and accessing learning materials but social networks like Facebook for peer discussions.

The separate uses of the two types of tools is seen in the paper by Deng (Deng & Tavares, 2013): "The students' perceived benefits of Facebook over Moodle can be understood by the immediacy of the responses they got since 'everyone is there' and so 'you ask a question, then you get a response quickly.' This gave the students a strong sense of community within the group, and acted as an impetus for them to visit the Facebook Group, post and exchange more comments. As one student exclaimed, 'I know there are some listeners, they really enjoy my sharing. I find it very satisfying, then I will keep doing it.'"

1.3 Photography, audiovisual technologies, and e-learning

The relationship between change and technology in retrospect appears in many instances to have been a smooth transition, for example the selling of music from long-playing record to tape cassette to CD to online downloads/cloud storage. However, many technologies have not had a clear path, as the final use of a new technology is often reliant on not one but several innovations and the wide availability of these to a mass market.

The current use of computers in teaching combines many means of access: the lecturer either loads a presentation onto a fixed computer via a flash drive or plugs in a laptop and the presentation can be viewed on screen; students can access the Internet and download a version of the lecture notes, or link to supporting resources. Multiple technological changes have been needed to get to this position.

For most of the twentieth century the mainstay of teaching was the epidiascope, overhead projector, or slide projector. The production of lecture slides was a labour-intensive process that required lecturers to submit work usually two to three weeks in advance of the slides being required for a lecture (Table 1.2).

The key technological changes include the transition from electric to electronic typewriters to computers and computer monitors, the move from silver photographic technology to digital, and the development of data projectors that can be used in the classroom (Hackbarth, 1996). This is indicative of the multiple changes that have taken place that together affect how we now work, teach, and enjoy our leisure time.

These transitions have meant a big change in how presentations are developed; from having a high level of technical support not just for production but also design of presentations, to a situation now where everyone is expected to be able to create their own presentations. Whether or not they are actually taught how to present or how to create a presentation for a lecture or talk is a separate debate.

There is certainly no one who would like to go back to the long, laborious process of slide creation; however, ardent photographers are for retaining black-and-white processing and printing. The difficulty of this transition is an assumption that everyone will know how to structure and design a presentation (Johns, 1995; Simmonds, 1993), and that the time that would have been taken to develop and process the slides is now used to support the development of the lecture. Unfortunately, this is not the case, as many lectures are poorly designed and structured with little consideration for the viewer or the appropriateness of the final projection for clearly viewing the material. This was often the case when slides were created for presenters (Essex-Lopresti, 1998), and the situation has become worse, rather than better, with time.

Standards for audiovisual projection laid down over many years by Kodak are now ignored as data projectors are installed. The commonest fault is the size of screen being too small for the size of the room, so that the text is too small to read unless the problem of small screen size is taken into account by the creator of the presentation,

Table 1.2 Changes in lecture production processes

Traditional	Interim	Current
Typed script presented Retyped in clear font Photographed using lith film Film processed and any white spots removed with opaque Film and Diazo material exposed to ultraviolet light Diazo processed with ammonia Film cut and mounted into slides and then labeled for projection Slides organized with other lecture slides and presented	PowerPoint delivered to department Presentation slides converted into color transparencies using a slidemaker Film cut and mounted into slides and then labeled for projection Slides organized with other lecture slides and presented	PowerPoint projected directly onto screen using a data projector

i.e., the lecturer or teacher. The other issue has been the increase in ambient light, as blackout blinds have given way to blinds that let too much light into a room, so the room is suitable for text slides but not for video or photographs.

1.4 Implications for the active ageing

The examples above in my personal experience are not the only transitions that have taken place, but they give an idea of the changes. We also need to look more closely at their effects.

1.4.1 Successive changes

The change in technology has been cumulative, as each successive change has added onto previous changes, and alongside these have been trends in technological improvement, e.g., from cathode ray tube (CRT) televisions to liquid crystal displays (LCD) to the latest versions where televisions have become less and less bulky and cheaper to have larger and larger screens. Designers used to have to assume that a web page would be viewed on a monitor that could view 640 × 480 pixels.

At the same time as screens have been developed to be lighter and larger so they have also been designed to be smaller and lighter such that we have tablets with greater power than computers we used to have on our desktop. Portable computing is a reality, from mobile phones to tablets, and how we use them and support everyone to use and access their capabilities is key to preventing the continuing development of a digital divide in the older age group.

Successive developments in data storage now mean that data storage is very cheap compared to even just 10 years ago. In 1997, I had to buy an extra hard drive; then, a large 6 Gb drive costed £699; now I can buy 1 Tb for about £50. The need for more and more storage has increased, but the cost has rapidly decreased and is continuing to decrease.

1.4.2 Digital technology and the active ageing

The concepts around whether there is or is not a "digital divide" have further developed the nuances surrounding the concepts of digital native and digital immigrant, as in Kaufman's (2011) separation into digital natives, immigrants, aliens, and integrators:

> *Digital Natives are called by many names – Millennials and Gen Y. They are approximately ages 10–29 years and born into the digital age. The internet is their default and comfort level in terms of playing games, doing homework, searching for events/information/products, and sending messages.*
>
> *Digital Immigrants are approximately ages 30–60 years and are the early and late adopters of the Web 2.0 technologies. Most immigrants were born "before the existence of digital technology and adopted it to some extent later in their*

life." They were brought up with a variety of computer technologies, used them in different contexts and have varied levels of experience with social media.

Digital Aliens think strategic, need to experience digital value. They are approximately ages 45–70 years and are the late adopters and laggards of the Web 2.0 technologies.

Digital Integrators live digital, innovate strategic, need to integrate. They are approximately ages 25–60 years and are the innovators and early adopters of the Web 2.0 technologies. Some were "Digital Natives," while others were "Digital Immigrants." They live the Digital World—experimenting, testing, sharing and implementing new social technologies and evolving their digital knowledge and presence.

These are very much in the way that Kaufman describes them orientated toward marketing, but equally the adoption or non-adoption of smart technologies is influenced by how technology is perceived.

Research papers have reported that "although the use of computers and the Internet among older adults is increasing there is still an age-based digital divide… in 2002 about 34% of people age 65 + accessed the Internet compared to nearly 100% of 16–18 year olds" (Czaja & Lee, 2007; UCLA Center for Communication Policy, 2003).

The issue highlighted above is that there are not two groups of people defined as users and non-users (Gardner, Netherland, & Kamber, 2012) but a large number of variants. This is seen in the breadth of ages that the terms "elderly," "older age group," or "active ageing" encompass. We should not clump the active ageing into one group from the ages of 50–100, when we talk about teenagers or twenty-somethings covering only 7 years and 9 years, respectively.

1.5 Conclusions

As we move forward and use more technologies, it is important to learn from the past. It is important to understand that it is not one technology but multiple technological changes that have an impact, and that these changes are not sequential but more often occur at the same time.

The age of one computer per desktop, that at one time was seen as fanciful, is now one per lap or almost one per pocket as another technology, Wi-Fi, frees us from a single desk with an Ethernet cable. This freedom depends not just on Wi-Fi but miniaturization of components from hard drives to screens, such that the power of a full computer is accessible in a mobile phone or a tablet like the iPad.

It is this multiplicity of changes that are a challenge as they have come and continue to come in successive waves, even if many like stereo television rely on earlier technologies. What was the realm of the expert becomes available to everyone not visiting a village hall or lecture theater to see amazing stereo photographs from a slide projector.

Designers of the future need to be aware of the impact of technology on society as a whole and individuals. Innovations and new ways of doing things have a positive impact, but developments need to be for the good of all, such that the active ageing are included in the future of textile and technology-led innovations and are not left behind.

1.6 Sources of further information and advice

For the active ageing I thoroughly recommend a number of books that remind us of how technology was developing before the baby boom. One example is *A computer called Leo*, about an early Computer used by Lyon's Coffee shops, by Georgina Ferry (2003). Looking at much older changes, *The Book Nobody Read*, by Owen Gingerich (2004) , considers how many actually read Copernicus's work including making annotations in its margins; and *Leonardo*, by Michael White (2000), looks at what Leonardo Da Vinci was actually doing as a scientist.

Books examining the impact, opportunities, and challenges of digital technologies on the active ageing, the world, and how we engage communities in these changes include *Digital Habitats: Stewarding Technology for Communities*, by Etienne Wenger et al. (Wenger, White, & Smith, 2009); and *Information and Communication Technologies for Active Ageing: Opportunities and Challenges for the European Union*, by Cabrera and Malanowski (2009).

References

Cabrera, M., & Malanowski, N. (2009). *Information and communication technologies for active ageing*. Amsterdam, Netherlands: IOS Press.

Czaja, S. J., & Lee, C. C. (2007). The impact of aging on access to technology. *Universal Access in the Information Society*, *5*(4), 341–349.

Deng, L., Tavares, N.J., (2013). From Moodle to Facebook: Exploring students' motivation and experiences in online communities. *Computers & Education*, *68*, 167–176. http://dx-.doi.org/10.1016/j.compedu.2013.04.028. Accessed 31.07.13.

Essex-Lopresti, M. (1998). Slide presentations. *Journal of Audiovisual Media in Medicine*, *21*(3), 104–105. http://informahealthcare.com/doi/pdf/10.3109/17453059809065495. Accessed 06.08.13.

Ferry, G. (2003). *A computer called LEO*. London: Fourth Estate.

Gardner, P., Netherland, J., & Kamber, T. (2012). Getting turned on: Using ICT training to promote active ageing in New York City. *The Journal of Community Informatics*, *8*(1), 1–16. http://www.ci-journal.net/index.php/ciej/article/view/809/930. Accessed 06.08.13.

Gingerich, O. (2004). *The book nobody read*. New York: Walker & Company.

Hackbarth, S. (1996). *The educational technology handbook*. Englewood Cliffs, N.J: Educational Technology.

Johns, M. (1995). Design for slides. *Journal of Audiovisual Media in Medicine*, *18*(3), 121–128. http://informahealthcare.com/doi/pdf/10.3109/17453059509021636. Accessed 06.08.13.

Kaufman, I. (October 24, 2011). *Are you a digital alien, digital immigrant or digital native? ... Marketing to the Digital WHO ... SocialMediaToday*. http://socialmediatoday.com/irakaufman/381667/marketing-digital-who-native-immigrant-alien. Accessed 31.07.13.

Prensky, M. (2001). Digital natives, digital immigrants Part 1. *On the Horizon*, *9*(5), 1–6. http://dx.doi.org.ezproxy.derby.ac.uk/10.1108/10748120110424816. Accessed 31.07.13.

Rudi, A. (2011). Hybrid learning: How to reach digital natives. *Learning Solutions Magazine*. http://www.learningsolutionsmag.com/articles/765/. Accessed 31.07.13.

Simmonds, D. (1993). Standards for medical graphics. *Journal of Audiovisual Media in Medicine*, *16*, 56–61. http://informahealthcare.com/doi/pdf/10.3109/17453059309064823. Accessed 06.08.13.

Steeples, C., Unsworth, C., Bryson, M., Goodyear, P., Riding, P., Fowell, S., (1996). Technological support for teaching and learning: Computer-mediated communications in higher education (CMC in HE). *Computers & Education*, *26*(1–3), 71–80. http://dx.doi.org/10.1016/0360-1315(95)00082-8. Accessed 31.07.13.

Toth, V. T. (2012). *Programmable calculators Casio FX-501P*. http://www.rskey.org/CMS/index.php/exhibit-hall/7?manufacturer=Casio&model=FX-501P. Accessed November 30.

UCLA Center for Communication Policy. (2003). *Surveying the digital future*. http://images.forbes.com/fdc/mediaresourcecenter/UCLA03.pdf. Accessed 06.08.13.

Wenger, E., White, N., & Smith, J. (2009). *Digital habitats*. Portland, OR: CPsquare.

White, M. (2000). *Leonardo*. London: Little, Brown.

Whiting, S. (2009). *Paul Brown's eHuman.com anatomy slides*. Wednesday 31st July SFGate. http://www.sfgate.com/bayarea/article/Paul-Brown-s-eHuman-com-anatomy-slides-3250669.php. Accessed 31.07.13.

Clothing, identity, embodiment and age

2

J. Twigg
University of Kent, Canterbury, UK

2.1 Introduction: clothing, social identity and age

Design needs to be set in the context of the wider debates around the role of dress in the constitution of social identities. Much of the sociological literature on fashion and dress has indeed been concerned with questions of identity, exploring the ways in which clothing and dress express larger social structures such as class and gender. Since the time of Veblen (1899) and Simmel (1904), sociologists have analysed fashion in terms of systematic entrenched change, in which competitive emulation and striving between different social classes acts as the driving force of fashion, and in which styles diffuse down the social hierarchy from social elites (Bourdieu, 1984). Though this account has been modified since the 1990s by the recognition of the significance of other sources of fashion, for example, bottom-up influences like street styles (Entwistle, 2000; Evans, 1997; Polhemus, 1994), and by the recognition that the modern fashion system is more fragmented and plural than this suggests, the sense that social class is a significant determinant of dress remains strong. The links between gender identity and dress have similarly been widely theorised within sociological work. Many writers indeed regard gender as the central preoccupation of fashion (Davis, 1992; Entwistle, 2000; Kidwell & Steele, 1989; Tseëlon, 1995), and it is certainly the case that gender is one of the most marked features of dress codes across cultures. Clothing interacts with the body in ways that both obscure and point up sexual difference, thus acting to deliver gender as self-evident or natural, when it is in reality a cultural construct (Butler, 1990, 1993). Dress thus helps to reproduce gender as a form of body style, reinforcing the complex interplay between sexed bodies and gendered identities, and playing a critical part in the way in which femininities in particular are rendered, played out, resisted and understood (Holland, 2004). Similar, though less extensive, analysis has been applied to the links between dress and other 'master identities' such as race, ethnicity and sexuality (Holliday, 2001; Khan, 1993; Rolley, 1993).

Age, however, has not received the same level of attention. Sociology and cultural studies, reflecting the wider ageism of the culture, have tended to ignore age, sidelining its significance and treating it as something more relevant to the fields of medicine or social welfare. And yet age is indeed one of the 'master identities'; how we are perceived, who we socialise with, how we are judged and ordered socially, what deserts are deemed appropriate for us – are all crucially determined by our age or our location

Textile-led Design for the Active Ageing Population. http://dx.doi.org/10.1016/B978-0-85709-538-1.00002-X

in an age categorisation. We should not be surprised, therefore, to find age reflected in norms around dress (Twigg, 2006).

2.2 Age ordering

One way we can understand these norms is in terms of the long-established phenomenon of 'age ordering'. By this I mean the systematic patterning of cultural expression according to an ordered and hierarchically arranged concept of age. It can clearly be seen in relation to children, where at least since the late eighteenth century, and often before, children have worn distinctive age-related forms of dress. The degree to which childhood is marked out in this way has varied historically (Marshall, 2008; Paoletti, 2012; Paoletti & Kregloh, 1989). Such practices are associated in particular with the Romantic Movement and the new cult of childhood that emerged in the nineteenth century with its emphasis on children as social beings in their own right, expressed through distinctive dress. Cook (2004) traces this development through the early twentieth century with the emergence of retailing, particularly based in department stores, specifically aimed at children. More recently, the trend has been for children's clothes to be less, rather than more, distinctive, with the spread to young girls in particular of adult female styles, as part of a more general extension of consumption culture to this group.

In relation to old age, there has been a similar pattern of structured expectations, expressing norms about what is appropriate – or more significantly in this context, inappropriate – for people as they age. Though always subject to historical specificity, certain features have recurred in relation to dress and age: more covered-up styles, higher necklines, tighter-drawn linen, longer skirts. Colours tend to be darker or more sombre, with many older women in the past moving permanently into a version of mourning. For men a similar pattern of 'longer means older' (Lurie, 1992) has been obtained historically, with older men adopting the long robe in contrast to the short hose of the young. Colours tend to be darker and more sober; styles less showy. Such strictures focus in particular on sexuality. Old women wearing ultra-fashionable or sexually explicit dress have long been the mainstay of misogynistic imagery that draws on the Vanitas tradition (Tseëlon, 1995). In a more muted form it is found in the cultural trope of Mutton Dressed as Lamb (Fairhurst, 1998). The pressure to tone down, to retreat from being visible, was reported even in Holland's (2004) more recent study of women who had adopted radical, transgressive styles of dress.

One of the ways in which we can understand such patterning is through a recognition of clothes as ideological (Barnard, 1996), one of the ways whereby social groups establish, sustain and reproduce positions of power, relations of domination and subordination. Such cultural processes enable relations of inequity to appear natural, proper and legitimate. This perception has largely been developed in relation to class and gender, but it applies also to age, for there are aspects of the dress of older people that can be said to be ideologically formed, suggesting subordination and social exclusion. I am thinking here of the grey, drab, self-effacing, 'do not look at me' dress often associated with age. This reflects the wider ideology of ageism, with its devaluation of the old and its construction of them as a version of the Other (Bytheway, 1995; Gullette, 1997; Laws, 1995; Laz, 1998, 2003; Woodward, 1991). It acts to naturalise and obscure what are essentially social and political processes, underwriting at a visible bodily level of dress, the structural exclusion and poverty experienced by many older people.

2.3 Age-related clothing

There is thus clear historical evidence of the existence of age ordering but does this still apply? There is certainly a widespread cultural assumption today that such ordering has gone, or is at least in sharp decline. It is often asserted that older people no longer expect to dress in distinctly old ways, and that they are free to adopt mainstream styles like everyone else. Evidence from a study I undertook on clothing and age suggests that the situation is more complex, with evidence of both continuity and change with regard to age-related clothing norms. The study was funded by ESRC and based on interviews with women over 55, fashion editors and design directors of branded retail-clothing companies (Twigg, 2013a).

It was clear from comments of the older women themselves that they were fully aware of traditional norms in relation to dress and age, and that these reflected the established structures of meaning referred to above. Respondents, for example, acknowledged, and to a large degree accepted, the pressure to 'tone down' with age. Their comments were often marked by a sense of caution. One respondent, in her 70s, who had always been interested in fashion, explained how as an older women she now had to be 'careful':

> *You have to be careful when you get to my age. I always have to have something with long sleeves. I think one or two of the things are perhaps a bit too young for me. I always wear very long skirts. I don't like showing my legs.*

Many of the comments were shaped by a concern to avoid styles that exposed the ageing body to view. Another respondent explained:

> *I wouldn't show my knees, because you don't show your knees after you're 50. Some people do and they look awful.*

The need to adopt covered-up styles, with sleeves in dresses and longer skirts, was endorsed by most respondents. There was also a widespread consensus that very 'girly' styles were best avoided. One respondent explained:

> *I wouldn't wear a frilly dress … It would make me look silly. I'm too old. I would look really silly.*

There was thus clear continuity in their comments with the cultural values that had shaped earlier forms of age ordering, in particular the adoption of quieter, more covered-up styles and the retreat from showy, ultra fashionable dress that exposed the body to view and made claims to sexual attention.

At the same time, however, it was clear that respondents felt that the situation of older people with regard to dress had indeed changed. A number remarked on the difference in their attitude from that of previous generations, contrasting their experiences with those of their mothers and grandmothers.

> *When my grandmother was older, she wore a pinny and her hair up in braids. And she was the same age as me! I'm 62. And when she was 62, she was walking round in this paisley pinny and her hair up!*

She was determined that she was not going to give up more youthful styles or adopt age-coded dress:

> *I say, 'What day do I wake up and I really just want a Crimplene skirt?' 'What day do I wake up and do that?' I don't think I'm ever gonna wake up and do that … I don't think our generation ever is. I think that generation's gone… We won't change a lot. We'll still be in our jeans, and we'll still be in our tops.*

The sense that this was a generation who would never give up wearing jeans was widely shared, and was reflected in the views of fashion editors who endorsed the view that the current generation was 'different', and that magazines and designers needed to recognise this. The fashion editor of *Women & Home*, a magazine aimed predominantly at middle-class women in their 50s and 60s, explained:

> *We've actually changed the magazine quite radically to move with the times, because women of 55 are not the same as women of 55 a few years ago. And obviously everybody knows this, and we've taken this onboard.*

Questions of appearance and dress were recognised to be more important than in the past for everyone, including older people, so that a magazine like *Yours*, predominantly read by working-class women in their 60s and 70s, now covers fashion in a way that it would not have done 20 years ago. The editor explained:

> *That's one of the reasons why we do fashion. Because it's saying you're not disappearing, you're out there in the community. Whether you're a Gran, whether you're a carer, you might still be working, whatever you're doing, wanting to make the best of yourself is part of that. And that's what fashion helps you do.*

Fashion was thus seen as having an important role in integrating older women into the mainstream of modern culture in a way that was not the case in the past.

Representatives of the fashion industry also agreed that expectations had changed. The design director for George at Asda commented:

> *When I first started working, 30 years ago, there was a point in time when people, the majority of people, would switch into that way of dressing, into classic dressing,*

> *because they felt that was appropriate to their age. But that is gone … This is a massive change, I mean it's a huge change in my lifetime.*

Even Edinburgh Woollen Mill, the most strongly age-related of the companies in the study, was conscious of the changing market, and was actively moving younger in its styling and presentation. As the marketing manager explained:

> *We're sort of, over the years, discovering that the 55-year-old lady is now demanding more than ever before, in terms of the type of fashion that's she's looking for… I don't think people who are in their 50s see themselves as being 50. They actually see themselves as being a lot younger. … Our customer was telling us, you know, we might be 55, we might be 65 but we actually don't wanna look like grannies, you know, we don't feel like grannies in our head. We're looking for something younger, slightly more fashionable.*

This sense that older people are moving younger in their clothing styles is widely expressed in the media, carried forward in the popular phrase that '60 is the new 50, or 40'.

2.4 The changing cultural location of older people

The shifting nature of clothing expectations clearly reflect larger social changes that have occurred in the circumstances of older people in the late twentieth- and early twenty-first centuries. Sometimes termed the 'reconstitution of ageing' thesis, this argues that the nature of later life has changed under the impact of a series of interconnected developments that have occurred across Western societies over the last 50 years (Basting, 1998; Blakie, 1999; Gilleard & Higgs, 2000; Gullette, 1997; Phillipson, 1998). Rising life expectancy and declining family size have together produced a dramatic demographic shift, so that between 1901 and 2003 the proportion of the UK population aged over 50 increased from 15 to 30 percent, and is projected to rise to 41 percent in 2031 (Tomassini, 2005). Similar figures are found in the US and other Western countries. With this have come changes in the nature of the life-course. The normative life-course entrenched in the early years of the twentieth century (Anderson, 1985) and persisting in the West broadly until the 1980s has been destabilised. This was based on a relatively secure pattern of education, followed – for men – by full-time, lifelong work in the same occupation, often the same organisation, accompanied by marriage, family formation and eventually retirement. Women followed a slightly different trajectory. But in the late twentieth- and early twenty-first centuries, this has been replaced by longer and more fragmented periods of education and training, with the possibility of looping back; more 'flexible' forms of work and portfolio careers; higher and more persistent rates of female labour market participation; greater fluidity in personal relationships, with partnering and re-partnering; and earlier, more flexible and longer retirement.

Together, these changes have thrown into the air many of the old cultural assumptions about the life-course, underpinning a sense that later years are no longer the clearly defined or scripted period of life they once seemed. Traditional age ordering within society has been weakened, with stages of life, and behaviour appropriate to them, less clearly prescribed. Öberg and Tornstam (1999, 2001) suggest that modern cultures are marked by a declining salience of age ordering at all stages of life, with adulthood an increasingly undifferentiated period between childhood and frail old age, so that later life is better conceptualized as an extended plateau of late middle years, which only ceases with the interruption of ill health and disability.

Under conditions of post- or late-modernity, it is argued, there has also been a shift in the locus of identity from production to consumption. Traditional formations such as class, gender – and to a degree age – have become less central, as identities become less socially entrenched, more fluid and more the product of agency and choice (Bauman, 2000; Giddens, 1991). As a result, patterns of consumption are increasingly significant in the narrative of the self and the formation of social identities, as they come to perform an integrative function within an individuated culture, acting to integrate people within a common culture of lifestyle. This impacts on the constitution of old age, which moves from being a relatively fixed state to one more marked by flexibility and optionality. Here, consumption offers older people the possibility of counteracting the cultural exclusion traditionally associated with age, as they join younger cohorts in pursuing the same cultural trends. This dream of integration, of course, rests on the possession of resources. Though in the period from the 1960s the incomes of some pensioners in the UK have risen, many remain in poverty. This has produced a dual trend, towards greater heterogeneity among older people combined with greater similarity in relation to the general population. Jones and colleagues (2008) suggest that this dual pattern extends to cultural identity also, with some elders joining younger cohorts in their spending patterns and self-identity, while others remain tied to older patterns of deprivation and exclusion.

2.5 Baby boomers

In the accounts of such developments, the baby boomers tend to feature prominently. Indeed at a popular level they have come to epitomise these social changes. The media in particular are fascinated by the boomer generation, and frequently feature articles on them as a key cohort in the potential revision of ageing. Such media accounts often focus on themes of defiance and resistance. There is a widely held assumption that the ageing of this cohort will be 'different': they will not want to adopt the 'frumpy' age-related dress assumed to be the pattern of previous generations. Surely the women who pioneered the miniskirt, journalists repeatedly say, will not want to sink into knitted tops and pleated skirts; they will insist on retaining 'youthful' styles and engagement with mainstream fashion. It is notable how often discussion of changes regarding age are framed in terms of the distinctive qualities of the 1960 generation (though many of the features described belong more to the early 1970s – understood here as the Long Sixties; Marwick, 1998). At a visual level, the 1960s were peculiarly youth oriented. This is particularly true of the fashions of the period, exemplified in the UK by the designs of Mary Quant and the appearance of models like Twiggy. They valorised a prepubescent look, with knock knees, panda eyes, flat barred shoes, short skirts and smock shapes. The media in presenting issues around clothing and age recurrently return to this period, granting it an iconic status, which is then attached to the baby boomer cohort in general.

Within academic discourse too the baby boomers are often presented as a 'special' generation (Biggs, Phillipson, Leach, & Money, 2007; Gilleard & Higgs, 2000, 2007). Edmunds and Turner (2002) describe them as a 'strategic generation in aesthetic, cultural and sexual terms…the first […] to live through a time when mass consumer revolution transformed popular taste' (quoted by Phillipson, 2007). Gilleard and Higgs in particular have presented them as cultural pioneers, the generation who invented youth culture and went on to develop lifestyle-oriented consumption culture, and have carried those attitudes into their later lives. More recently they have become the butt of more hostile accounts in which, with their free university education, access to rising house values and favourable pensions, they are presented as a greedy, overprivileged cohort whose expectations stand in the way of younger ones (Willetts, 2010).

There are, however, analytic problems in this focus on the baby boomers. The underlying demography is not as straightforward as is sometimes assumed, and varies significantly between different Western countries (Phillipson, 2007). In the US, the home of the 'baby boomer' terminology, there was a sustained surge in births from the mid-1940s to the mid-1960s. The UK had a distinctive pattern of two separate peaks, one in the late 1940s, the other in the mid-1960s. Baby boomer terminology is usually associated with the first, formally known as the 'bulge'. In some European countries there was no 'boom'. 'Baby boomer' is also problematic for being an implicitly classed term. As we noted, many of the special advantages of this group – pensions, housing, university education – only apply to sectors of the middle class. The imagery that underlies the account is one of prosperous, middle-class couples in good health – the Saga vision – and it implicitly erases the experiences of pensioners on low income and with poor health, though many of this cohort in fact fall into that category.

Baby boomer is thus a category heavily shaped by consumption imagery. It does reflect some of the reality of cultural change, but in distinctive and selective ways that isolate and amplify certain features. In particular it reflects the values and personal situations of women journalists, themselves often in their 50s and 60s, who are the source of much writing on boomer lifestyles, and who present the issues of dress and age in terms of a defiant account in which women are refusing to sink into old age or to give up their Manolas. Having said this, it is clear that the growth of consumption and the shift towards a lifestyle that is associated with this have reshaped aspects of the lives of older people. For those who are able to engage in this sphere, consumption does offer new forms of integration with the mainstream, allowing older people to counteract the cultural marginalisation that traditionally marks old age, and with this to assert their essential unity with the mainstream, younger culture.

2.6 Casual dress

The idea that older people are now dressing younger is perhaps best exemplified by their widespread adoption of casual dress. The rise of casual dress has been one of the most marked developments in the dress code over the last 20 years. It now presents the predominant form of dress for most of the population, or at least for those who are not employed in middle-class occupations in offices or formal settings; and even for them, it is still the main form of dress outside work. Tee-shirts, jeans, fleeces, jogging pants, sweatshirts and combat trousers are now the mainstay of most people's dress. Much of the sense that older people are now part of the clothing mainstream comes from this adoption of casual dress. What most people mean when asserting that older people now wear the same clothes is not that they are wearing the dress of the young, in the sense of the fashions of the teens and twenties, but that they are drawing on the same lexicon of casual dress that has come to dominate majority wear.

'Casual dress' is an umbrella term for a range of garments that have become established within the current clothing lexicon. Many have their origins in sport. In this they are part of a long-established process whereby new items enter the clothing code from sport, and become established as day-to-day wear, eventually to become entrenched as standard, even formal, wear. The process was originally analysed by Laver (1937, 1950) in his historical account of the development of men's tailoring, in which the informal hunting dress of the English sporting gentlemen in the late eighteenth and early nineteenth centuries turns into the tailored coat, and eventually becomes the formal wear of white tie and tails. More recently the process can be observed in relation to anoraks, which entered the clothing system in the 1950s as smart ski wear, but by the early twenty-first century have become the principal form of outerwear for large sectors of the population. The current diffusion of dress from exercise wear into casual dress, in the form of track suits, trainers, layered garments and fleeces, is a recent version of this long-established pattern. Walking clothes, indeed outdoor wear in general, are another prominent source of casual wear making its way into the general pool of dress.

Some of these sources of casual dress are gendered in origin, and to a degree remain so. This is particularly true of the sportswear sector. For example, walking clothes for women, as other chapters explore, are often designed around men, and sized down for women, with only superficial, cosmetic changes, typically in colour. This can present problems of fit and stylishness since women's bodies, both younger and older, are formed differently from men's and are not just smaller versions. The style and atmosphere of outdoor wear shops are also often gendered, with a heavy emphasis on equipment and on a rugged approach to the body, exemplifying the no-nonsense approach to dress that is part of traditional masculinity. The shops are often dark, with limited changing areas, masses of stock and few mirrors. They are very different in feel from shops aimed at the female market, and can represent somewhat alien territory for many women.

It is clear, however, if we return to the general category of casual dress, that older people are indeed adopting these more casual styles. We can, however, ask how far such adoption represents a change in the social location of age and how far a change in the meaning of certain styles of dress. I have argued elsewhere (Twigg, 2013b) that what we are seeing in part is a new form of fashion diffusion, in which styles originate in the high-fashion youth market, and are then diffused to the mainstream, until they eventually reach the older market. It is not, therefore, that older people are moving younger, so much as styles are diffusing to the older. Some elements in the adoption of casual dress by older people can be explained by this process. Garments like anoraks or fleeces have already taken on something of an age code. We can, however, also understand these changes in terms of development of a larger undifferentiated element of the market, a mainstream territory of casual dress that is not high fashion but that is widely worn by all ages. This does, to a degree, support the contention of Öberg and Tornstam (1999, 2001) that what we are seeing is the declining salience of age ordering at all stages of life, with adulthood an increasingly undifferentiated period between childhood and frail old age.

2.7 Adjusting the cut

Aspirations towards cultural integration and unity, however, encounter difficulties in relation to physical changes that occur with age. Though older people may want to wear the same dress as the majority, they may be frustrated in doing so by issues of fit that arise from bodily change. As the body ages, it alters: waists thicken, stomachs expand, shoulders move forward, busts in women lower (Goldsberry, Shim, & Reich, 1996). These shifts are evidenced in the systematic data produced by Size UK, a government-sponsored project that scanned a sample of the general population in pursuit of better fit for the garment industry (Bougourd & Treleven, 2010). As part of this, they developed a new tailor's dummy for the 'mature' figure, representing the distinctive body type of this age group. Designers thus need to adjust the cut of their clothes to reflect these changes. This can, however, present problems for retailers, for adjusting the cut to make it fit better can have the effect of ageing the garment, writing into its very structure information about the sort of body that is meant to inhabit it. Adjustments for the older figure – or its cultural interpretation, such as longer skirts or the addition of sleeves – can in particular detract from the fashionability of the garment, and may limit its marketability, certainly to younger age groups, but even paradoxically to older ones also.

Such ambivalences set up problems for brands in signaling that they are relevant to customers – that they have garments designed for them and their bodies – and at the same time avoiding the negative connotations of being directly labelled as for the old. Fashion is profoundly – perhaps inherently – youth oriented. At its High Fashion core, the system is centred on youthfulness. These values are to a significant degree shared by older customers who have internalised ageist standards, seeking youthful fashions in the attempt to look more youthful themselves. This, however, presents a dilemma. Many want styles that are associated with youthfulness, and with this, attractiveness, success, fashionability, but in forms that are adjusted to their bodies and to the social and cultural interpretation of these.

Retailers thus face a dilemma: how to signal that their garments are aimed at and designed for the older figure, without labelling them negatively in a market that prizes youth and associated fashionability in dress with youthfulness.

2.8 Conclusion

The nature of later years is changing, and we should not be surprised to find this reflected in clothing codes. Older people are adopting dress that reflects the greater cultural continuity they experience in their lives. This is particularly marked for younger older people, and for those who by virtue of good health and relative affluence are able to maintain lifestyles integral with the younger mainstream. As a result these groups are looking for clothes that are attractive and fashionable, and that do not mark them off in any distinctive way from the rest of the population. Casual dress is one of the fields through which they are increasingly achieving this, as the Design for Ageing Well project shows.

Acknowledgement

This chapter draws on an ESRC-funded project, Clothing, Age and the Body RES 000 22 2079, published as *Fashion and Age: Dress, the Body and Later Life,* Bloomsbury 2013.

References

Anderson, M. (1985). The emergence of the modern life cycle in Britain. *Social History, 10*(1), 69–87.

Barnard, M. (1996). *Fashion as communication*. London: Routledge.

Basting, A. D. (1998). *The stages of age: Performing age in contemporary American culture*. Ann Arbor: University of Michigan Press.

Bauman, Z. (2000). *Liquid modernity*. Cambridge: Polity Press.

Biggs, S., Phillipson, C., Leach, R., & Money, A.-M. (2007). The mature imagination and consumption strategies: Age and generation in the development of a United Kingdom baby boomer identity. *International Journal of Ageing and Later Life*, *2*(2), 13–30.

Blakie, A. (1999). *Ageing and popular culture*. Cambridge: Cambridge University Press.

Bougourd, J., & Treleven, P. (2010). *UK national sizing survey – Size UK*. www.fashion.arts.ac.uk/sizeuk.htm. Accessed 11.04.12.

Bourdieu, P. (1984). *Distinction: A social critique of the judgement of taste*. London: Routledge Kegan Paul.

Butler, J. P. (1990). *Gender trouble: Feminism and the subversion of identity*. London: Routledge.

Butler, J. P. (1993). *Bodies that matter: On the discursive limits of "sex"*. London: Routledge.

Bytheway, B. (1995). *Ageism*. Buckingham: Open University Press.

Cook, D. T. (2004). *The commodification of childhood: The children's clothing industry and the rise of the child consumer*. Durham, US: Duke University Press.

Davis, F. (1992). *Fashion, culture and identity*. Chicago: University of Chicago Press.

Edmunds, J., & Turner, B. S. (2002). *Generations, culture and society*. Buckingham: Open University.

Entwistle, J. (2000). *The fashioned body: Fashion, dress and modern social theory*. Cambridge: Polity.

Evans, C. (1997). Street style, subculture and subversion. *Costume, 31*, 105–110.

Fairhurst, E. (1998). "Growing old gracefully" as opposed to "mutton dressed as lamb": The social construction of recognising older women. In S. Nettleton, & J. Watson (Eds.), *The body in everyday life*. London: Routledge.

Giddens, A. (1991). *Modernity and self-identity: Self and society in the late modern age*. Palo Alto: Stanford University Press.

Gilleard, C., & Higgs, P. (2000). *Culture of ageing: Self, citizen and the body*. London: Prentice Hall.

Gilleard, C., & Higgs, P. (2007). The Third Age and the Baby Boomers: Two approaches to the social structuring of later life. *International Journal of Ageing and Later Life*, 2(2), 13–30.

Goldsberry, E., Shim, S., & Reich, N. (1996). Women 55 years and older: Part I Current body measurements as contrasted to the PS 42-70 data. *Clothing and Textiles Research Journal, 14*(2), 108–120.

Gullette, M. M. (1997). *Declining to decline: Cultural combat and the politics of midlife*. Charlottesville: University Press of Virginia.

Holliday, R. (2001). Fashioning the queer self. In J. Entwhistle, & E. Wilson (Eds.), *Body dressing*. Oxford: Berg.

Holland, S. (2004). *Alternative femininities: Body, age and identity*. Oxford: Berg.

Jones, I. R., Hyde, M., Victor, C. R., Wiggins, R. D., Gilleard, C., & Higgs, P. (2008). *Ageing in a consumer society: From passive to active consumption in Britain*. Bristol: Policy Press.

Khan, N. (1993). Asian women's dress: From burqah to bloggs. In J. Ash, & E. Wilson (Eds.), *Chic thrills*. Berkeley: University of California Press.

Kidwell, C. B., & Steele, V. (Eds.). (1989). *Men and women: Dressing the part*. Washington: Smithsonian Institute Press.

Laver, J. (1937). *Taste and fashion*. London: Harrap.

Laver, J. (1950). *Dress: How and why fashions in men's and women's clothes have changed during the past two hundred years*. London: John Murray.

Laws, G. (1995). Understanding ageism: lessons from feminism and postmodernism. *The Gerontologist, 35*(1), 112–118.

Laz, C. (1998). Act your age. *Sociological Forum, 13*(1), 85–113.

Laz, C. (2003). Age embodied. *Journal of Aging Studies, 17*, 503–519.

Lurie, A. (1992). *The language of clothes*. London: Bloomsbury.

Marshall, N. (2008). *Dictionary of children's clothes: 1700 to present*. London: V&A Publishing.

Marwick, A. (1998). *The sixties: Cultural revolution in Britain, France, Italy and the United States*. Oxford: Oxford University Press.

Öberg, P., & Tornstam, L. (1999). Body images among men and women of different ages. *Ageing & Society, 19*, 629–644.

Öberg, P., & Tornstam, L. (2001). Youthfulness and fitness – identity ideals for all ages? *Journal of Ageing and Identity, 6*(1), 15–29.

Paoletti, J. B. (2012). *Pink and blue: Telling the boys from the girls in America*. Bloomington: Indiana University Press.

Paoletti, J. B., & Kregloh, C. L. (1989). The children's department. In C. B. Kidwell, & V. Steele (Eds.), *Men and women: Dressing the part* (pp. 22–41). Washington: Smithsonian Institute Press.

Phillipson, C. (1998). *Reconstructing old age: New agendas in social theory and practice*. London: Sage.

Phillipson, C. (2007). Understanding the baby boomer generation: Comparative perspectives. *International Journal of Aging and Later Life*, 2(2), 7–11.

Polhemus, T. (1994). *Streetstyle: From sidewalk to catwalk.* London: Thames and Hudson.
Rolley, K. (1993). Love desire and the pursuit of the whole: Dress and the lesbian couple. In J. Ash, & E. Wilson (Eds.), *Chic thrills.* Berkeley: University of California Press.
Simmel, G. (1904/1971). *'Fashion', on individuality and social forms: Selected writings.* trs D. C. Levine, Chicago: University of Chicago Press.
Tomassini, C. (2005). *Focus on older people: Demographic profile.* London: Office of National Statistics.
Tseëlon, E. (1995). *The masque of femininity.* London: Sage.
Twigg, J. (2006). Clothing, age and the body: A critical review. *Ageing & Society, 27*, 285–305.
Twigg, J. (2013a). *Fashion and age: Dress, the body and later life.* London: Bloomsbury.
Twigg, J. (2013b). Age ordering in dress: Fashion, the body and age. In H. Thomas, R. Root, S. Black, A. de la Haye, J. Entwistle, & A. Rocamora (Eds.), *The fashion studies handbook.* London: Bloomsbury.
Veblen, T. (1899/1953). *The theory of the leisure class: An economic study of institutions.* New York: Mentor.
Willetts, D. (2010). *The pinch: How the baby boomers stole their children's future.* London: Atlantic Books.
Woodward, K. (1991). *Aging and its discontents, freud and other fiction.* Bloomington: Indiana University Press.

3 Attitudes to apparel amongst the baby boomer generation

P. Borcherding[1], *J. Bubonia*[2]
[1]University of Cincinnati, Cincinnati, OH, USA; [2]Texas Christian University, Fort Worth, TX, USA

3.1 Introduction

The collective baby boomer generation takes its name from the accelerated and substantial growth, or "boom," in birth rates that occurred in Western countries affected by World War II. This rise in birth rates occurred for many reasons, but most economists attribute the increase to the strong economic growth period that followed the war, which provided individuals with a sense of prosperity and security in which to raise a family. Prior to 1946, birth rates were much lower and had remained fairly steady. An increasing number of younger couples were deciding to start families and choosing to have more children than was typical for the time period. Before the war, the average family had 1.9 children; this number rose to 2.5 after the war. The steady economic growth led to expansion in manufacturing and business, as well as housing, which provided a more affluent lifestyle for middle-income families.

The baby boom generation is commonly considered to be inclusive of individuals born between the years 1946 and 1964. The name and significance of this generation has grown steadily throughout their lifespan—particularly in the United States and the United Kingdom—because of their market size, discretionary income, spending power, and love of consumer products. There are approximately 77 million baby boomers in the US and 76 million in the UK. This baby boom generation is responsible for over half of consumer spending currently in the US and 80% of leisure and travel expenditures. By 2015, baby boomers will make up 45% of their respective countries' population (US Department of Labor, 2012) (Lee, 2001). The leading edge of this generational group is now 65 years of age. Extensive study has been initiated to investigate the needs of this 65- to 80-year-old actively ageing cohort, including many studies detailed in this publication. Statistics show that the average baby boomer will live to be 83 years of age, so it is imperative that product markets research and anticipate the needs of this user group. This data needs to guide manufacturers to develop affordable consumer products targeted specifically for this generational cohort.

History shows that the impact of this generation reaches far beyond housing and consumer products. The US Department of Labor released a report in June 1997 stating the baby boom generation had "significantly changed demand for public services throughout their life" (DeFrancesco, 1999). The influence of this momentous group began with their entrance into kindergarten. Elementary schools were ill equipped to handle the number of entering students causing overcrowded classrooms and often

Textile-led Design for the Active Ageing Population. http://dx.doi.org/10.1016/B978-0-85709-538-1.00003-1

promoted makeshift teaching spaces in locker rooms, libraries, and storage rooms. Foresight allowed school districts to quickly anticipate classroom requirements, and many opened new intermediate and high schools specifically to meet the needs of this generation. This immediate response that led to gratification was not lost on these young individuals.

Baby boomers developed expectations that their needs would be met whenever necessary, in both social services as well as the product market. They attended college in record numbers, often being the first in their family to go to or graduate from college. Historically, the baby boomers became the most educated group of the twentieth century to ever enter the workforce, where they challenged labor markets, redefined work practices, and even, in the 1990s, redefined work attire. They were the first generation of predominately dual-income families, which allowed them to reevaluate family life priorities.

This generation is currently redefining what it means to be "empty nesters" and retirees. Baby boomers have always been rule breakers, and as they age they will keep demanding products to meet their individual needs, and they will change the way we live and the products we use (Smith & Clurman, 2007). Baby boomers say their happiness and satisfaction in retirement will come from involvement with family, friends, and remaining fit. They have developed expectations that both the commercial and social sectors will be ready to meet their needs (i.e., libraries, parks, recreation centers, and community living environments). They have also experienced expanded apparel product categories in the past which were introduced just to meet their lifestyle (i.e., junior sportswear, contemporary departments, lifestyle retailers, and women's athletic shops).

Boomers are also very accustomed to technology with their use of gadgets and social media. Initially they may have taken a little longer to embrace the technology boom but they are closing the gap with younger generations. More than half (55%) of older boomers (ages 56–65) now watch online videos, research and order products online, and participate in texting, social networking, and Skype or FaceTime with family members.

The first wave of baby boomers has already entered what is being referred to as the "active ageing" life stage. Most experts agree this group will approach it just as they have all other life stages—by redefining it to match the way they want to live, not the way they are necessarily expected to live. This generational cohort led industry to experience unprecedented growth in the consumer products market as manufacturers rushed to produce goods to meet their every want and need. This has included apparel and textiles as well—from natural and synthetic fibers to engineered textiles—and the demand is on the rise for manufacturers to meet these consumer needs. Currently, the market has been quite slow to react and target the needs of this growing demographic. Consumers in this age group are frustrated in their search for apparel that is designed for their bodies, expressive of their style, and functional for their needs.

Throughout this book there have been references to this cohort as the "active ageing," which is a very appropriate name for them. The World Health Organization defines Active Ageing as "the process of optimizing opportunities for health participation and security in order to enhance the quality of life as people age…allowing people to realize their potential for physical, social and mental well being throughout the life course" (World Health Organization, 2002, p. 12). This definition perfectly describes the way baby boomers plan to live the next stage of their lives. They are approaching life with high expectations, travel plans, bucket lists, and athletic goals for a healthy lifestyle. In the past, sociologists and marketers have referred to this time in a person's life as their "golden years." This term has a sense of finality to it, and as such, has been rejected by this generation. Today the baby boomers have introduced another lifestyle stage into this spectrum—the Silver Generation—ages 65 to 80—and they are making the most of these silver years. They may also be approaching this period with more physical changes than they had anticipated, but are firm believers that these will not stop them. Baby boomers are assuming the apparel and textile market will address their needs just as they have throughout their lives.

3.2 The baby boomers and the growth of marketing

This generation dominated consumer markets in so many ways throughout the second half of the twentieth century. Their impact on the textile and apparel industry will be examined in the next section, but first it is essential to look at the growth of marketing throughout this time period (Litman, 1950). This generation's desire for products specifically directed toward them paralleled with the growth of consumer-directed marketing practices makes for an interesting shared history.

The baby boom generation was the first to grow up experiencing the vast changes in companies' marketing approaches from the late 1950s to the present day. The marketing approach most common prior to 1950 stemmed from the early beginnings of the science of marketing, and can be summed up by the eminent scholar Jagdish N. Sheth, professor of business administration at the University of Illinois at Urbana–Champaign, who wrote in a faculty working paper in March of 1982,

> *Traditional marketing [before 1950]... stemmed from two axioms: (1) that Marketing was essentially an economic activity and therefore was a subset of the discipline of economics and (2) that the initiator of marketing activities and programs was the marketer and not the consumer in the marketplace. ...New schools of thought have emerged since the 1960's as these two fundamental axioms were questioned.*

Most scholars throughout the following two decades determined that the marketing of a product was not exclusively an economic activity but in actuality a function of business and was therefore housed in schools of business throughout academia. Furthermore, they determined that the marketing of a product, to a consumer really addressed the concept of value—price, product and distribution—and the significance of that combination of values was explicit to the targeted consumer.

Many scholars (Drucker, 1974; Kotler, 2005; Levy & Zaltman, 1975) expanded the first axiom of economic exchange to include the consideration of human behavior within uncontrollable variables such as politics, culture, social behavior, and economics. They saw the importance of considering the impact of a person's interest in the product or particularly the "value" of the product to the targeted consumer. Other scholars and those practicing marketing (Dichter, 1964; Engel, Kollat & Blackwell, 1968) refuted the second axiom that Sheth highlighted, and they proposed that the consumer was more powerful than the marketer and that the consumer ultimately determined the marketing mix and success of the product. Apparel and textiles is an excellent example of this school of thought. Prior to the 1960s, fashion designers (the marketers of fashion at the time) dictated the styles of clothing and fabrics that were popular. Numerous lower-priced merchants scrambled to duplicate runway fashions with little regard to consumer acceptance. These fashions were often developed in inferior or inappropriate fabrics, thus frustrating consumers. The onset of the 1960s and the movement of fashion trends evolving from the streets—from the Beatniks to the London Mod scene—changed the entire industry. These trends were being led by the baby boomer generation (Owram, 1997). The industry learned that in order to effectively market goods they must first understand the consumer. This understanding allowed them to ultimately develop products to meet the needs of the consumer while creating marketing plans that were more specifically directed (Sheth & Gardner, 1982).

The evolution of marketing as a discipline and science was aided by one of the largest consumer populations ever to come into the marketplace, and business soon realized that it was vitally important to study and understand the needs of this consumer group. The baby boomers were the first generation to have products designed and marketed to them from the day they were born. They were raised in a world that was beginning to develop and exchange products specifically to meet the needs of consumers in developed countries. In the US, infrastructure was laid and suburban neighborhoods multiplied offering a lifestyle more focused on the activities of families. Retail complexes, dubbed shopping centers, opened up a new world for families. Manufacturing and integration of technology allowed for the development of expansive product lines catering to consumers that were targeted by further marketing efforts. Qualitative research provided marketers with information regarding the individuals residing in these suburban areas that led to the newest of the schools of marketing thought—lifestyle market research or psychographics. This marketing concept became very important as a means to identify product opportunities.

The last half of the twentieth century was dominated by the gradual understanding of the power of marketing by the consumer products industry. Brand development allowed companies to create identities and they later discovered that they were building relationships with their consumers, thus courting brand loyalty. The focus in the 1960s was on price and value of a product. The 1970s tied the product to the user's lifestyle, focusing on what value meant to them. The 1980s ushered in the era of product quality and total quality management. The twentieth century ended with the concept of product saturation and an over-stored environment that was beginning to be threatened by the digital world and the growing power and influence of the consumer in product development. Now companies were seeking ways to interact with the user and to better understand the features they desired in their products, how they wanted these products delivered, and even what brand identity was desired.

As the twenty-first century began, and the baby boomers entered midlife (Whitbourne & Willis, 2006), marketing departments in all product industries began to add a new component to their consumer research: future needs forecasting. John Naisbitt's ground-breaking novel *Megatrends* (1982) ushered in a new vision for anticipating future scenarios and making intelligent marketing decisions based upon these predictions. The book, copyrighted in 1982, sold nine million copies in 58 countries (Iwata, 2006). This approach begins with sociocultural, economic, and technology research and connects otherwise isolated manifestations to each other to help determine future trends in the marketplace. Other marketing agencies, particularly the BrainReserve founded by forecaster Faith Popcorn, began to develop what BrainReserve calls strategic trend-based marketing (faithpopcorn.com). It is now considered vitally important for all consumer industries to have an understanding of the future lifestyles of their consumers in order to anticipate their product needs.

3.3 Baby boomers and their interaction with apparel and textiles

Cultural, social, and industrial changes shaped the apparel and textile industry in the second half of the twentieth century. The rationing of fabric throughout World War II led to a pent-up desire for new apparel and accessories. The movement of baby boomers to the suburbs offered a new lifestyle, and with it came the desire for more casual apparel for leisure activities. As early as 1957, *American Fabrics*, an American textile industry publication, noted that there were new and significant changes happening to the American family lifestyle that were impacting their choices of fabrics and apparel. The addition of electricity into homes, the advent of television, the nationalization of the long weekend, and the dissemination of luxury to the masses were contributing factors that led to the exodus of families to the suburbs (p. 42, Spring 1957, Vol. 40). The suburban lifestyle brought with it an increased demand for casual clothing to meet the needs of busy social schedules as boomers participated in activities within the community with family and friends, which left less time for household chores.

As young married couples began to raise families, there was a significant growth trend in the children's market. Department stores had long carried children's clothing, but now allocated significant space devoted to not only infant and toddler clothing but their accessories and furniture as well. Layette departments were given prime space and promotions were planned around special events in the lives of these baby boom children: new births, religious ceremonies such as baptism, first communions, bar mitzvahs, holiday apparel, and of course—their first entry into school. These departments were well stocked with children's apparel made of cotton and polyester blends in both knit and woven fabrics. Most homes had both a washing machine and dryer, and the demand was on the rise for all apparel to be "wash and wear" for the busy boomer household.

School apparel departments were opened in major department stores throughout the US and the UK. School uniforms were offered for sale where applicable, but if that was not required locally, proper pants, shirts, skirts, and dresses were merchandised to answer the needs of school dress codes. The term "back to school" was coined during the baby boomer academic years as they returned to the classroom each year. Today, tremendous marketing efforts are made during the July and August months by manufacturers and retailers whose business relies on students returning to the classroom, from preschool to university. Organized sports were expanded to include both girls and boys. This led to the development of specific apparel for individual sports as high school athletics grew in importance. Sports apparel was designed to give the wearer a specific edge in competition, such as—supportive footwear, protective elements of dress, support jerseys for contact sports, and stretch fabrics for gymnastics, cheerleading and dance squads. The boomer teenage years influenced the opening of junior and varsity departments in major department stores, tending primarily to casual sportswear and denim apparel. This provided an additional marketing opportunity for the junior apparel market. Collaborations between teen magazines and department stores featured many favorite apparel manufacturers and even fiber organizations such as the Wool Secretariat, the trade association for the wool industry worldwide, and Cotton Incorporated, the cotton trade association in the US (Cotton Incorporated, 2012).

Their college years brought about the collapse of formal dress codes for college campuses, and specific school-related fleece sportswear came into vogue. The growth of athletic scholarships, for both women and men, increased the number of teen athletes during the 1970s and 1980s. Athletic clothing and shoes became acceptable as everyday casual attire, no longer worn just for participating in sports activities. The boomer families were discovering that the traditional department store was not meeting all their needs, and the prevalence of retail options grew steadily, with the rapid growth of apparel store chains such as The Limited and Gap in the US and Topshop and Next in the UK. These chains had a narrow lifestyle focus, and as the boomers quickly moved their purchases to these stores, the brand connection began.

The baby boomers grew up with a variety of fabrics and fibers that offered an alternative to the natural fibers that require some special laundry care. Nylon was the first synthetic fiber offered to the general public and was manufactured early in the first half of the 1950s for women's stockings, sleepwear, and intimate apparel, as well as outerwear jackets and some active sportswear. Nylon was quickly followed by polyester, which became essential when blended with cotton for easy-care fabrics for men, women, and children. The introduction of acrylics brought easy care into the knitted apparel market, previously dominated by wool. Fabrics composed of blends of wool with nylon also began to appear, making wool garments more affordable and some even washable. Blends of Orlon and acetate began to appear even in the couture collections of Yves Saint Laurent, Pierre Cardin, and of course the designer who had the most impact on this generation during their formative style years—Mary Quant.

The apparel and textile companies soon realized the advantage in designing and marketing specifically to this new generation of young women in particular. The American sportswear company Junior Sophisticates, and their young designer Anne Klein, began to market specifically to these new young consumers and urged the department stores to give them designated space for their collections. Thus, the junior department began and soon increased in size and valuable real estate in stores. Other companies soon joined the junior market, such as Anne Fogarty designs, Sportwhirl, and the sportswear companies Bobbie Brooks and the Villager. At the same time, numerous children's wear companies were growing and expanding their offerings—Carters, Healthtex, Tiny Town Togs, Florence Eiseman—as well as the special offerings found in some department stores. These departments grew at a rapid pace throughout the 1960s as the consumers put trust in their favorite local department stores.

As the 1960s ended and the leading segment of this generation was in college or career, the apparel and textile marketplace began to change again. Ted Polhemus noted, in his 1994 book *Street Style*, that "style tribes" began to emerge in the late 1960s and early 1970s in both the US and the UK. He describes it as "encoded within style's iconography are all those ideas and ideals which together constitute a [sub] culture. Like-looking is like-thinking and in this sense the members of a style tribe have a great deal in common." This certainly describes the succession of fashion trends that began to diversify and signify this generation, while addressing the different consumer lifestyle groups in the population. This time period began with the Mods and the Rockers influence from the UK to the Preppy and then Hippie in the US. The antiwar sentiment for the Vietnam War catapulted denim into the single most desired fabric of this generation. What began as a protest fabric to upset "establishment" dress codes evolved into the most important fabric to this entire generation. Denim blue jeans are considered iconic to the baby boom generation and are by far *still* a favorite item in their wardrobe.

In the late 1970s and 1980s, baby boomers became parents and focused on providing their children with healthy and active lifestyles. In the process, they too took an active interest in living a healthy and active lifestyle of their own. They embraced running, aerobics, bicycling, and downhill skiing in the 1980s and 1990s. Advanced fiber and fabric technology such as seamless and stitchless manufacturing, 3-D stretch fabrics, microfibers, and breathable materials, were developed to address the comfort and functional needs of boomers participating in these sports. The twenty-first century trends of walking, yoga, Pilates, and weight training all led to new apparel trends and allowed for a 24/7 approach to active wear dressing. The boomers were beginning to appreciate active sports that were not quite as jarring to their bodies but rather burned calories by strengthening and stretching muscles and building stronger bones. The growth of the active wear market will continue as the baby boom generation moves through their silver years (Swanko, 2011). They show no intention of slowing down, and research shows that the majority of the baby boom generation thinks of themselves as nine years younger than their chronological age (wgsn,.com, 2011). Boomers will continue to find activities to challenge and engage them. They expect the apparel market to address the changing needs of their bodies as markets have always done for them in the past. Psychologically, they are the "me" generation that feels good about themselves. It is important for all of their active apparel to meet their needs.

3.4 Market implications

Synthetic fibers made life easy for the baby boomers in regard to their apparel needs. During the oil crisis of the late 1970s, the synthetic fibers fell out of importance for a short time. All of the textile excitement shifted to natural fibers; the comfort and feel against the body, and the aesthetics of beautifully knit and woven cotton, wools, and silks dominated the industry. The introduction of microfiber technology reignited the use of synthetics blended with natural fibers as well as garments made completely of manufactured fibers. The range of fabrics was greatly improved, and the multiple options of blended and textured yarns offered so many options for this baby boomer generation. Women were entering the business world and moving up in the corporate ladder at a rapid pace, and they relied on the industry to meet their needs once again. Executive shops that catered to working women were opened in many stores as well as more casual options that designers such as Liz Claiborne offered to working women.

In the spring of 1994, Du Pont made the strategic decision to expand the nylon business. Du Pont superbly positioned nylon to enter a dramatic new phase of production for the future. Nylon 6 and 6,6 are leading the industry in sustainable fibers by their lower cost of manufacturing, waste-free production, and energy efficiency. Polyester, the fiber accounting for 65% of all synthetic fiber production today, was first developed to be blended with natural fibers to provide strength and resistance to both wrinkles and garment wear. The development of microfiber technology provided a new life for polyester in popular athletic apparel as well as fine garments. Polyester fleece not only has applications in athletic garments but is also used in interior components of outdoor garments. This fiber is also found in products ranging from plastic bottles to rope to home furnishings and is currently the most recycled textile fiber in textiles.

Spandex, often better known by the brand name Lycra, is a synthetic fiber utilized in a wide variety of apparel today. This fiber was originally developed for the active wear market, which steadily grew as the baby boomers entered school and engaged in activities. Today spandex is found in all types of apparel products due to its stretch and recovery qualities, which provide comfort to garments while accommodating a variety of body types. In fact, a large majority of manufacturers today blend spandex with cotton to make fabric for the ever-popular denim market.

Jeans are a wardrobe staple for baby boomers today as with most generational groups because of their comfort and fit. This is an example of the remarkable tenacity of the textile products developed particularly for the baby boomer generational market. These product innovations continue to deliver market share for apparel and textile producers even though many of these manufacturers no longer focus on serving the baby boomer target market. This is not a well-thought-out plan for apparel and textile manufacturers. This actively ageing market cannot be ignored, because they possess the greatest spending power of any generation, and this will continue for at least the next couple of decades. The Design for Ageing Well research study investigated the application of smart textiles and wearable technologies in clothing that enhance the well-being of the active ageing to create a new generation of monitoring and feedback devices with the potential to enhance the autonomy and independence of this rapidlygrowing consumer majority, see Chapter 21.

3.5 Current lifestyle trends for the baby boomers and product needs for the future

In the current landscape of marketing, product design, technological advancements, and the expansion of the global market have intensified competition and consumer expectations. High turnover rates of production as well as swiftly trending pop-up stores and online applications test the quality, transparency, and loyalty of brand–consumer relationships. As baby boomers enter a new stage in their lives, this demographic is once again positioned to refashion and redesign apparel and textile products, services, and marketing industries. Elements of technology, health, and living choices will integrate and evolve, allowing boomers to influence and mold the future of design. The baby boomer generation created design trends for their lifestyle since the day they were born.

As technology intensifies the marketplace for socializing, consumption, and lifestyle enhancements, baby boomers will continue to embrace new services, applications, and operating systems to improve their quality of life. Finding comfort in connectivity and the availability of information, the meaning of legacy, and the way in which it is communicated will continue to evolve and manifest throughout the lives of this active generation. The availability of information offers a wide range of tools for boomers to use in their leisure and post-retirement endeavors of entrepreneurship. Connectivity fuels the encouragement of peer interaction and collaboration. This very connectivity with the marketplace has enriched product design as well. Today multiple customization options for apparel and footwear allow the consumer to personalize purchases to their precise needs. This level of interaction will become essential for product managers and marketers to connect with this cohort in much the same way they presently connect with younger generations. However, the specifications for design details must change to allow for adaptations to meet the needs of baby boomers in regard to sizing, fit, and donning and doffing in an effort to provide smart wearable options (see Figure 3.1). Branded manufacturers will earn brand loyalty as a result of online retailing and consumer participation. This will lead to increases in online consumption levels and product design—a bar that boomers have raised before. Brand honesty is communicated through the promotion of practical end uses of products versus an emotional connection to the purchase.

Currently boomers are breaking the mold by continuing to work long past the "age of retirement" that attracted past generations. Statistics show that 83% of them intend to keep working after retirement and about half of those polled expressed an interest in working in a new profession. The majority of these adults will return to

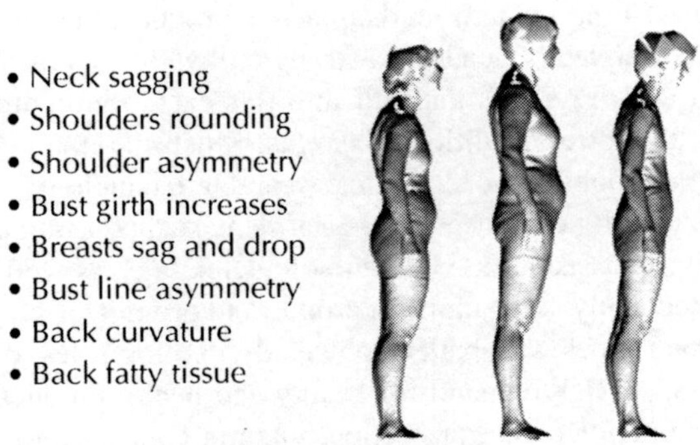

Figure 3.1 Changes to the body due to the ageing process.

work because they want to stay active, not for financial reasons. For some, the current economic landscape has enticed them to continue to earn an income to enhance retirement savings or to assist family members in need. The key is that they will continue to earn, adding to discretionary income, and this is new ground for marketers to consider. As this generation maintains its active lifestyle, they will also continue to purchase products without hesitation because they have the money and desire to do so.

Baby boomers are more accepting of cultural changes than their parents' generation. They are quickly beginning to close their gadget and social media gap with younger generations, and this will also lead to their knowledge and activity base. Those connected to the Internet find social groups, activity calendars, and products that interest them, all from the comfort of their home. They are traveling in groups to new cities and countries, spending their money at home and abroad. Shopping trips often include three generations together in stores, each seeking to fulfill their own product needs. Boomers are typically more frustrated than other generations when shopping for apparel items. They have special needs for fit and styling which are challenging to find in today's marketplace. Through the assessment of current economic and technological standings, windows of opportunity begin to show. History has demonstrated the strong influence of the boomer generation and its impact on culture, the economy, and the product industry (Tortora & Eubank, 2010). Before determining which products will be most effective to target the boomers, recognition of key spending habits and retail consciousness must be identified. Cooperative spending, beneficial bundling, and experiential equity will dictate the marketing and product development of future products.

Seen in various aspects of the boomer consumer lifestyle, cooperative spending is an increasingly appealing notion. Purchasing products that also enhance the lives of others, often referred to as the "Toms effect," will become more important. Products that offer multiple uses will be in high demand. Apparel that meets both the needs of casual and active sportswear is a clear example. Areas of growth and opportunity such as wearable technology will allow the user to not only remain active but also monitor their activity to insure their health. Travel is expected to increase with this generation, therefore creating more demand for all products associated with travel and leisure.

A recurring trend in the modern marketplace is products that offer multiple benefits. These products frequently generate a strong motive for purchase and often allow for a higher retail cost. Products that fall into this category commonly share some technological link that offers additional benefits. Much like the outerwear jacket or handbag that offers a mobile phone pocket, wearable technology will offer the consumer information directly evidenced on the garment, such as a blood pressure level or blood sugar reading. Massachusetts Institute of Technology, as well as other research centers, have created many adaptations to products to promote ageing adults' activity. They have developed shoes with battery-powered vibrating soles to improve balance and help prevent falls. High demand for beauty and health products that offer additional benefits of ultraviolet light protection, vitamin C, and antioxidant ingredients will continue to grow. Apps developed to help older shoppers choose food products based on their health and medical history will increase in popularity among this generation group.

Experiential equity is increasingly more popular with the boomer generation. Initially, a product readily marketed to the emergent young adult markets would not have any significant presence in the minds of older adults. Today however, as generations work and interact, the likelihood of the boomers knowing about a product and having the desire to purchase it is quite high. The newest active apparel suited to spinning is shared among the class, and current fashion trends are found on multiple sites on the Internet. Furthermore, they want all of this without sacrificing fit—for their bodies.

Most ageing adults experience a gradual shift in their posture, causing the shoulders to round forward, the back to curve, and often creating an asymmetry to the shoulders, bust, and hip area measurements. Serious bone and disc deterioration can result in osteoporosis and degenerative back problems. Common difficulties caused by the ageing process are joint issues that can range from occasional pain to debilitating and crippling arthritis. Extensive research studies of female body scans have confirmed many of these changes (McCann, 2010; Na & Ashdown, 2008). An additional significant adjustment that occurs for both men and women is an expanse in girth through the torso that often includes the hip measurement as well. Women will also experience both an increase in bust size as well as a drop in their bustline. These changes become more significant as people age, and will of course vary by individual. The common link between all of these physical changes is the need for modification in the pattern development of garments when designing for the ageing body. Development of garments to fit this generation will fit those of the future as well. Companies need to recognize and meet the needs of this generation—again.

3.6 Conclusion

This generation is the most highly educated cohort to ever enter their silver years. They are healthier than any generation group that came before them and they are in tune to how their bodies are changing. In the past couple decades many boomers have become overweight as a result of reduced activity or lack thereof. Subtle body changes that happen over time due to ageing are what many boomers are fighting to overcome as long as they can. This generation is projected to live an average of six years longer than their predecessors, so their bodies will need apparel products to adapt to their needs more than ever before.

The size of this ageing population will put a strain on every country's social services. An overwhelming number of ageing adults are choosing to stay in their own homes or at least live in their own space rather than move into a group home of any type. Many of these adults will live alone, often for the first time in their lives. They need to take care of their physical health while tending to their emotional and mental well-being.

Studies have shown that the older a person is, the more that they rely on others to stay physically active. Classes for aerobics, walking, Pilates, and yoga are among the fastest-growing group activities in the US. Hiking and biking tours are among the largest tourist draws in the UK and other parts of Europe. Actively ageing adults want to stay active, and this will be a vital part of their life—and they want to do so with others. The opportunity for apparel and textile manufacturers to meet the needs of this market is huge. Increased social activity of an individual expands the groups they identify with, which leads to an increased importance of their clothing. Boomers continually seek out appropriate and desirable apparel for all of their social needs, whether casual, active, or formal.

Currently the industry is simply not responding to this target market's needs, and the manufacturers producing garments that could appeal to and meet the requirements of this consumer group are not marketing to this generational cohort. According to Age Lessons, a consultancy focused on the boomer, its research shows that 45% of boomers believe they are overlooked by marketers. Other product markets have appealed to the boomers lifestyle trends and have tapped into the spending power of this market by providing products they desire. Apple maintains and builds their "cool" identity without neglecting the boomers. L'Oreal uses technology and youthfulness to appeal to this generation. Skype directly targets boomers through usability and connectivity. Textile and apparel manufacturers need to study the user needs of this generation to develop a carefully crafted marketing strategy that connects the user to their brand and they will reap the benefits for many years to come. Follow these leaders one more time.

References

American Fabrics. (Spring, 1957–Summer, 1968). New York: Reporter Publications.

Cotton Incorporated. (2012). *Boom-ing Business: Analysts Suggest Retailers Focus on Baby Boomers*. http://www.wwd.com/markets-news/textiles/boom-ing-business-analysts-suggest-retailers-increase-focus-on-baby-boomers-5960252. Accessed 11.07.12.

DeFrancesco. (1999). *Baby boomers - Drivers of change*. Department of Economic and Community Development, State of Connecticut government. http://www.ct.gov/ecd/cwp/view.asp. Accessed 24.07.12.

Dichter, E. (1964). *Handbook of Consumer Motivations*. New York: McGraw-Hill.

Drucker, P. (1974). *Management: Tasks, responsibilities, practices*. New York: Harper & Row.

Engel, J. F., Kollat, D. T., Blackwell, R. D. (1968). *Consumer Behavior*. New York: Holt, Rinehart and Winston Marketing Series.

Iwata, E. (2006). *Naisbitt turns lust for life into mega book career. USA Today*. http://usatoday30.usatoday.com/money/books/2006-09-24-naisbitt-usat_x.htm.

Kotler, P. (2005). *The role played by the broadening of marketing movement in the history of marketing thought*. Journal of Public Policy & Marketing *24*(1): 114–116. http://www.jstor.org/discover/10.2307/30000565?Uid=3739728euid=. Accessed 20.06.12.

Lee, S. J., Nelson, L. P., Lin, J., Tom, F., Brown, R. S., Jones, J. A. (2001). Today's dental student is training for tomorrow's elderly baby boomer. Special Care in Dentistry *21*(3): 95–97. doi:10.1111/j.1754-4505.2001.tb00233.x.

Levy, S., & Zaltman, G. (1975). *Marketing, society and conflict*. Englewood Cliffs: Prentice-Hall.

Litman, S. (1950). The beginnings of teaching marketing in American universities. *Journal of Marketing*, *15*(2): 220–223.

McCann, J. (2010). *Design for ageing well*. Miami, FL, USA: Smart Fabrics 2010 Conference.

Na, H., & Ashdown, S. (2008). Comparison of 3-D body scan data to quantify upper body postural variation in older and younger women. *Clothing and Textiles Research Journal, 26*(4), 292–307.

Naisbitt, J. (1982). *Megatrends*. Warner Books.

Owram, D. (1997). *Born at the right time: A History of the Baby Boom Generation*. Toronto: University of Toronto Press.

Polhemus, T. (1994). *Style tribes*. London: Thames and Hudson.

Sheth, J. N., Gardner, D. M. (1982). *History of Marketing Thought: An Update. Faculty Working Paper 857*. Urbana-Champaign: University of Illinois. http://www.jagsheth.com/wp-content/uploads/2014/02/history_of_marketing_thought_an_update.pdf. Accessed 20.07.12.

Smith, J., & Clurman, A. (2007). *Generation ageless. How baby boomers are changing the way we live today and they're just getting started*. New York: Harper Collins.

Swanko, K. (2011). *Designing activewear for the lively baby boomer market*. Textile Insight May/June: 22. http://textileinsight.com/2011index.html. Accessed 11.07.12.

Tortora, P., & Eubank, K. (2010). *Survey of historic costume* (5th ed.). New York: Fairchild.

Whitbourne, S., & Willis, S. (Eds.), (2006). *The baby boomers grow up: Contemporary Perspectives on Midlife*. New Jersey: Lawrence Erlbaum Associates.

World Health Organization. (2002). *Active ageing: A policy framework*.

The importance of colour in textiles and clothing for an ageing population

4

C. Johnston
Royal College of Art, London, UK

4.1 Introduction

I have approached this subject from a point of view that I believe that colour is relevant, important and to be enjoyed by anyone, at any stage of our lives.

I have worked as a designer in the fashion and textile industry for over 30 years, specialising in colour and printed textiles, for international fashion and interior manufacturers and retailers. Throughout my career, colour has been one of the most important and enjoyable tools of my trade. Although I have worked with colour for a considerable time, it is a subject that continues to fascinate, puzzle and interest me. It surprises me how little is really known about colour, both by the general public and the colour specialists in science and the arts.

Colour is a major part of our lives; we use it and need it for many daily purposes. Colour can be a language to keep us safe – by identifying colours we make appropriate decisions, think about crossing roads and changing plugs. Every day we make many decisions about colour, the clothes we wear, and the food we eat. We are so familiar with colour that we often take its value, influence and presence for granted.

Colour and how we perceive it is physically different for every individual, therefore it is not surprising that it is viewed as a very personal subject; it is like an unexplored fifth sense – there is still a huge amount to learn and discover about colour. The Society of Dyers and Colourists has an excellent website that provides extensive information on colour science and theory. This chapter looks at the process that colour designers use to work with retailers to colour products.

4.2 Attitudes towards colour amongst the active ageing

In preparation for this book, a colour workshop was organized with a group of mature individuals mostly over the age of 60. The aim was to take part in research to explore personal ideas and choices of colour for clothing in order to ascertain if there was a common consensus of colour choice that could be linked to age. There is scientific evidence that our colour perception weakens with age, but no information on why there would be a change in personal preferences.

Textile-led Design for the Active Ageing Population. http://dx.doi.org/10.1016/B978-0-85709-538-1.00004-3

The input from this group demonstrated the incredible diversity of each individual's choices and views about colour. During the workshop, participants used colour to express their personal preferences. The group had very strong personal opinions about colour, but very little common choice or agreement about their likes and dislikes. Some individuals liked neutral calm colours and some preferred bright strong colours. Everyone had an opinion.

This individuality should not be surprising, as surely, as we age we know ourselves better. We become more confident in what we enjoy, what pleases us, and what suits us best in terms of our environment and the decisions about how we present ourselves. Our personal colour choices for the products we buy to wear, and surround ourselves with, are two of the ways we express ourselves and our individuality throughout our lives.

The way we get to know what works is simply by trying things out, experimenting and finding the best solutions, and of course these lessons are often learnt by making mistakes. Our colour choices express whether we want to stand out and be different, or merge with the crowd.

The opportunity to choose and select colours is important for every age, gender and ability. It seems to boil down to confidence. This reminds me of the excellent and provocative poem by Jenny Joseph, written in 1961, that starts: 'When I am an old woman I shall wear Purple with a red hat…' expressing a desire to challenge the conformity and expectations of behaviour from people in this period of their lives. The poem blends humour with a serious social comment (Joseph, 1997). Why would society find it more comfortable for people to conform or disappear after a certain age?

The attitudes of the affluent baby boomer generation of the postwar era have to some extent broken down the stereotypical expectations of age and image, with healthier lifestyles, fashionable clothing and cosmetic surgery; people's age has often become less visually obvious.

The desire to stay looking young and fashionable has been cleverly exploited by international high street brands across products, at all market and price levels. Publications are working with older models to reach this market; *The Guardian* has a weekly fashion spread in the Saturday colour supplement for 'All Ages'. People living in the developed consumer world have more choices than ever in how they spend their money; competition is fierce and colour is a main consideration when people buy products.

4.3 The colour selection process for clothing

The process for the selection of colour for products is in general a mystery to the public. This chapter aims to explain how some of the major high street clothing retailers decide which colours they choose and develop into products on a seasonal basis. I am not suggesting that this is the best or most effective way of selecting and bringing colour to the customer; I simply aim to give an insight into how manufacturers and retailers decide on the colours.

In most developed countries the consumer takes it for granted that clothing appears in store on a seasonal basis slightly ahead of the actual season. Whatever the climate in Europe and the United States we will find autumn clothes arriving in stores in August and spring/summer clothes in stores in January. The reason is that retailers are aiming to achieve maximum sales at full price in order to realise the best profit margin.

To tempt us to purchase, the retailers will do everything in their power to attract our attention whether it is through marketing and press coverage, the drama of the shop window or the quality and value of the product. It is a tough, competitive field, and having the right product in the right colour, at the right time, is essential for success and ultimately for survival of the business.

Getting the customer to cross the threshold of the shop is the first challenge to the retailer, and although a visually attractive shop front and environment has a major impact on this action, colour has a strong influence on our decision to enter and engage, whether we are aware of this or not. The choice and range of the colour of a product in store can have a major impact on the sales, and ultimately the profit and loss for a company. The wrong colours can mean poor sales, drastic reductions on price and profit, as well as be potentially damaging for the visual identity and reputation of the brand.

There is also more growing concern in this more environmentally conscious era about the amount of waste that can result. Although there are better recycling systems in place especially for textiles, unsold clothing ultimately means further processing, or alternative methods of disposal, involving both cost and the unnecessary use of precious energy.

How do clothing brands and shops decide which colours to use on their products? This chapter aims to demystify this process and give an insight into how many of the high street companies go about this. It may seem to the public that these decisions are rather arbitrary and made on a whimsical ad hoc basis, but this is far from the truth. The colour process can take two years, from inspiration to the shop floor.

4.4 Colour forecasting

In the UK, there are many commercial companies that professionally predict and present colour forecasts and textile and clothing trends; this forecasting industry goes back to the late 1960s and early 1970s when mass market fashion started to boom and become more affordable and accessible.

Deryk Healy Studios and Nigel French Enterprises were two of the earliest and successful UK-based companies in this field. There were similar companies internationally such as Dominic Peclers and Nelly Rodi in Paris and Pat Tunsky in New York.

The forecasting industry has continued, and some companies have developed into large-scale consultancies advising international companies such as Trendunion in Paris headed by Li Eedelkoort who work with fashion and textiles as well as with a broad range of industries, including automotive, packaging, mobile phones, paint and home decoration.

There are currently fewer international large-scale forecasting businesses focused uniquely on colour. Over the last 20 years, many high street retailers and manufacturers have learnt from these original forecasting companies and set up their own in-house design teams that develop a specific design identity; colour is developed in conjunction with the product design. The commitment and investment in design will depend on the company, its size and complexity, but since the early 1980s, most high street clothing brands began to find it more effective to employ and develop their own in-house design teams.

4.5 Classic and changing colours

It is important to point out that when the colours of clothing are chosen by most retailers, approximately 75–80% of colours are constant and unchanged on a seasonal basis. In menswear collections the percentage is even higher. In most clothing the classic colours – black, white, navy and cream – are constant standard colours and represent the major financial commitment. There are also standard neutral colours in various tones of grey marl, denim and chambray along with beiges, browns and khaki that many companies will maintain and repeat for all seasons depending on whether the clothing has a more formal or casual emphasis.

It is not this unchanging part of colour selection that most concerns the commercial managers of companies who put the money behind the colour, but the 'icing on the cake', the seasonal fashion colour statements that need to be suitable to do the job of signalling newness, drawing in, enticing and attracting the customers to part with their money. It is this small but very significant percentage of colour that is so important to get right, and inevitably it takes longer to assure everyone involved in the process, especially financial management and buyers, that decisions made in the selection of these colours are sensible. The following section aims to explain how colour develops from the designer's initial inspiration to a commercial product on the shop floor.

4.6 How the colour selection process starts: designers and inspiration

Colour researchers and designers who work on seasonal colour have many ways of finding colour inspiration; there is no set formula, method or process.

It is a case of constantly looking for colours that excite the eye and set the mind in motion. Designers are trained to constantly look, and be visually aware, to keep a sharp eye open for stimulation that will be useful. They also tend to be obsessive collectors of objects and visuals that will feed the need for fresh ideas. They become adept at using eye and brain coordination as visual antennae to pick up ideas and explore potential design opportunities. This is not a nine-to-five job but a continual process of hunting and gathering.

Like many other designers, I constantly squirrel away all sorts of curiosities that I think might have future use. Watching films, going to exhibitions and travelling to interesting places are all fruitful hunting grounds for colour.

The digital camera and mobile phone camera provide instant capture and wonderful immediacy for inspiring imagery. Each designer has his or her own process and method of research. The selection process starts with personal inspiration and goes through a system, which enables the colours to be developed and perfected to be relevant for the intended brand or product.

4.7 Sharing information: the case of the British Textile Colour Group

In the UK there are groups of designers who meet to present their seasonal colour research to one another. One of these is the British Textile Colour Group (BTCG). I have been lucky to be a member of the BTCG for many years.

BTCG is a nonprofit-making organisation, with a membership of approximately 30 designers predominantly from fashion and textile design backgrounds. All members are currently involved in aspects of the industry that relate to colour. They may be part of a large organisation or independent design consultants, but each member has to be approved for membership by the whole group.

The group meets in London twice a year to discuss and share thoughts and ideas for the season two years ahead of the actual timing. For example, the meeting to discuss spring/summer 2014 would happen in spring 2012 and the discussion for autumn/winter 2014 would be in autumn 2012. This is a continuous process, keeping the development of colour exciting, dynamic and relevant.

At the BTCG meeting, each designer is required to present to the whole group a general reflection on current and possible future directions of colour for the season as well as six themes of around eight colours each, with visual objects and a short written description to back up and explain why these colours have been presented. Three members will form the rotating editing group to look for common themes, messages and colours.

All members then have the opportunity to vote for their favourite boards. From the common consensus the editing group will take the strongest themes to form a palette of Pantone-matched references and relevant images. These palettes are made available to all members but are strictly confidential and not for publishing or sale.

The BTCG palette is subsequently presented one month later at an international colour group meeting called 'Intercolour', and the process is repeated, but with a global representation. The members of Intercolour are similar in profile to members of the BTCG but represent global colour expertise. The Intercolour meeting will result in another joint palette, but with international input, (Intercolor, 2014).

The 'Intercolour pack' is in turn brought back to the UK and shared with all the members of the BTCG, and each member is given access to the pack for a period of time to study the information in detail. An interesting and important point is that there are some members of Intercolour who are involved in developing the colour trend presentations for international fabric and yarn trade fairs, which are shown at a later date.

4.8 How colour palettes are used

Essentially each BTCG member will use this information in his or her own way and for his or her own purposes, but it provides a starting point to define colour for a given season. In the subsequent two years there can be many influences and reasons for the palette to change and evolve. For example, a designer working with one of the large high street retailers would likely use a combination of their initial research, the BTCG palette and the Intercolour pack to discuss with the fashion designers and commercial teams which colours are going to give the required newness and be most suitable for the products they are planning to sell. Other members of the colour forums have a variety of uses for the colour information such as design consultancy, journalism and education. A particularly interesting part of the process is the early sharing of colour research with other experts.

In such a fiercely competitive field this openness may seem contradictory. However, the breadth and thoroughness of the process takes the outcome away from just being a personal point of view, giving the research greater credibility and, most importantly, a stronger sense of authority. It is this authority that gives the retailers and buyers greater confidence to take the risk of investing in and committing to colours.

Subsequently, design teams, buyers and commercial managers will travel to international trade fairs such as Premiere Vision in Paris, where fashion fabric manufacturers present their ranges for a year ahead of the market. The presentations of colour trends at these fairs include the influences from members of the Intercolour meetings. This will provide confirmation and assurance to the retailers that the colours that are being developed are relevant and suitable, but will also provide an opportunity to consider if there are alternatives that can be included.

Of course it can be dangerous for retailers to be over-influenced by the trends at trade fairs and result in the high street looking too slavishly similar, which does happen quite frequently. There is a need to carefully balance what trends forecasters propose with a company's own identity and competitive edge.

Therefore, ideally, it is best to be aware of the main colours forecasted, but decide as a team of designers and buyers what is right for your own company and customer.

4.9 From colour palette to product

Materials and colour have a strong influence on each other, and refining and developing the colours to make sure that they work to enhance the character of the materials is an extremely important part of the process. A colour that looks good on lamb's wool knitwear might not be so attractive on a cotton T-shirt, so a lot of adjusting and tweaking of the colours and tonalities is needed to make the appropriate colour for the product.

The finessing and perfecting of colours has improved over the last 20 years through the development of digital colour measurement systems, which have enabled colour technologists and dyers to help the designers achieve much greater speed and accuracy in colour matching. This also enables colour accuracy to be achieved in communications from one side of the globe to the other. If a colour is selected in the UK, it can be rapidly (almost instantly) digitally matched and replicated anywhere in the world where the materials and the manufacture occur.

During the selling season, retailers constantly monitor sales of products, and at the end of a season they will review the successes and failures and the colours will be assessed along with all other aspects. This data is essential for the retailer to build its knowledge and like any individual learn from the successes and failures for future decisions. This whole process of colour inspiration through to the products we see in the high street shops is a continuous flow. Although the process described here can take two years, current systems of production are constantly speeding up and facilitating more colours and more choice.

Undyed garments or fabrics can be held by manufacturers at the start of the season. When sales results show the most popular colours, the manufacturer can quickly match or adjust and dye the garments. This system can reduce risk and waste. Direct selling through the Internet is also changing the way we shop, and the ageing population is accessing this choice with confidence.

4.10 Conclusion

This chapter has explained a system that has developed over many years for high street retailers to choose colour and deliver products for a specific season, for people of many ages. I suspect that this process may already be outdated as emerging technology is likely to transform the process. As where, how and why we make and manufacture globally is rapidly changing, no doubt our methods of making colour choices will need to adapt and evolve in response.

The rapid rise of direct Internet sales has already had a major impact on how we shop and on the profile of the UK high street. There is currently great excitement around the accessibility of 3D printing. As schools acquire this technology, a new generation will learn skills to individually make the things they need and desire. I look forward to the day (which may not be so far off) when the consumer is more interactive and involved in the process of choosing colour, materials and shapes for themselves.

It is already possible to have an app on a mobile phone that can read colour data and store it digitally for our own inspiration and personal use. The data could be sent to a manufacturer or a 3D printer and applied to clothing or any product. It will be interesting to see where and how we will get the knowledge and confidence to make these decisions. We will no doubt make mistakes, but we can learn from our mistakes and enjoy the personal involvement in making colour.

References

Intercolor, (2014). Intercolor, international and interdisciplinary platform of coolur experts. http://www.Intercolor.nu.

Joseph, J. (1997). *Warning*. (1st ed.). Souvenir.

Further reading

Scully, K., & Cobb, D. (2012). *Color forecasting for fashion* (1st ed.). Laurence King Publishing.

The adoption and nonadoption of new technologies by the active ageing

5

D. Bryson
University of Derby, Derby, UK

5.1 Introduction

The initial idea for this chapter was to look at barriers to adoption of new technologies, which is too negative a way to approach the topic as it assumes that the active ageing are not adopting technology. This is very far from the truth, as more and more active ageing are using what are termed new technologies. The reality is that perhaps the active ageing are more sophisticated in their approach in opting for tools that work best for them in their particular circumstances—in other words, there is a needs-driven agenda.

This becomes a two-way street: "If I do not need it, why should I have it or have to use it?" and "Yes I need it, but I want it to work how I want it to work." The active ageing, while classed as digital immigrants, are also those who have adopted and used a wide variety of different technologies throughout their lives. They have had to adapt and change not once but many times already. Indeed, it may occasionally be "Enough is enough, if the old technology still works why change?"

Going back to basic marketing principles, what are the drivers behind the active ageing adopting new technologies? What are the benefits for end-users? Why should I spend my limited income on this?

A secondary strand that has caused a backlash against technology is where it has been imposed on the active ageing as being "a health benefit" or an assurance that "this is to keep you safe," with the imposing of the health agenda and the monitoring and assessment of health and well-being taking the form of compliance rather than user uptake.

5.2 Technological use by the active ageing

In looking at the literature, there appears to be two approaches to computer use, with a tendency for papers to be looking more at "reasons of senior citizens for not using the Internet" (Peacock & Künemund, 2007) rather than the second strand looking at "how and why active agers use the Internet."

Peacock reported on reasons for senior citizens not using the Internet in 2007 as part of a European survey, but also noted the trend toward greater use by younger active agers (see Figure 5.1).

Textile-led Design for the Active Ageing Population. http://dx.doi.org/10.1016/B978-0-85709-538-1.00005-5

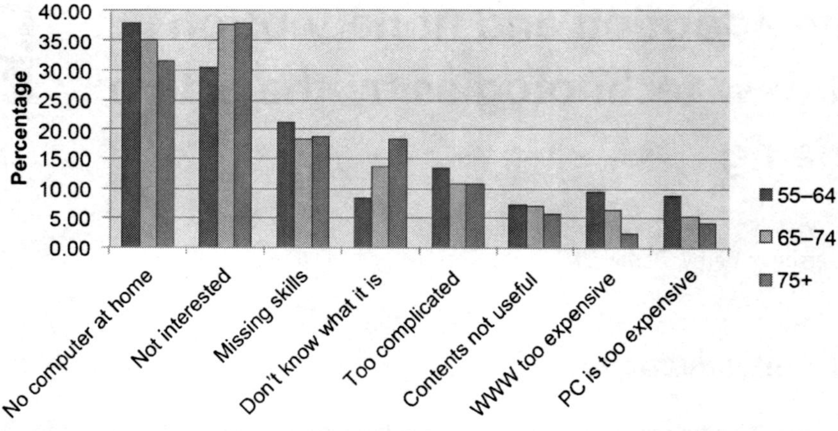

Figure 5.1 Graph of reasons of senior citizens in Europe for not using the Internet in percentages for different age groups.
Created from data in Peacock and Künemund (2007).

Table 5.1 Reasons for not having access to the Internet in the UK (Ofcom, 2011)

Reason for not having access by over 65s	%
Lack of interest—no need for the Internet	31
Age—too old to use the Internet	24
Need—did not want a computer	17
Training—did not know how to use the Internet/a computer	15
Cost—it was too expensive	3

In the United Kingdom, demographics from the independent regulator and competition authority for the UK communications industries (Ofcom, 2011), OfCom, found that older people and those in lower socioeconomic groups are less likely to have Internet access. However, use by older people is increasing, with 55% of those aged 65 to 75, and 26% of those over 75, having home Internet access. The reasons given for not having access are seen in Table 5.1.

5.3 Internet access in care and nursing homes

The lack of Internet access in care and nursing homes has been recognized as a serious issue with reports via blogs (Ayres, 2012) and the national media highlighting not only the benefits but also the lack of availability. "Elderly people are increasingly being 'excluded from modern society' because care homes are failing to offer access to the internet…" (Bingham, 2013).

Table 5.2 Percentage of care and nursing homes with Internet access

Region	No. with residents Internet access	Total no. of residential care homes	Percentage with resident Internet access
Channel Islands	14	54	26.0
England	3201	17,282	18.5
Isle of Man	6	23	26.0
Northern Ireland	30	468	6.4
Scotland	196	1285	15.0
Wales	120	1144	10.5

Figures taken from Care Homes and Nursing Home UK Guide (http://www.carehome.co.uk).

In looking at national minimum standards for care homes for older people (Welsh Assembly Government, 2004), Internet or broadband access is not included, and using figures from the Care Homes and Nursing Home UK Guide (http://www.carehome.co.uk), it can be seen that access to the Internet is seen as a very low priority in care homes, with just over 18.5% of homes in England including Internet access as an advertised service to users (Table 5.2). These figures indicate a severe lack of Internet access for older people with an attitude of why would older people need it or unfortunately not even thinking about providing access as a basic need.

This is in contrast to the data on Internet access for households and individuals from 2011 (Office for National Statistics, 2011), where the key points were:

- 45% of Internet users used a mobile phone to connect to the Internet.
- Six million people accessed the Internet over their mobile phone for the first time in the previous 12 months.
- The use of wireless hotspots almost doubled in the last 12 months to 4.9 million users.
- 21% of Internet users did not believe their skills were sufficient to protect their personal data.
- 77% of households had Internet access.

The differential between the two sets of figures, with 77% of households having Internet access but at maximum 25% of care and nursing homes having access, shows how the digital divide is not at the forefront of thinking for those providing the basic necessities for older people. The Care Quality Commission puts great stress on the ability of people to check and feedback on the quality of their care in the booklet "What standards you have a right to expect from the regulation of your care home" (Care Quality Commission, 2013), but the booklet does not include access to the Internet as a basic right so residents can e-mail to make a complaint.

The director of carehome.co.uk has called on homes to provide this "crucial access" (McCardle, 2013). "Carehome.co.uk has found that only a fifth of care homes – approximately 3400 of the 20,000 across the country – have given residents web access. Davina Ludlow, director of carehome.co.uk, is now calling on care homes to change this and help residents reap the potential health benefits of being online."

5.4 Internet access, leisure activities, and the active ageing

Technology has been used to promote active ageing, with regard to overcoming isolation, by enhancing social connectivity as well as social and civic participation (Gardner, Kamber, & Netherland, 2012). Näsi, Räsänen and Sarpila (2011), in looking at the issue of Information and Communication Technology (ICT) inequality in Finland, found that there was a clear delineation between users and nonusers correlating with their leisure activity patterns: "Seniors who have a variety of different leisure activities are by far more likely to be active users of the Internet than those less active during their day" (p. 173).

The drive, then, is their need to learn about activities and ways to support their ability to live independently. However, reports and views divide into conceptual frameworks looking at independent living in terms of healthcare needs (Institute for Prospective Technical Studies, 2006) and research into smart homes and telecare rather than looking first at end-user needs for the technology. This is still a trend even in recent literature toward what is done to older adults—e.g., phrases like "specially designed technological intervention" (Berkowsky, Cotton, Yost, & Winstead, 2013) rather than working with the older population to find out what they want, i.e., empowering rather than disempowering.

Many campaigns are working on overcoming digital exclusion. For example, Digital by Default 2012 (The Intelligent Business Company, 2011) looked at people in social housing, a sector that includes almost half of all single-parent families and a fifth of all pensioners, and Spring Online, an award-wining digital inclusion campaign (http://digitalunite.com/spring-online).

Other reports suggest the number of people online aged 65 years and older as remaining static to be between 25% and 35% (Nominet Trust, 2013). The report "Ageing and the use of the – internet: current engagement and future needs" (Milligan & Passey) emphasizes that the way older users engage with the Internet is different and is more needs- and benefit-driven rather than the fashion and social communication orientation in younger users. The executive summary highlights two areas in general: why older users are turning to the Internet and how technology is or is not being supported and promoted to this age group. The first of these is more appropriate to look at in this chapter:

- Older people are seeing potential uses that apply to their specific needs and circumstances.
- Any digital divide is more likely to be associated with specific technologies or certain practices, rather than with all technologies and all Internet uses and practices.
- Finding criteria through which to view digital appropriateness for older people, rather than focusing on comparisons across different age groups, is likely to provide more valuable indicators of digital trends and shifts.

This is partly the researchers' finding: If they are looking at health and well-being benefits and the money is going to that end of the market, then that is how those involved in that intervention will perceive how the technology can be used, missing out on its other, far more interesting uses.

5.5 How do the active ageing adopt new technologies?

The division of users into digital natives, immigrants, and aliens has been at the core of thinking for some time (Prensky, 2001a,b), from the more blunt commentaries (Alderman, 2013) to those analyzing the reality that not all young people really are digital natives (Chung, 2013) just as not all of the active ageing are digital immigrants. Others have rightly redirected thinking to those who can afford technology (Joy, 2012) and those who cannot, which applies equally to the old and the young. In a longitudinal study of ageing, Gilleard (Gilleard & Higgs, 2008) found that PC ownership decreased, moving through the age cohorts of the active ageing: "Those born nearer the end of the first half of the twentieth century are more likely to use the Internet than those born closer to its beginning." This reinforces the difficulty of viewing the active ageing as one group, whether for clothing or wearable technology.

There are a number of initiatives to support the wider use of technology and specifically Internet-based technologies by older people (discussed earlier). A number of these projects are in development as co-design processes and others are more intervention-based.

The SAAPHO Project is designed to create "…a usable system that covers health, safety and social needs of end-users taking into account their preferences in the design of the system behavior and architecture" (Doménech, Rivero, Coll-Planas, Sainz, Reissner, & Miralles 2013). The tools employed were a mix of healthcare surveillance using sensors, participation services, and security, again sensor-based in the participant's home for carbon monoxide, temperature, and motion detection. Older people were involved at the beginning of the project, but to assess what was being proposed rather than starting with a blank slate. Even the definition of active ageing they use—"Active Ageing refers to the optimization process in health, social engagement, and security opportunities as people age"—reveals the social care-style approach.

Another project, ePAL, takes a more proactive approach, aiming at developing a roadmap for extension of professional active life of the retiring and retired knowledge workers in developed countries, specifically Europe. The valuing of the older person as a contributor of knowledge and experience is essential for their feeling they can contribute. The project looks at three perspectives: technological, societal, and organizational (Afsarmanesh, Msanjila, & Camarinha-Matos, 2012).

There is a need for positive approaches to the older age group that ask them what they would like and facilitates its implementation; that ask them what they feel they can contribute to society or would like to contribute; and consider remedies and technologies as treatments for old age. Obi and colleagues, in their paper on "Promoting ICT innovations for the ageing population in Japan" (Obi, Ishmatova, & Iwasaki, 2013), recognize the societal changes that are needed to facilitate these innovations, though many are infrastructural needs for governmental development. Lifelong education, social participation, and pro-elderly employment are included alongside more traditional health and protection issues:

1. Lifelong education: encouraging senior citizens to participate in the knowledge-based society through distance learning, and lifelong education as a part of e-participation/e-inclusion.

2. Social participation: promoting positive e-participation to civil society through volunteer activities; expanding the ICT-utilized social networks for elderly people and creating a community volunteering culture.
3. Pro-elderly employment: creating new pro-elderly people employment systems and markets, such as teleworking and three-day working week (Obi, 2013).

There is a sense in some research projects where older people are participants that they are speaking out, and when allowed to view their own opinions are doing so. For example:

"There's strong evidence to suggest that integrating information technologies into people's daily lives can significantly enhance their quality of life. However, such integration requires that technologies be designed with human agency in mind. That theme is one of the key findings of our research with older adults, expressed by them as the belief that they wish to be on the active user side of the digital divide and not simply subjects of pervasive monitoring technologies. This desire to be an active user and not a passive subject of home-based technology expresses a larger theme: that overcoming the digital divide among older adults requires deep consideration of the ways in which technology mediates, influences, and is shaped by human relationships" (Shankar, Camp, Connelly, & Huber, 2012).

Similarly, the New Dynamics of Ageing SUS-IT project found that social processes were key not just for the initial use of computers but their sustained use. As explained in the briefing paper "Older people and digital engagement":

"Some older people reveal *exceptional tenacity in attempts to remain digitally connected,* often persisting in the face of many obstacles posed by age-related changes (e.g. changes in physical ability, memory, support, and/or technology problems/changes). Awareness of the interrelationship between social inclusion and digital inclusion from the perspective of personal experience may be an important part of the explanation for this." (Sus-IT, 2011).

5.6 Wearable technology and the active ageing

In order to get a clear picture of the adoption or nonadoption of wearable technology by the active ageing population, it is important to look at the directions in current research. In analyzing the research literature on wearable technology and the active ageing, it is clear that the predominant approach is a medical or care-based one: the idea that wearable technology, like a monitoring device, will be good for the health and well-being of the individual. Table 5.3 below shows the number of references returned from Scholar with the keywords "wearable technology active ageing" and how many of the references had particular keywords. The references, and by implication the research, is about the health and well-being agenda, not a better life and social interaction of the active ageing through the use of wearable technologies.

The titles of papers exemplify this approach: "Health and aged care enabled by information technology," "Intelligent technology for an ageing population: The use of

Table 5.3 **A scholar.google.co.uk search using keywords "wearable technology active ageing."**

Keyword	Number of references
Medical	5340
Sensing	3380
Care	5190
Measuring	3460
Health	5620
Healthcare	3270
Intervention	2740

The search produced 7200 references; then the smaller number of references when you remove a keyword, such as "medical," showed the number of the references with a particular keyword.

AI to assist elders with cognitive impairment," "Monitoring technology for the elderly patient," "HealthGear: a real-time wearable system for monitoring and analyzing physiological signals," "Special section on smart wearable devices for human health and protection, guest editorial," and "Applications of smartphones for ubiquitous health monitoring and wellbeing management." These are just examples of a trend that sees the active ageing only in terms of their health. This is also where the market can make the most money with the added value of medical interventions compared to solutions that work for all and therefore are mainstream with lower profit margins.

5.7 Tablet technologies and the active ageing

The use of tablets has grown substantially since the introduction of the iPad in 2010; though not the first tablet-style device, it has certainly been the device that has made the greatest impact in education, entertainment, and work-based portable computing.

Even here a quick glance at the research shows that the first idea is how it can be used to promote a healthy active lifestyle rather than the social and fun aspects. The fact that it is a powerful computer enabling easy Internet access without a fixed cable and heavy computer, something that can at one moment be used to read a book and then check the TV schedule or e-mail, look at finances online, or talk via video to relatives across the other side of the world or down the road, seems neglected.

The titles of papers and applications, as in the previous section, suggest a lack of imagination: "Exploring technologies to foster wellbeing and vitality among older people," "Ageing well: can digital games help older adults?," and "Motivating and assisting physical exercise in independently living older adults: a pilot study" (Silveira, Daniel, Casati, & de Bruin, 2012). The latter paper includes a description of how the elderly were taught how to use the iPad. Can we assume in this day and age that even the older of our active ageing have not used a computer when in many cases they have been involved in their development.

There is a tendency, along with the application of technology in health, to treat the elderly as if they are children in terms of the graphics (Silveira, Daniel, Casati, van het Reve, et al., 2012) see Figure 5.2 and in the way the elderly are represented.

As the iPad is intuitive enough to use at a basic level by children with autism and other learning disabilities, surely our active ageing if given iPads to use and take home will soon pick up on what it could do for them. Indeed, if given and not inflicted on the active ageing, the long-term benefit is likely to be greater.

The real aim should be getting the iPad or iPad mini and broadband/WiFi into as many homes of the elderly as possible, which will do more to alleviate issues of the elderly without interventions. This is primarily because a lot of apps designed to support cognitive function, exercise, and healthy lifestyle are being designed for everyone to use, not just the active ageing.

The issues that researchers often look at, while logical at first glance, are ultimately very short-sighted. An example is a paper looking at "How Unfamiliar Words in Smartphone Manuals Affect Senior Citizens" (Ishihara, Kobayashi, Takagi, & Asakawa, 2013). Do any of us actually read our smartphone manuals? If smartphones are so smart, they should not need manuals. It is better for video guides to be built into phones so users do not have to understand the specialized vocabulary.

These issues are raised through organizations like Respecting Elders: Communities Against Abuse (RECAA). As expressed by Kim Sawchuk (Sawchuk, 2013), "RECAA does not consist of older adults who are 'actively ageing,' the new governmental mantra that has gone global (WHO, 2002) with the realization that the world's population is ageing at an unprecedented rate. The women and men of RECAA are ageing as activists and providing a welcome alternative view of spirited oldness" (p. 48).

"RECAA members also understand that they are living in a new communicational social ecology of the digital, and more specifically Web 1.0 and 2.0, which is exerting transformative pressure upon their organization. They are keenly aware of this shift toward mediatization in the communications environment. They may be older, but they

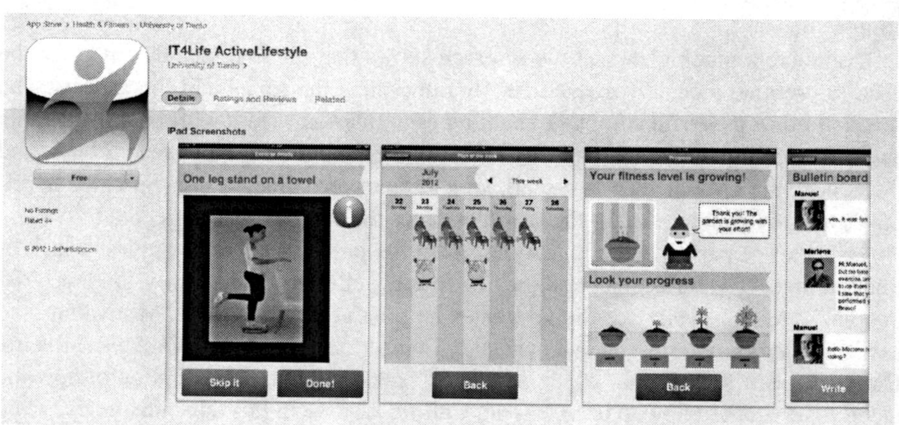

Figure 5.2 Screen capture of IT4Life ActiveLifestyle iPad application. (http://www.lifeparticipation.org/active.html.)

do not live in a bubble. They keep in touch with children and grandchildren over mobile phones and computers. Yet their digital skills vary, and those who have a computer do not necessarily know how to maintain it so that it runs without problems. Although they are in their 70s and above, they are deeply intrigued by what technologies can do for them and what they can do with them. They are curious and critical. One RECAA member (aged 92), who was recently given an iPad, stated bluntly, 'Kim, I like this iPad thing, but you know, I have been keeping a list of what I think they could improve.' "

5.8 Social media, communities, and the active ageing

The blogosphere is more active and aware to the potential of social media as an element in developing social wellness. Looking at the Council for the Third Age's (C3A) Age Well programs, Roan Yong (Yong, 2012) suggests three areas that help to strengthen communities:

1. Social media connect the seniors' activity with that of their family and friends.
2. Social media capture stories and promote knowledge sharing.
3. Social media provide space for conversation and community-building.

In comparing this section with the previous two, there is much more meaningful use of technology where either e-learning or social media are being looked at to support the active ageing, yet the technology is the same. There are still elements of technology being applied to the elderly, but less so. This shows the importance of emphasizing the social representation of the Internet rather than purely its use in promoting active lifestyles.

For example, from the 2012 paper by Gardner (Gardner et al., 2012): "ICT's have the potential to contribute to individual and community empowerment and capacity building and can be an effective antidote to social exclusion among older adults. Results from this study suggest comprehensive, community-based ICT training programs (such as OATS), can support the ongoing engagement and re-engagement of older adults within society by building, maintaining and restoring their place within their various communities (including city, neighborhood, family, and communities of choice based on interest or identity)."

Research by Cotten (Cotten, Anderson, & McCullough, 2013) provides evidence about the impact of Internet use on loneliness, concluding that "Using the Internet may be beneficial for decreasing loneliness and increasing social contact among older adults in assisted and independent living communities."

5.9 Conclusions

In many ways the relationship between society, the Internet, wearable technology, and the active ageing is symptomatic of the perception that old age is a disease or condition to be treated. The adoption or nonadoption ends up being an issue of compliance to treatment protocols, rather than using them to promote greater social freedom and interaction and the development of communities of active ageing, in a situation where the active ageing are also being segregated from the rest of the population through lack of transport and basic infrastructure.

5.10 Sources of further information and advice

There is a wide variety of books that look at changes in adoption and nonadoption of digital technologies, including *Digital Habitats: Stewarding Technology for Communities* by Etienne Wenger et al. (Wenger, White, & Smith, 2009) and *Information and Communication Technologies for Active Ageing: Opportunities and Challenges for the European Union* by Cabrera and Malanowski (Cabrera & Malanowski, 2009). Other sources on the Internet, including those talking to and engaging with active agers, cover projects like Sus-IT http://sus-it.lboro.ac.uk and outputs like "What older people want: A catalog of co-designed ICT concepts" (Frohlich, Lim, Woods, & Ahmed, 2012).

References

Afsarmanesh, H., Msanjila, S. S., & Camarinha-Matos, L. M. (2012). Technological research plan for active ageing. *Information Systems Frontiers*, *14*, 669–692. http://link.springer.com/article/10.1007/s10796-011-9295-z.

Alderman, T. (2013). *The fractious world of digital natives, immigrants and aliens*. Huffington Post, The Blog. http://www.huffingtonpost.com/tom-alderman/the-fractious-world-of-di_b_2821090.html. Accessed 18.09.13.

Ayres, S. (2012). *Should all care homes provide internet access for residents?* http://-shirleyayres.wordpress.com/2012/01/19/should-all-care-homes-provide-internet-access-for-residents/. Accessed 30.09.13.

Berkowsky, R. W., Cotton, S. R., Yost, E. A., & Winstead, V. P. (2013). Attitudes towards and limitations to ICT USE in assisted and independent living communities: Findings from a specially-designed technological intervention. *Educational Gerontology*, *39*(11), 797–811. http://dx.doi.org/10.1080/03601277.2012.734162.

Bingham, J. (2013). *Most care home residents 'excluded from modern society' by lack of internet access*. The Telegraph. http://www.telegraph.co.uk/finance/newsbysector/mediatechnologyandtelecoms/digital-media/10250664/Most-care-home-residents-excluded-from-modern-society-by-lack-of-internet-access.html. Accessed 30.09.13.

Cabrera, M., & Malanowski, N. (2009). *Information and communication technologies for active ageing*. Amsterdam, Netherlands: IOS Press.

Care Quality Commission. (2013). *What standards you have a right to expect from the regulation of your care home*. http://www.cqc.org.uk/sites/default/files/media/documents/standards_to_expect_carehome.pdf. Accessed 30.09.13.

Chung, J. (2013). *The digital native fallacy: Teaching new skills while learning from their strengths*. Huffington Post, The Blog. http://www.huffingtonpost.com/jared-chung/the-digital-native-fallac_b_2774480.html. Accessed 18.09.13.

Cotten, S. R., Anderson, A. A., & McCullough, B. M. (2013). Impact of Internet use on loneliness and contact with others among older adults: Cross-sectional analysis. *Journal of Medical Internet Research*, *15*(2), e39. http://dx.doi.org/10.2196%2Fjmir.2306.

Doménech, S., Rivero, J., Coll-Planas, L., Sainz, F. J., Reissner, A., & Miralles, F. (2013). Involving older people in the design of an innovative technological system promoting active aging: The SAAPHO project. *Journal of Accessibility and Design for All*, *3*(1), 13–27. http://jacces.org/jacces/ojs/index.php/jacces/article/viewArticle/76.

Frohlich, D., Lim, C., Woods, S., & Ahmed, A. (2012). *What older people want: A catalogue of co-designed ICT concepts*. Guilford: University of Surrey, Digital World Research Centre. http://sus-it.lboro.ac.uk/SusIT_DWRC.pdf. Accessed 30.11.13.

Gardner, P. J., Kamber, T., & Netherland, J. (2012). "Getting turned on": Using ICT training to promote active ageing in New York city. *The Journal of Community Informatics*, *8*(1), 1–16. http://www.ci-journal.net/index.php/ciej/article/view/809. Accessed 30.11.13.

Gilleard, C., & Higgs, P. (2008). Internet use and the digital divide in the English longitudinal study of ageing. *European Journal of Ageing*, *5*(3), 233–239. http://link.springer.com/article/10.1007/s10433-008-0083-7.

Institute for Prospective Technical Studies. (2006). *User needs in ICT research for independent living, with a focus on health aspects*. ftp.jrc.es/pub/EURdoc/eur22352en.pdf. Accessed 30.09.13.

Ishihara, T., Kobayashi, M., Takagi, H., & Asakawa, C. (2013). How unfamiliar words in smart-phone manuals affect senior citizens. In *Universal access in human-computer interaction. Applications and services for quality of life (pp. 636–642)*. Heidelberg, Berlin: Springer.

Joy, O. (2012). *What does it mean to be a digital native?* http://edition.cnn.com/2012/12/04/business/digital-native-prensky. Accessed 18.09.13.

McCardle, L. (2013). *Carehome.co.uk reveals over 80 per-cent of care home residents are missing out on 'crucial' internet access*. http://www.carehome.co.uk/news/article.cfm/id/1560819/plea-to-get-residents-online-as-research-reveals-thousands-are-missing-out-on-crucial-internet-access. Accessed 30.09.13.

Milligan, C., & Passey, D. (2011). *Ageing and the use of the – internet: Current engagement and future needs*.Nominet Trust. http://www.nominettrust.org.uk/sites/default/files/NT%20SoA%20-%20Ageing%20and%20the%20use%20of%20the%20internet_0.pdf. Accessed 30.09.13.

Näsi, M., Räsänen, P., & Sarpila, O. (2011). ICT activity in later life: Internet use and leisure activities amongst senior citizens in Finland. *European Journal of Ageing*, *9*(2), 169–176. http://link.springer.com/article/10.1007/s10433-011-0210-8.

Nominet Trust. (2013). *Ageing and the use of the internet*. http://www.nominettrust.org.uk/knowledge-centre/articles/ageing-and-use-internet. Accessed 30.09.13.

Obi, T., Ishmatova, D., & Iwasaki, N. (2013). Promoting ICT innovations for the ageing population in Japan. *International Journal for Medical Informatics*, *82*, e47–e62. http://dx.doi.org/10.1016/j.ijmedinf.2012.05.004.

Ofcom. (2011). *Communications-infrastructure-report 2011*. http://maps.ofcom.org.uk/broadband/downloads/ofcom-uk-broadband-speed-report-2011.pdf. Accessed 29.09.13.

Office for National Statistics. (2011). *Internet access – Households and individuals*. http://www.ons.gov.uk/ons/rel/rdit2/internet-access--households-and-individuals/2011/stb-internet-access-2011.html. Accessed 30.09.13.

Peacock, S. E., & Künemund, H. (2007). Senior citizens and Internet technology. *European Journal of Ageing*, *4*(4), 191–200.

Prensky, M. (2001a). Digital natives, digital immigrants – Part 1. *On the Horizon*, *9*(5). http://www.twitchspeed.com/site/Prensky - Digital Natives, Digital Natives, Digital Immigrants -Part1.htm. Accessed 18.09.13.

Prensky, M. (2001b). Digital natives, digital immigrants, part II: Do they really think differently? *On the Horizon*, *9*(6). http://www.twitchspeed.com/site/Prensky - Digital Natives, Digital Immigrants - Part2.htm. Accessed 18.09.13.

Sawchuk, K. (2013). Tactical mediatization and activist ageing: Pressures, push-backs, and the story of RECAA. *MedieKultur, Journal of Media and Communication Research*, *29*(54), 47–64. http://ojs.statsbiblioteket.dk/index.php/mediekultur/article/view/7313/6765. Accessed 24.10.13.

Shankar, K., Camp, L. J., Connelly, K., & Huber, L. (2012). Aging, privacy, and home-based computing: Developing a design framework. *Pervasive Computing*, *11*(4), 46–54. http://dx.doi.org/10.1109/MPRV.2011.19.

Silveira, P., Daniel, F., Casati, F., & de Bruin, E. D. (2012). Motivating and assisting physical exercise in independently living older adults: A pilot study. *International Journal of Medical Informatics*, *82*(5), 325–334. http://dx.doi.org/10.1016/j.ijmedinf.2012.11.015.

Silveira, P., Daniel, F., Casati, F., van het Reve, E., & de Bruin, E. D. (2012). *ActiveLifestyle: An application to help elders stay physically and socially active*. http://www.lifeparticipation.org/images/ActiveLifestyleApp.pdf. Accessed 30.11.13.

Sus-IT. (2011). *Older people and digital engagement*. Briefing Paper 1 . http://sus-it.lboro.ac.uk/SusIT_Briefing_Paper_1.pdf. Accessed 30.11.13.

The Intelligent Business Company. (2011). *Digital by default 2012: The case for digital housing*. http://www.housing-technology.com/dbd.php. Accessed 30.09.13.

Welsh Assembly Government. (2004). *National minimum standards for care homes for older people*. http://wales.gov.uk/subsites/caresocialservicesinspect/publications/nmsolderpeoplecare/olderpeopleche.pdf?lang=en. Accessed 30.09.13.

Wenger, E., White, N., & Smith, J. (2009). *Digital habitats*. Portland: CPsquare.

World Health Organization, (2002). *Active ageing: A policy framework*. http://whqlibdoc.who.int/hq/2002/WHO_NMH_NPH_02.8.pdf. Accessed 08.05.14.

Yong, R. (2012). Cultivating communities. http://roanyong.com/category/knowledge-management/cultivating-communities/ Accessed 24.10.13.

Part Two

Understanding and researching apparel needs amongst the active ageing population

Qualitative and quantitative methods applied in active ageing

6

B.R.M. Manning, S. Benton
University of Westminster, London, UK

6.1 Introduction

This chapter is set against the novel background of the integration of electronic and textile technologies within a multi-layered garment system, which would create an attractive ensemble for the active elderly that also functioned as a mobile multi-functional communications platform. Whilst its ultimate success would depend on gaining an in-depth understanding of the psychology of ageing and its attendant needs and behavioural characteristics, it would equally rely on the effective transfer of this knowledge to the two separate research teams, each with entirely different ways of working (Duranti & Goodwin, 1992).

From the clothing team's perspective, their design process involves a close interactive relationship between client and designer evolving through a series of iterations to achieve a mutually satisfactory result. By contrast, the engineering team's approach is based on agreement of a functional requirement specification at the outset. This then proceeds through a series of structured technical design iterations before the result is generally presented to the client for acceptance trials.

The potential dichotomy that exists between these two entirely different ways of working and their associated disciplinary cultures highlights the key issues that can bedevil effective elicitation and transfer of acquired knowledge. Whilst they are inherently rooted in language, the major stumbling block lies in how each recipient perceives and extracts meaning of the content that is presented to them.

Similarly this issue not only forms the core of the problem to be faced in eliciting relevant information from an appropriately large sample of the active elderly, but is then also magnified by the need to translate and transfer the findings back to the two teams (De Fina, Schiffrin, & Bamberg, 2006). As the elicitation process is inherently a questioning one, it is beset by the twin problems of unambiguously conveying the intended meaning and of obviating any latent bias.

6.2 Meaning and interpretation

The basic problem with interpreted meaning is that the content is inherently judged in the context of personal experience. Although this is obviously unique to each individual, the greater percentage of it is common to all within any given society.

Textile-led Design for the Active Ageing Population. http://dx.doi.org/10.1016/B978-0-85709-538-1.00006-7

This is then supplemented to varying degrees by experiential knowledge generated within and across groups with similar shared interests. Beyond this there remains a significant measure that is highly personalised to each individual, and comprises a complex mix of interactive experiences of the real world coupled with internalised reasoning, value judgements and beliefs (Berger & Luckmann, 1967).

The key issue then is how best to cope with the inherent ambiguities that all too easily occur in the interpretation of meaning behind questions presented as part of any elicitation process (Gauker, 2003). This problem is somewhat of double-edge sword, as it applies just as much to delivering its results – or even the contents of this chapter!

Probably the best starting point for all such exchanges is to anchor them on generally accepted factual information. In essence such 'a posteriori' knowledge will have been derived from previously established and tested evidence, and subsequently received via educational systems and the media. A second, more philosophical source, is 'a priori' knowledge, which is based on inferred reasoning and moulded by a range of psychological reactions, evolved responses and resultant preferences.

Unfortunately meaning derived from these two knowledge streams can be modified by contexts in which the recipients are instinctively operating to radically alter the perceived meaning of the intended content (Ogden & Richards, 1923). These contextual modifiers broadly centre on:

- Linguistic context
- Situational and environmental contexts
- Racial and gender contexts
- Cultural and belief contexts

6.2.1 Linguistic context

Word usage and phrasing can all too easily alter the way the recipient interprets the meaning away from that which was intended (Searle, 1979).

What is presented to any recipient is interpreted in terms of their existing knowledge and experience of using the particular language, their perceptions of the focus of the dialogue and its implied intent. Unfortunately word selection and usage together with phrasing can all too easily alter the way the recipient interprets the meaning away from that which was intended. Moreover this can then be further distorted by errors in correctly identifying the concepts and objectives involved.

6.2.2 Situational and environmental contexts

These are cues that tend to lie hidden within the presented content that are assumed to be known and can be intuitively recognised by the recipient. These tend to be unstated references back in time to figures, issues, circumstances or events that those familiar with a particular topic domain are expected to be aware of or recall. Once again, this is not necessarily the case, and missing such important nuances can radically skew or alter the perceived intent.

6.2.3 Racial and gender contexts

Built-in societal constructs that lead to stereotyping, stigmatisation and prejudice are a ready source of negative reaction and bias that once again can predispose recipients to misconstrue meaning or intent. This can be particularly true where clothing designs and appearances are involved, and run counter to attitudes and behavioural norms.

6.2.4 Cultural and belief contexts

A further basis for misinterpretation or offence can stem from ignorance or failure to recognise potential issues that can be aroused by inadvertently misunderstanding the significance of various customs, habits, and values exhibited in today's multicultural society.

6.2.5 Latent ambiguities

Although avoiding the pitfalls that can lead to misinterpretation of meaning in any form of discourse is particularly important, it is especially so when the aim is to elicit information. Forming appropriately worded questions or subtle prompts, to draw out thoughts, opinions and perceptions from recipients, is therefore crucial to obtaining useful information.

The snag is that the likelihood of total unanimity of experience, knowledge, and thought processes between the parties concerned is vanishingly small (Stonier, 1997). Hence, all researchers need to be well aware that, however careful they are, their results will inherently be subject to a degree of experimental error. This should be recognised, and ideally quantified in their findings, to avoid being taken pedantically as absolute certainties by the world at large.

Like it or not, this position was neatly summed up by Donald Rumsfeld, the US Secretary of Defense from 1975 to 1977, when he came out with the following comment:

> 'There are known knowns.
>   These are things we know that we know.
> There are known unknowns.
>   That is to say, there are things that we know we do not know.
> But there are also unknown unknowns.
>   There are things we do not know we do not know.'

6.3 Knowledge acquisition

As the psychosocial and associated socioeconomic domains do not lend themselves to direct quantitative measurement, the collection, collation, and analysis of information has to rely on qualitative elicitation techniques (Marshall & Rossman, 1999). These broadly are split between:

- Direct face-to-face contact
- Survey questionnaires

Although these two approaches are completely different in both their objectives and the techniques used, they are mutually supportive (Figure 6.1). In essence almost always survey questionnaires need to be preceded by in-depth investigation that reveals the key factors that needing further probing (Creswell, 2006).

The common objective of both these approaches is ultimately to add to the fund of readily accessible knowledge of specific domains, or amend it in the light of new information obtained. Other than in exceptional circumstances, the start point for all investigations is to use existing knowledge resources as the base from which to establish and anchor the search for relevant new information and insights in a defined topic area (Shadbolt, O'Hara, & Crow, 1999).

Depending on the current level of knowledge and expertise available, either of the two main techniques or a combination of both may be deemed appropriate. Where the focus is on experience-based opinions and preferences relating to a topic that is well supported by an evidence base, the use of survey questionnaires can suffice. However, when new knowledge has to be unearthed from individual sources of expertise or tacit knowledge, interviewing will be essential. The 'cross-over' combination point comes where additional knowledge and insights are necessary to set up an effective survey.

6.4 Qualitative research methodologies

6.4.1 Interviews

By definition, face-to-face investigations revolve around either group discussions or individual interviews focussing on various aspects of the field in question with a set of appropriately knowledgeable individuals whose expertise, experience, and opinions need to be teased out.

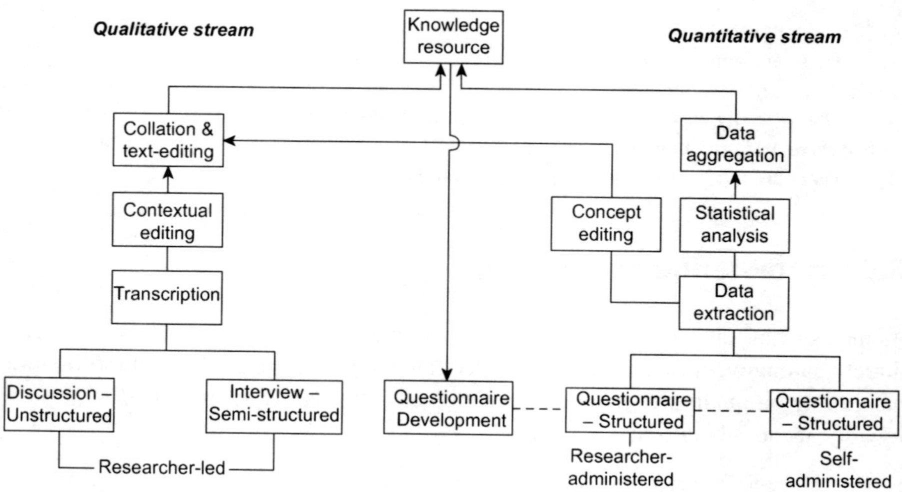

Figure 6.1 Knowledge processing.

Although these forms of research are of necessity time consuming and therefore costly, they are the only effective way of amassing sufficient information from which to establish a well-rounded, in-depth knowledge of the topic domain.

Whilst a focus-group approach can rapidly canvas a wide range of opinion in a relatively short time, it is inherently difficult to handle and keep on track, and may end up skimming the surface of the topic. By contrast, one-to-one interviews can yield considerably more valuable detail, albeit consuming far more time and effort.

Regardless of the approach taken, interviewing is essentially an art, whose basic skills can be learnt, but one that needs to be honed by continued practice. Although it should ideally be seen either as a relaxed conversation or as a convivial discussion, it needs an agreed focus that sets up the basic arena and nominal boundaries for the discussion.

Behind this, interviewers need to have a pre-planned framework within which to guide their gentle probing, and that allows them to steer the conversation back to the core topic and to cover all the key areas of interest. Although it is up to each interviewer to determine the degree to which this structure is imposed, they need to exercise careful control over 'out-of-area' excursions, where a fine balance is needed not to cut off valuable insights into 'cross-border' issues, whilst not losing sight of the main objective (Foddy, 1994).

Ideally this should be digitally recorded in situ to maximise the information gathered for subsequent review, collation, and editing. Whilst much of this will need to be copied across from voice into text, video input is equally important, as it can hold a host of behavioural cues that will also need to be noted and appropriate comments edited into the session record to ensure nothing is lost or misconstrued.

However, it must be remembered that this is only the first part of the research process, as the combined session results then have to be analysed to capture the overall 'rich picture' that emerges. This inevitably forms a complex pattern of concepts, interactions and inter-relationships that have to be documented and referenced back to the appropriate source evidence. Once peer-reviewed and published, it can then be added to the knowledge base of the topic domain.

6.4.2 Group processes

Although one-to-one sessions will potentially yield considerable in-depth knowledge and expertise, they also are likely to contain some measure of personal bias and strongly held opinions, which may need to be challenged. By contrast, group processes aim both to explore and contrast opinions, beliefs and attitudes held by its members. This produces information and insights as it stimulates memories, ideas, and experiences in participants that would otherwise be lost without this interaction.

6.4.3 Focus groups

These are discussion sessions led by a facilitator through a structured series of stages that address a set of questions that are designed and presented to provoke a positive contribution from the members. Obviously this needs careful management to ensure that all concerned can contribute their views, the key primary and secondary areas of interest are addressed, and a balanced set of conclusions at the end of each stage is achieved.

Whilst there are several variants of this basic theme, they generally involve some form of pairing – either involving two facilitators or dividing the group into halves – that enable comparison between opposing viewpoints, and ultimately leading towards consensus based on emerging shared perceptions.

The former case uses a 'dueling debating' technique used to draw out and examine contending arguments. The latter is a two-way approach where one half watches the other debate the issues, and then responds by critiquing these results and offering additional insights. Both methods have the advantage that they can lead to different and more subtle conclusions than those obtained by the simpler 'unipolar' one.

6.4.4 Design workshops

These are effectively variants of the focus group approach that incorporate a practical hands-on approach to the development of new concepts and approaches to design issues. The key benefit is that the group members can see, touch, and feel aspects of a design as it evolves, and provide feedback on its functionality, usability, and its ultimate acceptability.

This technique is particularly appropriate to products and services influenced by, or based upon, architecture and art, as it allows behavioural responses to be observed in context, and most importantly to pick up the visual cues to what could never be said – and hence might be lost.

6.5 Survey techniques

By contrast, research surveys use questionnaires to reach out and collect statistical input from a larger cohort of relevant individuals with interests or expertise relating to the topic area under investigation with the overall objective of obtaining profiles of their views at various levels of detail.

Creating a questionnaire that can collect meaningful data that reveals the impressions, opinions, and attitudes toward newly emergent concepts is essentially a complex bridging exercise between researchers and their audience. This depends on being able to translate quite subtle and frequently complex ideas into questions that are easily understood and unambiguously phrased.

This depends on developing a framework architecture that enables the accumulated topic knowledge to be reconstituted into a narrative sequence of concepts from which a logical series of sets of questions can be derived. These sets not only can directly test reactions to these ideas but also predicted consequences that have been derived from them.

The resultant statistics obtained are particularly important in that they provide a firm basis for well-founded decisions on ongoing and future research and application developments. Although primarily focused on gaining quantitative perspectives, questionnaires can also be used to probe for valuable additional qualitative input in similar ways to those used in interviewing.

In this case, questions are 'open' in that they ask the respondent to enter additional thoughts or comments on the preceding item in free text within a boxed space. However, overuse tends to inhibit responses by increasing the need to reflect and recall prior content rather than respond to prior prompts. As a result, this can break up a sense of continuity in working through the questionnaire.

Common to both approaches is the need to recognise and respond to the physiological dynamics involved in questioning and exploring issues. This is particularly so when there is no direct contact with the respondents, who might otherwise wish to seek guidance in interpreting the intended meaning behind some questions.

The design of questionnaires therefore needs to minimise potential ambiguities that could arise in the phrasing of questions and also make them concise and explicit. Then in the same way that any effective discourse needs to flow effortlessly, the sequencing and style of the questions need to keep the respondent engaged and interested right through to the end of the process.

This raises the thorny issue of how long and how many questions it is reasonable to pose. A balance therefore has to be struck between achieving all the research goals without disturbing the concentration of the reader and thereby undermining the authenticity of their responses.

As ever, this will invariably lead to compromises between the ideal and the practical, and calls for a pragmatic solution. This will need to be based upon a careful judgement as to the sampled population's motivation and tolerance, as much as the cognitive and emotional demands made by the questions. If this leads to constraints to the intensity and numbers of questions asked – then so be it – as quality should always trump quantity.

Much of the answer lies firstly in prioritising, and then varying, the ways in which questions are posed. This variability is essentially built-in, in terms of the form of response presented. Responses can typically range from a simple 'yes–no' option, through rating scales of various types and formats, to multiple-selection choices. Beyond this, a greater degree of sophistication can be introduced by using a matrix of multiple choices that includes a combination of rating and ranking procedures that expand the level of detail obtained within a single overarching question.

Where identification of consumer preferences in product selection are especially key to the viability of any design, the inclusion of sets of visual options into the mix is vital. This is particularly necessary in complex co-design situations such as those pioneered in the NDA Design for Ageing Well project in terms of a selection of clothing styles, shapes and cuts, and should ideally extend into all aspects of respondent interaction with any built-in wearable technologies.

These preferences tend to be influenced not just by the available functionality, look, and feel of the end product but also the attitudes and behavioural traits of the bulk of its potential consumers. Establishing such underlying consumer psychological profiles relies on a further variant of the basic question response formats, which uses a short ten-question form of psychometric test focussing on five very broad domains using paired sets of questions that probe opposing 'poles' of the respondents' perceptions of themselves.

Ultimately the key factor determining the success of the whole process, and any other research survey for that matter, centres on ensuring the 'buy-in' and preparedness to complete questionnaires from as large a cohort as possible. The aim must therefore be to ensure the experience is as relevant, interesting, and as satisfying as possible to take part in.

Although the analysis phase is relatively straightforward, it has to await the bulk of the expected returns to appear before work can realistically commence. Usually the broad picture begins to appear quite quickly section-by-section; however, so do some of the inevitable oddities that require investigation to resolve or delete.

The analysis itself ultimately entails a range of techniques dependent on the type of qualitative or quantitative data and providing the results in the most appropriate format to suit the particular audience involved.

Whilst broadly a lower-cost exercise than interviewing, it can be a time-consuming process that can easily overrun planned expenditures. From the outset, project plans need to take this into account by building in sufficient contingency to cover effects of delays and other factors beyond their direct control.

Locating sources and sufficient respondents to ensure a statistically appropriate cohort that takes account of nonresponders; mailing out questionnaires and chasing slow responders; and finally collating, resolving anomalies, and analysing the returns prior to generating the closing report can all introduce substantial delays and added costs.

6.6 Direct contact information-gathering techniques

As discussed previously, qualitative knowledge acquisition in psychosocial scenarios must centre on interactive dialogue with those who have in-depth expertise, practical experience, or tacit knowledge of the topic domain. Gathering such information will depend on establishing an appropriate mix of one-to-one or one-to-many interviews.

In the former case, this will involve essentially unstructured in-depth interviews or conversations; in the latter case, the role of the interviewer changes to that of a facilitator prompting an audience to respond to a range of sometimes provocative questions or statements (Kvale, 1996).

Although both situations will need a common underlying structural framework within which to operate, the in-depth interview will be broadly unstructured, responding in tune to the flow of the conversation, whereas the focused discussion will rely on a semi-structured approach, always in the end drawing the interviewee back to the core aspect of the session.

6.6.1 Unstructured interviews

As all interviews are effectively conversations that range across a topic area within implied or agreed boundaries, the simplest is essentially just that. Even though they have no predetermined questions or sequence and seem to just go with the flow, subtle guidance by the interviewer is needed to get the most out of the discussion.

Achieving this requires careful in-depth preparation to ensure the interviewer has sufficient detailed and relevant knowledge to be able to respond and steer the session appropriately. If not, the participants are likely to be drawn into a series of lengthy and often disconnected and extended diversions that obscure or obliterate potentially valuable input.

This is where the highest level of interviewing skills are needed to put the interviewee at ease and promote conditions in which all manner of valuable and often unexpected insights emerge and are captured. Particular care needs to be taken at all times to avoid biases slipping into the discussion and inadvertently skewing the results. Even then the problem is that there is still no certainty that the process will yield anything of significant value.

6.6.2 Semi-structured interviews

The next step is to inject a measure of structure, whilst maintaining the same level of freedom and adaptability. In order to achieve a this mix of open and close-ended questions from a pre-planned set should be inserted at appropriate moments in a way that does not disturb the overall flow.

Depending on the interviewee's responses, the interviewer can adapt the thrust of the discussion to clarify or elaborate on points made, whilst at the same time taking due account of any concerns or sensitivities that may emerge. This demands much the same level of interviewing skills as in the unstructured approach, coupled with the ability to steer the conversation to ensure everything is covered without blocking access to unexpected 'treasures'.

This should ensure optimum value is obtained in each interview, whilst providing a framework within which to collate, compare, and contrast inputs and which can allow for insertion of additional questions into the core set. Moreover, keeping options open throughout ensures that no opportunities are lost to glean as much as possible out of each session, and thus maximise the return on the investment of time and effort.

6.6.3 Structured interviews

In effect, a structured interview is a researcher-administered survey questionnaire. As with indirect self-administered questionnaires, it uses a predetermined sequence of fixed-response questions that require answers to be chosen from a set of alternatives.

Although some open-ended questions can be included in both questionnaire formats, they tend to break up the thought process, as they require a more considered response. Although this works reasonably well when the presence of an interviewer ensures a response, they are otherwise often skipped or answered with the briefest of comments.

The main research benefit of standardised formats is that responses can be compared and contrasted over time and between different cohorts, which increases data reliability. Moreover, the results obtained are far less dependent on interviewer skills and are therefore potentially cheaper.

6.6.4 Interviewing techniques

In setting up the interview, it is important to set the scene by confirming both the purpose and expected duration of the session, together with an explanation of how the content of the discussions will be recorded and subsequently used. The interviewees should be told why they had been selected, and their consent formally obtained for the use of their input, subject to anonymity or other constraints that they feel necessary.

The interview process itself draws on the capacity of people to respond and express themselves within a highly personalised context, which can be charged with varying types and degrees of emotion. As a complex blending of behavioural and cognitive processes, its aim in accessing individual views and feelings should be seen by all the parties as an opportunity to create a shared source of knowledge and insight. Thus all interviews will challenge the ability of the interviewer both to productively control the process as well as to minimise any untoward attitudes, behaviours or reactions.

The interview needs a relaxed, conversational ambience to work well. The interviewee should feel comfortable and set at ease, and the whole dialogue should be conducted as naturally as possible to minimise stress.

The flow of the conversation should dictate both the questions asked or omitted, as well as the sequence. Whilst the aim should be to stay within the topic boundaries, the odd excursion outside may be decidedly fruitful and yield valuable information – but should be deftly guided back to the central area of interest.

The art of interviewing is actually more about listening than posing probing questions – after all, it is the input that the interviewee provides that is the crucial component. A key skill in building and maintaining momentum throughout the process depends on effective summarisation and clarification at appropriate points in the conversation, as it helps reduce and manage any insidious misperceptions and misunderstandings.

Similarly, questions should be concise, unambiguous, and to the point, and slipped into the conversation appropriately without interrupting the flow. Leading or biased questions should be avoided at all cost, and personal opinions should not be allowed to surface if controversial views are expressed.

The aim should be to finish the session on time, whilst making sure that everything has been covered sufficiently, without seeming to force a closure before the flow has begun to slow naturally. Before drawing to a close, it is important to summarise the main findings and ask the interviewees to confirm their agreement, and then move on to pose an open question that enables them to reflect on answers given and issues raised.

The interviewee should be offered the opportunity to review a transcript of the session, in case they have anything positive to add subsequently; and then of course thanked for giving their valuable time to assist the project to move forward.

6.6.5 Processing information gathered

Where in the past interviews would have been recorded by a stenographer or in handwritten note form, digital voice or video has now become the norm. However, the process of post-interview editing essentially remains the same in that the transcript is reviewed and edited by the interviewer to include 'off-line' contextual comment.

The advantage of digital systems is not only that there is a permanent record of all aspects of each event, but also that speech-to-text software such as *Demon* and *Siri* can provide transcripts of each of them for subsequent processing and analysis. This allows the input from this knowledge acquisition process to be collected from a large number of source events and brought together into a single digital content resource.

The primary issue is then how to identify and extract the key 'pearls of wisdom' from the mass of content. In essence this comes down to a 'super-indexing' task that involves highlighting and then marking-up the text in line with an evolving set of classification rules (Wertz, Charmaz, & McMullen, 2011). This approach enables indexers to develop a codification framework that both identifies concept clusters and associated links, as well as accommodating ongoing finds from other sources, such as responses to other open-questions that can be collected from relevant surveys (Patton, 2002).

6.7 Qualitative analysis techniques

Although Text Data Mining is the broad heading for this highly complex topic, a number of underlying techniques use different variants of the super-indexing approach to analyse and then display the various structures and relationships that exist within complex evolving knowledge resources. These techniques typically include:

- Grounded theory
- Cognitive mapping
- Concept maps
- Mind maps

In many ways these are subsets of the wider field of digital libraries and complex techniques based on techniques developed within computational linguistics.

6.7.1 Grounded theory

This is a qualitative research methodology developed for use in sociology as a means of formulating hypotheses that can be deduced or inferred from the analysis of textual data. As such their use of theory is somewhat of a misnomer, as a scientific theory is one that has undergone extensive and rigorous testing (Glaser & Strauss, 1967).

In this case, key points identified in the data are tagged and given arbitrary codes extracted from the relevant piece of text. These coded anchor points are grouped into sets of similar concepts, and then in turn brought together to form broad groups of categories that form the basis of a theory of the underlying logic that underpins the knowledge that emerges.

Unlike the following approaches, Grounded Theory does not include a system-specific mapping process, but could use any of these generic methods or one of several commercially available software products.

6.7.2 Cognitive mapping

Originally developed as an operational research technique focused on goal-centred corporate governance and strategic problem-solving, it is used to analyse qualitative input surrounding management issues and options. In this context it usually forms into a three-level hierarchical structure where the top level goals are supported by strategic direction underpinned by potential options. However, it is equally applicable as a generic analytical tool for both operational and qualitative research.

Although starting from the same form of textual analysis as Grounded Theory, Cognitive mapping uses Kelly's Personal Construct Theory as its base. Its aim is to number and extract key concepts only in terms of short mono- or bipolar phrases, ideally incorporating an active verb to give a sense of action and direction. Kelly's bipolar approach is important, as it can add richness by capturing the 'shades of grey' and nuances that can be otherwise be lost (Ackermann, Eden, & Cropper, 1990).

This process continues until the whole dialogue has been processed, after which the complete set of extracts has to be mapped to create a diagrammatic version of the content of the original dialogue (Figure 6.2). This is created using arrow-headed links between concepts to create logic chains, with negative signs placed close to the arrowhead to indicate onward negative impacts or effects.

Significant additional insights can be revealed via bipolar formatting, in the form of 'this' rather than 'that' indicated as 'X … Y' within the text or where an opposite pole is implied. Equally having to seek a 'best fit' of consequential chains, where no

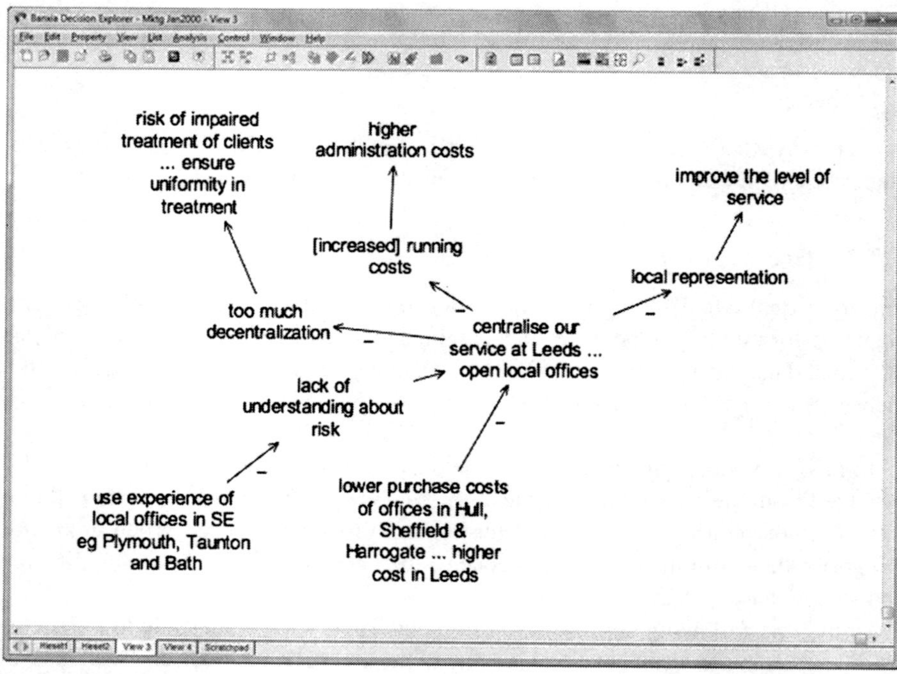

Figure 6.2 Cognitive map.

concept can be left out, gaps in supporting logic can be revealed together with unexpected clustering that can indicate hidden biases or preferences.

6.7.3 Concept maps

These were originally developed out of research into the psychology of learning and the development of knowledge structures in childhood, and are graphical tools for organising and representing knowledge. They have subsequently been increasingly used in computing and other high-technology areas to represent, organise, and visualise knowledge as a structured set of interlinked concepts within a domain (Novak & Cañas, 2008).

Once again, the core building blocks are concepts, which are labelled appropriately within a variety of boxes or other shapes, which may be used to differentiate between types of objects or events. Concepts also include 'propositions', which are statements that enlarge on various aspects of the basic objects or events. All relationships are then specified by link words or short phrases superimposed on connecting lines between individual concepts (Figure 6.3).

When used in academic research settings, maps are usually developed downwards and outwards, from a focus question via more general high-level concepts through to the more specific details in a hierarchical fashion. As this complexity increases, the structure tends to form into discrete segments, where cross-links are used to relate similar concepts together. These cross-links often represent creative leaps on the part of the knowledge producer in the creation of new knowledge.

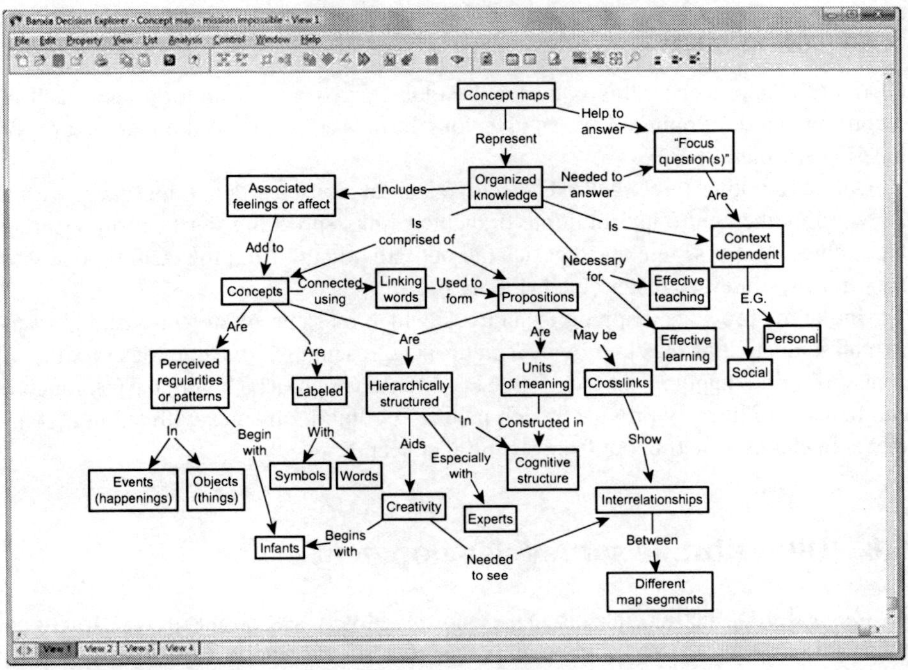

Figure 6.3 Concept map.

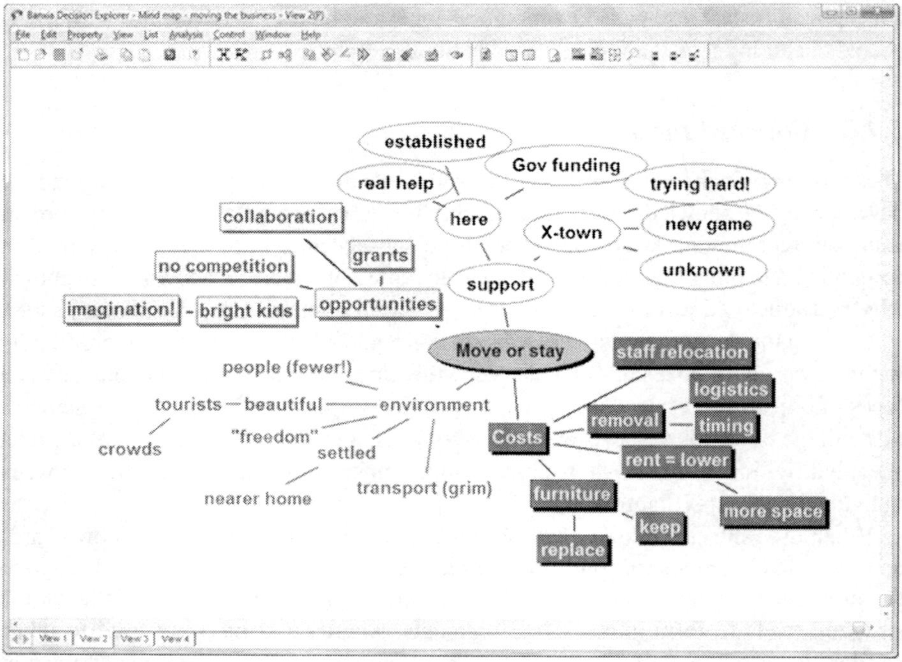

Figure 6.4 Mind map.

6.7.4 Mind maps

Of the four approaches, this is by far the oldest and simplest pictorial approach to recording and developing ideas, and has long been used for note-taking at lectures or meetings (Meier, 2007).

Unlike the other three methods discussed so far, it uses a radial approach, with a keyword as the centre and multiple branching links spreading out to form separate hierarchies. Major segment categories fan out individually from the central node into sub-branches in ever greater detail (Figure 6.4).

Unlike the previous mapping techniques, which rely purely on words and phrases to map semantic linkages to provide a pictorial representation of a complex mixture of concepts, mind mapping is a free-form sketch-based approach (Figure 6.5). As such it can be used to literally provide a 'rich picture' by linking images with textural comments in chains radiating out from a central concept.

6.8 Quantitative survey development

As touched upon earlier, questionnaires are effectively self-administered structured interviews, where the same general procedures should apply. Here the responses are closed questions that limit responses to choices between alternative attributes or

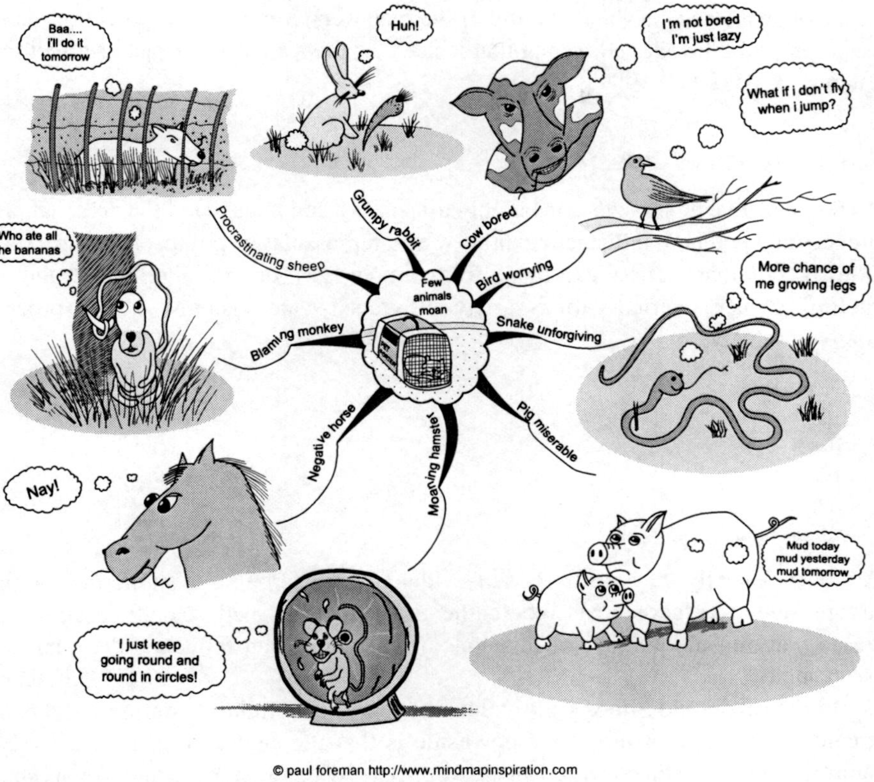

Figure 6.5 Image-oriented mind mapping.

perceptions, with the results effectively translated into quantitative measures for subsequent statistical analysis (Oppenheim, 2000).

However, single open questions are often injected at the end of natural clusters, that invite the respondent to provide additional input in the form of comments or ideas. As these are non-quantifiable, they have to be extracted and sifted either to be transferred across to add to the knowledge base, or to be used to amend and improve future questionnaires (LGDU – Wales, 2009).

An additional issue is that, in the absence of anyone to raise questions with, wording needs to be simple and focused unambiguously on a single point of interest. Ideally they should be set within natural 'clusters' of interest as part of the overall sequence.

A key problem with phrasing questions appropriately is how to avoid bias, especially where they may seem to be loaded in some way that begs a particular response. The aim should be that the whole sequence should remain neutral throughout in order to avoid inadvertently leading towards a chain of responses that skew the results.

From a psychological perspective, the questionnaire needs to be a comfortable fit with the background and shared frame of reference of the target audience, as it can affect the way that they interpret the meaning of the questions. Therefore it is important to ensure that the respondents are appropriately selected and briefed at the outset to optimise the accuracy and validity of their answers. Similarly, due attention should be given to the layout and format that is used, as this can also distract or affect their interest (BMJ, 2004).

6.8.1 Types of questions

As the objective of surveys is invariably to measure and understand the personal preferences and potential influences exhibited by a representative group, closed-end questions predominate. These can range from simple 'yes' or 'no' alternatives, through multiple-choice matrices, to a variety of graded-scale systems or rank-ordering options, namely:

- Dichotomous
- Likert
- Slider
- Ipsative
- Stapel
- Multi-choice

Although visually they are fairly self-evident, they can be quite subtle in how they are presented (Figure 6.6). Where the more sophisticated scaled questions are posed, varying the number of divisions or categories can influence the quality of the response.

Whilst Likert and Slider scales allow the respondent multiple options, and hence greater granularity of input, the downside is that the centre point is effectively a neutral response, which is often an easy opt-out. By contrast, by using even numbers

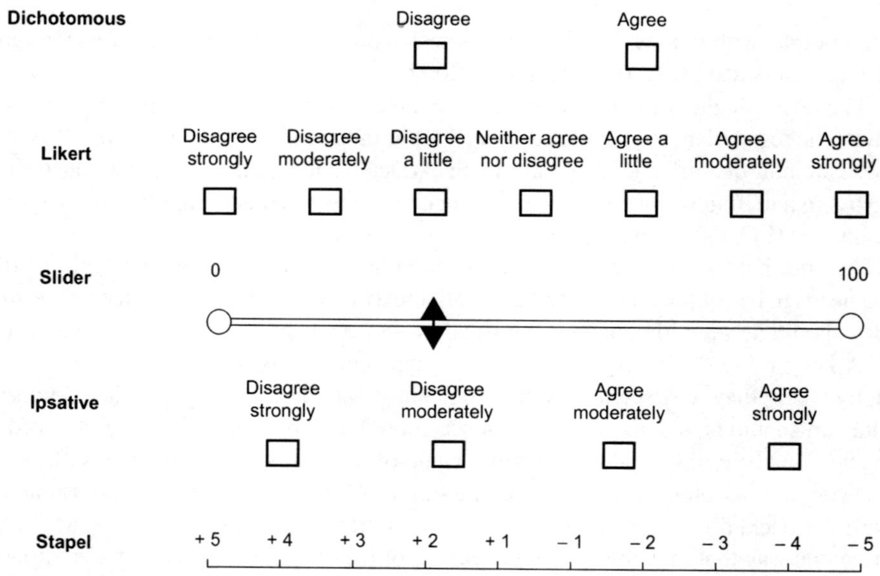

Figure 6.6 Basic scaling formats.

Multi-choice		Daily	Weekly	Monthly	Less
	Walking	☐	☐	☐	☐
	Hiking	☐	☐	☐	☐
	Running	☐	☐	☐	☐
	Biking	☐	☐	☐	☐
	Swimming	☐	☐	☐	☐
	Aerobics	☐	☐	☐	☐
	Team sports	☐	☐	☐	☐

Figure 6.7 Multi-choice scaling format.

of options, an Ipsative scale effectively forces the respondent to make a non-neutral choice, and the Stapel version extends the number of options to five in each direction.

Beyond this, multi-choice scaling (Figure 6.7) provides a variable range of options, ranging from a single selection per row, as shown, to ranking where the respondent is asked to insert numbers rating preferences.

Pictorial representations of different design options for both the arts and technological domains are key to obtaining feedback from potential purchasers with regard to both 'feel' and functionality of the end result. Whilst the questions themselves will still follow one or other of the closed-ended formats outlined above, they are more likely to seek a single preference or a ranking response.

However, greater value can be obtained by linking the visual component with one or more questions focussing on different aspects of significant design interest, such as that shown relating to pocket fastening (Figure 6.8). Others might include colour and shade, or a range of other preferences.

6.8.2 Attitude profiling

One of the key aspects of surveying is the need to establish a broad profile of the behavioural attitudes and likely prejudices of the respondents. This is particularly important to ensure that the overall mix is representative of the group as a whole, and not skewed towards a predominant point of view (McColl et al., 2001).

The methodology used is a very condensed form of psychometric test, where respondents are asked to rate their opinion of their personality traits for each line item within a set of 10 paired perceptions (Figure 6.9).

This Personality Inventory scores each entry in the range of 1 (disagreement) to 7 (agreement) and pairs each question with another, whose score is reversed and the totals used to provide an attitude profile covering the following five criteria:

- Extraversion
- Agreeableness

Q. Which type of pockets do you prefer?

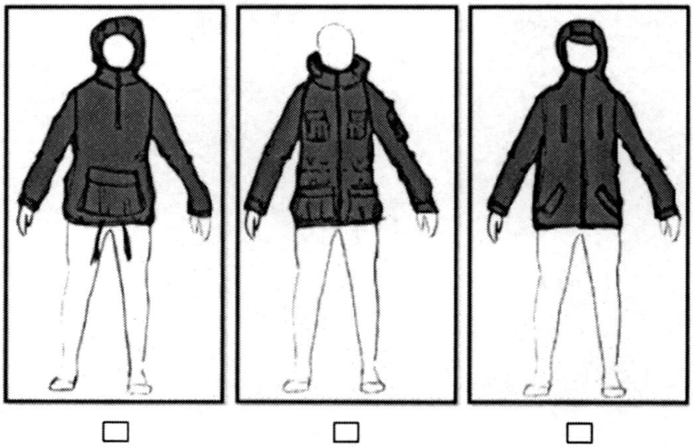

Q. Which type of pocket fastening do you prefer?

- Press studs ☐
- Velcro (patches) ☐
- Velcro (full length of pocket) ☐
- Zipper (small pull / water repellent) ☐
- Zipper (large pull / water repellent) ☐
- Buttons ☐

Figure 6.8 Preference scaling.

- Conscientiousness
- Emotional stability
- Openness to experiences

Whilst the meaning of most of the categories is obvious, Extraversion needs some explanation as this is a term specifically used in psychology. In essence it is the opposite of introversion, as it is the other pole of the extraversion–introversion spectrum, and tends to be seen as an outgoing, talkative, energetic behaviour trait.

6.8.3 Question sequencing

As a broad principle, the flow of questions should begin with the simple to answer, non-sensitive one, through subtle shifts of focus and interests, to the more sensitive and taxing. In essence this careful change of emphasis to the more focused later sections should move from general factual detail and background behavioural traits at the start, to more specific viewpoints and attitudes towards the end.

Good design should aim to capture interest from the start and maintain it throughout. Rapid 'go to' shifts between sections of clustered topics should be avoided, as this easily disconcerts the respondent and disturbs their concentration and commitment to complete the task.

Personality inventory

I see myself as:	Disagree strongly	Disagree moderately	Disagree a little	Neither agree nor disagree	Agree a little	Agree moderately	Agree strongly
Extraverted, enthusiastic	☐	☐	☐	☐	☐	☐	☐
Critical, quarrelsome	☐	☐	☐	☐	☐	☐	☐
Dependable, self-disciplined	☐	☐	☐	☐	☐	☐	☐
Anxious, easily upset	☐	☐	☐	☐	☐	☐	☐
Open to new experiences, complex	☐	☐	☐	☐	☐	☐	☐
Reserved, quiet	☐	☐	☐	☐	☐	☐	☐
Sympathetic, warm	☐	☐	☐	☐	☐	☐	☐
Disorganized, careless	☐	☐	☐	☐	☐	☐	☐
Calm, emotionally stable	☐	☐	☐	☐	☐	☐	☐
Conventional, uncreative	☐	☐	☐	☐	☐	☐	☐

Above are a number of personality traits that may or may not apply to you.
Please indicate the extent to which *you agree or disagree with that statement.*

Figure 6.9 Personality inventory.

Pre-planning is vital, especially where research needs to range across different aspects of a complex domain. In this instance, this needs to be a two-stage process, firstly mapping out all elements and inter-relationships that need to be covered; and then using this to work out how best to create a flow of questions that is as seamless as possible.

6.8.4 Administrative processes

The first issue to resolve is how best to identify and reach out to a wide section of the public that is likely to have interest in and experience of the domain and will then most probably respond positively. Once an appropriately sized listing, complete with all necessary contact information, has been assembled, the modes of contacting them have to be selected from:

- Post
- Telephone
- Internet

Whilst each has its respective merits and disadvantages, at least two will need to be used, providing paper or electronic means. However, telephone interviews are really only relevant to short and relatively generalised surveys, or where they result as a by-product of chasing up respondents for their replies!

From an administrative perspective all contact details that could identify respondents must be removed and coding used to anonymise all ‘downstream’ information processing in accordance with the Data Protection Act requirements.

6.8.5 *Quantitative analysis*

In essence this is a continuation of the administration process insofar as it involves creating a spreadsheet structure using any one of a number of standard statistical software packages, where each column will hold all responses in question order sequence per respondent. All calculations required to complete the analysis have then to be added, and the tables and display formats created to produce the required outputs, before being tested with dummy data to validate the process prior to going live.

Thereafter all incoming data has to be inserted into the appropriate cells against the coded reference for each respondent and cross-checked to obviate transcription errors. As statistical software automatically updates the analysis process incrementally as new input data are processed, any obvious anomalies can usually be recognised by reviewing the output as it evolves.

Whilst these software packages normally include a wide variety of graphical output formats, and in some cases will accept pictorial inputs such as those displayed earlier, it may be beneficial to reconfigure the output using purely graphical design-oriented software for reporting purposes.

A typical example of this type of approach is shown in Figure 6.10, where 80% of respondents preferred the pocket design shown above, and then preferred the large pull, water-repellent zipper closure over other alternatives.

Another approach that focuses on female buying preferences uses a set of 11 stacked histograms to shown the relative likelihoods of purchases being made from a range of sources (Figure 6.11).

Although there is a considerable range of other possible types of display formats, space limitations prevent their being presented in this chapter. The considerable potential to identify subtleties within the data sets can be seen in terms of the higher likelihood of purchases being made from High Street sports shops and sports warehouses.

6.9 Research ethics

All research that involves the use of information derived directly or indirectly from, or about, a specific individual has to be managed according to strict guidelines set by the professional bodies of the disciplines concerned and supervised by the academic institution leading the project.

The central ethical criterion is therefore that the research must ‘do no harm’ to the individual concerned, and is therefore particularly pertinent to qualitative research that spans the medical and sociological domains. Whilst there are minor differences between the two – particularly in regard to clinically sensitive issues – the generic approach follows the principles laid down in the British Sociological Association’s *Statement of Ethical Practice* (BSA, 2002).

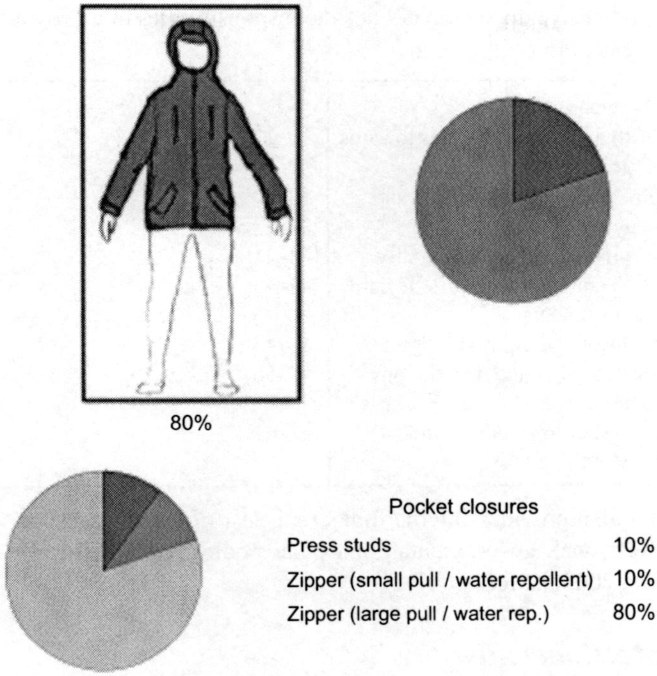

Figure 6.10 Pie chart for a preferred pocket design.

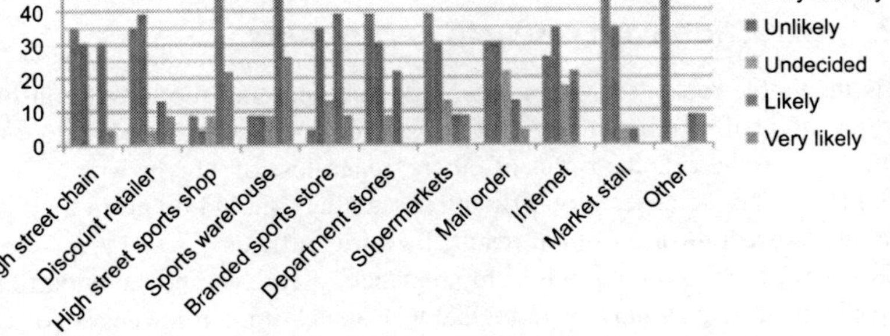

Figure 6.11 Stacked histograms for buying preferences.

This comprehensive document defines the responsibilities of all researchers within the following categories:

• Professional integrity	5–9
• Relations with and responsibilities towards research participants	10–12
• Relationships with research participants	13–30
• Covert research	31–33
• Anonymity, privacy and confidentiality	34–41
• Relations with and responsibilities towards sponsors and/or funders	42–44
• Clarifying obligations, roles and rights	45–48
• Pre-empting outcomes and negotiations about research	49–56
• Obligations to sponsors and/or funders during the research process	57–61

The document also provides Internet links to a host of useful websites from a wide range of professional, governmental, and other bodies having ethical concerns and interests (SRA, 2003; ESRC, 2007).

6.9.1 Professional integrity

As evolution of scientific knowledge is based on mutual trust in the validity of results obtained within the research community and beyond, it is vital that findings are accurately and truthfully reported. In the particular context of qualitative research, every effort must be made to avoid ambiguity, whereby content might be misinterpreted, either inadvertently or otherwise, to compromise the evidence presented.

Researchers need to carefully consider both the effects of their involvement and any consequences that may impact on those whose input is central to their work, and to ensure they safeguard them against any likely impacts. This must include ensuring that the proposed work is worthwhile; that the techniques used will be appropriate; and that the researchers do not stray outside the boundaries of their professional competence (MRC, 2005).

6.9.2 Interaction with research participants

As it is inevitable that researchers enter into personal and moral relationships with those they study, they must ensure that the advancement of knowledge does not entitlement them to override the rights of others, regardless of who or what they may be. Although this may be especially difficult, researchers should do all in their power to combat and correct misuse of their results by third parties.

A key part of the implicit relationship between researcher and study participants is that the physical, social, and psychological well-being of interviewees is not adversely affected by the research process. As a result, the protection of the rights, interests, sensitivities, and privacy of all concerned must be paramount.

One further aspect that should be guarded against is the likelihood a build-up of stress due to vulnerabilities that may not be obvious due to a variety of hidden factors.

These can arise as perceived uncalled intrusions into participants' private and personal worlds; distressing self-knowledge; or even power balances.

At the outset, the basis for the research should be explained in appropriate detail, and in terms meaningful to participants. This should also include how the results will be disseminated, together with the source of funding and other external influencing factors. At this point formal confirmation that informed consent to the intended process has been freely given should be duly confirmed, together with the fact that participants are aware that they always have the right to refuse to continue to participate at any time.

Part of this briefing must cover all aspects of data-gathering. This is particularly necessary where audio or video recording may be used, as interviewees need to be aware of their rights under any copyright or data protection laws.

At this stage, the positions of both parties need to be made clear and agreed upon with regard to future use of the material, with the formal consent given by the interviewees, together with copyright clearances. At this time it is also important that contractual and/or legal interests and rights in data, recordings, and publications are fully understood and agreed.

Finally, agreement should be reached on whether interviewees should have access to review content with the option to alter the content, withdraw statements, provide additional information, or be able to amend/correct interpretations.

6.9.3 Covert or third-party input

Whilst not generally relevant, in some circumstances input may be needed where there is doubt or concern over the behaviours exhibited by the interviewee, which lead to the view that some action or reference to a third party should be taken. This should only be taken in extreme circumstances, after reference to the ethics committee. In such a case all records should be removed from the study and securely stored pending any subsequent clinical or legal needs.

6.9.4 Anonymity, privacy and confidentiality

Guarantees of confidentiality and anonymity given to research participants must be honoured, unless there are clear and overriding reasons to do otherwise. All records that could identify interviewees or other parties in the study involved should be kept securely in accordance with Data Protection legislation requirements to ensure confidentiality.

In some cases it may be necessary to decide whether it is proper or appropriate even to record certain kinds of sensitive information, either during or subsequent to an interview. In such circumstances it is always best to err on the side of caution, and always make judgements in the best interests, and for the well-being, of the participant.

6.9.5 Quantitative survey development

Whilst there is a common interest with external agencies in advances in knowledge, ethical considerations, truthfulness, and rigour must always trump commercial or other interests. As such this must be central to any contractual arrangements.

Equally, although these agencies will have rights of access to the findings, it must be accepted that they have no rights to amend or alter the findings in any way. That said, that does not preclude their rights to challenge or disagree with the findings in open court.

6.10 Qualitative research aspects of co-design

The key problems with successfully breaking totally new ground with a dispersed multidisciplinary team are not solely confined to the obvious coordination and communication issues. They actually centre on the need to bridge the yawning gaps between disciplines.

These are made worse by the entirely different design processes and skill sets that are involved, which then lead to a lack of cross-disciplinary understanding. As if this is not a difficult enough hurdle to cross, setting this in the context of ageing adds a further third dimension of potential ambiguity enveloping all the parties (Manning, 2012).

This scenario sets a particularly difficult problem for any project leader involved with creating a 'total team' ethos from the outset; coordinating and unifying the efforts of the two design groups; establishing a common understanding of the evolving design needs of the elderly; and then successfully managing contending issues as they arise.

On the face of it, such a three-segment co-design as this needs to reconcile the seemingly irreconcilable! However, the secret lies in achieving a balanced mix of appropriate psychology and the management of change, all geared to overcoming sets of potentially competing preferences, processes, and procedures. For this to succeed, the motivation of all concerned needs to be focused on palpable and achievable benefits for all concerned that outweigh their perceived disadvantages.

In view of the critical dependency on translating these views into a representative reality, a front-end behavioural psychology profiling work package needs to operate as a two-way interface between a reference cohort of older people and the twin design groups (Figure 6.12). An integral part of this work package needs to include an interactive link with the care professions, who are likely to have a pro-active relationship with the intended beneficiaries.

Its key role is to act both as an analytical communication channel and moderator for co-design coordination. Central to the effectiveness of this linkage is the need to gain a detailed understanding of the impacts of ageing in terms of physiological, psychosocial, and socioeconomic constraints to mobility and well-being, and then to set this in the context of the potential assistance that a combination of clothing and electronics developments can deliver.

The issue that then arises is how best to gain and transfer this knowledge most effectively within the usual constraints of time and available funding. Whilst the basic principles have been outlined earlier, their application invariably requires balanced judgements to be made in terms of size and range of input sources to be sampled and the practicalities involved in finding them.

Although the first task is to build a research topic-specific knowledge resource that represents the views and expertise of as wide a cross-section of the chosen domain as

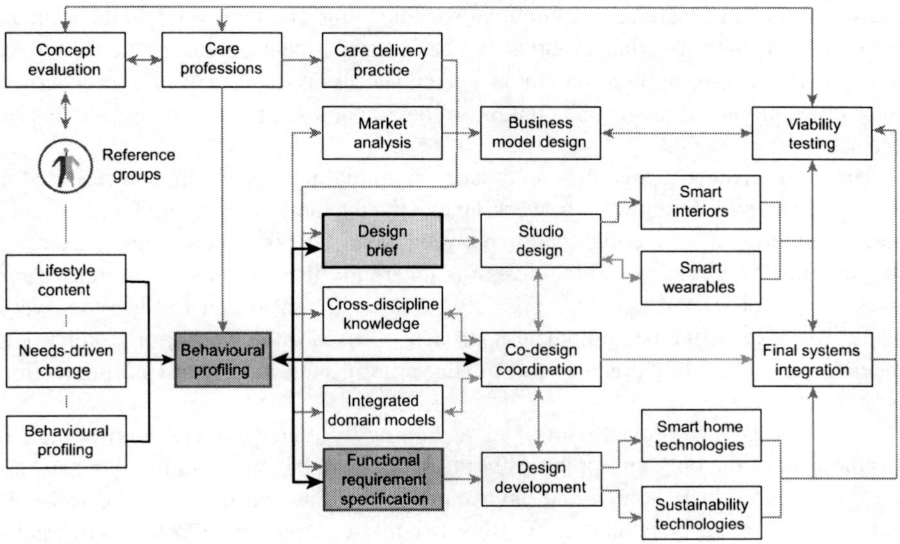

Figure 6.12 Systems co-design process.

possible, equal consideration should also be given to the size and sources of the group whose opinions will subsequently be sought. Unfortunately this is usually not as easy as it sounds. Even though there is a myriad host of organisations that focus on the pursuits and interests of the elderly, finding individuals from within them, whose interests match those of the research topic, can be extremely arduous and time consuming.

Once a sufficiently large set of individuals who are prepared to take part have been identified, they need to be selected as either potential interviewees or survey respondents, and then invited to join the appropriate group. This approach is normally better than asking them to choose a group to join, as the survey group has to be large enough to be statistically valid, whilst allowing for drop-outs.

Whilst this constraint does not directly apply to the smaller interviewee group, its size still needs to be sufficiently large to establish an adequate level of views, expertise, and knowledge from which to develop an appropriately detailed survey questionnaire.

Although not directly part of this group, a further, more pro-active trials cohort needs to be recruited to advise on and take part in the overall product development, test, and evaluation phases of the project. This should be a relatively small group, some of whom will need to be available to take an active part of the project over a fairly protracted period – albeit on a part-time basis. Whilst part of their role should ideally be to provide an independent cross-check on the initial design input, resourcing constraints may mean that some members of the original two groups may need to be recruited into the trials group on purely practical and pragmatic grounds.

6.10.1 Coordinating the co-design process

The role of a front-end behavioural psychology profiling work package is to provide input to these two product design work packages to enable them to establish their design briefs and functional requirement specifications and then validate them on an ongoing basis with the trials group as the basis for the management of the co-design process. This is especially important as overall success is dependent on all concerned fully grasping the problems and contexts of ageing and seeing how their joint efforts can satisfy these needs.

However, achieving this rests on creating a mutual understanding and respect for the significantly different ways of working that the two design teams are familiar with, and engendering a spirit collaboration between them. This is somewhat of an exercise in 'squaring the circle'! Clothing design is inherently interpersonal, with the wearer integrally involved throughout – providing critical appraisal and feedback at every stage. By contrast the technological approach is based on delivering a product that meets a predefined functionality and quality criteria with minimal or no interaction with its end-user.

In the broader context of arts and technology-oriented 'smart' environments, clothing is not the only end-point, as living spaces can benefit from the same co-design approach; smart homes and environments together with sustainable technologies can be equal beneficiaries. Further benefits can accrue from the twin tracks of building and maintaining an ongoing cross-disciplinary knowledge resource together with a reference library of increasing sophistication of the various domain models that emerge.

However, it should be recognised that whilst achieving these benefits, together with those that can benefit both the care professions and their elderly service users, there is little likelihood of this proliferating unless more work is done to develop viable business models that are suitably attractive to investors.

Although Design for Ageing Well, as pioneered by the New Dynamics of Ageing project, has considerable potential, its focus on 'active ageing' in terms of walking and hiking in all reasonably acceptable weathers is after all only a limited market opportunity. The real prize for all concerned is to use it to promote a sea change in attitudes toward to developing the massive 'hidden' market of older people by fully integrating smartphone and associated technologies into clothing.

6.11 Future trends

The biggest problem in all academic disciplines, and particularly in qualitative research (Naisbitt, 1982), is that it is perpetually:

> *Drowning in data yet starving for knowledge*
>
> *John Naisbitt*

Moreover, the question that follows from that is 'who has got what, and where is it'; closely followed in turn by 'who knows about it anyway!' Unfortunately, these are not only pitfalls waiting to trap the unwary, as even the best product or service can be undone downstream by a disinterested or demotivated potential customer base.

6.11.1 Bridging gaps

Psychology and qualitative research have much to offer. This is not only about finding and compiling knowledge gained through surveys but then 'translating' and transferring information between different parties involved in the design process and beyond. Equally important, it is about understanding the factors that engage and motivate the decisions of potential customers and those that make up the complete supply chain.

This is particularly evident in the context of mismatches between the design of many products and services and the attitudes and concerns of the recipients – and in the case of assistive technologies, their families as well. This is often made worse by the way in which it has been marketed, or 'prescribed', and dispensed by care services (Hogg & Wilson, 2004). In this latter case, the elderly and carers alike report of having systems 'inflicted' upon them regardless; of being patronised and having their wishes and concerns dismissed; of the obvious lack of in-depth expertise in what can help and what cannot; and of very limited ranges of inappropriate and incompatible products (Allen, unpublished). It is therefore hardly surprising that this sets up an underlying negative reaction triggered by such a raft of thoughtless and unhelpful behaviours.

Could there be a better example of how not to respect and relate to a potential customer base? Whilst obviously at the extreme end of the spectrum of 'not-buy–buy' reactions, such unwanted effects can be largely avoided by recognising the critical decision factors involved in obtaining anything at any point in one's life (Figure 6.13).

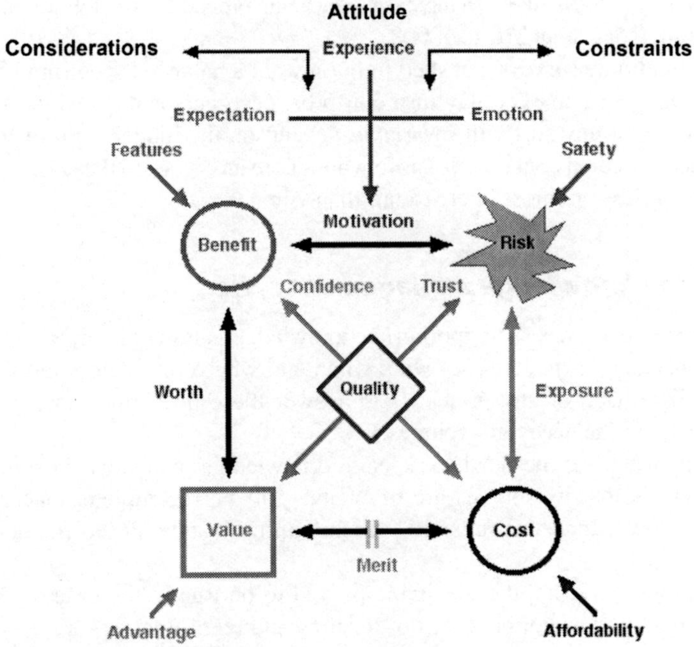

Figure 6.13 Critical decision factors.

In essence, these can be distilled to four major factors that are ultimately centred on the required quality of the goods or services under consideration (Manning, Benton, & Altemeyer, 2011), namely:

- Benefit
- Risk
- Value
- Cost

All four are then paired and bound to their immediate neighbours by a predominant characteristic and cross-linked with each other in terms of a common goal of quality. This goal-seeking instinct ultimately jointly depends on achieving acceptable levels of trust and confidence that a purchase will meet and deliver personally desired quality criteria.

Even then, most personal decisions are taken on the basis of merely balancing the perceived benefits against possible downside risks that may be involved. Often this excludes any effective cost or value judgements of the relative merits of alternative options. This seemingly odd behaviour is driven by attitudes derived from past experiences, which either sets desires battling against emotions or conversely starts their reinforcing each other (Correia de Barros, Duarte, & Cruz, 2011).

Whilst this may seem reasonably self-evident to those involved in clothing design, this is not yet so within the electronics domain. Although the co-design approach developed in the Design for Ageing Well project points the way ahead, much needs to be done to carry its benefits across this gap. This is especially so as micro-miniaturisation of devices and complete systems is likely to increase their being embedded in clothing in the future (Harms, Amft, Roggen, & Tröster, 2008).

Although attitude surveys can shed light on what a balanced spectrum of mindsets that go to make up a market may turn out to be, each personal decision to accept or reject offerings of any sort will invariably depend on the hidden sum of life experiences of the individual concerned. Those who ignore any clues to these key factors and press on regardless are most likely to fail than win.

6.11.2 The knowledge management cycle

Unsurprisingly, easy access to appropriate knowledge is immediately recognisable as a specific instance of the generic 'who, what, and where' problem identified earlier. The immediate need of any project is to answer these three questions in sufficient detail to seek out the necessary sources.

However, whatever the need for specific knowledge or detailed information, the core problem is that its sources are often likely to be incomplete, inaccessible, or worse – unknown. Indeed, Rumsfeld's dictum unfortunately all too frequently holds true!

As with most projects, the inevitable proved to be true in the case of Design for Ageing Well activities, both in terms of locating and recruiting appropriate cohorts of the active ageing to take part in various aspects of the process. Exactly the same issue arose both with regard to locating relevant textiles and devices offering the required functionality and also finding useful information to support and enhance the design process.

Equally, once underway the same problem applied to sifting the flood of incoming information once again in answering the question of how to locate and interlink the clues within the data needed to reveal the pearls of wisdom that are being sought. Whilst dealing with survey data is a relatively straightforward exercise using readily available statistical analysis software, the same cannot be said for textual input. Here the only option is to go through it line-by-line with a highlighter marking up or super-indexing relevant content for subsequent extraction and collation into a project knowledge resource.

Until now considerable effort has been put into processing vast amounts of textual material using a variety of text data-mining techniques that depend on considerable computing power (Feldman & Sanger, 2006). However, somewhat slower progress has been made using statistical natural language processing techniques (Manning & Schutze, 1999). Ideally what would be a great leap forward would be the development of more simplistic applications of these techniques, which would greatly assist those working at the 'coalface' text-based qualitative research.

Similarly, little or no real attempt seems to have been made to consolidate and maintain comprehensive current reference sources spanning a wide range of fields, which makes locating ageing condition-related or other topics and specialist interest groups extremely difficult. By the same token, keeping abreast of current or emergent new technology can be equally awkward, especially when any reference to inherent functionality sought is often hidden behind journalese or branding, especially where fashion and technology meet.

References

Ackermann, F., Eden, C., & Cropper, S. (1990). *Cognitive mapping – A user's guide. Working paper 90/2 from University of Strathclyde, Management science: Theory.* Method and practice working paper series 1990. http://www.banxia.com/pdf/de/GettingStartedWithCogMapping.pdf.

Allen M. C. *Notes on knowledge exchange partnership [KEP] – Care in business programme seminars.* Coordinated by the University of Swansea, Centre for Innovative Ageing, Unpublished.

Berger, P., & Luckmann, T. (1967). *The social construction of reality: A treatise in the sociology of knowledge.* First Anchor Books Edition. 240 p.

BMJ. (2004). Hands-on guide to questionnaire research: selecting, designing, and developing your questionnaire. *British Medical Journal, 328,* 1312.

British Sociological Association. (March 2002). *Statement of ethical practice for British Sociological Association.* http://www.britsoc.co.uk/media/27107/StatementofEthicalPractice.pdf.

Correia de Barros A., Duarte C., & Cruz J. B. (2011). Designing objects or systems? The role of design in assistive products' delivery. Include 2011 – The role of inclusive design in making social innovation happen, Conference papers, Royal College of Arts http://ubi.academia.edu/AnaCorreiadeBarros/Papers/522952/Designing_objects_or_systems_The_Role_of_Design_in_Assistive_Products_Delivery.

Creswell, J. (2006). *Qualitative inquiry and research design: Choosing among five approaches*. Sage.

De Fina, A., Schiffrin, D., & Bamberg, M. (2006). *Discourse and identity*. Cambridge: Cambridge University Press.

Duranti, A., & Goodwin, C. (Eds.). (1992). *Rethinking context. Language as an interactive phenomenon*. Cambridge: Cambridge University Press.

ESRC. (2007). *Research ethics framework*. http://www.esrcsocietytoday.ac.uk/ESRC InfoCentre/Images/ESRC_Re_Ethics_Fram e_tcm6–11291.pdf.

Feldman, R., & Sanger, J. (2006). *The text mining handbook*. New York: Cambridge University Press, ISBN: 978-0-521-83657-9.

Foddy, W. H. (1994). *Constructing questions for interviews and questionnaires: Theory and practice in social research* (New ed.). Cambridge, UK: Cambridge University Press.

Gauker, C. (2003). *Words without meaning*. MIT Press.

Glaser, B., & Strauss, A. (1967). *The discovery of grounded theory*. Chicago: Aldine.

Harms, H., Amft, O., Roggen, D., & Tröster, G. (March 13–17, 2008). *SMASH: A distributed sensing and processing garment for the classification of upper body postures*. Submitted to the third International Conference on Body Area Networks (BodyNets08), Tempe, AZ. http://www.wearable.ethz.ch/research/groups/textiles/smash.

Hogg, G., & Wilson, E. (2004). *Does he take sugar? The disabled consumer and identity*. British Academy of Marketing Conference . http://www.csrs.ac.uk/Publications/BAMdisabled.pdf.

Kvale, S. (1996). *Interviews: An introduction to qualitative research interviewing*. Sage Publications.

Local Government Data Unit – Wales. (2009). *A guide to......Questionnaire design*. Cardiff ISBN:978-1-907073-02-1. http://www.dataunitwales.gov.uk/SharedFiles/Download.aspx?pageid=79&fileid=447&mid=459.

Manning, B. (July 2012). *'universAAL': Open interoperability in ambient assisted living*. Report and research development outline to the knowledge exchange partnership project board – Swansea University. Appendix 3 – Co-design.

Manning B., Benton S., & Altemeyer B. (2011). *Sustainable social innovation by design: Breaking down the boundaries*. Include 2011 – The role of inclusive design in making social innovation happen, Conference papers, Royal College of Arts.

Manning, C., & Schutze, H. (1999). *Foundations of statistical natural language processing*. Cambridge, MA: MIT Press. ISBN: 978-0-262-13360-9.

Marshall, C., & Rossman, G.B. (1999). *Designing qualitative research* (3rd ed.). Thousand Oaks, CA:Sage.

McColl, E., Jacoby, A., Thomas, L., Soutter, J., Bamford, C., Steen, N., et al. (2001). Design and use of questionnaires: a review of best practice applicable to surveys of health service staff and patients. *Health Technology Assessment*, *5*(31).

Medical Research Council (MRC) (2000, updated 2005). Good Research Practice. London: MRC. http://www.mrc.ac.uk/Utilities/Documentrecord/index.htm?d=MRC002415.

Meier P. S. Mind-mapping. Social Research Update Issue 52: Autumn 2007 (ISSN: 1360-7898) Published by the Department of Sociology, University of Surrey. http://sru.soc.surrey.ac.uk/SRU52.pdf.

Naisbitt, J. (1982). *Megatrends: Ten new directions transforming our lives*. (p. 178). Warner Books/Warner Communications Company. ISBN: 978-0-446-35681-7.

Novak, J. D., & Cañas, A. J. (2008). *Theory underlying concept maps and how to construct and use them*. Technical Report Institute for Human and Machine Cognition CmapTools 2006-01 Rev 2008-01. http://cmap.ihmc.us/publications/researchpapers/theorycmaps/theoryunderlyingconceptmaps.htm.

Ogden, C. K., & Richards, I. A. (1923). *The meaning of meaning*. New York: Harcourt Brace & World.

Oppenheim, A. N. (2000). *Questionnaire design, interviewing and attitude measurement* (New ed.). London, UK: Continuum International Publishing Group Ltd.

Patton, M. Q. (2002). *Qualitative research and evaluation methods*. Thousand Oaks, CA: Sage.

Searle, J. (1979). *Expression and meaning*. Cambridge: Cambridge University Press.

Shadbolt, N., O'Hara, K., & Crow, L. (1999). The experimental evaluation of knowledge acquisition techniques and methods: history, problems and new directions. *International Journal of Human-Computer Studies*, *51*, 729–755.

Social Research Association (SRA). (2003). *Ethical guidelines*. London: SRA. www.the-sra.org.uk/documents/pdfs/ethics03.pdf.

Stonier, T. (1997). *Information and meaning. An evolutionary perspective*. XIII, 255 p. 23,5 cm.

Wertz, F. J., Charmaz, K., McMullen, L. M., Josselson, R., Anderson, R. & McSpadden, E. M. (2011). *Five ways of doing qualitative analysis: Phenomenological psychology, grounded theory, discourse analysis, narrative research, and intuitive inquiry* (1st ed.). (pp. 16–18.) The Guilford Press. (Print).

Effective communication in product development of smart wearable clothing for the active ageing population

7

D. Taylor[1], *M.W. Timmins*[2]
[1]University of South Wales, Newport, UK; [2]Consultant, UK

7.1 Introduction

The merging of technical textiles and wearable electronics demands the collaboration of disparate disciplines with their quite different study backgrounds, working cultures, and terminologies. For products to be developed to be fit for their purpose, and to be willingly worn by the end-users, there needs to be a common language, and it must be understood by all those involved in the product development team. This language must also be communicated throughout the product sourcing and development chain. Once the product has been developed, potential buyers have to be made aware that the product exists, informed of the 'benefits and functionality' of the product attributes, and that it suits their customer needs. This is clearly in the 'promotion' area of an organisation's marketing strategy. The need here is to use a clear and readily understandable plain language to explain the product benefits and features to end-users. Experienced marketing specialists will be quite familiar with traditional routes to market, but the merging of textiles and electronics will require new strategies.

There is, however, a complication. When considering a functional layered clothing system with built-in electronics, this hybrid product has no obvious retail home, and thus matters of marketing strategy are difficult to formulate. The garment system could be thought of as an item of clothing, in which case a sportswear outlet may be appropriate. Alternatively, the item may be regarded as an ICT product, in which case an electrical/computer outlet would be appropriate. In each case, staff in these outlets would need to understand a new language – technical staff need to know about the attributes and properties of the clothing, and sportswear clothing outlets need to develop a technical language. The use of a mobile phone as a central processing device serves merely to complicate matters further, since if these mobile phone outlets were to be considered for use, both clothing and technical language would need to be learnt by the staff. It may also be the case that the ultimate user, as in the case study, the active ageing walker, would not normally use of any of these three outlets. Owners of these outlets are well aware of their customer base, and recognising that active ageing walkers do not form a substantial part of it, may be reluctant to stock these innovative hybrid products.

Textile-led Design for the Active Ageing Population. http://dx.doi.org/10.1016/B978-0-85709-538-1.00007-9

The content of this chapter is concerned with developing common understanding between stakeholders in multidisciplinary projects where the range of specialisms is particularly wide. For the purposes of this chapter, the project under consideration is 'Design for Ageing Well', one of 35 projects under the general heading of New Dynamics of Ageing (NDA). Stakeholders are taken to include specialists employed by the project, advisors, industry collaborators and those for whom the outputs are intended: the 'users'. What is particularly significant, and unique within the NDA network, is that this collaborative, cross-disciplinary research has been design-led. The overarching project aim was to assist the confidence of active older people (60–75 years) when they are out walking.

The project aims were to develop, through co-design engagement, a garment system that is electronically enabled to provide the wearer with (1) information about their environment – locality, distance travelled and route taken; (2) biometric data, including heart rate and (3) the ability to communicate with other walkers. The system would also use accelerometer algorithms to detect a fall and if necessary, call emergency services. The project, led by the University of South Wales (originally the University of Wales, Newport), involved three work packages: 'WP1, Behaviour', linking the Universities of Salford and Westminster; 'WP2, Clothing', linking the University of South Wales and Brighton and 'WP3, Technology' at the University of Ulster. The project demanded the production of sample garments and work on subsequent introduction of the system to the market including manufacturers, buyers and potential users. The design-led approach has been key to the adoption of visual strategies to translate and explain complex language, both in breaking down the barriers between the disparate disciplines and industry collaborators, and in communicating with end-users.

7.2 Communication complexities in product design

There is what may be called a 'continuum of complexity' in the iterative product design development processes, and in particular complexities in the design communication aspects. At one end of the scale is the designer working alone to identify the need, manufacture the product and then exploit it in the open market. This approach, although increasingly rare, can still be found in the computer software business and certain high-end bespoke clothing businesses. Direct communication with the client, throughout the iterative stages of development, would be the norm here. Next on the continuum, a further level of communication complexity exists where the designer identifies the need, but has other people or organisations produce the item concerned. Communication of what is required in terms of function and quality can become problematic even at this level, particularly if the item is produced on the other side of the globe. Price and Lewis have discussed the issues pertaining to the communication of 'working design flats', and supporting technical detail presented in Chapter 22. It is worth noting that in this model there is little direct involvement of the user as a stakeholder in the design process.

A further level of complexity can exist, where the design of a product is the output of team working, drawing not just from the skills of one designer but of more than one – possibly several. An example relevant to the Design for Ageing Well (DfAW) project could be drawn from the fashion/apparel industry, where the accepted approach to apparel design and style selection is typically driven by trend prediction in combination with data from the previous season's sales. This approach is also production system led, with little room for real involvement of the end-user – the customer. The apparel design and production process has been researched by Hussey (2007). Her findings demonstrate that there is very little space in the process for real integration of end-user needs or the concept of co-opting the user into the design process, as summarised in Figure 7.1.

In Hussey's model, 'customer requirements' are taken as the requirements of the brand or retailer, rather than the end-user. This is also well demonstrated by Mckelvey and Munslow (2003), in their discussion to explain the design process. Implicit in the Hussey model is that the stakeholders – designers, developers, manufacturers, and retailers – will have at least some idea of the issues confronting the design/production process, and will be aware of the vocabulary of the activity, and thus technical terminology will to some degree be understood by all. Complexity may exist in the Hussey model, but there is still 'space' at this more complex end of the continuum of complexity for a further set of considerations, by trying to make the end-user central to the design process. These considerations lead to the idea of 'co-design'.

Many similar models have been constructed for ICT products, and in an early paper Löwgren (1995) analyses the software development process in terms of design methodology, arguing that such an approach, rather than an impersonal engineering methodology, provides greater ease of use of the finished product.

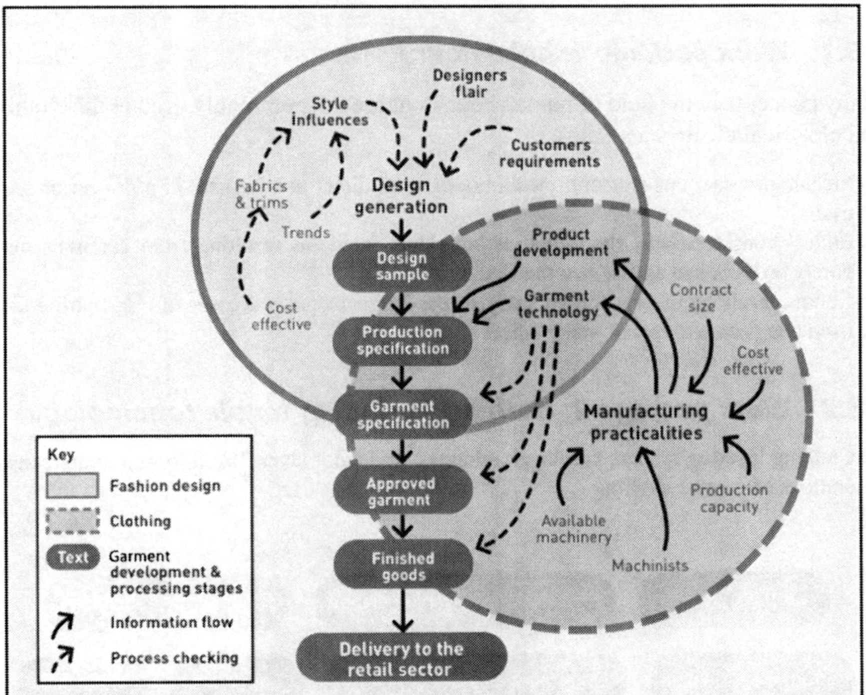

Figure 7.1 Hussey's model of the apparel design and production process.

7.3 Understanding the terminology of different disciplines in product design

Where different project work packages are located in different establishments, stretched lines of communication, travel difficulties, and varying and incompatible administrative systems and even working practises, are the norm. In Cummings and Kiesler (2005), the issues both of multidisciplinary and multi-institutional projects are well explored. Although the paper is based around 62 science-based projects and has no textile aspects, and the range of disciplines is less than the range in the DfAW project, the paper reflects accurately the DfAW situation. The authors' implicit conclusion, that larger numbers of institutions/disciplines lead to harder work needed to ensure successful outcomes, is also recognised in the DfAW project.

Drawing from a wide range of disciplines (Figure 7.2), it is hardly surprising that some of the terminology used across the DfAW project work packages will be unfamiliar, even confusing, to the project's stakeholders. The issues here divide into three groups. Firstly, there is a set of terms specific to a subject, for which there is no real equivalent. These jargon terms, not infrequently initials, need to be explained, and in the DfAW project it was obvious that there was reluctance in some cases for design team members to ask for clarification of poorly understood expressions. Examples of such terms include:

7.3.1 *Work package 1: behaviour*

Many concepts in the field of human behaviour are not commonly used in the clothing and electrical/electronics fields.

- Meaningful user engagement: medium/long-term direct involvement in a project or set of trials
- Ethical considerations: the effect on individual freedoms resulting from decisions made purely on technical and/or cost matters
- Communications theory: formal study of the interpersonal processes of transmitting ideas from one person to one or many others

7.3.2 *Work package 2: clothing (including textile terminology)*

- Clothing layering system: baselayer, midlayer, and outer layer. Taken from military considerations of combat clothing

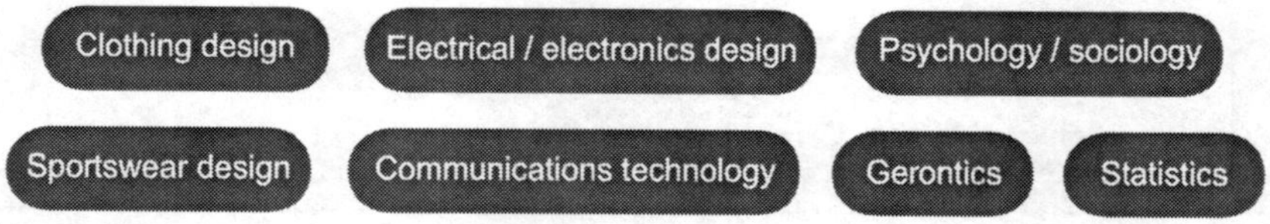

Figure 7.2 Expertise involved in the Design for Ageing Well project.

- Lay: layers of fabric in a pile ready for cutting out to form garments
- Breathability: the ability of a fabric to transmit water vapour from one of its surfaces through the material to the other surface
- Colourway: one of several different colour combinations in the same pattern printed on fabric
- Crocking: the loss of dye colour due to rubbing or abrasion
- Regain: the ratio of the weight of water in a fabric to the oven-dry weight of the fabric
- Body-mapping: the engineering for stitch constructions in 'seam-free' knit

7.3.3 Work package 3: electronics/systems

- Bus: a system that transmits data between components inside a computer-based system
- Chipset: a group of integrated circuits designed to work together
- Algorithm: a procedure or formula for solving a problem
- Bandwidth: the amount of data that a communication channel can handle in a given time
- Open source: software supported by a community of software users, rather than a commercial software company. Usually free

7.4 Terms with different meanings between specialisms

For all ordinary purposes, however, this type of specialist terminology is solved by consulting a good dictionary, or an online glossary. It is more problematic, however, when the same word can have different meanings between specialisms, or even the same meaning has different descriptors in the different topic areas. The context of the term's use is of course vital and will frequently quickly establish meaning, but a few examples will serve to make the point:

- Thread: in systems/engineering/technology activities, 'thread' is the serrated spiral at the end of a bolt, on to which a nut can be fixed. In the textiles field, it is the long filament forming cloth when it is woven or knitted. In communications, the term is used to describe an interconnected sequence of messages, often ordered as a list from the first communication between one party and another (or many parties).
- Processing: in textile terminology, 'processing' is treating a fabric to give it additional properties, such as colour, waterproofing, or fire resistance. In ICT it is the term used to describe the background manipulation of data by a computer system as it calculates changes to information.
- Loom: for the textile industry, a 'loom' is one of the fundamental pieces of textile machinery, used to produce woven fabric. In electrical/engineering matters it is the pre-prepared electrical wiring harness to be installed in electrical equipment and vehicles (and potentially in electronically enabled garment systems).

As outlined above, the context may well reveal the intended use, but confusion can occur, as it can when the same idea is called different things across specialist topics. Again, here are some examples:

- Denier and gauge: these terms relate to measures of the fineness or coarseness of a thread and wire, respectively, although they are arrived at differently. Denier is defined as the weight in grams of 9000 m of yarn, and gauge from the number of holes of gradually decreasing size in a steel plate that a length of wire must be drawn to arrive at the required thickness.

- Toile and breadboard: when making clothes out of expensive fabric, it is by no means uncommon for the designer to make a trial garment out of a relatively cheap material such as calico. This is done so that any modifications to fit and size can be done before cutting expensive fabric. This trial garment is a 'toile'. An electronics/systems designer will produce a trial circuit from individual components or subassemblies to test for functionality before producing the final design, and he/she will call this trial circuit a 'breadboard'. The concept is identical, but the terminology very different.

7.5 Visual approaches to developing a common understanding

Perhaps the main reason for developing a common project language across such a complex project as DfAW is to achieve a common level of understanding amongst the wide range of specialist knowledge required. Explanations in verbal forms as above are a vital part of developing this common understanding, but graphics in the form of diagrams and illustrations – even video and animation – have a valuable role in reaching a common understanding of a topic. Considerable study has been carried out over many years on this topic of information graphics, which now takes its place as a recognised area of academic study (McCandless, 2010). Not all approaches to developing understanding need the complexity demonstrated in the McCandless' publication, however; but the use of graphical approaches in DfAW and other New Dynamics of Ageing projects has been useful, and in some cases essential. In DfAW, in collaboration with Sizemic, data derived from body shape/size scanning, when converted into a computer-based image, formed the basis of basic blocks developed to fit users representative of the age group. Patterns were adapted from these blocks, by technical design experts, to coordinate throughout the various garment layers, for the development of trial garments. Examples of the visual 'common language' approach abound in other projects too; Table 7.1 shows a small sample of the use of this approach in other projects in the 35 separate NDA programmes. Some of these visual approaches are used within the projects, and some are themselves project outcomes.

7.6 Bringing different disciplines together in co-design

It was noted earlier that in the standard approach to developing new clothing products, the user likely to wear the garment is often given scant attention, and indeed the closest the clothing designer gets to the potential wearer is often the buyer of a clothing retail chain.

Some attempts at opinion-gathering can be carried out by using focus groups, but these groups have limited use by their nature and very short lengths of time during which opinions can be sought, typically an hour or so.

Table 7.1 **Examples of visual communication approaches in new dynamics of ageing (NDA) projects**

NDA project name	Example of use of graphics to develop understanding
Envision	Animated representations of the stress in limbs and joints as a result of various day-to-day activities likely to be carried out by older people
Transitions in kitchen living (TiKL)	Design drawings of more appropriate age-related kitchens
Older people's use of unfamiliar space (OPUS)	Maps, plans, and pictures used to develop a PDA-based navigation tool
Tackling ageing incontinence through theory, tools, and technology (TACT3)	One of the outputs of this programme was the great British Toilet Map
Safety on stairs	Using research data to produce plans and drawings for building examples of appropriate staircase design with use by older people in mind

In co-design, the end-user is firmly in the frame, and makes a major contribution to the design process. The general ground rules are well explained in Sanders and Stappers (2008), but broadly the co-design process used in DfAW was characterised by:

- Assembling a design team of designers, industry stakeholders, and end-users
- Recognising that the team is hierarchy free
- Meeting of the whole team regularly, sometimes over long periods
- Focussing on an identified project facet at each meeting
- Carefully recording visually and verbally what was said and what was decided

It is clear that the co-design process tends to be resource-hungry, particularly where more than one user is part of the team. This was certainly the case in the DfAW project, but with such a ground-breaking project, involving many formal disciplines, it became even more apparent as the project progressed that the co-design approach was the only appropriate route. Only by this process could cross-discipline communication channels, mutual understanding, and a common language be achieved.

Implicit in the Hussey model there is a degree of homogeneity of what is being researched, with project team members having broadly the same background and a common understanding of the topic. This was certainly not the case in the DfAW project, and in any project where different work packages are widely spread geographically, and where the disciplines are widely different with no common language across the activity. Couple these factors with a determination to deeply involve the end-user, and it is safe to assume that communication problems will emerge. These three factors are a feature of the DfAW project, and needed to be addressed before all stakeholders could understand what their colleagues were saying.

7.7 Using visual communication to help develop a common language in the Design for Ageing Well (DfAW) project

As with most projects, good communication channels are an important way of developing familiarity with new ideas and concepts. Table 7.2 shows the main channels used in the DfAW project, and comments on their usefulness in developing a common language and enhanced understanding.

Table 7.2 **Channels of communication in Design for Ageing Well**

Communication channel	Comments on usefulness in assisting in generating common project language
Project banners	Six banners communicated a strong corporate image in introducing the project aims and work package descriptors to industry and other stakeholders. The corporate image (developed by Taylor) was common to newsletters (2–6) and in all project communication
Co-design workshops	An essential method of developing a shared language. Each workshop dealt with a different topic, and misunderstandings could usually be quickly corrected
Information seminars	Another important tool, these seminars dealt with a range of topics (such as statistics) and were intended to update project team members to a common level of understanding of the topic
Newsletters	Intended to keep stakeholders up to date with the project progress, these newsletters usually contained one or two common language-building items
Information sheets	Intended specifically to address the need to develop a common understanding, these sheets were produced on an ad hoc basis as the need arose. The style varied from drawings and diagrams, concept maps, and written text materials
Informal out-of-session discussions during session breaks and evening meals	Useful in sharing ideas, but not likely to be widespread amongst all of the project team
Routine management meetings	Not essentially designed for common language development. Occasionally provided space for informal common language development
Web-based Wiki	Intended to produce a repository of relevant content and information generated by and shared by the stakeholders to encourage common language development
Social media	Specifically, a Facebook page intended to create an ongoing diary of the project in an informal way and encourage common language development between the stakeholders

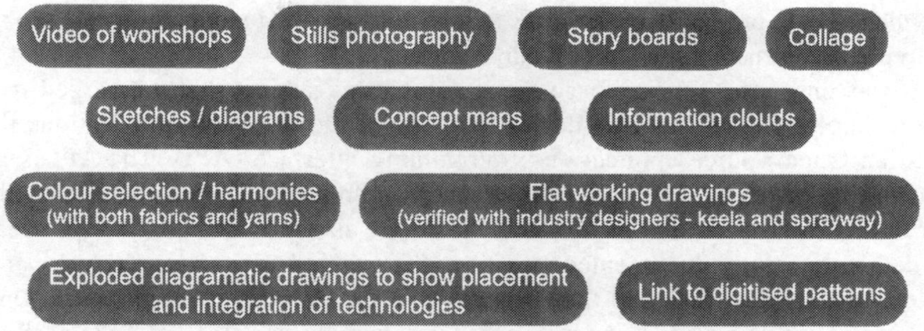

Figure 7.3 Methods of visual communication in the Design for Ageing Well project.
Source: From McCandless (2009).

As noted in Section 7.5, common understanding can often be approached by means of visual and graphic methods. In some cases graphic approaches are essential, such as preparing paper patterns that are laid over fabrics to provide an accurate guide to cutting out the material to be used. Even though considerable use of computer-aided design and advanced fabric cutting methods are becoming very common in factory-based clothing production, for optimal 3D fit and articulation (needed in technical clothing), a garment pattern drafted by traditional means (on a pattern table with reference to the tailor's dummy and user trials) has to be 'plotted' onto the computer, as the first stage in these processes, where it may then be manipulated on screen. In the DfAW project, considerable use has been made of visual communication for the purposes of recording data, informing stakeholders and planning approaches to market. Figure 7.3 shows the principal methods that have been used.

7.8 Case study: communication between disciplines

It is one thing to understand the definitions of words and phrases used by specialists, but quite another to comprehend the thinking behind their use, even where understanding is enhanced by visual approaches. This case study shows some of the difficulties that might be encountered. Much of the material in this case study is taken from Taylor and McCann (2012) and its use is gratefully acknowledged.

As background to this case study, a 'Shimmer' is a proprietary device that collects data from electrical sensors and transmits the data to a processing unit for analysis and display. In the DfAW case, some of the sensors are in the Shimmer unit and others in the garment system base layer next to the skin of the wearer. The processing unit (and some of the display) is a smartphone, in this case an HTC Desire.

The Shimmer device, whilst not deemed to be particularly large at $50 \times 30 \times 20$ mm, was seen by the clothing design team to be potentially too bulky for inclusion in a tight-fitting skin layer garment, where it could cause significant discomfort to the wearer. An ongoing discussion developed during several design and development meetings over a period of several months, with the clothing design team questioning the selection of this item. The debate focused on whether it was possible to either rebuild the unit, originally rectangular, to give it a flatter and more ergonomic profile, or to use an alternative smaller or flexible device.

It was only after several discussions around this subject that it emerged from the technology team that this device was one of the only available options that used an Open Source application-programming interface (API). The API is the gateway by which a software developer can programme the device and extract data for further processing. In this instance, the device also provided continuous stream of data with a fully documented structure that enabled the software developer to extract the relevant data and then process it in the smartphone application. Once this issue was understood, the rationale for selection was clear, and the clothing design team was able to accept the limitations of the hardware and work around the problems.

The misunderstanding was an example where no one was to blame, but the discussion was very clouded for some time. The introduction of a new member to the design team, with complementary knowledge and skills from architecture, at a late stage in the research and development process, revived the same discussion with its associated frustrations to do with the different ways of working between the clothing and technology work packages. This problem in communication, due to the merging of different cultures and terminologies, is indicative of the overall challenges in bringing wearable devices to market in clothing.

7.9 Case study: communication with textile industry designers and manufacturers

From the early stages of both co-design training and design iteration workshops, well-known clothing brands were invited to submit samples for comparative review or, ideally, to come into the academic environment to present ranges to users in the workshop environment. Regular workshops were designed to give active older participants the opportunity to show and describe some of their own outdoor garments and to discuss some design prototypes for new garments presented by the trade. Members of the DfAW design team were present to enable clear communication. After working for about six months in the DfAW co-design team, user group members had begun to understand the language and concepts in use across the project. Fruitful interactions between stakeholders in a design-based project are more likely when a common understanding is based on a common language.

This case study outlines a situation where the user group had a sufficient level of mutual understanding with the design team and a sportswear company's designer to be confident enough to provide real consultancy advice that was implemented in future garment production.

During research trips to the international sporting goods trade show (ISPO 2010 and 2011), held in Munich, the clothing-based project team spent time with project lead users inspecting and trying ranges of garments. What became apparent during liaison with the senior designer from the collaborating company was that, prior to exposure to the project, none of the garments had been designed with any real thought for how an older user may want or need to interact with them (Table 7.3). One of their senior designers was invited to attend one of the co-design workshops held in Newport, and this led to ongoing involvement, sanctioned by the particular company, throughout the co-design process. This organisation also invited users to visit their company showroom.

Subsequently, the DfAW design research team next met the same senior apparel designer from the company at 'Outdoor 2010', the European summer outdoors industry trade show held in Friedrichshafen in July 2010. At the meeting, the team were shown with production samples of the same garment that had been discussed at the user show-and-tell workshop in Newport. The garment design had been modified as a direct result of the DfAW design team's observations at ISPO and the active ageing participant workshop held in Newport in May 2010, with the key modifications to the design shown in Table 7.4.

The making of these changes was directly influenced by the experiences of the garment designer in working with an individual active ageing participant, the participant 'show-and-tell' co-design workshop, and facilitated by the designer changing the fabric specification to a different brand with similar properties but with the flexibility to work to user needs rather than marketing and brand specification needs. The process demonstrated a step towards a co-design methodology for the designer and allowed the opening of a path to addressing the needs of a new market segment.

Table 7.3 Feedback on design features from the collaborating company

- Small and difficult to grasp Velcro closures on cuffs.
- Small and difficult to use 'poppers' on outer zip storm flaps – the participant was unable to successfully fasten the poppers.
- Double-ended zippers (allowing the bottom of the zip to be opened upwards), which were difficult to actually fasten – the participants had difficulty sliding the leading edge of the zipper through both pullers and were often unable to fasten the zips.
- Storm flaps over pockets that were very challenging to push hands through – the participants, having donned the garment, had real problems locating the pocket, reaching under the storm flap, and then locating the zip pull tag to open the pocket.
- No garments in the female range, which meant garments were cut to the wrong shape for a female figure.

Table 7.4 Key modifications to the design

- Removal of outer front zip storm flap, replaced by an inner flap with a drain channel, thus removing the need for any fiddly small poppers to close the storm flap.
- Replacement of zip with a water-resistant version.
- Removal of two-way zipper function to facilitate easier fastening.
- Removal of outer storm flaps from outer pockets, and replacement of standard zip closures with a water-resistant version.

7.10 Case study: communication with retail

This section addresses some of the issues that have been identified in relation to bringing the attributes of smart clothes and wearable technologies to near market. In particular there is the challenge of addressing the rapidly growing – but neglected – ageing consumer. The project team, through co-design–led user engagement, and with findings from Generation Sport's 'Best Ager' market research, became well aware that the active ageing have difficulty in finding suitable products and are dissatisfied with their treatment and overall experience within the retail environment. Understandably they are unlikely to go into a 'hard core' outlet with a climbing wall in the entrance! They want age-appropriate size, shape, and colour. As Taylor and McCann neatly express in the abstract to their paper:

'To be a success in the market place, and specifically, to be attractive to the Active Ageing cohort, the benefits and functionality must be clearly articulated otherwise the target end user will see no reason to invest in the end product'.

In other words there must be a common language, and it must be understood. The product 'benefits and functionality' must be clearly communicated to inform potential consumers that the product exists, and that it suits their needs. Clear and readily understandable language is required to explain the product features not only in terms of textile attributes and garment features but also in terms of the function and usability of embedded wearable electronics. This is clearly in the 'promotion' area of an organisation's marketing strategy, but confusion can often prevail.

As already discussed, the project's layered clothing system with built-in electronics has no obvious retail home, and thus matters of marketing strategy are difficult to formulate. The garment system could be thought of as an item of clothing, in which case a sportswear outlet may be appropriate. Alternatively, the item may be regarded as an ICT product, in which case an electrical/computer outlet would be appropriate. In each case, staff in these outlets would need to understand a new language: technical staff need to know about attributes and properties of the clothing, and sportswear clothing outlets need to develop a technical language. The use of a mobile phone as a central processing device serves merely to complicate matters further, since if these mobile phone outlets were to consider being used, both clothing and technical language would need to be learnt by the staff. It may also be the case that the ultimate user, the active ageing walker, would not normally use of any of these three outlets. Owners of these outlets are well aware of their customer base, and recognising that active ageing walkers do not form a substantial part of it, may be reluctant to stock these innovative hybrid products.

The situation is further complicated when the question of 'Who actually is the client/customer for a DfAW clothing system?' is posed. At least two possibilities exist. The first, and perhaps the most obvious answer is the older walker himself/herself, and some of the issues in this case are outlined in the earlier part of this section. The second possibility is that the purchaser of the garment system is someone else, perhaps a son or daughter of the potential garment system user, who is concerned that the user may lose their way or otherwise come to harm while out walking. They buy for reassurance that the older walker will be safe when using the garment system.

In these kinds of examples, different promotion and sales techniques will be needed, depending on the target group.

7.11 Case study: communication with wearers

Wearers of the electronically enhanced garment system will also need to develop their own understanding of its capabilities, use, and maintenance. Whilst the users who were part of the co-design team became familiar with the project common language, this will not be the case with those purchasing from a retail outlet. The purchasers will in any case need rather different information, possibly along the following lines:

- *Garment functionality*: this includes the capabilities and limitations of the garment system as a whole, use of the mobile phone 'central processing unit', and collecting, analysing, and presenting data.
- *Garment operation*: switching on and off, charging garment energy sources, smartphone, and Shimmer power supplies, and their replacement/renewal when needed.
- *Garment maintenance*: day-to-day surface cleaning, washing/dry cleaning, storage when not in use, and end-of-life disposal.

It is common practice in the garment sector, particularly in connection with garment cleaning, to use standard symbols as a common language to indicate, for instance, recommended wash temperatures, advisability of ironing, and other maintenance topics. It is tempting to think the same visual approach could be used in the case of the electronics part of the DfAW layered garment system. In this particular field there are also commonly accepted symbols for on/off switches, play/stop a recording, and many others. None of these visual identifiers, however, can describe the fine-grained detail needed for complete confidence in using the DfAW garment system. This aspect was discussed in a paper by Taylor (2011), where he experimented with a range of visual approaches to a common language for garment labelling. A decision on this point has yet to be made, but it may yet prove to be the case that operation of the garment may be sufficiently complex to need some form of written manual.

7.12 Conclusion

Throughout the DfAW programme two linked key themes have underpinned the project. The first theme has been the importance of adopting a design-led approach with a focus on visual, written, and verbal communication at every stage. Emphasis on this theme has helped to capture spoken and visual information to be translated by designers as facilitators. Without this approach, the DfAW project, involving many academic specialisms across five geographically distant universities, a lay user group, and with strong industry links, would have been extremely difficult to manage. Even with this emphasis, the project has at times had its communication channels stretched to the limit. However, great care taken at the beginning of the project to ensure the robustness of those channels, particularly in their role in transmitting a common language, has meant that confusion has been kept to a minimum.

The second underpinning theme has been the strict adoption of a co-design process, as outlined in Section 7.6. The thorough (albeit slower) nature of the co-design process has been key to developing understanding across academic disciplines, between industry stakeholders and, above all, empowering users to have confidence in expressing their views, formulating their needs and wants, and their horizons widened – both in terms of smart clothes and in being more confident when they are out walking.

The clothing layering system has been designed, and clothing design and electronic content specified. Preliminary clothing samples to meet these design parameters have been produced, not in experimental university workrooms but in the production facilities of collaborating companies, thus providing the opportunity to produce real garments under real production conditions. These samples have been evaluated in the field by knowledgeable individuals – the user group that formed part of the co-design team.

The next step is bringing the layered garment into the users' hands. Some of the considerations in what is largely a matter of a garment system producer's marketing strategy are outlined in Section 7.10. These considerations will apply in particular to the parts of the marketing strategy that relate to promotion, selling, and pricing, and will need to be addressed for successful product takeup.

References

Cummings, J. N., & Kiesler, S. (2005). Collaborative research across disciplinary and organisational boundaries. *Social Studies of Science*, *35*(5), 703–722.

Hussey, C. (2007). *Investigating clothing sector interrelationships: A study focused on industry specific technology use (Ph.D. thesis)*. University of Northumbria.

Löwgren, J. (1995). Applying design methodology to software development. In G. M. Olson, & S. Schuon (Eds.), *DIS '95 proceedings of the 1st conference on designing interactive systems: Processes, practices, methods, & techniques, 1995. ACM New York* (pp. 87–95).

McCandless, D. (2009). *Information is beautiful* (1st ed.). London: Collins.

McCandless, D. (2010). *Information is Beautiful*. Harper Collins.

Mckelvey, K., & Munslow, J. (2003). *Fashion design: Process, innovation & practice*. Malden, MA: Blackwell Science.

Sanders, E. B., & Stappers, P. J. (2008). *Co-creation and the new landscapes of design*. [www] http://www.seainsideproject.com/wp-content/uploads/2011/12/CoCreation_Sanders_Stappers_08_preprint.pdf. Accessed 10.12.13.

Taylor, D. (2011). Bridging methodological gaps through cross-disciplinary dialogue for design of smart clothes wearable technology for the active ageing. Online. In *Proceedings of Include 2011 – Helen Hamlyn Centre, 18–20 April, 2011, Helen Hamlyn Centre, Royal College of Art*.

Taylor, D., & McCann, J. (2012). Embedding design in the process of communication of functionality in smart clothes and wearable technology to the active ageing consumer. In *Proceedings of the 88th Textile Institute world conference, 16–17 May, 2012, Textile Institute*.

Anatomical and physiological changes with age: implications for apparel design

8

D. Bryson
University of Derby, Derby, UK

8.1 Introduction

In a paper in 2007, I looked at the idea of "Unwearables" (Bryson, 2007) as a way of looking at clothing design and its lack of address to the realities of the body's actual shape and the way we function. In this chapter I want to expand on this by looking at how we change anatomically and physiologically with age and how this affects wearability and unwearability in the active ageing and why "Things aren't the way they used to be." (John Gorka, 2013).

The actual market size for some clinical conditions is such that they are major areas for medical product development as well as areas for apparel and smart clothing development, e.g., with 160,000 incontinence shoppers (Chemist + Druggist, 2013) in the UK, the incontinence market is worth over £68 million with a 12% growth in the value of the pads subcategory. This and similar markets will only increase in size as more people are expected to buy their own products rather than be provided with them by the National Health Service (NHS) (C and M Research, 2012).

Given that the active ageing population covers such a wide span of ages, there are changes and variations over the age groups. In addressing their bodily needs there is no one set combination that will determine why a piece of apparel is either wearable or unwearable, and the combination of reasons will change depending on circumstances such as age, health, and day of the week, as well as from person to person. Clinical conditions have consequences in terms of changes to the body whether physical, neurological, or psychological. These may be of prime importance or overridden by what are seen as other more important factors such as comfort, price, fashion, and usability.

8.2 Anatomical and morphological changes

Our genes and whether we have good genes or not so good genes along with environmental factors will determine what will happen with the growth, development, and ageing of our bodies, and these processes cannot be stopped—only delayed. Some of the changes are visible, while some of them like closure of sutures to the cranium are not.

Textile-led Design for the Active Ageing Population. http://dx.doi.org/10.1016/B978-0-85709-538-1.00008-0

The aim in this chapter is to look at functional changes and limitations, which may affect clothing choice; for example, how easy or hard donning and doffing clothing may become.

8.2.1 Body proportions, stature, shape, and build

Chumlea (Chumlea, Garry, Hunt, & Rhyne, 1988) reported in his study following a population of 220 (122 female, 98 male) over 6 years that the mean annual decrease in stature was 0.4–0.5 cm/year, which is a decrease per decade of 4.0–5.0 cm, with a standard deviation of 2.0–4.0; and also that this change is approximately constant with age and sex. Changes in weight, except for the lightest who recorded increases in weight, were small and negative.

These changes in stature are alongside changes in musculature, with a gradual loss of muscle fibers beginning at the age of 50 and continuing so that by the age of 80 approximately 50% of fibers are lost from limb muscles (Faulkner, Larkin, Claflin, & Brooks, 2007). Faulkner et al. recognized that age-related changes in the musculoskeletal system have a significant impact on the everyday activities of the elderly, and that the impact is not limited to those with average activity, as those who are athletic see their performance decline after approximately 40 years and, by the age of 80, the decrease in peak performance in events has decreased by 35–65%.

8.2.2 Bones and joints: movement and mobility

Functional limitations due to joint problems like osteoarthritis are declining as earlier diagnosis and treatment is leading to a steady decrease in upper and lower body limitations. Freedman (Freedman & Martin, 2000) reported a decline from 5.1% to 4.3% in upper body limitations between 1984 and 1995, and from 34.2% to 28.5% in the lower body in Americans 70 years and older. This change shows the general trend but is also accompanied by an increase in chronic conditions as the population gets older; as Daley (Daley & Spinks, 2000) reports, "Forty percent of those aged 60–74 years have a chronic disorder that may result in some functional limitation compared with 65% aged 75 and over."

8.2.3 Skin, hair, and nails

Skin changes with age, becoming thinner due to a decrease in elasticity of the collagen and reticular fibers in the dermis. This results in the skin becoming more fragile, such that at the age of 70 and beyond the skin can appear almost as thin and friable as paper. As the skin becomes more fragile it is more liable to tear, leading to difficulties where the elderly have to be nursed in to pressure sores, especially where there are also damp and shearing forces, e.g., sliding down the bed.

The difference between younger skin and older skin is that after elastic deformation of the skin, there is an exponential recovery back to the original thickness. In a young adult, the recovery is complete within a few minutes, while in an elderly person, full recovery often requires over 24 h (Daly & Odland, 1979).

Hair or lack of hair in the ears is a distinct feature that distinguishes males from females. Men, even as early as 30 years of age, start to grow coarse hair in the nostrils, in the eyebrows, and around onto the cheek below the lower eyelid. They also grow a beard-like tuft of hair on the tragus of the ear. In comparison, old women have no tuft of hair at the tragus even if they grow hair more easily on their face.

Thin skin can lead to increased incidence of tears, and if inappropriate dressings are used, these can lead to further tears. Nonadhesive dressings are preferred; e.g., Steristrips and Melolite, in a study from 1998 by Edwards (Edwards, Gaskill, & Nash, 1998). The implication for clothing design is the need to avoid sharp edges or areas that could tear the skin, e.g., making sure that zips and closures are smooth and work easily.

Skin changes are accompanied by changes in hair. As we age, less melanin, which is produced in the hair follicles by melanocytes, leads to graying. This can occur as early as 30 years of age but is very variable depending on genetic factors. Everyone has some hair loss with age, but alongside this the rate of hair growth also slows.

8.2.4 Morphological changes

There are several clinical conditions affecting the body's shape, such as osteoarthritis of the knee joints, osteoporosis, and a condition that can start very much earlier—hallux valgus, commonly called bunions.

Osteoarthritic changes in the knee joint can lead to genu valgus, where men and women walk along with what appear to be bow legs. This occurs due to uneven wear across the whole surface of the lateral condyles of the joint.

Osteoporosis can lead to fractures of the vertebral bodies, primarily in the thoracic region of the spine, resulting in height loss and a stooped or hunched posture, termed kyphosis. This can have an effect on clothing, as described by the National Osteoporosis Foundation: "After becoming shorter by an inch or more in one year's time, some people realize there is a problem with the spine. Because of height loss and changes in the spine, clothes may start to fit poorly" (National Osteoporosis Foundation, 2013).

Clothes fitting poorly due to kyphosis can have an impact on body image, so clothing that is designed to be flattering and comfortable for women with osteoporosis and ways of combatting bad fit have been suggested by a number of organizations and specialists. For example, "Clothing, Body Image and Osteoporosis: Finding Clothes to Suit a Changing Body Shape after Vertebral Compression Fractures" (National Osteoporosis Society, 2011), where clinicians create clothing blocks using the shape produced for an osteoporotic woman using a body scanner (Maas, 2008); and "Bone Loss and Posture: Six Tips to Look Your Best" (David, 2008).

This issue is not new and affects a large number of women, so projects have been undertaken to create styles of clothing that function for osteoporotics. As expressed by Vann in 1999:

"Imagine you've shrunk in height, stooped over and have no waistline. For the 23 million women in this country (US) who suffer from osteoporosis, a disease that causes bones to thin, weaken and break easily, the simple act of getting dressed is a daily ordeal. Body changes from osteoporosis include loss of as much as 3 inches in height, rounding of shoulders, a larger waist and protruding belly, all of which can make clothes shopping a nightmare. Jackets and blouses pull across the back and shoulders. Collars gape. Shirts and dresses do not hang properly" (Vann, 1999).

Hallux valgus affects the joint of the big toe and can start in early teens but can get so bad that it is difficult to find shoes that fit the feet. However, clinicians and designers have been working on prevention and shoes that will cope with bunions as they stretch and are padded to prevent chafing of the bunion (Meanfeet, 2010). Shoes with a split separating the big toe and second toe have been developed from shoes native to Japan called jika tabi (Japonista Sole, 2013).

8.3 Physiological changes

8.3.1 Thermoregulation

Kenney, in a minireview on ageing and human temperature regulation, summarized the state of knowledge in his 2003 (Kenney & Munce, 2003) paper as follows:

- Baseline temperature levels and circadian temperature rhythm in healthy older adults are not different from younger individuals.
- Ageing is associated with a decreased vasoconstrictor response during cold exposure and skin blood flow due to decreased vasodilation.
- Active muscle loss affects thermogenesis, so resting heat production decreases by 20% from the age of 30 years to 70 years.
- Reduced sweat gland output, as older men and women sweat less during passive heat exposure.
- Less redistribution of blood flow from kidneys and visceral organs.

Studies (Natsume, Ogawa, Sugenoya, Ohnishi, & Imai, 1992) have found that "thermal sensitivity is decreased with advancing age and that thermal perception in the elderly, especially to cold, is less sensitive in the summer." The result of decreased or absent sweating due to impairment of thermoregulatory function has been linked with increased mortality during heat waves (Foster, Ellis, Doré, Exton-Smith, & Weiner, 1976; Kilbourne, 1996).

8.3.2 Healing processes

Normal skin exhibits characteristic changes with age that have implications for wound healing. It has been reported that in healthy individuals ageing leads to delays in epithelization of 1.9 days (Holt, Kirk, & Regan, 1992), though the actual physiological process remains unchanged. As the age of the skin is more related to solar exposure and less related to chronological age, the change in healing is an effect of the skin's underlying physiological status.

The final results of healing in skin and other parts of the body with age are the same qualitatively as for young subjects (Eaglstein, 1989) but more likely to be delayed. As impaired wound healing represents a major clinical and economic problem, prevention and protection of the skin and other parts of the body against injury could have a significant role to play.

It is the presence of underlying clinical conditions, especially vascular and neurological diseases, e.g., diabetes, nutritional deficiencies, and decreased blood flow due to venous insufficiency, that leads to failure of wound healing. These may lead to leg ulcers and loss of limbs. The cost of chronic venous insufficiency has been estimated in the UK to be between £400 and £600 million (Simka & Majewski, 2003). Multilayer high-compression elastic bandaging systems are seen as the most effective method of compressions for treating venous leg ulcers (Rippon, Davies, White, & Bosanquet, 2007).

8.3.3 Touch and physical sensation

Touch is a useful example of change with age, as we rely on touch for managing many everyday clothing-related tasks from putting on clothes at a gross level down to buttons and other closures at a smaller scale.

Along with the other changes in skin, the spatial acuity of the skin of the fingertips deteriorates with age, with Stevens (1992) reporting changes in the minimum separable distance as a mean threshold of 1.95 mm for the young (18–33 years), 2.68 mm for middle-age (41–63 years), and 5.03 mm for the elderly (66–91 years). Other researchers have found that this decline in discrimination is reversible and can be restored by suitable stimulation (Dinse, Kliebel, Kalisch, Ragert, Wilimzig, & Tegenthoff, 2006).

Age-related changes were also found over the body surface by Stevens and Choo (1996): "Declining acuity with age was found to characterize all regions to one degree or another, but the hands and feet turned out to be far more vulnerable than the more central regions, including the very acute lip and tongue. Deterioration of acuity in the great toe (averaging 400% between youth and advanced age) and fingertip (averaging 130%) may adversely affect such diverse activities as braille reading, grasping, and maintaining balance" (Stevens & Choo, 1996).

The changes in touch are also accompanied by a decline in thermal sensitivity, with the greatest decrease in the extremities, especially the foot (Stevens & Choo, 1998).

8.3.4 Dexterity and proprioception

Proprioception is the sense of knowing where one's body is located in space, and consists of joint position sense and kinaesthetic movement sense. These components are important for generating smooth, coordinated movements and the maintenance of body posture and balance. As we get older, there is a decrease in joint position sense (Kalisch, Kattenstroth, Kowalewski, Tegenthoff, & Dinse, 2012), and it takes longer to detect movement—i.e., the feedback mechanism between moving and understanding where we have moved to is slower.

We rely on proprioception to understand where our body is located, so balance is impaired, and as we age it is more difficult to stand on one leg compared to when young, which may affect dressing as will similar decreases in dexterity, the fine-movement control that would enable us to easily manage buttons and other closures.

These abilities may be further degraded due to arthritis of the joints with related lack of mobility, dexterity, and lack of strength in the hands and fingers, leading to clothing problems. These were expressed as follows by Reich in 1991:

"Proper fit of clothing was a significant problem for females. Hands and wrist problems were significantly related to difficulty in manipulating fasteners. Difficult clothing design features included turtlenecks, long and set-in sleeves, and back closures for women and pullover sweaters, t-shirts, and ties for men" (Reich & Otten, 1991).

8.4 Factors affecting wearability and unwearability

Considering what an active ager might or might not wear, and which clothing is wearable or unwearable, is not straightforward, as the range of ages that are covered, and the effects of self-actualization, means there are different views and opinions across the age range.

8.4.1 Fashion, comfort, and fit

In looking at clothing preferences in elderly female customers in 1981, their preferences suggested little influence of contemporary fashion trends, but 61% expressed a definite need for special sizing (Richards, 1981). There has been little change in how elderly perceive clothing, with Guzel reporting in 2013 that "the elderly gave importance to the comfort of clothes and to the availability of the elastic and pocket in the models and often preferred the shirt and trousers as clothes type came out. It was determined that people aged 65 and above needed ergonomic clothes facilitating their movements and economical ones in terms of price" (Güzel, 2013).

Looking at baby boomers aged 45 to 54 in Canada, Spadaro in 2012 found that "the clothing they preferred did not change along with the natural changes their body undertook as they aged. This suggests that there is a need to change the fit model at the production level to accommodate the changes that take place as women enter late adulthood" (Spadaro, 2012).

Thomas and Peters, looking at adults over 65 years, found that they are "still interested in fashion and that looking fashionable for themselves and others is important" (Thomas & Peters, 2009).

Clarke and associates, looking at women aged 71 years to 93 years, found that the women chose their clothing styles to compensate for age-related health issues and that the clothing choices "were influenced by their changing lifestyles and constrained by a lack of desirable and affordable clothing options for the older female body" (Clarke, Griffin, & Malhila, 2009). They also found that "Expressing a preference for classic or traditional styles, the women also reported that they used clothing strategically to mask or compensate for bodily transgressions that had occurred over time as a result of the physical realities of ageing, including weight gain, altered body shapes, wrinkles and sagging or 'flabby' arms and necks, referred to respectively as 'bat wings' and 'turkey wattles'" (Spadaro, 2012).

8.4.2 Changes in body, body image, and wearability

What is worn or not worn or in many circumstances bought or not bought depends on availability of the desired product but more often on the match between what is required and what most meets those requirements. The change in how active agers view their bodies or the "body image," has a significant effect on their world view. When young we appear invincible, but as we get older it is more likely that we experience one or more parts of our body, to all intents and purposes, "letting us down"; this can be changes in eyesight (glaucoma or cataracts), increased risk of cancers with age, or the change in skin thickness and appearance.

In 2011, Oh looked at body image in older people. "Ageing-related physical changes may lead older people to feel growing discrepancy between how old they look and how old they actually feel. Older people maybe particularly susceptible to developing problems in body image because physical changes associated with the ageing process generally are perceived as socially undesirable and unattractive" (Oh, 2011).

8.4.3 Market segmentation

The value of the active ageing market has long been recognized but continues to be neglected. The implications of the ageing body are unrelated to age as many of the above anatomical and physiological changes are dependent on genetic and environmental factors, so the market is not clearly segmented by age.

If you consider the likelihood of a teenager, a 20-something, an over 30, and an over 40 agreeing or having the same needs in terms of fit and fashion, this will lead you to the conclusion that this is an impossibility. Yet in terms of sector, the active ageing, which may cover from 50 or 60 all the way up to 95 or older, are almost invariably looked at as one market. While the active ageing sector will all be experiencing anatomical and physiological changes as the transition into old age towards senescence takes place, the degree to which these changes will be impacting on their choice of clothing and lifestyles varies considerably.

As commented on in 1988 by Schewe, "Yet chronological age, especially the use of the retirement age of 65 as a demarcation point, is most inappropriate. It merely describes the market; it does not foster a powerful understanding of the many underlying motivations. More important, our ageing population is highly heterogeneous, a virtual crazy quilt of desires, psychological and physiological capacities, lifestyles, and roles" (Schewe, 1988).

8.5 Conclusions

This chapter has placed an emphasis where possible on the natural processes of ageing and included those that impact most directly on clothing design, trying to get away from the notion of ageing as a disease. Clothing industries need to look at the positive impact they can make to support an ageing population, as expressed by Larkin in the strapline for her article in 2011: "As an industry, we need to help overturn the spreading notion that ageing is a disease, and put forth a positive wellness model that supports people in living as fully as possible at any age, regardless of health conditions."

The factors that affect wearability—fit, comfort, and fashion—are really no different for active agers compared to the rest of the population, except that most clothes are not designed for their needs and body shapes, and the older market has become more healthcare-driven, where clothing for the older person is seen as a medical device. This has been recognized as the "medicalization of ageing" (Larkin, 2001), and increasing inactivity with old age seen as inevitable, whereas many of the anatomical and physiological changes can be slowed or mitigated through practice and exercise (Goggin & Morrow, 2001) as an ever-increasing number of active agers engage and enjoy their old age.

References

Bryson, D. (2007). Unwearables. *AI and Society*, *22*(1), 25–35. http://dx.doi.org/10.1007/s00146-006-0072-3.

C and M Research. (2012). *UK incontinence market to be driven by NHS cutbacks*. http://www.companiesandmarkets.com/News/Healthcare-and-Medical/UK-incontinence-market-to-be-driven-by-NHS-cutbacks/NI6107. Accessed 06.08.13.

Chemist + Druggist. (2013). Incontinence. Category focus. http://www.chemistanddruggist.co.uk/incontinence Accessed 06.08.13.

Chumlea, W. C., Garry, P. J., Hunt, W. C., & Rhyne, R. L. (1988). Distributions of serial changes in stature and weight in a healthy elderly population. *Human Biology*, *60*(6), 917–925. http://www.jstor.org/stable/41464085.

Clarke, L. H., Griffin, M., & Malhila, K. (2009). Bat wings, bunions and turkey wattles: Body transgressions and older women's strategic clothing choices. *Ageing and Society*, *29*(4), 709–726. http://dx.doi.org/10.1017/S0144686X08008283.

Daley, M. J., & Spinks, W. L. (2000). Exercise, mobility and ageing. *Sports Medicine*, *29*(1), 1–12. http://link.springer.com/article/10.2165/00007256-200029010-00001.

Daly, C. H., & Odland, G. F. (1979). Age related changes in the mechanical properties of human skin. *Journal of Investigative Dermatology*, *73*(1), 84–87. http://www.nature.com/jid/journal/v73/n1/pdf/5616244a.pdf. Accessed 06.08.13.

David, J. L. (2008). *Bone loss and posture: 6 tips to look your best*. http://www.minnesotamedicine.com/CurrentIssue/PulseDesignMarch2008/tabid/2481/Default.aspx. Accessed 21.08.13.

Dinse, H. R., Kliebel, N., Kalisch, T., Ragert, P., Wilimzig, C., & Tegenthoff, M. (2006). Tactile coactivation resets age-related decline of human tactile discrimination. *Annals of Neurology*, *60*, 88–94. http://dx.doi.org/10.1002/ana.20862.

Eaglstein, W. H. (1989). Wound healing and aging. *Clinics in Geriatric Medicine*, *5*(1), 183–188. http://www.ncbi.nlm.nih.gov/pubmed/2645996.

Edwards, H., Gaskill, D., & Nash, R. (1998). Treating skin tears in nursing home residents: A pilot study comparing four types of dressings. *International Journal of Nursing Practice*, *4*(1), 25–32. http://onlinelibrary.wiley.com/doi/10.1111/j.1440-172X.1998.00066.x.

Faulkner, J. A., Larkin, L. M., Claflin, D. R., & Brooks, S. V. (2007). Age-related changes in the structure and function of skeletal muscles. *Clinical and Experimental Pharmacology and Physiology*, *34*(11), 1091–1096. http://dx.doi.org/10.1111/j.1440-1681.2007.04752.x.

Foster, K. G., Ellis, F. P., Doré, C., Exton-Smith, A. N., & Weiner, J. S. (1976). Sweat responses in the aged. *Age and Ageing*, *5*(2), 91–101. http://dx.doi.org/10.1093/ageing/5.2.91.

Freedman, V. A., & Martin, L. G. (2000). Contribution of chronic condition aggregate changes in old-age functioning. *American Journal of Public Health*, *90*(11), 1755–1769. http://www.ncbi.nlm.nih.gov/pmc/articles/PMC1446390/.

Goggin, N. L., & Morrow, J. R. (2001). Physical activity behaviors of older adults. *Journal of Aging and Physical Activity*, *9*, 58–66.

Güzel, S. (2013). Clothes preferences and problems of consumers aged 65 and above. *The Macrotheme Review*, *2*(5), 168–181. http://macrotheme.com/yahoo_site_admin/assets/docs/14MR25Gu.22265009.pdf.

Holt, D. R., Kirk, S. J., & Regan, M. C. (1992). Effect of age on wound healing in healthy human beings. *Surgery*, *112*(2), 293–297. http://europepmc.org/abstract/MED/1641768.

Japonista Sole. (2013). *New generation shoes that will revolutionize how the world walks: Tabi sneakers*. http://www.japonista-sole.com/la-feet-tabi-shoes-boots-sneakers/cat_14.html. Accessed 21.08.13.

John Gorka. (2013). http://johngorka.com.

Kalisch, T., Kattenstroth, J.-C., Kowalewski, R., Tegenthoff, M., & Dinse, H. R. (2012). Age-related changes in the joint position sense of the human hand. *Clinical Interventions in Aging*, *7*, 499–507. http://dx.doi.org/10.2147%2FCIA.S37573.

Kenney, W. L., & Munce, T. A. (2003). Invited review: Aging and human temperature regulation. *Journal of Applied Physiology*, *95*(6), 2598–2603. http://jap.physiology.org/content/95/6/2598.short.

Kilbourne, E. M. (1996). Heatwaves and hot environments. In E. K. Noji (Ed.), *The public health consequences of disasters*. Oxford: Oxford University Press.

Larkin, M. (January/February, 2001). The 'medicalisation' of aging: What is it, how it harms, and what to do about it. *Journal on Active Aging*. http://www.baypath.org/Themedicalizationofaging.pdf. Accessed 21.08.13.

Maas, S. (March, 2008). *Designing Duo*. Minnesota Medicine. http://www.minnesotamedicine.com/CurrentIssue/PulseDesignMarch2008/tabid/2481/Default.aspx. Accessed 21.08.13.

Meanfeet. (2010). *Hallux*. http://www.meanfeet.co.uk/hallux-information.asp. Accessed 21.08.13.

National Osteoporosis Foundation. (2013). Osteoporosis and your spine. http://www.nof.org/articles/18 Accessed 21.08.13.

National Osteoporosis Society. (2011). Clothing, body image and osteoporosis: Finding clothes to suit a changing body shape after vertebral compression fractures. http://www.nos.org.uk/~/document.doc?id=805. Accessed 21.08.13.

Natsume, K., Ogawa, T., Sugenoya, J., Ohnishi, N., & Imai, K. (1992). Preferred ambient temperature for old and young men in summer and winter. *International Journal of Biometeorology*, *36*(1), 1–4. http://link.springer.com/article/10.1007/BF01208726.

Oh, K. (2011). Body image in older people as it relates to exercise and dieting. *Journal of the Korean Society of Clothing and Textiles*, *35*(6), 678–686.

Reich, N., & Otten, P. (1991). Clothing and dressing needs of people with arthritis. *Clothing and Textiles Research Journal*, *9*(4), 34–40. http://dx.doi.org/10.1177/0887302X9100900405.

Richards, M. L. (1981). The clothing preferences and problems of elderly female consumers. *The Gerontologist*, *21*(3), 263–267. http://dx.doi.org/10.1093/geront/21.3.263.

Rippon, M., Davies, P., White, R., & Bosanquet, N. (2007). The economic impact of hard-to-heal leg ulcers. *Wounds*, *3*(2). http://www.woundsinternational.com/pdf/content_119.pdf. Accessed 19.08.13.

Schewe, C. D. (1988). Marketing to our aging population: Responding to physiological changes. *Journal of Consumer Marketing*, *5*(3), 61–73. http://dx.doi.org/10.1108/eb008232.

Simka, M., & Majewski, E. (2003). The social and economic burden of venous leg ulcers: Focus on the role of micronized purified fraction adjuvant therapy. *American Journal of Clinical Dermatology*, *4*(8), 573–581. http://link.springer.com/article/10.2165/00128071-200304080-00007.

Spadaro, J. (2012). *Market analysis of apparel preferences and consumer satisfaction of female baby boomers in Canada*. Toronto, Ontario, Canada: Ryerson University, Master of Arts in Fashion. http://digitalcommons.ryerson.ca/dissertations/923/. Accessed 21.08.13.

Stevens, J. C., & Choo, K. K. (1996). Spatial acuity of the body surface over the life span. *Somatosensory and Motor Research, 13*(2), 153–166. http://dx.doi.org/10.3109/08990229609051403.

Stevens, J. C., & Choo, K. K. (1998). Temperature sensitivity of the body surface over the life span. *Somatosensory & Motor Research, 15*(1), 13–28. http://dx.doi.org/10.1080/08990229870925.

Stevens, J. C. (1992). Aging and spatial acuity of touch. *Journal of Gerontology, 47*(1), 35–40. http://dx.doi.org/10.1093/geronj/47.1.P35.

Thomas, J. B., & Peters, C. L. O. (2009). Silver seniors: Exploring the self-concept, lifestyles, and apparel consumption of women over age 65. *International Journal of Retail & Distribution Management, 37*(12), 1018–1040. http://dx.doi.org/10.1108/09590550911005001.

Vann, K. (January 26, 1999). Fashion project provides styles to suit those with osteoporosis. The Courant. http://articles.courant.com/1999-01-26/features/9901260751_1_competition-and-fashion-show-disease-strips-women-disease-suffer-extensive-body. Accessed 21.08.13.

Further reading

Arber, S., & Evandrou, M. (1993). *Ageing, independence, and the life course*. London: Jessica Kingsley in association with the British Society of Gerontology.

Johnson, M. L. (2005). *The Cambridge handbook of age and ageing*. Cambridge: Cambridge University Press.

Sinclair, D., & Dangerfield, P. (1998). *Human growth after birth* (6th ed.). Oxford: Oxford University Press.

Spirduso, W., Francis, K., & Macrae, P. (2005). *Physical dimensions of aging*. Champaign, Illinois: Human Kinetics.

Victor, C. (2010). *Ageing, health, and care*. Bristol, UK: Policy Press.

Thermoregulation and clothing comfort

9

K. Stevens[1], *M. Fuller*[2]
[1]University of Innsbruck, Dornbirn, Austria; [2]University of Leeds, Leeds, UK

9.1 Introduction: what is clothing comfort?

The concept of clothing comfort is incredibly difficult to define, and in practice, to achieve, due to continual changes in the requirements and productivity of the body and unstable environmental conditions. These factors are further complicated by different perceptions and individual preferences related to comfort, and by differing physiological responses related to environmental changes. Ageing contributes to this changed response to the environment and as a result, designing clothing that is comfortable for the active ageing can present a multitude of additional challenges.

In order to discuss how clothing may be specifically designed for the active ageing, it is first necessary to understand how clothing comfort is defined; how the human body reacts with, and in response to, changing environmental conditions; and how the clothing worn as the interface between the body and the environment can affect this.

Clothing comfort is both multidimensional and complex and what it involves has been a point of discussion for decades. It is generally agreed that it involves both thermal and non-thermal components, and changeable environmental and activity situations. Due to the wide physiological buffer zones within which the human body operates, it is believed that it is easier to quantify and express *discomfort* (e.g. hot, cold, tight, scratchy, etc.). As such, the following definition is widely accepted for comfort: 'freedom from pain and from discomfort as a neutral state' (Li, 2001); though this is a somewhat vague definition and does not adequately express the complex phenomenon that is comfort.

Taylor (1972) states that 'body comfort can only be achieved if the skin is kept in a dry condition and the body temperature is constant', while Saville (1999) suggests that it is possible to distinguish two aspects of wear comfort.

- *Thermophysiological wear comfort:* The heat and moisture transport properties of the clothing and the way that the clothing helps to maintain the heat balance of the body during various levels of activity.
- *Skin sensorial wear comfort:* The mechanical contact of the fabric with the skin; its softness and pliability in movement; and its lack of prickle, irritation, and cling when damp.

Like Saville, Rossi (2005) agrees that thermophysiological and skin sensorial elements play a large role in clothing comfort; however, Rossi suggests there may be additional factors involved.

Textile-led Design for the Active Ageing Population. http://dx.doi.org/10.1016/B978-0-85709-538-1.00009-2

- *Garment fit:* The fit, design, and style and of a garment. The tightness and weight of a garment will affect its thermal and skin sensorial properties, as well as the behaviour of the microclimate (areas within the clothing system where heat and moisture transfer take place).
- *Psychological factors:* The relationship between the construction and performance of a garment and its suitability for the situation, particularly when the clothing is intended to provide protection in some manner. Deficiencies could result in stress to the wearer, particularly if they do not feel the garment is adequate for the situation they are exposed to.

However, due to the complex nature of the human body, there is no simple way to determine the ideal balance of the above factors, to develop the ideal clothing solution. As evidenced by observation of our friends being comfortable in short sleeves on what we might feel is a cold day, or others opting for longer sleeves and a cardigan in what we might consider warmer weather, temperature reception and accommodation, and skin sensitivity, are hugely different amongst individuals. This is related to a number of factors including age, gender, ethnicity, etc. This is further complicated by constantly changing environmental conditions, and the implications associated with various levels and locations of activity, thus rendering it impossible to have one system that is perfect for all individuals, activities, and conditions.

9.2 Homeostasis and thermoregulation: maintaining a constant body temperature

Homeostasis is the ability of a system to remain in a stable state. In a human, a hugely complex organism, the physiological system of an individual must be controlled within narrow limits to function at maximum efficiency. This includes simultaneous regulation of various factors, including O_2 and CO_2 concentrations, pH, urea, blood composition (plasma, proteins, platelets, haemoglobin, etc.), ions, sugar, H_2O, and temperature. The control of these factors is essential, and failure to stay within specific limits will result in illness.

The human body contains billions of receptors that can detect even the smallest change in a physiological system. Receptors can distinguish various different types of stimuli, including chemical changes, light, temperature, tissue damage, and physical forces such as pressure and stretch. Receptors detect changes within systems, and this information is sent to the brain. Any deviations from normal parameters initiate responses to return the system to normal. These regulatory mechanisms are known as feedback loops, and are of paramount importance to health.

One of the most important aspects of homeostasis, and one that may be directly affected by the presence of clothing, is the regulation of body temperature (thermoregulation). Enzymes, proteins responsible for catalysing chemical reactions in the body, work best at 37.5 °C, and it is not surprising that the human body is programmed to operate very close to that temperature and will strive to maintain a stable core temperature despite wide variations in environmental conditions and heat generated through activity. To achieve this, heat production by the body and heat exchange with the environment must be in balance.

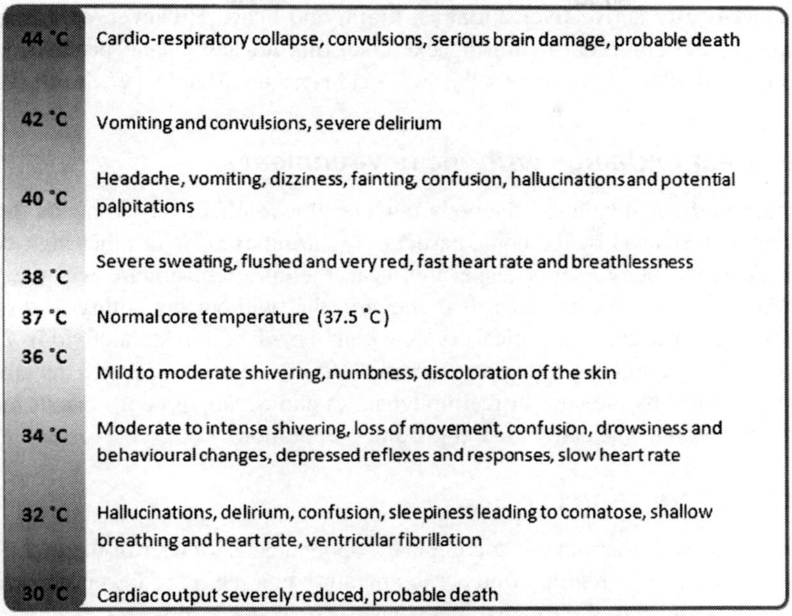

Figure 9.1 Adverse physiological reactions to a change in core temperature.

Increases or decreases in core temperature, even on a minor scale, can be very damaging to health, and can result in illness, with only a ±7 °C variation likely to result in death. Adverse reactions to changes in core temperature are displayed in Figure 9.1.

9.2.1 Heat production by the body

Heat is a form of kinetic energy that can be measured as temperature and expressed in units called calories. One calorie is the amount of heat energy required to raise the temperature of 1 g of water from 14 to 15 °C. Because 1 cal is a relatively small unit, kilocalorie (kcal) is often used to measure the body's metabolic rate and express the energy content of food (Tortora & Grabowski, 2000); 1 kcal = 1.163 W.

The body requires energy to function and this energy comes from food. As energy sources (carbohydrates, fats, etc.) are broken down and metabolised to produce adenosine triphosphate (ATP) to fuel muscles, heat is produced as a by-product. It has been suggested (Saville, 1999) that this process is relatively inefficient and only 15–30% is converted into useful work, with the remaining 70–85% of the energy being lost as heat. This is not entirely wasted energy, however, as much of it is used to maintain body temperature.

Heat production has a linear relationship with physical activity: the harder the physical activity, the greater the need for ATP within the working muscles and so the greater the by-production of heat by the body. Only a small portion, however, is lost directly through the skin, with the majority returning to the core via venous blood returning to the heart.

At rest, approximately 18% of the body's total heat production comes from energy breakdown in the muscles and skin, with the remainder being produced by the metabolically active liver, kidneys, brain, and heart. However, as a result of skeletal muscle contractions during exercise, this muscular heat production can rise to exceed 90% of the body's thermal load (Pascoe, Shanley, & Smith, 1994).

9.2.2 Heat exchange with the environment

To remain in thermal balance, the body must be able to effectively dissipate the heat that is being produced by the body, particularly during exercise or other high-energy situations. If the body cannot dispel this heat, the core temperature will rise, with potentially serious consequences. It is thought, that without the ability to dissipate heat, the body would reach critical levels within 4 h when at rest, and within 25 min while exercising at room temperature (Rossi, 2005). Heat exchange with the environment is governed by the laws of thermodynamics and occurs through various mechanisms: conduction, convection, radiation, and evaporation.

9.2.2.1 Conduction (K)

Heat conduction or thermal conduction is the spontaneous transfer of thermal energy through matter from a region of higher temperature to a region of lower temperature. The energy is transferred through direct contact. For example, if we touch a hot cup of tea with a hand, heat is conducted from the region of higher temperature (the cup) to the region of lower temperature (our hand).

Conduction between a human and the environment is a two-way process, and so heat can be lost or gained. For a human being, this is the heat exchange that occurs between the skin and surrounding surfaces with which it is in direct contact. Heat exchange via conduction is minimal, and at rest it accounts for no more than 3% of body heat loss, which could occur through clothing, jewellery, and between the soles of the feet and the ground (Tortora & Grabowski, 2000).

The amount of heat exchange via conduction is governed by the following:

- Temperature gradient (difference) between the two surfaces
- Surface area in contact
- Thermal conductivity (ease at which heat moves through a substance) of the two surfaces

An example of the capacity for substantial heat loss via conduction would be a naked human submerged in cold water. A large temperature gradient would exist between the surface of the skin and the water, and so the rate of heat loss would be very rapid as the energy moved from the areas of higher temperature (the skin) to the areas of low temperature (the water). The body surface area in contact with the water would be close to 95% if only the head remained above the water. The thermal conductivity of water is some 24 times greater than that of air, and so heat exchange via conduction is a much more rapid process in the presence of water.

9.2.2.2 Convection (C)

Convection is the transfer of heat by the movement of a fluid (liquid or gas) between areas of different temperature. Warm air is less dense than cold air, and so convection currents can form in the presence of a temperature gradient. When currents are produced only by temperature-derived density differences in the fluid, it is known as *natural convection*. When the convection currents are due to an external factor such as a pump or fan, this is *forced convection*. The faster the fluid is moved, the faster the rate of convection.

The exchange of heat by convection between a body and its environment depends on:

- The temperature gradient between the two (this determines the amount of heat absorbed or donated by a given mass of air that comes into contact with the skin).
- Relative movement of the fluid with which the body is in contact.

Despite the name, a radiator produces a good example of convective heat transfer: if a hand is placed some distance above the radiator, the hot air rising from it can still be felt. Within clothing, wind may enter the clothing system and warm air within the clothing may be replaced with cold air due to convective currents.

9.2.2.3 Radiation (R)

Thermal radiation is radiant energy produced solely as a result of an object's temperature. Unlike convection and conduction, radiation occurs without direct surface contact and can even happen through a vacuum.

Due to the nature of heat exchange via radiation, it is safe to assume that for a naked person outdoors, heat would be gained via radiation during the day from the sun, but lost at night via radiation to the sky. In a room of 21 °C, about 60% of heat loss occurs via radiation in a naked resting person (Tortora & Grabowski, 2000).

Factors affecting heat exchange via radiation include:

- Effective radiating surface areas
- Difference between the mean surface temperatures of both objects
- Emissivity and reflectivity of the objects in relation to the emissivity and average radiant temperature of the environment

9.2.2.4 Evaporation (E)

Evaporation is the process by which heat energy is used to transform liquids into gases. This process requires large amounts of heat energy, and by producing sweat that evaporates from the skin's surface, the human body can dissipate significant amounts of heat. The heat required to drive this process is known as the latent heat of evaporation (latent heat of vapourisation), and for water it is 2270 kJ kg^{-1} or 542.4 kcal kg^{-1}. Thus, if sweat is to be evaporated from the skin, this much energy will be lost to the environment.

The rate of evaporation is inversely related to vapour pressure, and so the higher the vapour pressure, the lower the potential for evaporation.

Evaporation provides the body's main defence against overheating during exercise. Under extreme conditions, a maximum of about 3 l of sweat can be produced each hour, removing more than 1700 kcal of energy if it all evaporates.

However, an evaporative threshold can be reached where the production of sweat exceeds its capacity to evaporate. Here, sweat that drips off the skin accounts for very little heat loss, and this may present a particular problem in high-humidity environments.

Evaporative heat loss that occurs due to evaporation of perspiration from the skin is termed *sensible perspiration*, which is excreted by sweat glands in sufficient quantity to appear as moisture on the skin. This will be discussed in greater detail later in this chapter. Humans also produce what is termed *insensible perspiration*, and this is perspiration that evaporates before it is perceived as moisture. This insensible water loss is outside of the body's control.

The rate of evaporation is dependent on:

- Area of wet skin open for exchange
- Forced and natural convection currents
- Vapour pressure gradient between the skin and the environment (Figure 9.2)

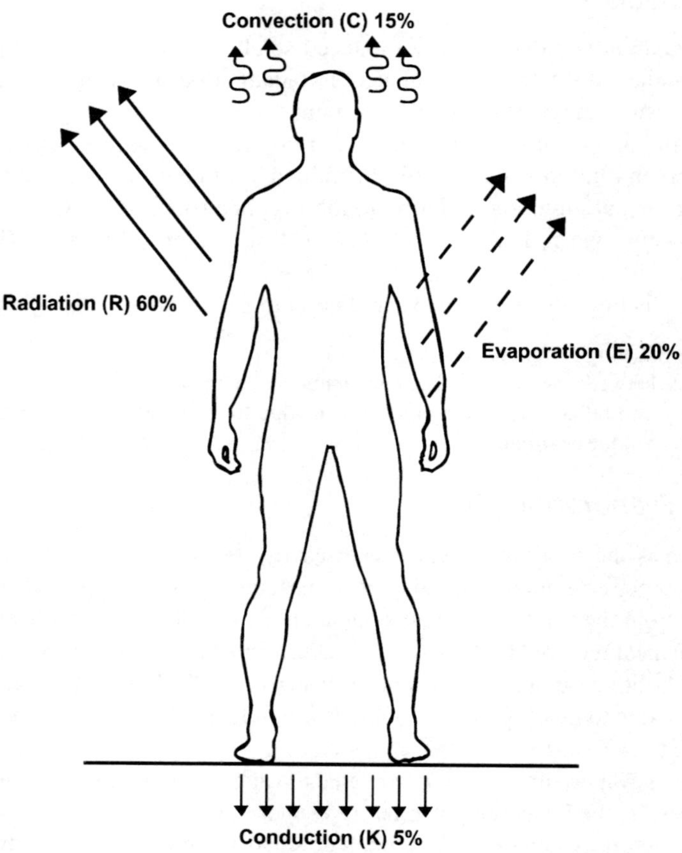

Figure 9.2 Avenues for heat exchange with the environment for a naked resting person at an ambient temperature of 21 °C.

9.2.2.5 Airway Heat Exchange

Heat exchange via the body's airways and respiratory system is also to be considered, particularly when exercising in low ambient temperatures: upon inhalation, the low-temperature air enters the respiratory system, cooling local tissues and blood, and thus adding to the body's heat loss. This cool air is warmed by convection and saturated with water vapour in the lungs and airways and then exhaled, where the heat is lost to the environment.

As an individual engages in exercise, the respiratory rate (the number of breaths per minute a person is taking) increases and so too does the volume of cold air that is taken into the lungs, and hence heat loss via airway heat exchange increases. When resting, or in comfortable ambient temperatures, this heat loss can be small, but when exercising or in low ambient conditions this method of heat loss can be considerable and it is thought that airway heat loss may amount to 8–10% of total heat loss, with around 3% of this from the warming of the inhaled air and a further 7% as heat exhaled from the lungs (Pascoe et al., 1994).

Airway heat losses are not under physiological control, and in extreme ambient conditions may be reduced by a simple covering of the mouth or special masks for heat and moisture retention.

9.2.3 Heat balance equation

The interactions between heat production and storage, and that lost to and gained from the environment, can be displayed by Fanger's heat balance equation:

$$M = S - W \pm K \pm C \pm R - E_{SP} - E_{IP} - E_{RP}$$

where

M is heat production via metabolism,
W is work carried out by the body,
S is heat storage by the body,
K is conduction,
C is Convection,
R is radiation,
E_{SP} is evaporation of sensible perspiration,
E_{IP} is evaporation of insensible perspiration, and
E_{RP} is evaporation from the airways.

9.3 Human thermoregulatory system

Core temperature (temperature of the body's internal organs) must be maintained at 37–37.5 °C. The shell temperature (temperature near the body surface, in the skin and the subcutaneous layer) is said to be 1–6 °C lower (Tortora & Grabowski, 2000). Skin temperature has been seen to vary with many factors, such as age, gender, ethnicity, etc., though average comfortable skin temperature is said to be 33.3 °C (Kim & Cho, 2002).

A naked human at rest should be able to maintain a comfortable body temperature in still air of 26–30 °C (Taylor, 1972). This range of environmental temperatures is known as the vasomotor range, and represents an interval at which the body can maintain normal core temperature by simply altering the amount of blood (and heat) to the skin, without the need to shiver or sweat. It is important that this vasomotor range exists; otherwise one would continually be shifting between shivering and sweating as a means of thermoregulation, which would be a wasteful of energy and body fluid.

The human body has a complex system of measuring body temperature and responding to temperature increases and decreases. As many biochemical reactions are temperature-dependent, the thermoregulatory system may often override other systems in the body.

9.3.1 Thermal reception

To maintain thermoregulatory function, the body must be able to sense temperature. As well as the brain and spinal cord, the skin plays a major role in thermoregulation. Located throughout the skin are millions of temperature receptors known as *thermoreceptors*, which can sense rises (warm receptors) and falls (cold receptors) in body temperature. Thermoreceptors are free nerve endings, about 1 mm in diameter on the skin's surface. Two distinct thermal sensations are perceived: cold and warm (Tortora & Grabowski, 2000).

The body has three times as many cold as warm receptors, and these cold receptors are located in the epidermis, near the surface of the skin. Due to the superficial location of these receptors, they are very quick to respond to change in temperature, and temperatures between 10 and 35 °C activate cold receptors.

Warm receptors are located a little deeper than their cold counterparts, in the dermis, and are activated at temperatures between 32 and 48 °C. However, the hypothalamus in the brain contains many more warm receptors than cold, and these are more sensitive than those on the skin. They are not stimulated until changes in environmental temperature or metabolic heat production affect the temperature of the brain by the flow of blood through it. The spinal cord is thought to contain thermosensitive structures and be equivalent to the hypothalamus as a sensor of temperature. Temperatures <10 and >48 °C stimulate mainly nociceptors (sensors that detect damage) rather than thermoreceptors, producing painful sensations (Tortora & Grabowski, 2000).

9.3.2 Thermal processing

The hypothalamus, aside from being sensitive to temperature, is also home to the central control of thermoregulation, the preoptic/anterior hypothalamus (PO/AH). Information from thermoreceptors on internal (core) and surface (skin) temperatures are relayed to the PO/AH, which then initiates the appropriate efferent response.

9.3.3 Other thermoregulatory components

In addition to thermoreceptors, the skin has 3–4 million sweat glands located all over the body. They are divided into two main types: apocrine and eccrine.

Apocrine sweat glands are located in the armpits, groin, ears, nipples, and navel and do not play a role in thermoregulation; they function in response to stress or sexual stimulation.

Eccrine sweat glands are simple, coiled tubular glands and are more abundant than apocrine glands. They are distributed throughout the skin, and their main function is to help regulate body temperature. Eccrine sweat glands are most numerous in the skin of the forehead, palms, and soles of the feet. Their density can be as high as 450 per cm^2 (Tortora & Grabowski, 2000). It is not the eccrine glands themselves, but evaporation of the sweat produced by the glands, that rids the body of the heat produced by metabolism. The release of such sweat (insensible perspiration) is usually indiscernible, yet even in cool weather an individual can lose up to 1.5 l of fluid per day. Production of sweat is controlled by the hypothalamus.

9.4 Thermoregulatory responses

The human body can detect minute changes in the temperature and chemical reactions taking place within the physiological system. Processing of this information within the brain will lead to a number of physical and behavioural responses, which will be explained in the following sections.

9.4.1 *Responses to falling temperature*

There is a natural behavioural response to the detection of a change in local ambient temperature: this is to adjust clothing, and it may be performed as a preventative or a resultant action. In the event of a detection of a fall in temperature, the behavioural response would be to put on more clothing to conserve heat.

Physiologically there are two responses to a detected fall in temperature. These are to increase heat conservation or to increase heat production.

9.4.1.1 Increased heat conservation

- Alteration of peripheral blood flow (vasoconstriction): Blood vessels on the surface of the skin narrow and constrict. This decreases the flow of warm blood to the surface of the skin and thus reduces the transfer of heat from the internal organs to the environment. Consequently, skin temperature decreases and can become close to that of the ambient environment. This can exacerbate the feeling of being cold and cause the body to initiate further responses.
- Piloerection: The standing up of hairs all over the body. The hairs trap still air in order to increase the boundary layer of the body and preserve heat. However, as humans are relatively hairless, this provides little discernible insulation.

9.4.1.2 Increased heat production

- Increase in muscle tone: An increase in muscle activity (tone) not associated with movement. This activity breaks down energy and heat is produced as a by-product.

- Shivering: Involuntary synchronous contractions of motor units. These are usually in the region of 10–20 contractions per second. At its maximum, heat production from shivering is 5–6 times than at rest; it is effective for short periods but uncomfortable.

9.4.2 Responses to rising temperature

Responses to rising temperature can again be behavioural or physiological. The behavioural response is the removal or adjustment of clothing. The physiological responses are:

- Alteration in peripheral blood flow (vasodilation): Blood vessels within the body dilate to allow an increased blood flow to the skin so that heat can be lost via convection to the environment. Consequently, skin temperature rises, which can aggravate the feeling of being hot. When exercising, this increased blood flow to the skin can limit the amount available for the muscles.
- Sweating: The rate at which sensible perspiration is produced increases so that there is moisture on the skin, in order that heat loss via evaporation can occur more readily (Figure 9.3).

9.5 Factors affecting thermoregulation

There are numerous non-thermal factors that can affect thermoregulation. These include:

- Age
- Gender
- Fitness level
- Illness/injury
- Intoxication
- Raised ambient CO_2
- Lowered ambient O_2

Of all the factors, gender and age present the largest differences in thermoregulation, and so will be discussed in greater detail.

9.5.1 Gender

Differences in thermoregulation occur between the genders due to morphological differences: size, body composition and skin surface area. Typically, women have smaller body mass, less lean muscle mass, larger amounts of body fat, and a greater proportion of surface area to body mass than men (Pascoe et al., 1994).

In cold environments these physiological differences can be considered advantageous to females. Females have average skin temperatures 1–2 °C lower than males, and so the thermal gradient between the skin and environment is smaller, thus reducing heat loss via convection. Also, the higher percentage of body fat seen in females provides increased levels of thermal insulation. However, this fat does not extend to the hands and feet, and so makes them more susceptible to cold injury. Furthermore,

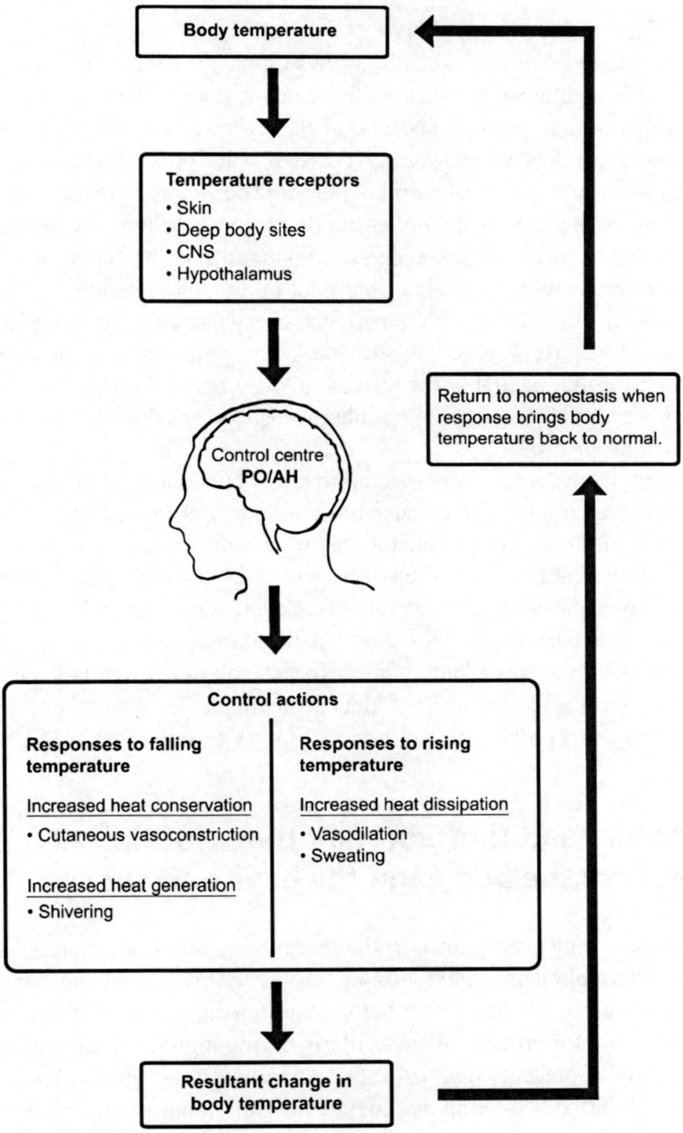

Figure 9.3 Body temperature control system.

this risk of cold injury is increased as females experience greater amounts of peripheral vasoconstriction (Pascoe et al., 1994).

In warm environments, these thermoregulatory differences can cause females to be at a disadvantage. Initially, the smaller muscle mass to fat ratio is less efficient and causes greater heat production. Women require a higher core and skin temperature to initiate sweating, and sweat production is lower than in males, and so at the onset of sweating they have higher heat stores and so heat loss via evaporation is significantly lower than in males.

9.5.2 Age

Like gender, it is the morphological and physiological changes associated with age that affect thermoregulatory responses. Children are at a greater risk of thermal stress and more susceptible to cooling because of their greater surface area to mass ratio. Pascoe discusses studies where greater decreases in core temperature occurred in younger members of a group of swimmers (between 8 and 20) who swam in water of 20.3 °C (Pascoe et al., 1994). After correction for area-to-mass differences, the fall in body temperature was still greater in the younger than the older swimmers. This is attributed to the increasing levels of body fat in the older swimmers. This factor is more pronounced in females, who experience greater changes in body composition after the onset of puberty. Pascoe suggests that with ageing and growing, boys are able to compensate for the inevitably higher heat loss by convection due to their increasing area-to-mass ratio by increasing peripheral vasoconstriction or by increasing their metabolic heat production.

Due to age-related decreases in metabolic rate, metabolic heat production is lower in older people. Ageing may also cause morphological changes in blood vessels that could lead to inefficiency in peripheral vasoconstriction (Pascoe et al., 1994). This leads to increased heat loss in cold environments with ageing. Older individuals also experience delayed onset of sweating and decreased sweat output, lowering their efficiency of evaporative heat loss. This can be particularly problematic: an older person will experience a higher core temperature before the onset of sweating is initiated, and will then sweat reduced levels of fluid, thus achieving little benefit from the sweating process and being at a greater risk of illness or injury from heat or cold.

9.6 Clothing and thermoregulation: clothing as a barrier between the body and the environment

Clothing can offer much assistance to the thermoregulation of a human; it can help to maintain a thermal environment, provide protection from cold and windy ambient conditions, and can be engineered to help move moisture away from the skin. However, when dressing for protection, particularly during high levels of activity, clothing can also potentially present significant thermoregulatory problems. The avenues for heat exchange described earlier in this chapter are true for uncovered objects, but since clothing mediates convective, radiative, and evaporative heat exchange, it makes heat exchange more complex. Furthermore, due to water being more thermally conductive than air, moisture in and on clothing is a concern due to its propensity to strip the clothing system and body of heat, potentially leading to physiological complications, as detailed earlier in this chapter.

However, the presence of fibres, fabrics, and clothing does not have to be a thermoregulatory problem, and they can in fact be engineered to support the thermoregulatory system, in terms of both heat and moisture management.

Thermophysiological comfort in clothing can be attained through achieving suitable levels of dry thermal insulation, heat and moisture management, and air permeability to suit the activity and ambient conditions. A layered clothing system is widely accepted in dressing for the outdoors, carefully tailoring fabrics and garments to support thermoregulation. This layering approach will be discussed, with specific reference to the active ageing population, after the interaction between the body and clothing has been addressed.

9.6.1 Thermal management

Thermal insulation, as provided by clothing, is a complicated area due to the complex nature of thermal conductance and the different proportions of air and fibre found in clothing. The presence of moisture further complicates this problem, and so this section shall deal with the transmittance and storage of dry sensible heat only. For further information please see Section 9.7, where the interaction of heat and moisture will be addressed.

Below 26–30 °C the body needs insulation provided by clothing to maintain homeostasis and thermal comfort. The thermal requirements of clothing are dictated by the ambient conditions and the activity level of the individual. In cold weather, it is desirable to store heat within clothing to remain in thermal comfort. As ambient temperature increases, the heat storage requirement decreases, and at ambient temperatures above 26 °C the thermal insulation required from clothing is nil. A similar trend is seen at different activity levels: during low levels of activity, when a person may be static for a long period of time, high levels of heat storage are required to remain thermally comfortable. As the level of activity and heat produced by the body increases, the necessity for storage of heat within clothing is eradicated. Under most conditions, thermal comfort can be maintained by tailoring the clothing to the ambient temperature and activity level, but this is not always possible. When a high-output activity is carried out in low ambient conditions (situations often encountered during outdoor sports), then the thermal requirements of clothing can be complex. During periods of lighter activity or rest, the participant will need to store heat and will require that their clothing provide thermal resistance; but during periods of intense activity where large amounts of heat energy are produced, it is essential that the clothing can release this to the environment.

Much theoretical and experimental research has been carried out in the field of thermal insulation, and it is generally agreed that factors affecting the thermal behaviour of clothing include the heat exchange with clothing (conduction, convection, radiation, evaporation), dry thermal insulation, transfer of moisture and vapour through clothing (e.g. sweat, rain), compressions (e.g. caused by high wind), pumping effects (caused by body movement), air penetration (e.g. through fabrics, vents and openings), and subject posture (Parsons, 1993).

9.6.2 Dry thermal insulation

Dry thermal insulation refers to the amount of heat clothing can store, and is governed almost solely by air. In a simplistic manner, in order to increase the thermal insulation of a garment or clothing ensemble, an increase in free air spaces that may potentially trap and warm the air must be achieved. For the purpose of insulation it is permissible that free air spaces may be engineered in a number of locations within a clothing system.

Air tends to adhere to solid surfaces, and this still air layer is known as the boundary layer, defined as 'a layer of fluid, close to the surface of a body placed in a moving stream that is distinguishable from the main airflow by distinctive flow characteristics of its own caused by friction' (Rees, 1971). Materials with a large exposed surface area will be a better restrictor of air movement, and it is because of this that fabrics made up of masses of fine fibres, such as pile, fleece, and down, make such good insulators. In clothing the boundary layer can be up to 6 mm thick (12 mm total between two surfaces) and can account for up to 0.8 clo (a clo is a unit of thermal insulation, where 1 clo equals the amount of clothing required to remain comfortable at 21 °C), although this may be reduced to about 0.2 clo with air velocities of $5\,m\,s^{-1}$ and above (Pascoe et al., 1994).

In addition to the boundary layer, and unarguably an area that is less at risk of disruption by environmental conditions, air can also be trapped within interstitial spaces between fibres and yarns for insulative purposes. If we consider a conventional 'warm woollen jumper', the first thing that may spring to mind would be a knitted garment, made with a thick yarn. Considering this structure, knitted materials contain far more structural air space than a woven material, and a multifilament yarn, with many small filaments making one larger fibre, contain even more free air spaces between the filaments. Furthermore, recent technology has allowed the engineering of yarn where the individual fibres are hollow to encourage increased insulation, mimicking the fur of a polar bear.

In addition to the air trapped within fibres or fabric, air layers trapped within a clothing ensemble also contribute to the thermal insulation afforded by the ensemble. These air gaps generally occur between the skin and the first layer of clothing, and from then on between each clothing layer in the ensemble. Air gaps can add considerably to the total insulation value of the garment, assuming that the air gaps are not so large that air movement can take place within them, leading to heat loss by convection. Because of this limitation, the fit of a garment has a great influence on its insulation value as well as the fabric from which it is constructed (Saville, 1999).

9.6.3 Effects of environmental conditions and motion

Physical activity and air velocity can have a considerable effect on effective thermal insulation within clothing and can have a profound effect on thermal values needed at various ambient air temperatures. Dry thermal insulation is governed primarily by air trapped within the fabric structure and between clothing layers; however, these air pockets are easily disturbed by local air movement and convection currents caused by wind and body movement, which can lead to considerable levels of heat loss. Therefore, basic insulation values, as can be found in literature, in practice apply only to a standing person in still air. For use in evaluation of physical activity and with varied ambient conditions, corrections are required.

Air flow within clothing is termed ventilation and can have both positive and negative effects upon the thermal comfort of the individual. It can lead to unnecessary and unwanted heat loss from the garment, directly affecting the thermal comfort of the individual. However, in some circumstances, often during high levels of activity, it is necessary for the individual to rid the garment of surplus heat, and avenues to achieve this are often engineered into the garment in the form of self-operational vents and openings, often found on the back, and under the arms, usually positioned to prevent water ingress in case of rain. Ventilation can occur either through the fabric, or through apertures such as collars, cuffs, and vents, and can be exacerbated by wind, body movement, and natural convection.

9.6.4 Air permeability

Air permeability is an expression describing the properties of a fabric that permit the passage of air through the fabric's interstices. The air permeability of fabrics for active wear is extremely important, as it will directly affect the thermal properties, and to some degree, the moisture management of a clothing garment or ensemble.

It has been found that there are many factors that can affect the air permeability of a fabric. Several early studies of air permeability of different materials showed the following factors to affect air permeability:

- Porosity: Defined as the ratio of free space to fibre in a given volume of fabric.
- Number, depth, and size of pores: These are governed by fibre, yarn, and weave characteristics. Studies have shown that when all of the above factors are kept constant, other parameters will affect the air permeability of a textile fabric.
- Fabric thickness: Air permeability will decrease as fabric thickness increases (although this is a function of porosity, as thickness will affect pore size).
- Yarn linear density or fabric count: As either of these is increased, air permeability will decrease.
- Yarn twist: An increase in yarn twist will lead to an increase in air permeability.
- Yarn crimp: As yarn crimp increases, so too does air permeability due to the increased extensibility of the fabric.
- Fabric construction: A study of fabric structures (Dhingra & Postle, 1977) found that air permeability was directly connected with fabric extensibility. Knitted fabrics were found to have higher air permeability than woven fabrics due to the inherent extensibility of knitted fabrics.
- Layering in garment ensembles.
- Moisture within garment ensembles: Absorption of atmospheric water can cause swelling of hydrophilic/hygroscopic fibres such as cotton and wool, which in turn changes fabric porosity and thickness, resulting in changes in air permeability.

9.6.5 The chimney effect

The chimney effect is a function of the air permeability of the fabric, the spatial design of the garment, and the size and location of openings such as collars, cuffs, and the hem. The chimney effect is an example of natural convection: air enters the garment either through the fabric or through apertures, disturbs the microclimate and boundary layer, ridding the garment of warm, still air pockets and replacing them with cold air. This cycle continues until apertures are securely sealed or activity ceases.

9.6.6 The bellows effect

The bellows effect is an example of forced convection and is caused by either the effects of body movement or, for outer fabrics with no air permeability, from the compression caused by wind on the outer surface. Here, the movement of the fabric changes the volume of air within the garment. This changes the pressure and causes convection currents to be formed. Warm air is then forced out through the fabric and through apertures and is replaced with cold air. An impermeable cover fabric worn over a pile insulating layer showed an 18% loss in insulation at a wind speed of 48 $km\,h^{-1}$ due to the effects of compression (Saville, 1999).

9.7 Moisture management

Moisture within clothing can be detrimental to comfort levels and can affect both thermophysiological wear and skin sensorial wear comfort.

Moisture within the clothing system decreases the stability of the thermal environment, leading to numerous physiological changes to maintain thermoregulation, thus affecting thermophysiological wear comfort. Cheung and Cheng (1995) state that 'dermal (cutaneous) health and thermophysiological comfort of people are affected by the wetness of their skin. When skin wetness increases, the skin becomes more susceptible to abrasive damage, more readily absorbs chemicals and is more prone to microbial growth. The intensity of sensations of wetness discomfort also increases.' Furthermore, wet clothing can produce negative psychological effects, as users might start to feel anxious about its effects, thus reducing concentration and enjoyment levels. It is therefore essential that clothing can effectively manage both internal and external moisture to provide total clothing comfort.

9.7.1 Internal moisture

Internal moisture is that which comes from within our clothing system, from the body as sensible (liquid) or insensible (vapour) perspiration, and it is this moisture that poses the more serious threat to comfort. When dressed in everyday clothing made from natural or synthetic fibres, even in conditions of light exercise, the fabrics can easily transmit all moisture vapour from insensible perspiration and the evaporation of sensible perspiration from the skin in a temperate climate. Any textile designed for protection from water or chemicals must contain a physical barrier to protect the user from the source. This barrier also inhibits the passage of moisture vapour through the fabric into the environment; the moisture vapour within the clothing will then condense and turn to liquid moisture. The passage of moisture vapour through clothing is dependent on many factors, and the ability of clothing to manage this moisture is vital to maintaining comfort.

9.7.2 Liquid moisture

Thermophysiological and skin sensorial comfort are incredibly vulnerable to the negative effects of liquid moisture – in particular the extent to which wet fabrics cause prickle, wet cling, and chafing – as well as the conductive heat losses. The interaction of liquids with textiles may involve one or several fundamental physical phenomena that are collectively termed 'wicking': wetting of the fibre surface, transport of the liquid into an assembly of fibres, adsorption on the fibre surface, and diffusion of the liquid onto the exterior of the fibres (Kissa, 1996). The course observed will depend largely on the wettability of the fibres (related to the chemical surface of the fibre) and the structure of the fabric (geometry and surface roughness).

9.7.3 Moisture vapour

As briefly mentioned, it is not the actual presence of moisture vapour within clothing that is problematic, but rather the condensation of the vapour back into liquid moisture. Condensation forms in clothing whenever the local vapour pressure rises to the saturation vapour pressure at the local temperature. Gretton (1999) stated that the possibility of condensation formation is enhanced when saturation vapour pressure is reduced by low ambient temperature and rain, or when venting of the clothing system is restricted. In clothing systems, with high levels of vapour present, it is essential that the temperature is kept constant, as any sudden or unexpected decreases in local temperature will cause vapour to condense within the clothing. The presence of condensation further lowers the microclimate temperature, causing the saturation of more vapour. Low to moderate levels of vapour can go unnoticed within clothing systems, and it is thought that only at values above 80% relative humidity would the body be unable to evaporate sweat from the skin's surface (Ren & Ruckman, 2003). There are numerous fabrics that claim to have the ability to transmit water from within (breathable), while still protecting from external moisture (waterproof), with perhaps the most well known of these being Gore-Tex, Sympatex, and eVent.

9.7.4 External moisture

External moisture is that which comes from the environment, usually in the form of rain or snow, and should be prevented rather than managed. This can be achieved by use of a selection of waterproof or water-resistant fabrics. In addition to the application of water-resistant fabrics, garments must also be designed in a manner that prevents ingress. This can be via sealable arm cuffs, adjustable neck fastenings, and intelligent hood design.

However, both fabrics and design have their limitations: waterproof membranes can in fact encourage internal moisture, even in dry conditions, which is counterproductive to the protection they provide and they can be a threat to comfort. As such, they should only really be used in extreme conditions, where prolonged precipitation is a certainty.

Water-resistant coatings (durable water repellence, DWR) can be extremely effective in all but severe conditions, but they may deteriorate with the abrasion from wear or washing and may need to be reapplied. The level of repellence available and the durability of treatments is dependent on many factors including chemical type, application methods, and properties of the fabric being treated.

It is very important to remember that every person's body is different, and despite extensive academic research in this area, there exists no universal system to remain comfortable in all conditions. The design of clothing to maintain thermoregulation is the same whether the user is young or old; sensible designs and good fabrics are essential. The difference between the young and the old, however, often lies in their activity levels; clothing that will maintain homeostasis of a young male will not maintain homeostasis of a female pensioner. It is thus down to designers to remember their target audience and for the wearer to remain aware of their own comfort. One area where thermal comfort is extremely important is in the design of outdoor clothing, and in this industry, fashion often follows function. In extreme environments when users' lives may be endangered by poorly designed clothing, a thorough understanding of the needs of the clothing users is vital. Clothing for the aged may not require such high weather resistance as outdoor clothing, but when the wearer may be more vulnerable to changes in the environment, considering thermoregulation in clothing design is still very important.

9.8 Thermoregulation and the traditional outdoor layering system: discussion

During exercise the body produces far greater heat than when it is at rest. Thus when exercising indoors, many people wear only thin clothing. However, when taking part in outdoor activities in potentially changeable weather, wearing such thin clothing all day might be unpractical or even dangerous. However, to begin climbing a mountain wearing all your clothing regardless of conditions could rapidly lead to overheating. Therefore, in outdoor pursuits such as hiking or mountaineering, it is important to wear an adaptable clothing system that can be changed to suit the activity levels of the wearer and the environmental conditions. For example, a hiker should be able to easily add layers when they stop for a break, and easily remove them when they start to exercise again. The primary advantage of a layering system over a single layer is this adaptability, allowing the wearer to remain comfortable in a large number of environmental situations, and when working to greater or lesser extents.

The layering system that most hikers and outdoor clothing wearers are aware of describes three layers that cover one's upper body, though it could equally be applied to the lower body. These layers are:

1. A baselayer, worn against the skin, designed primarily to transport moisture away from the body
2. A midlayer to keep the body warm, worn over the baselayer
3. An outer layer to keep the body and other layers dry

9.8.1 Baselayer

This is arguably the most important piece of clothing in the layering system as it is worn against the skin throughout the day and most of the body's interactions with the environment take place through this garment. Baselayers may be made of synthetic fibres (usually polyester or polypropylene) or natural fibres (usually merino wool or silk) and are usually quite tight-fitting to help transport moisture away from the wearer. They should not be too thick, as their primary purpose is to keep the body dry; it is not to insulate.

Water's high thermal conductivity means that wet skin can lead to overly rapid cooling. This may not only be unpleasant but could also be dangerous in certain situations. Thus, a baselayer should transport moisture from the body and into the environment or into other clothing layers. This transport of moisture along a concentration gradient is known as wicking, and is a vital criterion for a baselayer. Baselayers should dry quickly to ensure they do not become saturated. For this reason, highly absorbent fibres such as cotton are poorly suited to use in baselayers.

Because baselayers are worn for long periods, it is crucial that they are comfortable against the skin. Stiff fibres tend to be itchy, making coarse wools or thick synthetic fibres unsuitable. An inevitable downside of wearing a garment against the body for long periods is that it is a prime candidate for the growth of bacteria and can harbour smell. To counteract this, some synthetic fabrics are treated with an antibacterial finish that resists odour. Wool garments are inherently antibacterial.

9.8.2 Midlayer

The midlayer is the primary insulating layer in the three-layer system. Fleeces, pullovers, synthetically filled jackets, or other thick garments are all suitable insulating layers. Their role is not normally to block out wind or rain but to keep the body warm in relatively still or dry conditions.

As is the case with most insulating textiles, midlayers trap still air to keep the body warm. The thermal conductivity of air is very low, thus helping keep the body isolated from the cold environment. However, as the user's activity level increases and the body temperature changes from warm to hot, so too does the sweating rate increase. Thus, the midlayer should be as breathable as possible, allowing this sweat to escape. If sweat cannot escape the insulated layer, then it will become wet, and this will reduce its thickness and its subsequent ability to trap still air. A highly air-permeable midlayer is often the most breathable, but the disadvantage of this is that the midlayer is more vulnerable to wind blowing through the garment, thus disturbing the warm still air trapped in the garment. This is a somewhat unavoidable trade-off.

Traditionally, a '200 weight' fleece (fleece of $200\,g\,m^{-2}$) has been used for a midlayer, though in many conditions this was too warm for active use. As a result, a lighter '100 weight' fleece ($100\,g\,m^{-2}$) is often used instead.

A midlayer fleece should be cut relatively close-fitting to ensure that warm air remains trapped in the garment and in the gap between the midlayer and baselayer.

9.8.3 Outer layer

This layer is sometimes called a shell, and rightly so: it is the layer protecting the user from the environmental effects of wind, rain, and snow. Thus, waterproofness and windproofness are the most important properties of this layer. It is designed to be worn over either a baselayer or, in colder conditions, over a baselayer and midlayer. In fine weather, an outer layer is usually carried as an emergency layer, but may not be worn. In these conditions, having a lightweight shell layer is beneficial.

Many users obsess over the technicalities of waterproof fabrics, though in reality the performance of many waterproof fabrics is very similar. The crucial differences between outer layers are in the design of elements such as the hood and cuffs; if these let water in or make activity difficult, then the jacket impacts on overall comfort regardless of how waterproof or breathable the fabric is. Thus, the fit and design of an outer layer is every bit as important as the fabric from which it is made.

The increased protection of an outer layer has the drawback of reducing breathability in the clothing system. Thus, if this layer is worn when exercising hard, the wearer is likely to overheat, and sweat will build up inside the clothing system. This leads not only to a decrease in the thermal insulation of the midlayer but also causes discomfort as the baselayer becomes wet. For this reason, a waterproof shell should only be worn when necessary, and should not be worn in still and dry weather.

Waterproof-breathable shell garments usually require separate washing, and should be cared for using specialist products. In addition, the durable water repellence that is applied to these garments can degrade over time and may require occasional reproofing.

9.8.4 Criticism of the layering system

For many hill-goers, this simple three-layer system is very adequate and it is the one recommended by many shop staff, websites, and textbooks. However, the classic layering system has some drawbacks that can reduce comfort in certain situations. An early critic of the layering system was American mountaineer Twight (1999), who wrote:

> *'… marketing hype [has] created a consensual reality called the "layering system" – a reality that doesn't exist in the world of extreme alpinism.'*

A major concern with the layering system is how long it takes to adjust the layers. For example, when active in wet conditions, a baselayer and outer layer will be worn. However, when you stop, the outer layer must be removed to fit a midlayer underneath it; then when recommencing exercising, both layers must be removed before putting on the outer layer again. This is very slow and awkward. It also means that outer layers must be tailored to allow for numerous layers underneath them, making them baggy and therefore poorly suited to activities such as technical mountaineering. Another problem with the normal three-layer system is that the only windproof layer is the outer layer. This means that in mild but windy conditions, a baselayer and outer layer will be worn. However, a waterproof outer jacket has relatively low breathability, making this combination potentially sweaty and uncomfortable.

9.8.5 Alternative systems

There are numerous alternatives to the three-layer system, each with their own advantages and disadvantages. It is up to the user to determine the best system for their activity, environment, and ability. Some of these systems completely replace the normal three layers; others merely modify these layers

1. The pile and Pertex system, popularised by the UK company Buffalo, simplifies layering by replacing all three layers by a single garment. A deep pile fleece liner is sewn to a thin windproof nylon (Pertex) face fabric, making a windproof and very insulated garment. To avoid overheating, numerous zips can be opened that allow venting directly to the skin. If the activity level decreases, or the environment gets colder, then these vents can be shut. Though the outer fabric is not waterproof, the system dries very quickly and is designed to remain warm when wet. This system is popular in cold and damp conditions such as Scottish winter, though is sometimes criticised in warmer climates for its weight and inflexibility.
2. The use of a thin weather-resistant layer in addition to the three normal layers can greatly increase the flexibility of the system, particularly in windy or changeable conditions. These weather-resistant layers that are more breathable than a traditional waterproof outer layer but more protective than a normal fleece midlayer have become known as 'softshells'. Though not waterproof, they offer excellent protection in many circumstances and have gained widespread popularity in recent years.
3. The use of over-layering was popularised by Mark Twight in his book *Extreme Alpinism* (Twight & Martin, 1999), and mountaineers such as Andy Kirkpatrick popularised this system in the UK. The traditional layering approach is modified by the addition of an oversized synthetically insulated layer that can be worn over the top of the other layers. This layer is designed to be worn only when stationary or as an emergency layer, and its origin in mountaineering has led to this jacket being known as a 'belay jacket'. It is intended to let a mountaineer climb unencumbered by bulky layers, while making it easy to keep warm when stationary, simply by putting the jacket on over all other layers. By using synthetic insulation, moisture has minimal effects on the warmth on the garment.

Acknowledgement

This work was completed while both authors were based at the Centre for Technical Textiles at the University of Leeds, UK.

References

Cheung, K. Y., & Cheng, S. K. (1995). Comfort in clothing II. *Textile Asia*, 24–29.

Dhingra, R., & Postle, R. (1977). Air permeability of woven, double knit and warp knit outwear fabrics. *Textile Research Journal* (47), 630–631.

Gretton, J. (1999). Condensation in clothing systems. *World Sports Activewear*, 38–43.

Kim, J., & Cho, G. (2002). Thermal storage/release, durability, and temperature sensing properties of thermostatic fabrics treated with octadecane-containing microcapsules. *Textile Research Journal*, *72*(12), 1093–1098.

Kissa, E. (1996). Wetting and wicking. *Textile Research Journal*, 660–668.

Li, Y. (2001). The science of clothing Comfort. *Textile Progress*, 1–135.

Parsons, K. C. (1993). *Human thermal environments*. Taylor & Francis.

Pascoe, D. D., Shanley, L. A., & Smith, E. W. (1994). Clothing and exercise. I: Biophysics of heat transfer between the individual, clothing and environment. *Sports Medicine*, *18*(1).

Rees, W. H. (1971). *Physical factors determining the comfort performance of textiles*. Third Shirley International Seminar. Manchester.

Ren, Y. J., & Ruckman, J. E. (2003). Water vapour transfer in wet waterproof breathable fabrics. *Journal of Industrial Textiles*, *32*(3), p165–175.

Rossi, R. (2005). Interactions between protection and thermal comfort. In R. A. Scott (Ed.), *Textiles for protection* (pp. 233–260).

Saville, B. P. (1999). Comfort. In B. P. Saville (Ed.), *Physical testing of textiles* (pp. 209–243).

Taylor, M. A. (1972). Physical requirements. In M. A. Taylor (Ed.), *Technology of textile properties: An introduction* (pp. 212–231). Textile Book Service.

Tortora, G., & Grabowski, S. (2000). *Principle of anatomy and physiology* (9th ed.). Wiley.

Twight, M. F., & Martin, J. (1999). *Extreme alpinism: Climbing light, fast and high*. Seattle: Mountaineers Books.

Ageing populations: 3D scanning for apparel size and shape

10

J. Bougourd
University of Wales, Newport, UK

10.1 Introduction

The ageing population is growing rapidly and, while the World Health Organization (WHO) has alerted countries to the need to collect data about the changing natural ageing process, there is a paucity of interest in the apparel and three-dimensional (3D) scanning communities. This chapter outlines the concept of active ageing in relation to inclusive design and explores some of the ways in which both traditional and digital anthropometrics can reveal the varying size, shape, and posture of our growing inter-generational population.

Three-dimensional digital anthropometry has much to offer the apparel industries in the future but, at present, the integration of static 3D capture and its application into the design development process are of paramount importance. Some of those applications, drawn from a 3D sizing survey, are explored in two case studies.

10.2 Population

World population is heading towards 10 billion people. By 2050, it is expected to approach nine billion, with the five most populous countries being India, China, the United States, Nigeria, and Indonesia. India is expected to have overtaken China as the most populous country on Earth, with a predicted 1.7 billion (Walker & Mesnard, 2009).

While the growing global population will bring its own challenges, the growth of the ageing population is also unprecedented. Reports published by the United Nations (UN) Population Division suggest that, in 1950, some 200 million people were over the age of 60; by 2000 this had grown to 600 million; it is projected that, by 2050, it will be two billion (see Figure 10.1).

The number of people in the world who are aged 60 and over has doubled since 1980 and, by the mid-twenty-first century, there will be more people who are aged 60 years and over than children under the age of 15. Further, the older population is itself ageing, with those 80-plus being the fastest-growing group worldwide (UN, 2009).

It is expected that nearly all countries will experience this increase in longevity, and that there will be a corresponding decline in mortality and fertility, though the pace of change will vary between developed and developing nations (UN, 2009; as illustrated

Textile-led Design for the Active Ageing Population. http://dx.doi.org/10.1016/B978-0-85709-538-1.00010-9

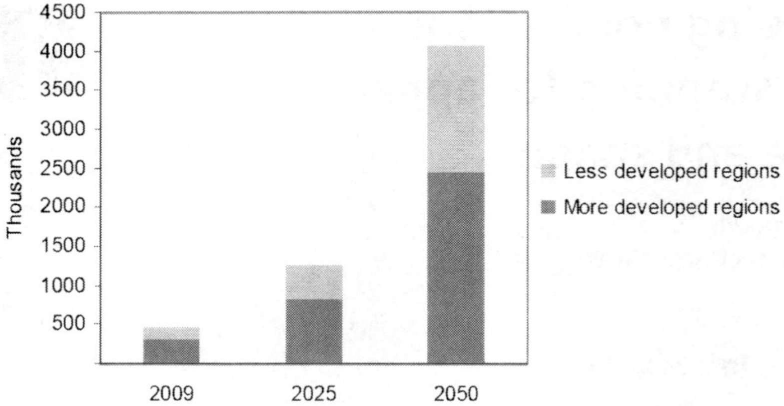

Figure 10.1 The UN projects huge growth in world population, particularly in the older population in the developed countries.
Source: UN, 2009, p. 26.

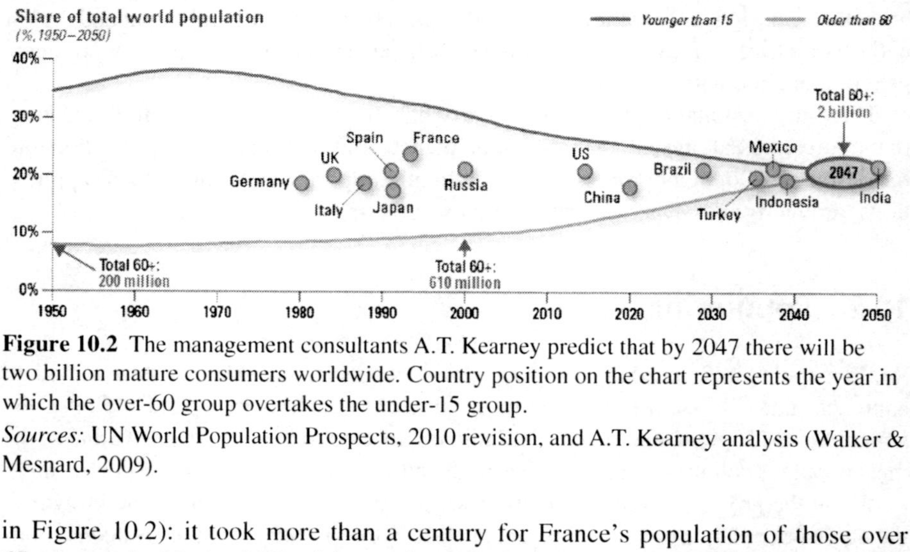

Figure 10.2 The management consultants A.T. Kearney predict that by 2047 there will be two billion mature consumers worldwide. Country position on the chart represents the year in which the over-60 group overtakes the under-15 group.
Sources: UN World Population Prospects, 2010 revision, and A.T. Kearney analysis (Walker & Mesnard, 2009).

in Figure 10.2): it took more than a century for France's population of those over 65 years to double (to 14% of the total), while it will take Brazil and China less than 25 years to reach the same level of growth (WHO, 2012a).

There are, however, mixed responses to these trends. Some media commentators voice negative and stereotypical views of older people that can influence the way that both societies at large and older people in particular think about age. Policy makers can in turn offer 'doomsday scenarios', as exemplified in a recent report that suggests that an ageing population is itself the problem, because of its impact on the health of an entire nation (De Filippo, 2012).

Taken in a historical context, population ageing is one of the most remarkable human success stories of any era, reflecting contributions made in public health, medicine, education, and economic development. But capturing and unlocking the full benefits of that success requires a change of perspective and institutional reform (WHO, 2012b).

10.3 Active ageing

UN population reports, the UN Action Plan on Ageing, and the WHO Framework for Active Ageing have gone a long way to changing our perspective on ageing. They have been influential in generating global, regional, and national debate. So much so that ageing has been described as a hot topic, singled out by a growing number of commentators in international, academic, and media circles. Evidence of this is the attention being given to ageing in 2012: the UN, the WHO, and the European Union (EU) are all highlighting events. An EU event, the European Year for Active Ageing and Solidarity between Generations (Age Platform Europe, 2012) was organised not only to celebrate the 10th anniversary of the UN Action Plan for Ageing but also to reflect the demographic challenge faced by all EU states.

The term 'active ageing' has, in the view of Hutchison, Morrison, and Mikhailovich (2006), become a catchphrase for national and international policy developments. In their review of active ageing, the authors suggest that different definitions of that phrase have emerged since it was first used, in the 1960s. Ageing, portrayed as successful or as productive, was regarded as generalised, or as failing to pay attention to the rights of older people; and, although a broader approach was taken by the WHO during the 1990s, it was not until the early 2000s that the term active ageing (put forward in documents published by the WHO, the Organisation for Economic Co-operation and Development (OECD), and the EU) began to be defined.

The OECD definition focuses on the transition from work to retirement; it has been suggested that this does not take account of environments that influence health and well-being across the span of a life (Hutchison et al., 2006).

The WHO offers a definition of 'active ageing' as:

> *... the process of optimizing opportunities for health, participation and security in order to enhance quality of life as people age. It applies to both individuals and population groups [allowing] people to realize their potential for physical, social, and mental well-being throughout the life course and to participate in society, while providing them with adequate protection, security and care when they need. The word 'active', when used by the WHO, refers to continuing participation in social, economic, cultural, spiritual and civic affairs, not just the ability to be physically active or to participate in the labour force.*
>
> *WHO, 2002, p. 12.*

In documentation associated with the European Year of Active Ageing, this WHO definition is extended to embrace the concept of 'inter--generational society' – one that creates an awareness of society for all ages and helps find innovative solutions that are sustainable and that serve all generations. While those statements support the findings of Hutchison et al. (2006) – that there are competing definitions and inconsistencies in the meaning of the phrase active ageing – it has not halted the drive to address issues associated with increasing longevity within the global population (Hutchison et al., 2006).

It is within the context of the broader EU understanding of active ageing – that of creating a society for all ages – that the issues in this chapter are discussed.

10.4 Design for all ages

Creating a society for all ages is not a new concept in design. It encompasses issues of Inclusive Design that are described by the Centre for Education in the Built Environment (CEBE) as a complex and evolving concept whose definition can include not only age, gender, and disability but also race, income, education, and culture; that is, design that 'includes and does not preclude' (CEBE, 2001). In the context of this discussion on the size, shape, and posture of the body, this concept can follow two routes.

1. The first can result in inclusive environments and products that encompass all user characteristics. An example would be a retail shopping environment where (in the context of population ageing) it was reported by Walker and Mesnard (2009) that – for those between 60 and 100 years of age – there was insufficient seating, shelves were too high, label print was too small, and packaging was too difficult to open.
2. The second route (relevant to this discussion) involves ensuring equal provision of products across the ages forming a population, using either a specified single user characteristic or a set of characteristics. This would imply the provision of fashionable, aesthetically pleasing, and appropriately fitting clothing for all ages within a population. It would include, of course, an emerging group of mature individuals who already aspire to wear the same clothes as the rest of the population (Twigg & Majima, 2014). However, while visible cultural connotations of 'age ordering' in clothing (frumpy styles and dull colours) may be disappearing, and a greater cultural integration of fashion needs and requirements may be emerging (Twigg & Majima, 2014), 'age ordering' remains an important concept - if we are to understand the changes that influence the size, shape, and posture of the adult body throughout its life span.

Age ordering for population projections – for health studies and anthropometric studies for design – is reflected in the way that international standards are drawn up. Two papers offer guidelines for making age divisions from zero to 100 years of age; for adults these are either every five years, or in multiples of five years (Ahmad et al., 2001; ISO, 2006). However, the terminology used to categorise older people varies widely and is suggestive of a somewhat pejorative view. This is exemplified in Tables 10.1–10.3.

With our longer life spans we are entering uncharted territory in which the challenges for individuals and societies are formidable, not least because we continue to cherish rather negative stereotypes of the ageing process. Those stereotypes have, if anything, grown more negative as life expectancy has increased. Survival to old age is perhaps less of an achievement and, as life has become more secure, the inevitability of eventual ageing seems more of an affront (Kirkwood, 2001).

Notwithstanding the importance of the International Organization for Standardization (ISO) standard for collecting, compiling, and comparing population data and the

Table 10.1 A schema for consideration of subject populations across the life span

0–2	Infant
3–12	Child
13–17	Adolescent
18–24	Young adult
25–44	Adult
45–64	Middle-aged adult
65–74	Young old
75–84	Old
85–99	Old old
100+	Oldest old

Source: Spirduso, Francis, and MacRae (2005).

Table 10.2 A schema for consideration of subject populations in older age groups

Age group (years)	Description
60–70	Old age
75–80	Mature age
90+	Venerable old age

Source: Nowak (2006).

Table 10.3 A schema for consideration of subject populations across adult life span

Age group (years)	Description
20–35	Young adulthood
35–45	Young middle age
45–65	Later middle age
65–75	Early old age
75–85	Middle old age
85+	Old old age

Source: Shephard (1998).

need to address terminology that is stereotyped and negative, it is the change occurring in the natural ageing processes that influences morphology.

The natural ageing process involves universal changes, independent of disease (Spirduso et al., 2005). The physical ageing process is one that progresses every day and through which we all travel. And, as with the varying pace of population ageing between nations, so there is variation between ageing in individuals and between their systems and organs. There is, however, evidence to suggest that there are correlations between our biological and our chronological ages, some of which can have a significant effect on anthropometric values.

10.5 Anthropometrics

Anthropometrics is the science of measuring individuals for the purpose of understanding human physical variation. Data about one, two or three dimensions enable us to capture and monitor physical changes throughout the life cycle and to size and shape items to fit their users.

There are two components:

- static (structural anthropometrics); and
- dynamic (functional anthropometrics).

Both represent the body and its action (IIE, 2012).

Static body dimensions are those collected from a series of standing or seated positions. These can include body heights, lengths, breadths, widths, postures, circumferences, contours, volumes, masses and shapes. Dynamic dimensions are those that involve the body in a functional attitude (IIE, 2012). These can include strength measurements such as grip, pull and push as well as the reach or range of the body in movement or motion of a physical activity where one, two, three or four dimensions could be captured.

The methods and equipment used to collect anthropometric data vary between traditional manual methods and emerging digital systems.

10.5.1 Traditional anthropometry

The development of traditional anthropometry and its specialist equipment has been the bedrock of our understanding of the size of the human body; despite fast-developing digital technologies, traditional anthropometry remains the primary method of capturing body dimensions for many practitioners. The WHO considered anthropometry '... as the single most portable, universally-applicable, inexpensive and non-invasive methods available to assess proportion, size and composition of the human body throughout the life cycle' (Onis & Habicht, 1996).

In reviewing anthropometric reference data in 1996, the WHO noted that there were little data on older generations and, because of wider geographic, ethnic and environmental variations, countries were encouraged to collect data on healthy adults 60 years and above (Onis & Habicht, 1996). The paucity of data prompted a number of studies – some longitudinal (multinational, national, see e.g. WHO, 2012c) and more restricted academic studies – that are increasing our understanding of the normal physical processes of ageing.

Specifically, those that underpin size, shape, posture and movement – the physical processes that decrease bone and muscle mass (with a corresponding increase in adiposity) that helps to characterise normal ageing and musculoskeletal health during the life span.

10.5.1.1 Bone mass

Skeletal mass is mostly formed in teenage years. Mineral density usually peaks at 30 years of age, after which there is a gradual decline, as bone mass begins to lose minerals faster than they can be replaced. The loss is more rapid for women following the menopause. A consequence of this change in bone mass (together with the shrinkage of the discs in the spine and in the lower limb joints) is a loss in the trunk and overall height with increased curvature of the hips and knees, a broadening of the feet and a flattening of the arches. In addition, as we age there may be a change in posture. Typical postures found in a population, pioneered by Florence Kendall and known as 'sway back', 'flat back', 'khyphosis lordosis' and 'military' (Kendall et al., 2010), are as a result of a departure from the natural spinal curvature, known as the neutral or ideal posture. Too much curvature of the spine brings about a posture associated with lordosis. Excessive lordosis can be referred to as 'saddle' or 'sway back' it can be manifested in the upper or lower sections of the spine, and in a combination of the frontal and side planes. A reduction in the normal amount of curvature is sometimes referred to as 'flat back'. Excessive curvature of the thoracic portion of the spine is particularly associated with the ageing process; as the spinal column becomes compressed and the back begins to bend over, three things can happen: the shoulders (typically natural, square, sloped and asymmetric) can become very rounded, the neck can project forward, and the breadth of 'across back' can increase (MedlinePlus, 2012).

A photographic system for measuring the angle of the forward head position (using the seventh cervical and tragus of the ear as landmarks) has been described by Nemmers, Miller, and Hartman (2009). Their results indicate that a forward head posture is directly linked with increase in age among women of 65–85 years and beyond, though the significant year of change is thought to be around 75 years. See Figure 10.3.

As indicated above, the loss of bone mass can lead to an overall loss of height. Most people begin to lose height after the age of 40, with an accelerated loss after 70–75 years, but the amount of shrinkage experienced by women can be twice that of men (5 cm, compared with 2.5 cm). Marked differences in height have nevertheless been reported in inter-generational studies between a 65–69 group and an elder group of 80–84. These older subjects also showed a thinner body frame than the younger age group, suggesting that, at the same time as we lose height, we lose fat-free mass muscle tissue (Perissinotto, Pisent, Sergi, Grigoletto, & Enzi, 2002).

10.5.1.2 Muscle mass

Overall, muscles decrease in strength, endurance, size and weight over the life span. The rate and extent of change varies between men and women (particularly in the strength of grip and flexion and extension of joints). For men it can begin in their twenties, and, for women, in their forties (MedlinePlus, 2012). Loss can accelerate after the sixth decade and, in some cases, can rise to a total loss of 23% by the age of 80. As this lean body mass declines there is normally a gain in fat mass which, in turn, has implications for both the overall body size and weight (muscle weighs more, and it is more compact than fat; fat weighs less than muscle, but can take up to four times as much space) (Spirduso et al., 2005). Overall body weight can increase between ages of

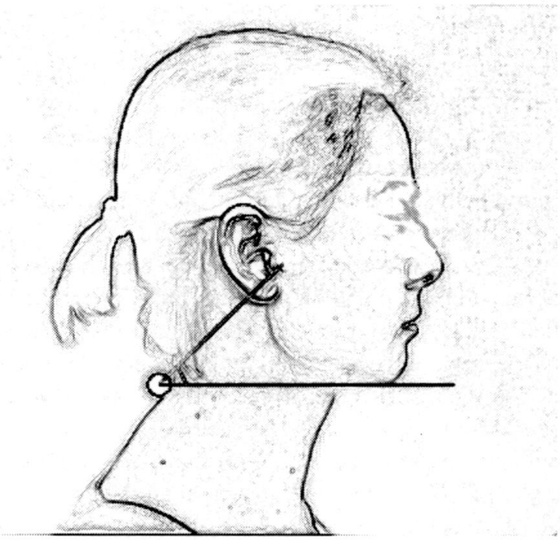

Figure 10.3 Forward head posture measured as an angle from a horizontal drawn from the seventh cervical vertebra (C7) across under the chin, radiating upward to a line drawn from C7 through the tragus.

20 and 45 years, peak during the fifth and sixth decades, showing a significant decline after 70–75 years.

10.5.1.3 Fat mass

Change in fat mass both between and within genders has a significant effect on body size and shape throughout the age span. There are two types of fat tissue: subcutaneous fat, which forms a layer beneath the skin; and visceral fat, which is found around the organs in the abdomen. The amount and distribution of fat can vary with gender and with age.

Adipose tissue, found in the upper body, is termed android, and is found mostly in men, while tissue in the lower body of women is termed gynoid, creating what is popularly referred to as a 'pear shape'.

However, after the menopause, fat mass in women shifts towards the upper body, creating a more male android shape. Fat tissue increases for both men and women through middle age when it is redistributed from subcutaneous to intra-visceral fat. This redistribution continues with advancing age (more slowly for women), after which there is a steady decline, creating further body, sizes, shapes and postures.

The body mass index (BMI) is described as a simple numeric measure, a person's thickness or thinness (size and shape). It is a commonly used indicator of optimal weight – how much a person's weight deviates from normal weight for a given height (although, since the population trend is towards an increase in weight, it has been used as a measure of obesity). The initial BMI classification system comprised four categories, but the WHO index has now been extended to accommodate not only the rise in obesity levels but also, as a result of differences observed between Asian populations, alternative cut-off limits (see Table 10.4).

Table 10.4 The international classification of adult underweight, overweight, and obesity, according to BMI

	BMI ($kg\,m^{-2}$)	
Classification	**Principal cut-off points**	**Additional cut-off points**
Underweight	**<18.50**	
Severe thinness	<16.00	
Moderate thinness	16.00–16.99	
Mild thinness	17.00–18.49	
Normal range	**18.50–24.99**	18.50–22.99 23.00–24.99
Overweight	**≥25.00**	
Pre-obese	**25.00–29.99**	25.00–27.49 27.50–29.99
Obese	**≥30.00**	
Obese class I	30.00–34.99	30.00–32.49 32.50–34.99
Obese class II	35.00–39.99	35.00–37.49 37.50–39.99
Obese class II	≥40.00	

The cut-off points reflect incorporation of data on Asian populations.
Source: WHO (2004).

Despite these adjustments to the BMI classifications there remain limitations. Results indicate that, for the elderly, it is healthier to have a BMI between 25 and 27, rather than, as indicated in the classifications, under 25 – suggesting the need for further adjustments to the classification limits for the older population (Keeler, 2010). BMI is also a poor proxy for body composition, as it cannot distinguish between fat mass and fat-free mass, and does not measure the percentage of body fat (Noor, Koon, & Hashim, 2005; Wells, Cole, & Treleaven, 2008) and, because of these inadequacies, it is not sensitive to age-related changes in body fat distribution (Price, Uauy, Breeze, Bulpitt, & Fletcher, 2006). It is recommended that, if traditional anthropometric measures are to be used, then the waist – hip ratio may be a better predictor of healthy and successful ageing (Price et al., 2006; WHO, 2008).

10.5.2 Digital Anthropometry

Important to the practice of inclusive design is the understanding of not only the complexities underlying natural body changes during life span but also the methods by which the size, shape, posture and movement that reflect those changes might be captured using digital technology.

Some of the specialist equipment used for traditional anthropometry is already designed with digital components (weight scales, electronic measuring 'tapes', stadiometers) and, whilst data from studies that use this 'low technology' equipment remain invaluable, the majority are limited to one-dimensional measurement that does not adequately describe the dimensional variability of the human shape.

The development of digital technologies continues to revolutionise our everyday activities. But it was the advent of commercially available 3D body scanners that, for the first time, provided opportunities and challenges – for researchers in health (Wells et al., 2008), kinanthropometry (Stewart, 2007), ergonomics (Sims et al., 2012) and anthropometry (Ball, Luximon, & Chow, 2012) – to expand our understanding of the physical shapes of the body and its parts (i.e. the head, face, hands and feet).

There are many devices able to capture the geometry of physical objects that can be called 3D scanners: e.g. computed tomography (CT or CAT scan and 3D digitisers, see e.g. Rapidform, 2012), but the technologies most commonly used to capture the surface shape of the human body are laser, projected light and, latterly, millimetre radio waves. Each has its own advantages: resolution level, automatic landmarking and measurement extraction, portability, cost, the ability to capture shape through clothing, etc. The launch of new systems (e.g. Cartesia, Kinect) or those being extended or enhanced (Vitronic – Human Solutions; Telmat, TC2) occurs frequently and, though regular reviews of systems are published (Hometrica, 2012) and guidance is given in some ISO standards, teams planning surveys are advised to benchmark available scanners (Bougourd & Treleaven, 2010; Simmons & Istook, 2003).

Initial interest in these systems was to find a quicker, cheaper and more efficient method of both capturing and extracting measurements; that is, to try to replicate traditional anthropometric methods of landmarking and measuring the body. There is, however, continued debate regarding issues of accuracy, repeatability and reliability:

- within traditional methods;
- between traditional and digital methods;
- within and between digital systems.[1]

The debate has thrown up new research questions and has prompted revision to some ISO standards. But the primary driver for 3D scanning in the apparel industry has been the quest for better-fitting clothes (and, in the context of this discussion, inter-generational apparel provision). The prospect of being able to quickly collect body-sizing data ultimately led to the utilisation of 3D scanners for several apparel applications. Scanners have now migrated into many North American and European retail stores where they are used to support made-to-measure (MTM), mass customisation and ready-to-wear (RTW) size, shape and fit solutions. Prime applications have been for the collection of anthropometric data for sizing surveys and research in university departments.

[1] See e.g. Daanen & Ter Haar, 2013; Lu and Wang, 2010; Milanese, Lovato, Piscitelli, Giachetti, & Zancanaro, 2011.

However, unlike the WHO (which recognised the need to map the process of natural ageing within populations), apparel companies have yet to fully acknowledge and address the changing size, shape and posture of an ageing population that is, as we have seen, growing worldwide. Clearly, the use of 3D scanners to help capture these dimensions for the design and development of MTM garments is an ideal inclusive design solution. This can also, to some extent, be the case with mass-customised garments. However, if the products being offered to the customer (as with in-store Best-Fit/size prediction or online shopping sites) are limited in size and shape, either to a specified market or to ages arbitrarily represented in a sizing survey database, it will not always be possible to create or to offer garments that meet the principle of inclusive design.

It is apparent in Table 10.5 that not all population surveys collected data from the same age groups. All surveys offer data collected from subjects more than 60 years of age (discussed at Section 10.5.1), but the surveys identify differing upper age limits. Some have a cut-off age at what is, for some countries, the traditional retirement age of 65 years (CAESAR), some a 65-plus category (SizeUSA, SizeMexico, SizeThailand, Sweden), while others have a cut-off age around 70 years (SizeChina – heads, SizeFrance, SizeSpain – women only, SizeGermany, E-Tailor Germany). Senior-specific studies have been conducted in Germany (female 50–80 years) and in France (male and female micro survey 70–100 years). Studies that have data for populations beyond the current 'significant age' of 70–75 years – when physical changes can accelerate – include two Japanese studies, SizeUK and Germany (female bra project). At the time of writing, a study is under way in Belgium and others are planned in Australia, China and Italy.

While 70–75 years has been proposed as an age of significant change – in the characteristics of bone, muscle and fat mass – the fastest-growing population is made up of individuals 80 years of age and above. It may be that, as reported by Kearny, people are 'living longer, staying healthier and becoming increasingly active and energetic', so that the significant age of decline will shift towards the ninth and tenth decades (Walker & Mesnard, 2009). Furthermore, it is suggested in a study of nonagenarians and centenarians that longevity itself may be a potential effect; that is, slimmer, shorter people may live longer, making long-lived people a particular, genetically selected, population with, perhaps, their own individual size, shape and posture (Ravaglia et al., 1997).

So, much effort is being put into the enrichment of our understanding of the physical shape of the ageing population. But the picture beyond the traditional retirement age remains patchy. If we are to provide comprehensive data on the projected global population, then the tremendous potential of 3D scanning technology needs to be exploited in a more systematic way to improve apparel design and fit; using, where applicable, one, two, three and four-dimensional anthropometrics. We are reminded here of the fact that, in 2008, the ISO proposed that the general requirements for anthropometric databases should incorporate five- and ten-year divisions to the age of 100 years (ISO 15535:2006). However, if we

Table 10.5 **Selected digital surveys discussed in text**

Group	Ages	Country, Survey, Year	No. Male/Female	Scanner/Postures/Measurements
1	18–65	USA, 1998–2000. CAESAR	2375, M & F	WB4, Cyberware/3/57 (+40 manual)
1	18–65	Netherlands, 1999–2000. CAESAR	1255, M & F	Vitronic/(As USA)
1	18–65	Italy, 2000–2001. CAESAR	801, M & F	(As USA)
2	16–70	Germany, 2001. E-Tailor	500, F	Vitus Smart/2/10
2	18–65	USA, 2002–2003. SizeUSA	10,500, M & F	TC2 NX?/2/200+
2	(18–65)	Sweden,* 2005–2007 or 2010	367, M & F	Vitus Smart/4/55 (+10 manual)
2	16–60+	Thailand, 2006–2008.	13,442	TC2 NX16/1/140
2	18–65+	Mexico, 2010. Size Mexico	17,364, M & F	TC2 NX16/1/200+
3	5–70	France, 2005–2006.	11,562, M & F	Vitus Smart/2/55 (+10 manual)
3	18–71+	China, 2006. Size China§	2500	Cyberware head scanner/Not known
3	12–70	Spain, 2007–2008¶	9159, F	Vitus Vitronic with; Antroscan/3/95
3	6–87	Germany, 2007–2008. SizeGermany#	13,400, M & F	Vitus 3D/4/43
4	50–80	Germany, 2002. Elderly women	1300, F	Vitus Smart/2/84 (+1 manual)
4	70–100	France, 2007. Seniors: 'micro-survey'	400, M & F	Vitus Smart/2/' varied'
5	7–90	Japan, 1992–1994. HQL	34,000, M & F	Voxelan, Hamano/not known/178
5	14–80	Germany, 1999–2000. Hohenstein	1500, F (bra)	Vitus Pro,/3/84
5	16–90+	UK, 1999–2002. SizeUK	11,000, M & F	TC2 NX?/2/130 (+10 manual)
5	18–89+	Japan, 2004–2007. Size-JPN	6700, M & F	Not known/Not known/217

* Hanson, Sperling, Gard, Ipsen, and Olivares Vergara (2009).
§ Ball et al., 2012.
¶ Alemany (2010).
# Preiss, et al., 2012.

are to take account of the reports referred to earlier, this standard may need to be revised upwards.

10.5.2.1 Dimensions in digital anthropometrics

Three-dimensional capture continues to advance and enhance our knowledge and understanding of the body. Three dimensions are data rich, offering individual, combined and interactive dimensional applications, although the full potential of these systems in addressing inclusive design is only slowly emerging.

Individual one-dimensional measurements that represent the body (e.g. length, girth) remain at present the foundation of sizing in the apparel industry. There appears, however, to be variation in the amount and type of digital measurement available from 3D scanners. Some manufacturers, such as Telmat and Human Solutions, advertise their measurement extraction software as compliant with both ISO 8559:1989 (Garment Construction and Anthropometric Surveys, which specifies 52 measurements) and ISO 7250:2008 (Basic Human Body Measurements: Definitions and Landmark, which specifies 56 measurements). While the two standards contain 14 measurements in common, and there is provision for additional measurements in ISO 15535:2006 (Section 7.4), little attention has been given to extending them to include other, essential and apparel-specific measurements. Other scanner manufacturers, such as TC2, offer their own apparel-specific measurement software, including measurements for intimate apparel. TC2 is able to extract over 200 measurements with the opportunity to considerably extend that number to nearer 400, if required.

It is also noted that ISO 20685:2010 (3D Scanning Methodologies for Internationally Compatible Anthropometric Databases) only makes reference to ISO 7250:2008.

In principle it would therefore appear that, if international, one-dimensional measurement standards are to be universally applied, they may need to be revised so as to take account of:

- variability found in the number of measurements collected in the course of national sizing surveys (see Table 10.5);
- the specific measurement needs of the apparel industry – all primary, secondary and product-specific measurements (e.g. ISO standards do not contain Top Hip or many of the other apparel-specific measurements, offered by TC2, or those proposed by the US standards body[2]);
- the need regularly to survey the growing population of those 60–100 years of age so as to identify any changes in the natural ageing process that need to be reflected in measurements. (For potential examples related to postural change see Figure 10.3, and, an illustration showing a sagittal view – Ashdown & Na, 2008).

Putting ISO standards to one side, the application of one-dimensional measurements extracted from 3D scanners has, as a process, been extremely successful. New and updated sizing systems have been produced for many company-specific and generic country requirements, ensuring a first stage of improved apparel fit.

To reiterate: in the same way as apparel companies have yet to address inclusive design, so have the 3D digital communities. While there is immense interest in 3D

[2] American Society for Testing and Materials (ASTM) – the relevant standard is D5219.

scanning for apparel applications, few are addressing older age needs. There are, however, some studies looking at inter-generational needs; for example, 60–75 year old apparel needs for walking groups[3] (NDA, 2012; see also Section 10.6.1). The discussion below looks at ways in which 3D capture is being generally applied and which could be applicable to all groups, irrespective of age.

One dimension. One-dimensional measurements are still used extensively in the apparel industry, and although it is acknowledged that they cannot fully describe variable population shape, they can, together with both two- and three-dimensional information, enhance existing practice and add new perspectives to the size and shape of the body. Song and Ashdown (2011) used 14 proportional measurements and a back seat angle to develop a new method to categorise body shapes for adult women (which can, they suggest, be used for all ages). While studies of shape classification have improved our understanding of ranges of shape, it is proposed that these existing body shape classifications are too limited (those offered by Connell, Ulrich, Brannon, Alexander, & Presley, 2006 – e.g. 'hour glass', 'pear' and 'geometric'; Simmons, Istook, & Devarajan, 2004 – e.g. 'diamond', etc.; also, what are known as 'drop values'; and some of those using principal component analysis), because they fail clearly to distinguish body shape characteristics (Song and Ashdown, 2011).

Two dimensions. Cross-sectional analysis of 3D body scan parts has been found useful to help to establish the shape of neck curves and waist suppression on block patterns for the elderly. In relation to garment fit, scans of minimally clothed bodies and scans of garments (made from a variety of materials) have been superimposed to assess air gaps (Daanen & Reffetrath, 2007) and degrees of fit (Ashdown & Mete, 2010). Silhouettes of whole body scans or scans of body parts have been used to observe age changes over the life span (see Figure 10.4), and to specify the shape of posture, using measured lateral distances from a vertical axis (see

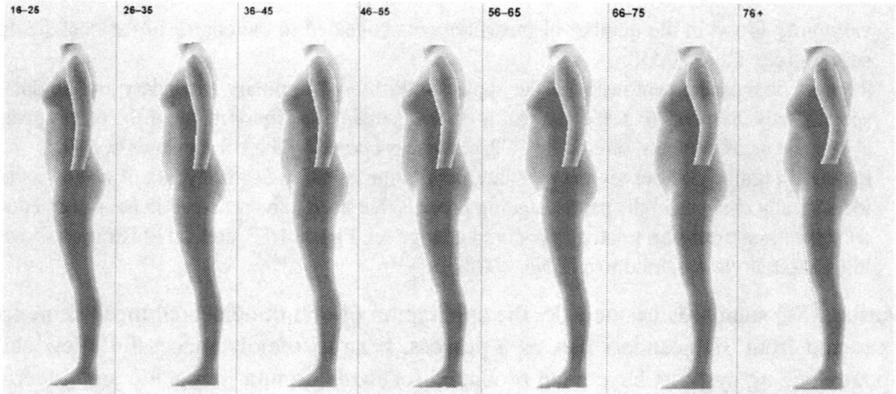

Figure 10.4 Profiles of the seven age groups for female subjects.
Source: SizeUK database.

[3] Design for Ageing Well is a co-design project for 60- to 75- year-old subjects, forming part of a UK research programme The New Dynamics of Ageing, 2009–2012.

Figure 10.5); alternatively, by determining compliance of angles and curves with a normal spinal posture.

Three dimensions. Three-dimensional digital avatars have been developed by companies such as TC2 for application in digital environments. These systems are designed to wrap 2D patterns on to a 3D form (e.g. Browzeware, Optitex and Assyst), and are used by manufacturers to help retailers to remotely assess apparel, size, fit and design. The creation of 3D digital footwear design and 2D pattern unwrapping processes led the way to apparel contour design and fit (e.g. Lectra, Optitex and an earlier system discussed in Krzywinski, Rödel, & Siegmund, 2005). The more ambitious application of freeform virtual apparel design has been realised for some classic apparel (e.g. shirts and jeans), but the wider challenge of applying virtual representations of some different materials is slowly being met. Other 3D morphological applications that are already commercially available include automatically generated 3D grading and block patterns, and physical mannequins (see case studies in Sections 10.6.1 and 10.6.2).

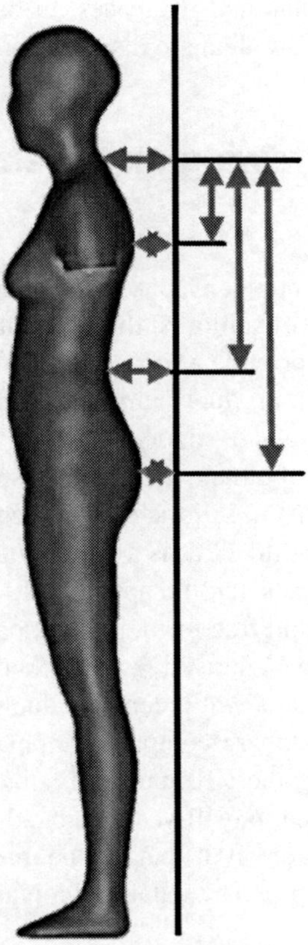

Figure 10.5 Extraction of postural data as lateral measurements from vertical datum.
Source: Kirchdoerfer, Mahr-Erhardt, & Morlock, 2011, p. 151.

Four dimensions. 3D body scanning with dynamic movement capture is in its infancy. There is a wide range of physical movements to consider. Those used in general activities; those associated with traditional dynamic anthropometrics for indoor and outdoor environments; and those with extreme movement found in some work and sporting activities. The last of these is discussed helpfully by Lotens (1989). In some cases it is possible to capture simulated movement in a range of static poses and to create dynamic models (Ruto, 2009). In the same way that static scans of clothed and unclothed bodies can be morphed and compared to assess ease for size and fit, so also it is possible to simulate dynamic positions to capture a range of movements where garments can be assessed for their suitability for that range of movements. Virtual walking, motion and morphing of the gait and postural variations of older people can be evaluated using traditional motion capture systems, but new techniques for the capture of motion, using scanning techniques, are becoming available. Telmat has added tracking with wireless, active landmarks that can be detected automatically.[4] XYZRGB captures movement by recording 360 degrees of information in real time, with colour and a high level of resolution (XYZRGB, 2012).

Notwithstanding the obvious benefits that these developments offer for capturing dynamic movement, current national databases do have extremely rich, static data. Some of the applications of those data are discussed, in two case studies.

10.6 Case studies drawing on the sizeUK national sizing survey

Sizemic, a fashion technology company, based in London, and its partner Institut TPC, a Hong Kong consultancy, have facilitated the application of information drawn from SizeUK data. The companies provide a 'one stop' sizing and fit solution for the fashion industry. They specialise in products and services based on 3D body data derived from 3D body scanning – from size survey data to size chart development and to fit mannequins to 3D pattern development. Sizemic is responsible for hosting the SizeUK data and has the exclusive rights to sell the data on behalf of the SizeUK consortium.

The technology partner, Institut TPC, is a garment technology-consulting firm that has pioneered an innovative and scientific approach to morphological grading and the development of both contour and free form garment patterns in a 3D environment.

Size surveys and 3D body data derived from 3D body scanning provide a wealth of measurement and shape data that are of potential value to retail, fashion, healthcare and other sectors. The Size UK database comprises anonymised personal details on each of 11,000 subjects representing the UK male and female population between the ages of 16 and 95. Data are organised into 10-year age bands (16–25, 26–35, 36–45, 46–55, 56–65, 66–75, 76–85 – see Figure 10.6) but can be mined according to selected target markets, including an 85+ group. The database contains 22,000 3D scans, over 130

[4] Telmat Industrie, the Symcad ST system.

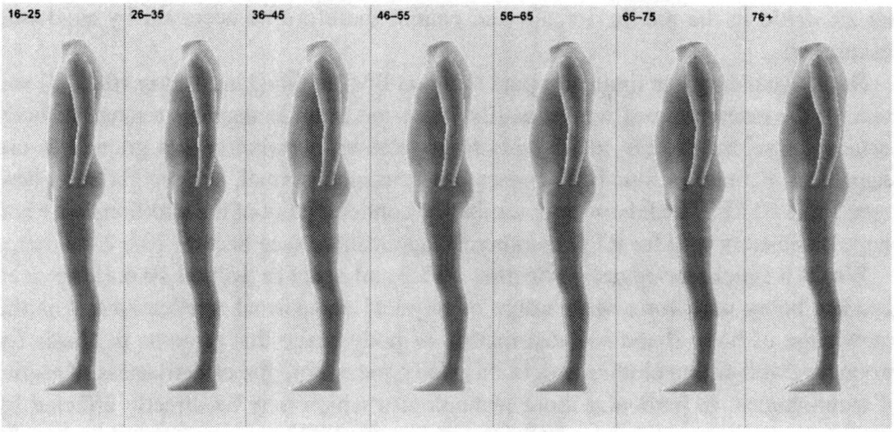

Figure 10.6 Profiles of the seven age groups for male subjects.
Source: SizeUK database.

anthropometric measurements per subject and market research information covering the results of 40 questions per subject on clothes, shopping and lifestyle habits.

Given that 3D shape is more important to achieving good fit than linear measurements – measurements do not adequately describe shape and people with similar measurements can have very different shape profiles – the case studies (Sections 10.6.1 and 10.6.2) seek to demonstrate the importance and added value of 3D shape data in terms of understanding inter-generational shapes, implementing results and maximising benefits.

In the context of this chapter the challenge is to deliver data in a form enabling:

- an *academic* apparel project, Design for Ageing Well, to be undertaken with a small group of walkers aged between 60 and 75 years (Section 10.6.1);
- retailers and clothing companies to utilise the SizeUK results and realise commercial benefits for market segments (Section 10.6.2).

10.6.1 Case study one: using SizeUK data for an academic project

This case study indicates how SizeUK 3D scan data were used to help to guide understanding of the size, shape and posture of 60- to 75-year-old walkers, recruited to support the co-design and development of a clothing-layered system comprising:

- a body layer with integrated technology to monitor vital signs;
- a mid insulating layer with optional warming devices;
- outer protective wear with multimodal user interfaces.

10.6.1.1 Academic considerations

The SizeUK survey was funded by the government and leading clothing retailers and manufacturers. Data collected are owned by the industrial partners and are not available in the public domain and cannot therefore be accessed by academic researchers.

Notwithstanding that limitation (and those on BMI outlined in Section 10.5.1.3) and because the research team was primarily interested in understanding a range of body surface shapes for the 60- to 75-year-old population, permission was granted to use shape data, defined as four BMI classes (underweight, normal, overweight and obese – see Table 10.4). This did not compromise the confidentiality of the traditional size and shape parameters used for RTW commercial applications (see Section 10.6.2, below).

Whilst it is acknowledged (in Section 10.5.2 and again in Section 10.6.2) that scan data are being used for a wide range of physical and virtual applications, it is the knowledge of body shape and distribution of body tissue that remains the basis for producing well-fitting clothes and, in this study, measuring the effectiveness of a suite of technologies; in particular those technologies which may be directly affected by varying body tissue spread (see Cleland et al., 2012). Therefore, since this was the first time that 3D shape data and inclusive co-design practice were to be used for RTW smart clothing, the scanning applications were confined to:

- aiding the recruitment and selection of participants;
- automatically generating one-dimensional measurements and traditional block patterns (e.g. woven, fabric).

10.6.1.2 Selection of participants

Following project ethics approval from the participating universities, a two-stage quota strategy (non-random) was used to select participants.

10.6.1.2.1 Stage 1: identifying national body shapes

The SizeUK database was mined to retrieve the 60- to 75-year-old population of men and women and, using proprietary shape analysis tools from University College London (UCL), average BMI shapes (controlled by average height for each class) were extracted for each BMI class. However, no underweight men were found in the data set and only four underweight women. So this BMI category was initially removed from the study.

10.6.1.2.2 Stage 2: matching participant body shapes

All walking clubs and groups within a 30-mile radius of the lead university were contacted, the project was explained, and members were invited to help with the study. One hundred people replied and submitted height and weight data, from which their BMI classes were calculated. On completion of that analysis, it was found that the project sample reflected that of the national data; that is, no underweight men. There was, however, one underweight female who was identified and included as a participant in the study. Four more males and four females were selected to represent each of the other BMI classes (i.e. a ratio of 4:1). All were invited to be scanned and their measurements extracted (electronic measurements, and a small number of supplementary manual measures) in a mobile anthropometric unit fitted with a TC2 NX16 scanner, a height gauge and a Tanita body composition monitor. Following confirmation of height and BMI class, the 3D scans were matched to the national representative BMI shapes.

In addition to the one underweight female, 12 participants were selected – three men and three women, one for each BMI class, plus six reserves. See Figure 10.7 for an example of the final matched female shapes.

All were invited to take part in the study as representatives of the national body shapes for the 60- to 75- year-old population. These participants also formed the nucleus of the project's User Reference Group.

10.6.1.3 Three-dimensional Pattern Block Development

The 3D body scans of the final seven participants were sent to Sizemic's technology partner Institut TPC. Their Parametrics Pattern Generation (PPG) software, with 3D to 2D real-time process, was used to automatically create the pattern blocks for trousers, shirts and upper body, hip-level bodices for females. See Figure 10.8 for the PPG process.

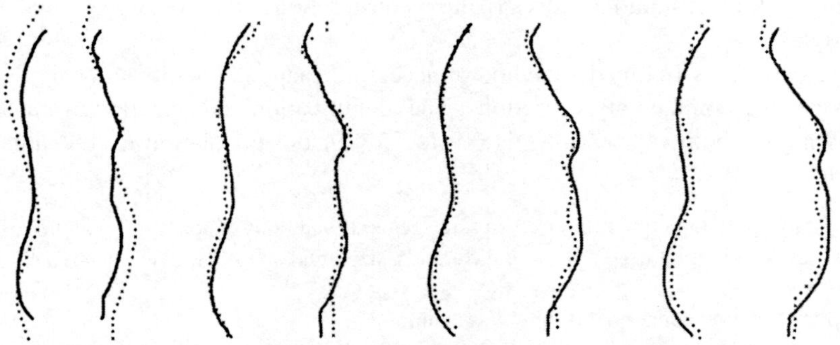

Figure 10.7 Silhouettes of morphed and matched 3D scans. The dotted lines indicate the silhouettes of representative BMI categories, drawn from SizeUK data; solid lines indicate silhouettes of matched selected sample from the study.

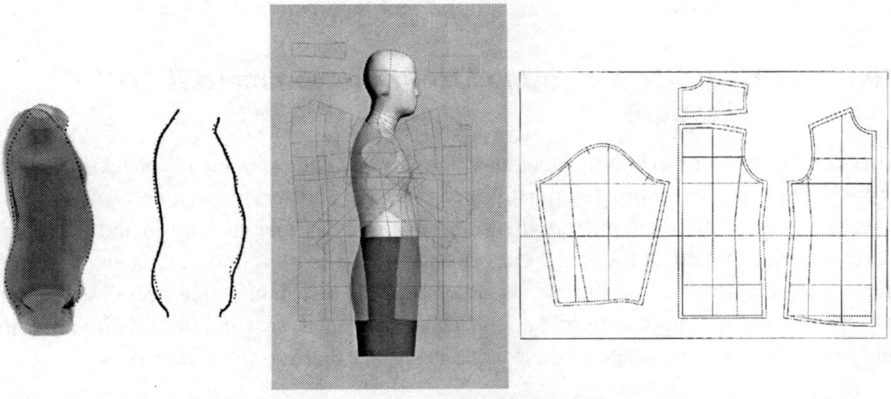

Figure 10.8 An example (left) showing a male participant example matched to morphed and matched SizeUK data, discussed in Figure 10.7. The central image shows shirt pattern unwrapped from a body scan by PPG. Image to right shows shirt pattern extracted from the participant's scan data.
Sources: SizeUK database; NDA project; Institute TPC.

On receipt, the full-scale patterns were checked and measured to ensure corresponding percentage of ease on key body measurement levels (bust/chest, waist and hips) for all male patterns and all female patterns. Minor discrepancies were adjusted prior to cutting, assembling and fitting the block toiles.

However, while the fit of the majority of garments was satisfactory, it was discovered that the size, shape and posture of some of the participants had changed since being scanned. Some of the changes were due, in part, to the participants being involved in the project (i.e. an instance of the 'Hawthorne effect'); others to changes in weight (i.e. affecting girths). Ideally, these participants would have been re-scanned, further measurements extracted and patterns regenerated. But, as both time and funding were limited, the existing block patterns were amended, ease percentages re-calculated, and the toiles refitted. Both electronic and manual body measurements and the final block patterns were used to create the garments in each of the garment layers. In some cases, sets of derived blocks with movement allowances (e.g. arm lift) were developed prior to using traditional studio practices (pattern construction and toiles) for the final co-design styles.

This case study outlined a method of accessing shape in a national database without breaching commercial restrictions. The identification and selection of nationally representative body shapes for this 60- to 75-year-old population have enabled the team to:

- begin to understand the differences in inter-generational body shapes;
- address the clothing needs of these differently sized and shaped walkers using a co-design process;
- explore the effectiveness of a suite of technology.

However, whilst this case study proved to be a successful route by which to access 3D data for a small, academic study of the 60- to 75-year-old consumer, the next case study considers how sizing data might be utilised by commercial designers, manufacturers and retailers of RTW garments and how they might help to realise better-fitting clothes for inter-generational and 'age-ordered' consumers.

10.6.2 Case study two: using SizeUK for commercial RTW applications

This case study seeks to demonstrate how Size Survey data, specifically SizeUK, can be utilised and exploited through a portfolio of products and services, an end-to-end solution, taking a specified client from measurement analysis, shape analysis and size chart development to practical tools like customised fit mannequins and 3D pattern development. The objective was to help clients understand the distribution and overall shape and size profile of their target customers, to improve the shape, size and fit of their garments and maximise the percentage of their target customer that can fit their clothes.

10.6.2.1 Commercial considerations

While SizeUK provides a rich and comprehensive database of measurement and shape data, it does not provide users with standard size charts based on UK garment size designations. For several reasons this was not an objective of the sizing survey project. Definitions of UK size designations vary in practice, as do the increments between sizes and the different types of size designation are used. It was also expected that the designers, manufacturers and retailers would want to create new size charts for their specific customer profile, rather than size charts based on broad national averages. Although this case study sets out five stages undertaken by a leading UK retailer to utilise the results and realise the commercial benefits of the SizeUK sizing survey, it is applicable to any company that is seeking to provide clothing, for any age group represented in the database.

10.6.2.2 Implementation – use of SizeUK by a leading fashion retailer

The client was a market leader in the fast-growing value segment of the UK clothing and accessories market, targeting a core audience of fashion-conscious women as well as men. They were one of the first retailers, outside the main group of SizeUK retailer partners, to purchase the data and one of the first to fully implement SizeUK survey results across their product range and supply chain in what has subsequently become a template for other clients. This case study describes the process and stages they went through and the benefits that the client subsequently enjoyed. They purchased data licences for both genders.

Due to the sensitivity of the information, it is not possible to provide detailed statistics to illustrate the commercial benefits. The client may not have shared detailed performance information with us anyway, other than anecdotal evidence that implementation of the results has had a beneficial effect. As far as possible, this case study describes those benefits and it is therefore reasonable to assume that others would enjoy the same experience. Some benefits are difficult to accurately measure because they cannot be attributed solely to the application of the data or a related initiative.

10.6.2.2.1 Stage 1 – data analysis

Database setup: after purchasing SizeUK data licences, the programme started with the client defining a target customer profile in terms of gender, age range, socio-economic groups, ethnicity and regions. Apart from gender and age, all the other profile criteria were not specifically defined. The overall gender and target age range was subdivided into two groups – a younger segment and an older one – in order to compare the two. This was partly to help the client decide whether to opt for one consolidated size chart or separate ones.

The initial measurement analysis was designed to show the:

- overall size distribution of their customers;
- overall size range from smallest to largest;
- frequency distribution by size and shape;
- market share by size;
- relationship between key measurements;
- effectiveness of their current size chart.

The results included statistical tables, measurement correlation charts, and both frequency distribution graphs and tables for the key measurements. This was carried out for each of the two age groups and the results reviewed and compared.

10.6.2.2.2 Stage 2 – size chart development

Sizemic's approach to size chart development is based on 3D shape data as opposed to linear measurements and traditional grade rules, and this differentiates the process. This is combined with a process of size chart optimisation to maximise the percentage of the sample population covered by the chart.

Size Tables. Based on the results of the initial measurement analysis, the next stage was to set the parameters for the size tables – the upper and lower limits of the chart and the increments between sizes. The tables were based on hip girth. The client had separate size tables for regular and larger sizes as well different age groups. Based on the results, they decided to go for one size chart.

Three-Dimensional Shape Analysis. Once the new size chart was established 3D shape analysis software was used to generate 3D average body shape (one for every size in the range), starting with the three main reference sizes (equivalent to sizes 12, 18 and 22 for women and 36-, 40- and 44-in chest girth for men). This showed the average body shapes as well as the variation in shape across the sample for each size. The 3D shape analysis was one of the main benefits of SizeUK because it enabled the client to properly understand their target customer's body shape – in many cases, as in this one, for the first time.

The 3D average body shapes provide a basis for 3D virtual parametric mannequins for each of the main reference sizes. These are then aligned and the virtual mannequin for the intermediate sizes generated through a process of interpolation. The result is a complete size range of virtual, parametric mannequins, each interrelated and aligned with the others.

Size Chart Optimisation. The client chose to go through a size chart optimisation process to maximize the percentage of the sample population covered by the chart and to produce a more refined size chart. The size charts were populated with measurements extracted from the 3D virtual parametric mannequins. Measurements were therefore the output of the 3D process, not the input, and shape was the driver. Most importantly, it meant that the measurements for each size in the chart related directly to a body shape in a way that measurement-based charts could not. Grade rules – the definition of the interval between sizes – were automatically generated from the 3D virtual bodies, demonstrating that the body does not grade in the same way as traditional arithmetical grade rules. (It is often impossible to create a body shape from a two-dimensional measurement-based size chart without amending some of the measurements.) After an internal company review process, the company then adopted the new size chart.

10.6.2.2.3 Stage 3 – customised fit mannequin programme

The client chose to opt for a customised mannequin solution based on their new size chart. This was a key stage in the programme because it provided a physical manifestation of the 3D shape data and, crucially, enabled the client to implement the results of the analysis programme across their supply chain, whereby all suppliers would be developing products on the same body shape and size.

The key stages in the development process of the mannequin were:

- three-dimensional virtual mannequin from all 3D average body shapes;
- polyfoam prototyping to test the body shapes against the original 3D virtual mannequin;
- prototype mannequin development produced from moulds developed from the same 3D data to ensure accuracy and consistency;
- mannequin production (see Figures 10.9 and 10.10).

This client rolled out the mannequin programme out across their supply chain and made it a requirement for their suppliers to purchase them including details on the programme in their supplier manual. There are big benefits in terms of cost and time saving for the supplier with a payback on the investment in less than one season.

10.6.2.2.4 Stage 4 – fit model measurement and shape comparison

We then scanned the client company's fit models and compared the measurements and shape to the size chart and the 3D average body shapes. Based on the results, the

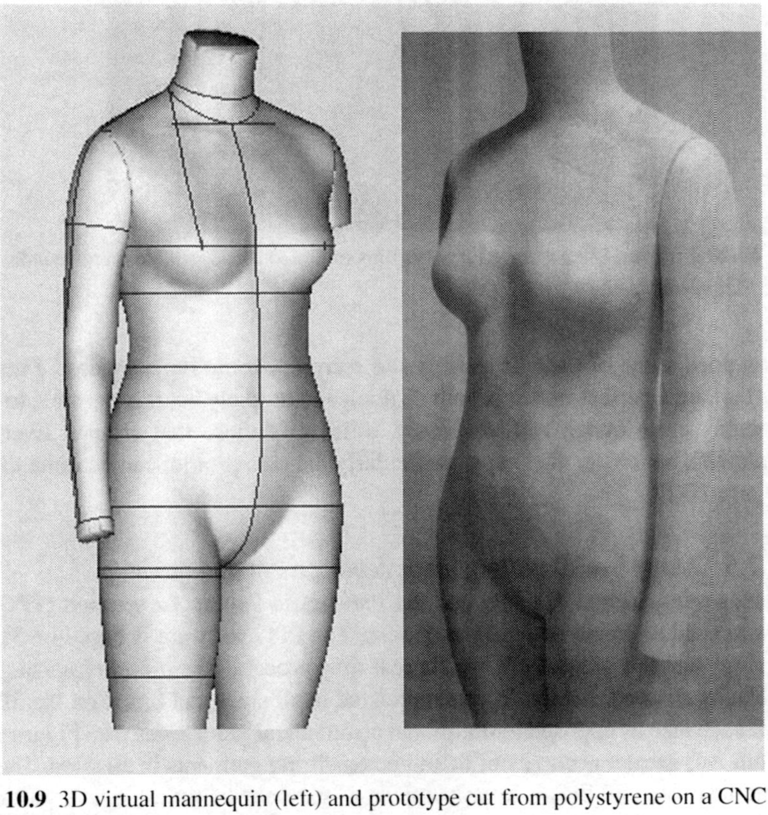

Figure 10.9 3D virtual mannequin (left) and prototype cut from polystyrene on a CNC rapid prototyping machine.
Source: Sizemic Ltd.

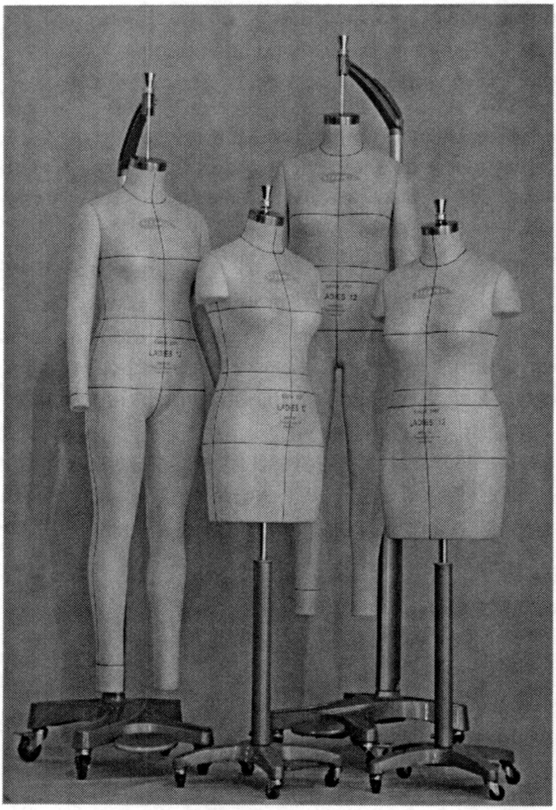

Figure 10.10 Finished linen covered mannequins mounted on adjustable metal stands.
Source: Sizemic Ltd.

client retained some of their fit models and recruited some replacements. Finding fit models that are a perfect match is both difficult and unlikely but it is possible to ensure a reasonably close match and, obviously, to identify those that are not. Even if the measurements are close, the shape, particularly the side profile, can be quite different (see Figure 10.11).

10.6.2.2.5 Stage 5 – 3D pattern block development and grading

The mannequins, virtual 3D data and the Parametric Pattern Generation (PPG) software were used to create new pattern blocks. The PPG software is based on 3D body morphology and is a uniquely 3D to 2D real-time process. The morphological grading of the blocks ensured that the fit was replicated in all sizes and based on the 3D body shape of each size as opposed to traditional arithmetical grade rules (see Figure 10.12). The result was greater accuracy of fit and better-fitting garments in all sizes. The client was able to make significant improvements to garment fit (in all sizes), particularly in their jeans and denim category. Simply generating the rise shape provided a 'quick win' fit improvement for their jeans.

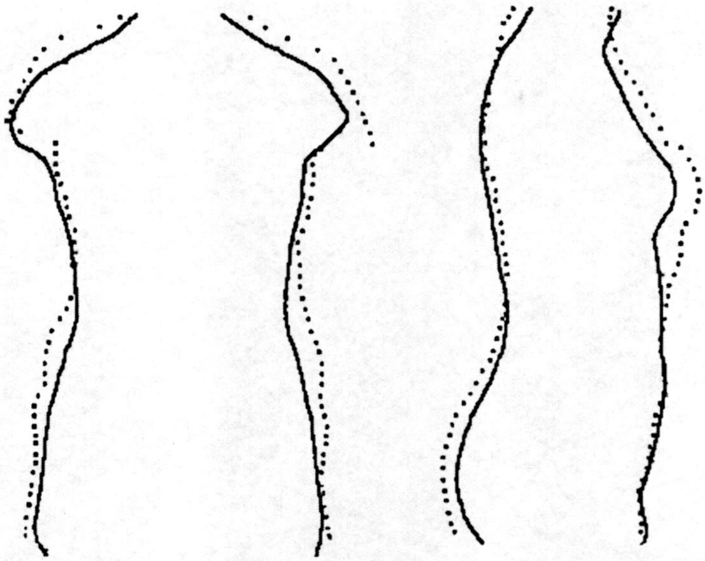

Figure 10.11 Client's fit model (dotted outline) matched with 3D average body shape (solid line).
Source: Sizemic Ltd.

10.6.2.3 Overall commercial benefits

The benefits the client enjoyed can be summarised as follows:

- returns percentage was reduced for reasons of sizing and fit – both in store and online;
- margins were improved due to reduced returns and better sell-through rates/lower markdowns, attributable to a more efficient size chart and improved sizing and fit;
- improved garment fit and consistency of fit across sizes and products has had a positive impact on customer loyalty and sales, though this is difficult to quantify as sales increases cannot be attributed solely to these products;
- analysis of the data enabled the client to understand demand in specific size categories and encouraged expansion of their size range;
- because suppliers were fitting on the same size and body shape (mannequin roll-out programme), the product development process became much more efficient. It also enabled them to establish a fit standard for all their products across their supply chain. Fewer iterations of samples (typically three to four per style before and one or two thereafter) resulted in shorter lead times and increased speed to market had a positive impact on sales;
- efficiencies achieved in the product development cycle had a cost benefit for the client. Fewer samples meant fewer fit sessions and lower fit model costs and saved staff time processing and managing sample approval;
- for suppliers there were also considerable benefits: a higher sample approval rate meant lower sample costs, lower sample delivery costs and a considerable time saving;
- the mannequin programme led to improvement in the overall level of quality control and provided and excellent quality assurance communication tool.

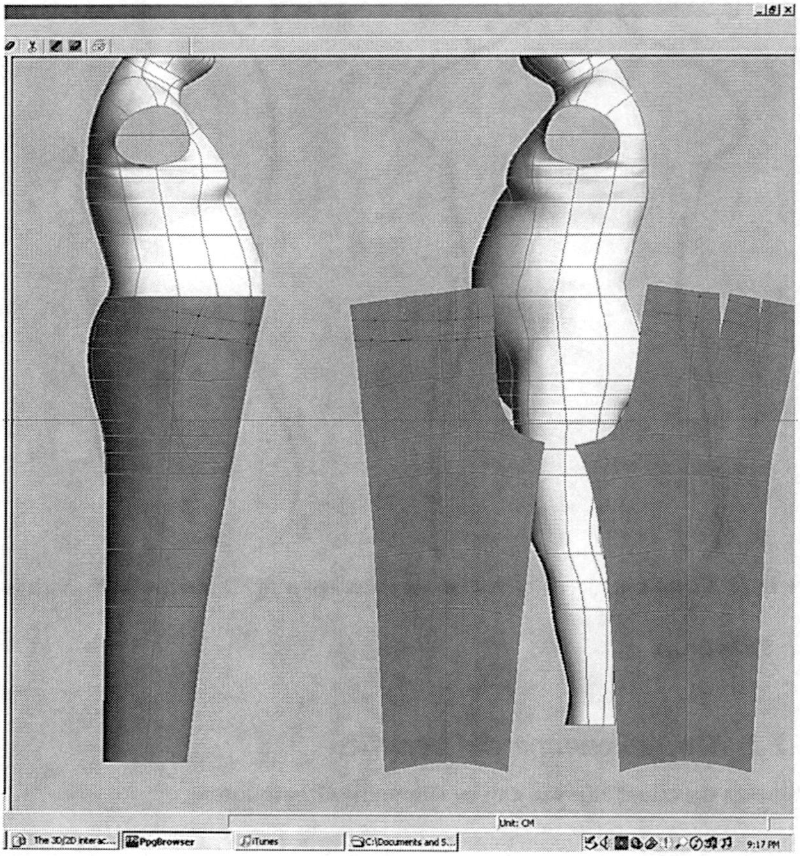

Figure 10.12 Morphological grading.
Source: TPC, Hong Kong.

Commercialising and exploiting size survey data in this way has enabled the client to enjoy tangible benefits that have improved the performance of the business, measureable through key performance indicators. These benefits can be divided into two distinct groups – those that had impact on the speed and cost of the product development process and those that were market oriented and had impact on sales, margin, customer satisfaction, buying and merchandising.

10.6.3 Utilising size and shape survey data

The case studies have demonstrated how national survey data can be mined and applied to consider the size, shape and posture of inter-generational communities and how accrued benefits can offer opportunities, not only to address the needs of growing market segments through an inclusive design strategy, but can also provide considerable commercial advantage.

10.7 Future trends

Ageing is one of the most remarkable human success stories of our era. If we are to unlock the full benefits of that success, there needs to be a change of perspective (WHO, 2012b). The implications of this perspective for the apparel industry include, in the context of this chapter, the need to:

- update ISO standards to take account of measurements required to create better-fitting garments for inter-generational communities;
- undertake comprehensive size and shape surveys at regular intervals so as to ensure that relevant data are available to reflect the natural ageing process, enabling designers, manufacturers and retailers to address emerging markets.

However, as indicated by the case studies, these immediate needs are directly associated with one-, two- and three-dimensional static capture and their applications for understanding the size, shape and posture of all populations to help improve and provide better-fitting clothes. Future inclusive design practices could, with the use of 3D avatars, include established 3D virtual design and product-fitting processes, as well as consider emerging technologies that offer the capture of 4D anthropometrics. For example, for the capture of those movements made while walking or the more ambitious dynamic model offered by DRAPE: Dressing Any Person (Guan et al., 2012).

3D scanning systems with apparel applications are in continuous development. An example is KINECT (and similar systems) that are relatively new and able to capture measurements, shape, motion, audio and image (KINECT, 2012). And although at present the resolution of the captured image is relatively low, it is already being trialled by consumers in retail environments for best fit of RTW apparel and has potential for use in the home, where body size, shape and posture can be captured at leisure. Future applications are seen as online RTW shopping (though of course restricted to age bands of available surveys and size ranges, determined by retailers) and, more importantly, the development of mass-customised apparel, where size, shape and posture can be individually addressed and where aspects of co-design protocol are already being practised.

A theme of this chapter is that longer life spans take us into uncharted territory in which the challenges, for individuals and for societies, are formidable. Part of that challenge for the apparel industry is to achieve an understanding of the changing shape, size, posture and movement during the natural ageing process within growing populations, and to fully exploit the advantages offered by 3D scanning technology.

Acknowledgements

The author would like to acknowledge the kind assistance of: all members of the project User Reference Group, University of Wales, Newport, UK; Mr Andrew Crawford, CEO of Sizemic Ltd, UK; Dr Anthony Ruto, director of Shape Dynamics UK; Dr Simone Morlock, Hohenstein Institute, Germany; Mr Patrick Robinet, IFTH, France; Dr Supiya Charoensiriwath, Thai National Electronics and Computer Technology Center, Thailand; Mrs Kerry King, director of product development, TC2, USA; M. Jean-Loup Rennesson, marketing manager for Telmat, France; Mrs Sandra Alemany, Universidad Politécnica de Valencia, Spain; and Mr Stroud Cornock, London, UK.

References

Age Platform Europe. (2012). *European year for active ageing and solidarity between generations*. http://www.age-platform.eu/en/age-policy-work/. Accessed July 2012.

Ahmad, O. B., Boschi-Pinto, C., Lopez, A. D., Murray, C. J., Lozano, R., & Inoue, M. (2001). *Age standardization of rates: A new WHO standard.* (GPE Discussion Paper Series: No.31).

Alemany, S., Gonzales, J. C., Nácher, B, Soriano, C., Arnáiz, C., & Heras, A. (2010). Anthropometric survey of the Spanish female population aimed at the apparel industry. In d'Appuzo (Ed.), *International Conference on 3D Body Scanning Technologies, conference proceedings*. Lugano, Switzerland: Hometrica.

Ashdown, S., & Mete, F. (2010). Development of visual fit assessment tool for apparel firms. In *Annual report*. Pennsylvania, USA: National Textile Center.

Ashdown, S., & Na, H. (2008). Comparison of 3D body scan data to quantify upper body variation in older and younger women. *Clothing and Textiles Research Journal*, *26*(4), 292–307.

Ball, R., Luximon, Y., & Chow, H. C. E. (2012). Anthropometric study on Chinese head. In *Asian workshop on 3D body scanning technologies, Tokyo conference proceedings*. Switzerland: Hometrica.

Bougourd, J. P., & Treleaven, P. (2010). UK national sizing survey – SizeUK. In d'Appuzo (Ed.), *International conference on 3D body scanning technologies, conference proceedings*. Lugano, Switzerland: Hometrica.

CEBE. (2001). *An introduction to inclusive design, Centre for education in the built Environment: Special interest group in inclusive design*. Cardiff and Salford Universities.

Cleland, I., Nugent, C., Finlay, D., Burns, W., Bougourd, J., Stevens, K., et al. (2012). Effects of accelerometer coupling within smart clothing on physical activity monitoring. *The Journal of Health and Technology*. http://dx.doi.org/10.1007/s12553-012-0036-1.

Connell, L. J., Ulrich, P. V., Brannon, E. L., Alexander, M., & Presley, A. B. (2006). Body shape assessment scale: instrument development for analyzing female figures. *Clothing and Textiles Research Journal*, *24*(2), 80–95.

Daanen, H. A. M., & Haar, T. (2013). Whole body scanners revisted. *Displays*, *34*, 270–275.

Daanen, H. A. M., & Reffeltrath, P. (2007). Function, fit and sizing, in Sizing in Clothing. In Ashdown (Ed.), Cambridge: Woodhead.

De Filippo, A. (May 2012). *The problem of population ageing in Italy*. http://www.lifeinitaly.com/lifestyle/aging. Accessed July 2012.

De Filippo, P., Reiss, L., Hirshberg, D.A.,Weiss, A. & Black, M.J. (2012). DRAPE: Dressing Any Person, Siggraph. http://.s.brown.edu/~pguan/publications/DRAPE2012.DRAPE2012.pdf.

Guan, P., Reiss, L., Hirshberg, D. A., Weiss, A., & Black, M. J. (2012). *DRAPE: Dressing Any Person, Siggraph*. http://.s.brown.edu/~pguan/publications/DRAPE2012.pdf.

Hanson, L., Sperling, L., Gard, G., Ipsen, S., & Olivares Vergara, C. (2009). Swedish anthropometrics for product and workplace design. *Applied Ergonomics*, *40*(4), 797–806.

Hometrica. (2012). *Organiser of conferences on 3D scanning*. http://www.3dbodyscanning.org/2012/home.html. Accessed August 2012.

Hutchison, T., Morrison, P., & Mikhailovich, K. (2006). *A review of the literature on active ageing*. Canberra: University of Canberra, Prepared for the Australian Government by the Healthpact Research Centre for Health Promotion and Wellbeing.

IIE. (2012). *Z94.2 – Anthropometry & Biomechanics*. Note published by the Institute of Industrial Engineers in Georgia, USA . http://www.iienet2.org/Details.aspx?id=2624. Accessed August 2012.

ISO. (1989). *Garment construction and anthropometric surveys – Body dimensions*. (Switzerland: ISO 8559:1989). Accessed November 2012.

ISO. (2006). *General requirements for establishing anthropometric databases.* (Switzerland: ISO 15535:2006). Accessed July 2012.

ISO. (2008). *Basic human body measurements for technological design – Part 1: Body measurement definitions and landmarks.* (Switzerland: ISO 7250–1:2008). Accessed November 2012.

ISO. (2010). *3-D scanning methodologies for internationally compatible anthropometric databases.* (Switzerland: ISO 20685:2010), Accessed November 2012.

Keeler, A. (2010). *Body mass Index and older people.* http://www.livestrong.com/article/336966-body-mass-index-older-people/. Accessed August 2012.

Kendall, F. P., McCreary, E. K., & Provance, P. (2010). *Muscles: Testing and function with posture and pain.* London: Lippincott Williams & Wilkins.

KINECT. (2012). *Motion sensing technology.* Discussed at http://en.wikipedia.org/wiki/Kinect. Accessed November 2012.

Kirchdoerfer, E., Mahr-Erhardt, A., & Morlock, S. (2011). 3D body scanning – Utilization of 3D body data for garment and footwear design. In d'Appuzo (Ed.), *International Conference on 3D body scanning technologies, conference proceedings.* Lugano, Switzerland: Hometrica.

Kirkwood, T. (2001). *Lecture 1: Brave Old World* (The Reith Lectures 2001: The End of Age, published in London by the BBC).

Krzywinski, S., Rödel, H., & Siegmund, J. (2005). Virtual product development for close-fitting garments of knitwear with Elastan yarns. *2nd international conference of textile Research Division, NRC, Cairo, Textile processing: State of the art and future developments,* (Vol. 2), 2, pp. 136–140.

Lotens, W. A. (1989). Optimal design principles for clothing systems. In R. F. Goldman, & B. Kampmann (Eds.), *Handbook on clothing.* (Research Study Group 7 on Bio-Medical Research Aspects of Military Protective Clothing, Chapter 17.

Lu, J.-L., & Wang, M.-J. J. (2010). The evaluation of scan-derived anthropometric measurements, *IEEE Transactions on Instrumentation and Measurement, 61*(9), 2048–2054.

MedlinePlus. (2012). Advice published online by U.S. National Library of Medicine as 'Aging changes in the bones – muscles – joints'. http://www.nlm.nih.gov/medlineplus/ency/article/004015.htm. Accessed August 2012.

Milanese, C., Lovato, C., Piscitelli, F., Giachetti, A., & Zancanaro, C. (2011). Photonic scanning for anthropometry. *Italian Journal of Anatomy and Embryology, 116*(1 Suppl.), 120.

NDA. (2012). Newsletters on design for ageing well, published online as part of The New Dynamics of Ageing project, at http://www.newdynamics.group.shef.ac.uk/design-for-ageing.html. Accessed August 2012.

Nemmers, T. M., Miller, J. W., & Hartman, M. D. (2009). Variability of the forward head posture in healthy community–dwelling older women. *Journal of Geriatric Physical Therapy, 32*(1), 10–14.

Noor, M. I., Koon, P. B., & Hashim, Z. (2005). *Strategy for the prevention of obesity – Malaysia.* Kuala Lumpur: Malaysian Association for the Study of Obesity.

Nowak, E. (2006). Anthropometry for the needs of the elderly, Chapter 60 of the *International encyclopedia of ergonomics and human factors* (2nd ed.). In W. Karwowski (Ed.). Danvers, Mass., USA: CRC Press.

Onis, M., & Habicht, J.-P. (1996). Anthropometric reference data for international use: recommendations from a WHO Expert Committee. *The American Journal of Clinical Nutrition, 64*(4), 650–658.

Perissinotto, E., Pisent, C., Sergi, G., Grigoletto, F., & Enzi, G. (2002). Anthropometric measurements in the elderly: age and gender differences. *British Journal of Nutrition, 87,* 177–186.

Preiss, A., & Botzenhardt, U. (2012). Size survey – Process chain and available products. In *Asian workshop on 3D body scanning technologies, Tokyo conference proceedings* (pp. 106–114). Switzerland: Hometrica.

Price, G. M., Uauy, R., Breeze, E., Bulpitt, C. J., & Fletcher, A. E. (2006). Weight, shape, and mortality risk in older persons. *The American Journal of Clinical Nutrition, 84*(2), 449–460.

Rapidform. (2012). *A note on scanning technology*. http://www.rapidform.com/3d-scanners/. Accessed August 2012.

Ravaglia, G., Morini, P., Forti, P., Maioli, F., Boschi, F., Bernardi, M., et al. (1997). Anthropometric characteristics of healthy Italian nonagenarians and centenarians. *British Journal of Nutrition, 77*, 9–17.

Ruto, A. (2009). *Dynamic whole body modelling and animation*. [Engineering Doctorate thesis]. University of London.

Shephard, R. J. (1998). Aging and exercise. In T. D. Fahey (Ed.), *Encyclopedia of sports medicine and science*. Internet Society for Sport Science. http://www.sportsci.org/encyc/agingex/agingex.html. (Accessed July 2012.

Simmons, K., Istook, C. L., & Devarajan, P. (2004). Female figure identification technique (FFIT) for apparel; Part II: Development of shape sorting software. *Journal of Textile and Apparel, Technology and Management, 4*, 1.

Simmons, K. P., & Istook, C. L. (2003). Body measurement technique: Comparing 3D body scanning and anthropometric methods for apparel applications. *Journal of Fashion Marketing and Management, 7*(3), 306–332.

Sims, R., Marshall, R., Gyi, D. E., Summerskill, S., & Case, K. (2012). Collection of anthropometry from older and physically impaired persons: Traditional methods versus TC2 3-D body scanner. *International Journal of Industrial Ergonomics, 42*(1), 65–72.

Song, H. K., & Ashdown, S. P. (June 2011). Categorization of lower body shapes for adult females based on multiple view analysis. *Textile Research Journal, 81*(9), 914–931.

Spirduso, W. W., Francis, K. L., & MacRae, P. G. (2005). *The physical dimensions of ageing*. Champaign, Illinois, USA: Human Kinetics, ISBN-13: 9780736033152.

Stewart, A. (2007). Kinanthropometry – the interdisciplinary discipline, editorial. *Journal of Sports Sciences, 25*(4), 373.

Twigg, J., & Majima, S. (2014). *Consumption and the constitution of age: Expenditure patterns on clothing, hair and cosmetics among post-war 'Baby Boomers'. Journal of Ageing Studies, January 2014, 30*, 23–30.

UN. (2009). *World population ageing 2009*. New York: United Nations Department of Economic and Social Affairs/Population Division.

Walker, M., & Mesnard, X. (2009). *What do mature consumers want?* Published by the corporation's Global Business Policy Council in conjunction with the A.T. Kearney Global Consumer Institute.

Wells, J. C. K., Cole, T. J., & Treleaven, P. (2008). Age variability in body shape associated with excess weight: the UK National Sizing Survey. *Obesity, 16*(2), 435–441.

WHO. (2002). *Active ageing: A policy framework*. World Health Organisation, Published online as whqlibdoc.who.int/hq/2002/who_nmh_nph_02.8.pdf. Accessed July 2012.

WHO. (2004). Appropriate body-mass index for Asian populations and its implications for policy and intervention strategies. *The Lancet, 363*, 157–163.

WHO. (2008). *Waist circumference and waist–hip ratio: Report of a WHO expert consultation, Geneva*. Published online at http://www.who.int/nutrition/publications/obesity/en/. Accessed July 2012.

WHO. (2012a). *Are you ready? What you need to know about ageing*. World Health Day 2012. http://www.who.int/world-health-day/2012/toolkit/. Accessed July 2012.

WHO. (2012b). *Global population ageing: Peril or promise?* World Economic Forum, Global Agenda on Ageing Society, collection of essays published on World Health Day 2012.

WHO. (2012c). *WHO study on global ageing and adult health (SAGE)*, Report on a study published online at http://www.who.int/healthinfo/systems/. Accessed August 2012.

XYZRGB. (2012). *Live body scanning*. Commercial Website at http://www.xyzrgb.com/3d_body_scanning.html. Accessed August 2012.

Part Three

Apparel design requirements for the active ageing population

The role of wearable electronics in meeting the needs of the active ageing population

11

D.C. Jones
Fibretronic Limited, Kowloon, Hong Kong, China

11.1 Introduction

This chapter will begin with a brief introduction to what is meant by the term 'wearable electronics', (or 'wearables') and how this technology area has evolved from a research curiosity into a growing sector of the apparel and soft goods industry. Wearable electronics represents a cross-fertilization of the garment industry and the consumer electronics industry. Whilst the term 'wearables' has more recently come to include body-worn gadgets, wrist bands, and smart watches, this chapter will focus on wearable electronics under the definition of textiles and garments with embedded electronic functionality.

The convergence of two traditionally separate industry sectors began with product designers of conventional electronic goods asking questions such as 'why do electronics need to be in hard plastic boxes?' and 'why cannot softer more tactile materials be used to explore the interfaces between man and machine?' This came at the same time as a new trend in the design of personal electronics, with products becoming more subtle, ergonomic, and in tune with the wearer and their lifestyle needs. Personal electronics began to turn into fashion accessories manufactured in different styles and colours to coordinate with clothing. This led product designers to push boundaries to find new form factors that could offer more integrated solutions for the end-user. Softer, more flexible materials such as rubber, foams, and fabrics were explored to create new form factors that opened up new product opportunities.

At about the same time, textile researchers were starting to explore how it might be possible to make intelligent fabrics with advanced functions: textiles that could sense or respond to stimuli, and in particular, fabrics with embedded functionality such as the ability to conduct electricity.

One of the first academic institutions to consider the combination of clothing and electronics was the Massachusetts Institute of Technology, through their wearable computing research program in the 1990s. The focus of this work was on the design of computing hardware in portable form factors, such as clothing and other body-worn accessories. In particular, the use of conductive yarns as a conduit for the wearable computer's user interface technology was groundbreaking and it helped to kick start the electronic textiles field. From a more commercial perspective, Philips Electronics was one of the first companies to dedicate a research team to explore

Textile-led Design for the Active Ageing Population. http://dx.doi.org/10.1016/B978-0-85709-538-1.00011-0

the possibilities of merging electronics and clothing. The 'visions of the future' project in the late 1990s saw Philips establishing a multidisciplinary team of electronic engineers and fashion, textile, and industrial designers to investigate potential consumer applications for wearable electronics. The team came up with a series of concept garments to show future application possibilities, such as digital suits with built-in keypads and communications capabilities for business and corporate clothing; electronic sportswear with built-in audio controls, GPS, and biometric sensors; and interactive clothing for children with embedded games and security systems. Although most of these concepts have now been commercially realised by various other companies, sadly they did not envisage any concept garments specifically for older consumers.

One of the first commercial examples to show how electronics and clothing could be joined together to make a hybrid product was the Levi Strauss ICD Jacket, which was developed in collaboration with Philips Electronics in 2001. In this limited-edition garment, conventional electronic hardware and wires were routed through the jacket lining to make a garment that could control an MP3 player and mobile phone. A control module or 'hub' linked the devices together so that they could be used as a network in the jacket.

The first commercial wearable electronic garment to use electronic textile components was the Burton Snowboards Analog Clone Jacket in 2002, which was developed by Softswitch Limited in collaboration with Sony to control their latest Walkman minidisc player directly from the jacket sleeve. This was followed quickly by the Amp Jacket in 2003, which was a collaboration between Softswitch Limited, Apple Inc., and Burton Snowboards. Both of these jackets incorporated a pressure-sensitive fabric keypad in the sleeve and textile cables to connect to the Walkman/iPod, allowing the audio player to be remotely controlled by interaction with the garment. These jackets were the first of their kind in that they used washable electronic textile elements that were permanently integrated into the garment construction during the cut and sew process.

Following on from these early explorations, wearable electronics has since developed into its own niche market. Over the last 10 years, many of the world's largest garment brands have added advanced functionality to their products by incorporating electronic features such as entertainment controls, lighting devices, communications accessories, sensors, heating panels, and other 'intelligent' features. In parallel, we can witness how over the last decade more and more everyday products have become 'intelligent' through some form of embedded electronic functionality. The miniaturisation of circuits and micro-components has meant that electronic systems have become covert and almost invisible, making them easier to embed in flexible substrates. Increasingly, today's consumers and those in the future will come to expect their clothing to have advanced functions, such as health monitoring capabilities or built-in communications and safety features, in addition to the basic properties of keeping them warm or covered. User-assistive technologies built into clothing will be of particular relevance to older consumer groups, not only from a medical perspective but in helping them to lead fuller and more active lives into old age.

11.2 Current applications and end-users

Over the last decade many apparel companies have developed and retailed a wearable electronic garment or accessory. Many of the early concept ideas have now been realized as commercial products sold by some of the world's best-known clothing brands. The pressure of low-cost competition for commodity products in the garment and textile industries has led many brands to look for ways of adding value and point of difference to their products by incorporating new technology. Wearable electronics has been used as a marketing tool to build brand equity via cutting-edge and innovation-driven products. Wearable electronic products are high-profile items and they provide a way for clothing and soft goods companies to differentiate their products from the competition, create a futuristic aura, and generate consumer interest. This in turn has helped to advance the sector and distil those technology solutions that are highly functional, can be easily applied, and can meet critical cost targets.

Initially, wearable electronics was confined to high-end, limited-edition ranges typically marketed to the young, tech savvy, and predominantly male demographic. More recently, wearable technologies have expanded into broader apparel categories in active sportswear and mainstream High Street clothing. However, the target demographic continues to be the younger, technology-hungry consumer. Unfortunately, wearable tech for the older age groups is still largely seen as a future solution to address the medical and health needs of an ageing population.

The main examples of wearable electronic products that have been integrated in garments, bags, and textile-based accessories can be divided into three main categories, which will be described in the following sections.

11.3 Communication and entertainment

- Switches and keypad controls to operate the music functions of audio players/iPods
- Smart phone controls with built-in microphones/speakers/keypads to receive phone calls
- Ergonomic and integrated two-way radio controls

The communication and entertainment category was the first area of development for wearable electronics. Early researchers envisioned a world where computer interfaces were integrated in clothing, but to date, commercial products have only considered much more simplified input devices for controlling personal electronics such as audio players and mobile phones. Switches and signal-carrying cables have been developed in different formats suitable for sewing and bonding permanently into garments, being washable and durable to apparel end-use conditions. Keypad and joystick controllers, typically with five or six functions, can now be easily integrated in high volume to a broad range of garment styles including outerwear, gloves, suits, bags, and trousers. These high-tech trims are used as remote interfaces to devices such as iPods, iPhones, and other smart phones, allowing the wearer to operate the music functions or make phone calls by interacting with the switches built into the clothing they are wearing. Most products make use of Bluetooth or other high-frequency radio for wireless transmission of signals between the embedded wearable components and the hardware. This technology has found application in outdoor sports clothing such as snow and ski jackets and gloves, running/cycling apparel, urban outerwear, suits, and uniforms. The functional requirement for the wearer is to be able to easily access the main control functions of their personal electronics whilst not interrupting their activity by having to remove the device from an inside pocket or bag. In some applications the purpose has been one of security, to be able to hide away a valuable item such as an iPhone, but still be able to discreetly operate the main functions using a remote controller exported to the garment instead. The main demographic for this application is a younger, technology-hungry consumer who wants to be both connected and mobile. These end-users have no difficulty in interpreting a garment with technology content, and this has been critical to the success of this application to date. Whilst older consumers may not want instant access to music from their iPod, there may be more relevant apparel applications focused around communication, connectedness, and security where some of these technologies could be applied.

11.4 Comfort and safety in the outdoors

- Heating systems for apparel
- Technical/safety lighting using high-brightness LEDs
- GPS/tracking components

One of the most rapidly growing application areas for wearable electronics is apparel-based heating systems designed for integration in outerwear, gloves, and footwear. The recent growth in this market has been mainly due to the availability of lower-cost solutions, together with technical improvements in battery technology to reduce size and extend operating times. To date, the main uptake markets for heating systems have been in outdoor sports applications such as snow and mountain apparel, motorbike clothing, wetsuits, and work wear. Wearable heating technology is relevant to a much broader range of end-users since the functional benefit to the wearer is very easy to communicate and understand. For older consumer groups, this technology offers an immediate opportunity to create garments that improve comfort levels for people in cold conditions, allowing them to remain active in the outdoors all year round.

For outdoor safety purposes, there is an increasing use of lower-power, high-brightness LEDs designed in modular formats for easy integration into garments and accessories. LED modules can be simply applied as garment trims, magnetic buttons, zip pullers, and eyelet-mounted units, which can be designed in a range of shapes and colours. For high visibility in low light and dark conditions, these products add a safety feature to active sportswear garments such as those for running and cycling. However, being seen at night is of importance not only for those who are exercising in the dark but has wider relevance to pedestrians, in particular children and older people. Improving visibility on poorly lit streets will help to improve the safety of older people as they cross roads and negotiate the urban environment during winter months.

Wearable GPS in garments has been attempted by various companies in products such as gloves and jackets. Other than as a way-finding device to prevent you from getting lost, GPS has more interesting functions that lend themselves well to active sportswear. Being able to measure speed, distance, altitude, and duration, and then combining this with data about your route, provides all the information you need to record your activity, whether this is walking, climbing in the mountains, or back country snowboarding. The data can be stored and downloaded later, or for information in real time it is necessary to have a small display screen built into the garment. This type of technology, whilst useful, is limited by the cost and complexity of integrating a display screen into a garment and has now been superseded by the recent proliferation of smart phone apps and wrist-worn smart watches for tracking route and activity. Wearable GPS technology is now being distilled down to simple chips that can be embedded in garments to be used as real-time monitoring devices, to track the movement and location of the wearer, and even to transmit this data by GSM to third parties. This type of application is of interest to parents of younger children to provide up-to-the-minute feedback about where their child is going and what they are doing. Parents can use software to track position and receive emails or mobile alerts if their child has crossed a boundary or is at a location that is not allowed or is dangerous. This technology is commonly used in commercial vehicle tracking and freight logistics, but due to privacy concerns it is questionable how far this type of application will extend into wider consumer markets. For vulnerable groups, including older people, the potential of this type of device as a safety tool cannot be overlooked.

11.5 Fitness monitoring, sports performance and health care

- Heart rate and respiration sensors for fitness and performance monitoring
- Heating systems for sports performance enhancement
- Wearable therapy devices

Skin sensors made from fabric and other flexible conductive materials have been integrated in textile chest straps, bras, and other intimate apparel for the purpose of monitoring heart rate and respiration rate. The target market for these wearable sensors is the active sportswear consumer who wants to monitor their fitness and performance levels during exercise such as running, cycling, and gym workout. Wearable sensing and monitoring technologies are predicted to be the fastest area of growth for the wearable electronics sector over the next decade. Next-to-skin clothing with fully integrated sensors will not only monitor fitness levels of the wearer but it is hoped that the next generation of garments will be able to collect a much broader range of physiological data. Information about health, activity, and well-being will be collected by an array of fully integrated sensors to give the wearer an intimate picture of how well they have slept, their stress levels in a particular situation, the status of a health condition, or whether they have met their activity target for the day. These technologies could include sensors for position and movement, temperature,

ECG and EEG, arterial oxygen saturation, blood pressure, glucose, perspiration, and other chemicals. Coupled together with wireless data transmission to smart phones or smart watches, this provides a very compelling application for a wide range of consumer groups.

The ageing population is quite clearly the main group that would have the most benefit from the healthcare and medical aspects of these technologies. If we set aside the desire to simply monitor a particular medical condition, it can be seen that there is a broader possibility of combining data relating to activity and wellness to allow a more objective assessment of lifestyle. It is known that consumers who use simple activity monitoring devices such as step counters, *do* tend to significantly increase their activity levels as a result. So for older people, passive, unobtrusive wearable sensor technologies could be a powerful tool that may allow them to optimize their chances of living a fitter and healthier life longer into old age.

In the performance sportswear field, heated garment systems are starting to be used as muscle-conditioning tools to enable elite athletes to maintain muscle temperature following warm-up. Keeping muscles at an optimum temperature prior to strenuous exercise not only improves an athlete's performance but reduces the likelihood of injury. For older people who want to continue to enjoy higher-impact sports such as running, this approach of using a garment to warm-up a critical muscle group prior to exercise could be of significant benefit.

In addition to passive monitoring technologies, there will also be a growth in wearable therapy devices using cold, heat vibration, light, and electrical stimuli integrated in soft and comfortable body-worn garments and supports. Designed to treat and alleviate specific health conditions and complaints, these portable devices will place simple therapy treatments in the hands of end-users to self-administer when required. These active therapy technologies will be wirelessly linked to remote control hardware such as smart phones and smart watches and follow preprogrammed settings that could be determined by health professionals. This technology area, as it develops, will be highly relevant to the active ageing market.

Of the currently available technologies discussed above, wearable heating systems stands out as the most immediately applicable to the active ageing consumer. The functional benefits are very easy to understand and communicate to the end-user, and it is an application more suited to the older demographic. A simple system can be designed that does not require a complicated interface or control screen for operation. The next section will take a brief look at a commercially available technology and how it could be applied to products for the older consumer.

11.6 Apparel heating systems

The HEAT*wear*™ system by Fibretronic Limited allows textile heating panels to be designed in different sizes and shapes for easy integration in all types of outerwear garments, gilets, gloves, and footwear. The textile panels are integrated during the garment manufacturing process so the technology is discreetly hidden from the end-user. The panels are breathable, soft, lightweight, washable, and do not affect the aesthetic properties of the garment. The power source is a separate lithium ion battery pack sized to fit inside a pocket or pouch on the garment. The battery size is tailored according to the size of the heat panel being used and the length of operating time required between recharging. The battery is mains rechargeable, or for glove and footwear applications the battery can be charged via a USB. An LED control switch on a textile extension cable allows the wearer to control the battery heat settings and turn the system on and off. The LED switch is permanently integrated in the garment in a location that is easily accessible for the user and where the indicator light can be easily seen. In a typical outerwear jacket application, a single heat panel is usually integrated in the lower back area of the garment positioned between the lining and the outer fabric. This lumbar region has been shown to provide the maximum benefit in terms of maintaining core body temperature. The LED control switch is usually integrated into the upper chest area, allowing the wearer to easily control the system without having to open up the garment and remove the battery from an inside pocket. For more extreme conditions, the system can be configured with multiple heat panels that can be distributed around the front and back of the garment, for example in motorbike clothing and for winter sports professionals.

When the body is cold, blood flow to the extremities is restricted in order to maintain the body's core temperature. Using the heating system to maintain core temperature will also help to prevent the hands and feet from quickly becoming cold. However, for the older user, in whom circulation may be a particular problem, it could also be beneficial to have heating capabilities incorporated in gloves and footwear. For the active ageing market the heating system is ideally suited for outerwear garments or midlayer vests and liners. This type of application would allow older people to extend the amount of time during the year that they can remain active in the outdoors. It would enable people to continue to participate in outdoor activities during colder months such as walking, golfing, gardening, or winter sports. By providing heating when needed, the system can be intelligently controlled to maintain an optimum temperature that is comfortable for the wearer. In situations where the wearer starts to get cold when standing still for some time, or in cold wind, it is simple to get a boost to core body temperature by activating the heating system. Examples might be when watching spectator sports or taking grandchildren to the playground, getting that heating boost to make the experience more enjoyable. During colder months of the year when there is less desire to be outside, having an active heating system that is simple and effective to use will considerably improve comfort levels and tolerance to cold weather conditions. This in turn will have an impact on physical and mental well-being. Getting out of the house and continuing to be active will reduce the tendency for older people to become more sedentary and isolated in the home during winter months.

For the active ageing market, the most affordable solution would be to offer the heating system in a sleeveless midlayer vest or liner that could be worn under any type of jacket or coat. For most users, the vest would only require a single heating panel in the lower back region and could also incorporate temperature sensors to intelligently manage the heating function and maximise the battery lifetime. With a standard system, the user could expect to get around 3–4 h of continuous heating if the battery is operated in constant mode. If the heating is activated only when the user needs it, then the operating lifetime of the battery can be extended to a whole day of use.

Further benefits of heating technologies are linked to the human body's ability to easily absorb far infrared radiation. Studies have shown that far infrared waves can actually penetrate two to three inches deep into muscular tissue and ligaments, helping to promote a wide range of additional beneficial effects including pain relief, arthritis relief, and relief of other ailments commonly found in the aged. When far infrared heat penetrates through the skin into muscular tissue and ligaments, it transforms from light energy into heat energy. The thermal effect within the deep layers of tissues cause blood vessels in capillaries to dilate, which improves blood circulation and promotes healing and wellness.

11.7 Commercial challenges of wearable electronics for active ageing

There are a range of challenges in introducing wearable electronics products to an older customer base. It is a consumer group that presently has little awareness of this new technology area and a group that may be, in general, more resistant to adopting new product innovations (perhaps with the exception of medical and therapeutic products). Wearable electronics has been mainly targeted at aspirational young consumers via lifestyle clothing brands, but these products may need to be marketed in a completely different way to an older demographic. For apparel and soft goods brands which will adopt these technologies, a key consideration is how to raise awareness amongst their older customers that these technologies are available and could be of benefit to them. The dilemma will be how to do this effectively whilst still appealing to the younger customer who is the driving force behind their marketing efforts. The question is whether a product is still cutting edge and cool if your parents and grandparents are using it? The solution may come down to a simple dividing line based purely on function. In general terms, those technologies that address comfort, health, and well-being will be of more interest to the older consumer, while those technologies that link to personal electronics, music, communication, and sports performance will be of greater interest to the younger consumer. Adding electronic functionality that does not bring anything useful to the garment other than as a novelty feature or gimmick will not be appropriate for the older consumer group. It is also important that the integrated electronic features add value to the finished garment to create a unique proposition rather than it making more sense for the customer to buy a stand-alone accessory that does the same job.

A further issue to address is the possible preconception by older consumers that 'wearable electronics' sounds too technical and that these products are too complicated to use. The responsibility falls on product designers and developers to consider carefully the form factors of critical components to avoid interfaces that are too fiddly and displays that cannot be read. The size and spacing of switches or buttons on a keypad need to be appropriate for those with reduced dexterity. A single larger multifunction switch combined with a multicolour indicator light, to show which function has been activated, may be more appropriate than a series of smaller single function buttons on a keypad. A switch designed with strong haptic feedback will give the user a clear activation response, and together with a variable-colour indicator light it will provide a simple interface that is easy to use, easy to see, and easy to understand. Software/firmware written to control the electronics should be developed perhaps in conjunction with feedback from user groups to determine only a limited number of key functions that are important to the operation of the device. Less is more, and simple operation makes for a more satisfying user experience. A further consideration is keeping the interface and control functions consistent between different styles in a garment range and critically from season to season. Unfortunately, the garment industry, driven by fashion, thrives on change. But for older people, having learnt how to use something new, keeping a consistency in the way a product operates and avoiding dramatic changes is more important.

It is critical for product designers and merchandisers to ensure that a garment's function and benefits can be immediately understood and simply communicated at retail. To date for wearable electronics, this last link in the chain, getting the product off the retail shelf and into the hands of the end-customer, has been the weakest. It is particularly important for wearable electronics that the customer has three main questions answered quickly when they look at a garment on the shop rail: what does the product do, how does it operate, and what are the functional benefits of using it? In most retail environments, a dedicated point-of-sale display promoting wearable electronic garments is not possible. This means that currently the only opportunity to communicate with the customer in store is via a hangtag, which of course has its limitations. It can only be hoped that as the market grows in the future, retailers will begin to dedicate separate rails in their stores to wearables and staff will be trained to explain them. Incorrect retail presentation of wearable electronics products still remains a major problem for those garment brands that do not have control over their own retail outlets.

Last but certainly not least are economic factors: how to make a garment with advanced functionality, but with a retail price tag that is affordable for an older customer. User groups of older people have highlighted understandable concerns about buying and investing in a sophisticated garment with integrated electronics that might go wrong or cannot be upgraded. They are also worried about aftercare. Since the garment is an electronic product, it is necessary to supply a user manual to advise the correct conditions of use together with cleaning instructions. Aftersales support and a manufacturing warranty are also required.

With regard to the final retail price of the garment, it is the cost multiplier structure in the garment industry supply chain that largely dictates the finished ticket price. For example, an electronic part supplied to a garment manufacturer for \$10 translates to \$40 of added cost to the final retail price tag. This four-times multiplier means that a technology-integrated garment is always going to be significantly more expensive than its standard counterpart. It becomes prohibitive to make inroads into higher-volume clothing categories that are usually more price sensitive. Since the garment industry is particularly sensitive to component supply costs, when designing wearable electronics it is important to avoid complex systems using expensive components that cannot be purchased in significant volume. Higher garment assembly costs can also be incurred if additional manufacturing steps have to be introduced to integrate the electronic components. In general, the increased care and precision necessary for handling and installing electronic components is more time consuming and therefore increases labor cost. Assembly costs can be reduced by supplying components that are simple to install and are compatible with garment production lines. An additional source of cost for the garment manufacturer is the requirement for more detailed and frequent quality control testing. Incoming QC testing is required on electronic parts prior to integration, followed by final QC testing of all finished garments before they leave the factory.

Currently, in most applications, the design and format of electronic systems is simply dictated by cost and ease of implementation rather than using the most elegant technology solutions. It is therefore the availability of simple and cost-effective components that will be the driving force behind future expansion of the wearables sector, rather than the most innovative of smart fabric technologies. From a commercial perspective, it can be seen that wearable electronics will never become commodity items. In marketing these products to the end-consumer, the focus needs to be less on the price and more on presenting value for money.

11.8 Implementation considerations

The projected size of the active ageing market in the future presents a considerable commercial opportunity for garment brands to make functional, assistive, and integrated clothing for a consumer group who will become increasingly comfortable with technology products. The key to unlocking the mass market potential is to get the correct mix of garment design, embedded functionality, easy-to-implement parts, and an acceptable retail price. An approach to getting the design and functionality elements right could be a strategy that brings older end-users together with (typically) younger designers and developers. Being able to gain direct feedback to inform the development process will ensure that concerns about usability, ergonomics, form factors, and value for money are addressed. Working closely with user groups could also be a way to get the retail communication and marketing approach right.

With regard to design and implementation of electronic technology for apparel, this is Fibretronic's particular area of expertise. Our focus lies in being able to supply an electronic part that is designed for volume manufacturing, which in turn is easy for our customers to handle and apply in garment production lines. Electronic parts need to be designed in simple formats that can be applied in the factory much like another garment trim item. Systems should be reduced to as few parts as possible to avoid overly complex integration procedures and added manufacturing cost. Where a system part is intended for permanent integration in the final product, it is important that it is designed to be compatible with the particular garment construction process – for example, sewing, thermal bonding, or ultrasonic welding. Parts that can be customised to suit different garment designs offer a significant advantage rather than a one size/shape fits all approach. Also it may be necessary to make parts in a range of sizes to fit the graded sizes of the garments, for example textile cables supplied in different lengths. Parts supplied in ready-to-apply packages with fully finished detailing, such as stitch-in moulded housings for keypads, offer a more complete solution. Fibretronic supplies garment factories with instruction manuals showing how parts need to be handled, installed, and tested. A QC testing regime is also designed and testing equipment supplied to allow the factory to test critical electronic functions on incoming component parts and finished products.

Another design approach is to make parts that can be applied to the garment at a late stage in the manufacturing process, for example attachable modules. Self-contained electronics modules that can be attached via snaps, Velcro, eyelets, or other fastening methods greatly simplify implementation and supply logistics for soft goods manufacturers. This approach minimises the involvement of the manufacturer in handling, processing, and testing of critical electronic subcomponents. It also gives garment designers the flexibility to apply the electronic technology to a broader range of products by simply adding the appropriate attachment point to the finished item.

11.9 Conclusion

Wearable electronics is a cutting-edge niche sector of the apparel industry that has been growing in momentum since the 2000s. Today, 'wearables' has become the latest buzz word and the term now encompasses a much broader philosophy that technology should be with us at all times, being portable, assistive, and immediately accessible. There is no reason why this connectedness to technology is not equally applicable to older consumers as younger ones, who are the main target market for companies selling wearable products. This chapter has aimed to highlight how wearable electronics could offer some solutions in response to the growing demand by older people to have a more active and healthy life for longer. Feedback from user groups has already been valuable in identifying technologies such as heating and communication systems in garments as well as health/fitness monitoring as key areas of interest for the active ageing customer. For heated clothing, technology solutions are already available to make effective heated clothing at an affordable price, but today's commercial products are mainly marketed at younger people. The next challenge is how to engage garment brands in designing, manufacturing, and marketing this type of cutting-edge product for older customers.

Looking to the future, it is clear that there will be a growing demand for products that keep us connected and that monitor our lifestyle to encourage us to make healthy choices. These products will be wearable and the biggest market sector will be the older consumer. The timing is ripe for the textile and garment industry to take advantage of new innovations in electronic components, smart materials, and intelligent fabrics to lead the way with a new generation of clothing designed to support us as we age.

Overview of the design requirements of the active ageing

12

M.W. Timmins[1], *J. McCann*[2]
[1]Consultant, UK; [2]University of South Wales, Newport, UK

12.1 Introduction

This chapter, and the next, focus on identifying and addressing the design requirements of the active ageing, in terms of the design of functional clothing that has the potential to promote healthy exercise, autonomy and independence. In particular, the application of smart clothes and wearable technologies is discussed in a case study involving cross-disciplinary research 'Design for Ageing Well: Improving the quality of life for the ageing population using a technology enabled garment system'. This study has begun to address the current lack of understanding of the actual and potential role of mobile/wearable technological advances in everyday lives of older people, with developments that have often failed to address their aspirations. Clothing is a major contributor to how people define and perceive themselves and is a necessary part of everyday life. The ageing population represents the new consumer majority, embracing the post-Second World War 'baby boomer' generation that has been accustomed to making design choices in the selection of their clothing and accessories throughout their lives.

However, transient fashion ranges do not generally cater to the physiological demands of the changing older body, resulting in clothing that is often uncomfortable due to inappropriate fit, styling, proportion and weight. Clothing may be difficult to take off and put on, and fasten, and has compromises in quality and aftercare. The performance sportswear sector, as an early adopter of textile and technology innovation, is targeted primarily at athletes and the youth market. Current advances in smart clothes and wearable technology may be utilised to promote health and well-being, but assistive technologies and monitoring devices for 'ill people' are normally developed with little aesthetic appeal and with technology user interfaces often difficult for the wearer to read, understand and use. Wearable technology, for everyday use, is not readily available to older users as little has been done to address older wearers' changing physical and cognitive limitations, and style requirements, when developing products and services.

The 'Design for Ageing Well' project has been concerned with adapting the concept of the functional clothing 'layering system' to the design requirements of the active ageing (60–75 years) with a focus on promoting healthy exercise and, in particular, the activity of walking. It has begun to address the gap between the clothing

Textile-led Design for the Active Ageing Population. http://dx.doi.org/10.1016/B978-0-85709-538-1.00012-2

and electronics markets, which have traditionally been separate, with technological advances not readily accepted by older users due to unreliable integration of technology into clothing, and with styling and the design of the user interfaces, and marketing, primarily targeting the youth market. Little textile-based wearable technology product development has been carried out to address the everyday design requirements of older wearers in terms of human factors such as sizing, fit, predominant posture, thermal regulation, moisture management, protection and the psychological 'feel good factor'.

Emerging wearable electronics and advanced textiles are often confusing to traditional clothing and textile designers, while electronics and health care experts are not normally conversant with textile and clothing technology. This collaborative research has evolved a shared 'language' to enhance communication between these sectors and older wearers. A cross-disciplinary team of researchers has applied a co-design approach, driven by end-user engagement with industry stakeholders, in the development of a smart clothing 'layering system' as a wearable technology platform. A spirit of cross-disciplinarity has been fostered in a holistic approach between partners in clothing, technology and behaviour, thus establishing and testing new co-design methodology. This chapter begins to look at the development of shared language to enable the elicitation of the design needs, perceived wants and aspirations of the active ageing population to inform the co-design of a functional clothing layering system. The iterative co-design development process and subsequent practical realisation of working prototypes is explained in the following chapters.

12.2 Defining smart clothes and wearable technology

12.2.1 Some existing definitions

The definition of *smart*, in respect to clothing, is still open for discussion and debate. 'Smart' is a highly relative expression that needs to be defined in the context of current research. Various definitions of 'smart clothes' and 'smart textiles' are concerned, predominantly, with electronic (e-textiles) as part of the 'wearable computing' domain. 'The underlying technology of smart textiles is multifaceted and integrating. It is best perceived as a platform that aggregates technological breakthroughs coming from fields as diverse as textile engineering, nanotechnology, microsystems, polymers and displays, communicating engineering (including ad hoc protocols), multi-sensor data fusion, and many others' (Kotrotsios & Luprano, 2011). Ultimately 'the essence of wearable computing is to embed computers into anything that we normally use to cover or accessorise our bodies' (Poslad, 2009).

Van Langenhove originally defined 'smart textiles' with three subgroups: 'Passive Smart' that sense environmental stimuli; 'Active Smart' that sense and react to the environmental condition or stimuli, and 'Very Smart' that can sense, react and adapt their behaviour to the circumstances. Active smart textiles require a sensor and an actuator and, for very smart textiles, a controlling unit (Van Langenhove, 2007; Van Langenhove & Hertleer, 2004). Technological advances permit power and signal pathways to be integrated into garments and accessories to facilitate applications with the potential to provide functionality that ranges from electrically heated and illuminating textiles to wearable electronics that deal actively with life-threatening situations, comprising a unit with cognition, reasoning and activating capacities. Textile-based sensors may monitor vital signs, such as heart rate, temperature regulation, respiration, and activity including movement and location, as well as social and emotional contact. Subsequently, over time, monitoring can provide sufficient data to detect behavioural changes and alert the user in providing information and feedback.

Systex, the European 'Platform for Smart Textiles and Wearable Microsystems', under 'Definitions of Smart Textiles', proposes Euratex's definition of 'smart textiles', as a result of the work done by TEG6 (Thematic Expert Group 6 (Smart Textile and Clothing)). Euratex, described as 'the voice of the European textile and clothing industry', states that 'In our everyday work with smart textiles different definitions can be found' as follows:

12.2.1.1 Smart textiles

A passive textile structure capable of responding to external stimulation, i.e. from pressure, temperature, light, low voltage current etc.

12.2.1.2 Wearable technology

Any electronic device that is small enough and light enough to be worn or carried on the body.

12.2.1.3 Interactive textiles

Wearable technology that is woven/sewn into clothing and operated/controlled by an integrated control panel or operation button.

12.2.1.4 Intelligent textiles

The concept of intelligent textiles is most often used as a synonym for interactive textiles, phase change materials, illuminating textiles etc. that are covered by the above definition of smart textiles when used in clothing or interiors (http://www.systex.org/content/definition-smart-textiles).

12.2.2 Examples of smart textile applications

Van Langenhove reminds us that we are surrounded by textiles in our everyday lives that are readily available in many constructions, widths, weights and finishes and that we are familiar with their use and aftercare. Textiles provide a large flexible and comfortable contact area with the body by comparison with 'hard' wearable technologies. In contrast, for example, a watch as a 'closed box' with protected technology is not appropriate if the technology is to be placed on many areas of the body. In describing 'smart textiles at a glance', at the Medical Smart Textile Symposium in Aachen (2012), Van Langenhove presented an overview of smart textiles that can absorb, reflect, shield, measure or generate in response to temperature, heat flux, electromagnetic fields, humidity, chemicals, gases, radiation, electric properties, movement, forces, mechanical strength, odour, acoustic and biological applications (Van Langenhove, 2012).

Topics within her presentation, with direct relevance to outdoor clothing, included conductive textiles as sensors for vital signs monitoring, GPS sensors for location, conductive fibres for heating, phase change materials to aid temperature regulation, and textile assemblies with the potential for self-adapting insulation/ventilation.

Communication is enabled between components, through conductive fibres, with the wearer by means of keyboards and displays, and with the wider environment through antenna or induction. Smart electronic functionalities demand the optimised consumption and distribution of energy. In terms of the storage of energy for the powering of devices, Van Langenhove referenced the development of flexible chemical batteries and capacity-based fibre batteries and energy sources generated from heat, motion and light. The introduction of flexible and stretchable electronics (www.stella-project.de) now offers potential for enhanced clothing comfort and fit.

In respect to the ageing population, discussion of the topic of smart textiles and wearable electronics is normally centred on medical devices, where the 'patient' is based in the home, in a care environment or in hospital. Smart functionalities described by Van Langenhove, within a medical context, include textiles to monitor stress (EMG) and pressure sensors, with double layer fabrics or Quantum tunnelling effects, to monitor movement on a hospital bed, in relation to sleep patterns or for the avoidance of bed sores. Electrotherapy textile electrodes, with varying current intensities, may be used to stimulate the skin, stimulate muscle contraction or enforce the delivery of electrically charged drugs. Cooling functionality is more complex with garment design detail, including ventilation through closures, making a major contribution to temperature regulation. Shape memory textiles, using Nitinol alloy fibre, may alter garment drape as in the 'Oricalco' shirt by Grado Zero Espace, with sleeves that could be programmed to shorten immediately as the room temperature heats up, but the fibre expense is unrealistic for commercial clothing applications (www.gzespace.com). Active dyes can be used in textile displays in response to stimuli such as heat, light, humidity and odours and used, for example, in wound dressings or for fire fighters to alert the wearer to dangerous gases. LEDs are now commercialised in textile-based displays (Philips Lumalive), and in using blue light for pain relief (Philips, Blue Touch), and for enhancing visibility in motor bike jackets (iLightX – Safety, Interactive Wear). Examples of applications other than for health and safety include shape memory properties as in the example of the shirt by Grado Zero Espace (Figure 12.1).

12.2.3 Nonelectronic smart textile attributes to enhance clothing functionality

Euratex recognises that the definition of 'smart textiles', when applied to clothing, embraces innovative textiles with nonelectronic textile properties. The clothing layering system, with appropriately designed user interfaces, can benefit from the application of a variety of nonelectronic smart textiles strategically placed throughout the

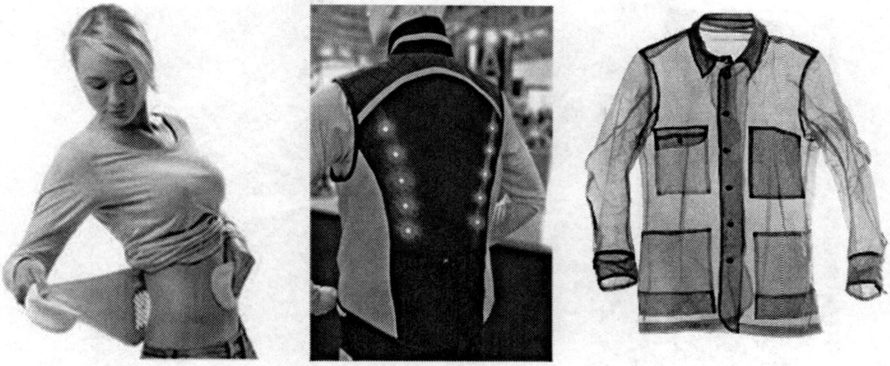

Figure 12.1 Examples of smart applications.

garment layers. The positioning (or body mapping) of smart textile properties demands expert knowledge in the selection of fibre and fabric constructions, to address the physiological demands of the human body, in highly specialised garment development.

Nonelectronic smart textile properties that may be selected for applications throughout the clothing 'layering system', and that have the potential to directly impact human comfort and the 'feel good factor', include: moisture management (wicking), thermal regulation (phase-change materials), waterproofness and breathability (hydrophilic and hydrophilic coatings and laminates), durable water repellency (DWR), visibility (reflectivity), stain repellency, (nanocoatings), antimicrobial properties (often silver encapsulation and finishes), improved blood circulation (far infrared rays, FIRs) sun blocking (ultraviolet protection), impact protection (shock absorbing materials, i.e. D3O (UK), three-dimensional spacer knit fabrics, Baltex (UK) and silicones combined with spacers or thermoplastics, Dow Corning). Additional examples of smart functionality include shape memory materials and thermo chromic dyes with chameleonic characteristics. Some of these properties, in relation to the needs of older people, are described in greater depth in 'Smart protective textiles for older people' (McCann, 2013) in *Smart Textiles for Protection.*

12.2.4 Integrating wearable electronics into the clothing system

In terms of bringing products to market, companies such as Philips are promoting medical devices, such as 'Blue Touch', to be sold in pharmacies, with an adjustable waistband that, in theory, will fit a range of body sizes, while vital signs monitoring devices have adjustable shoulder and chest straps. Interactive Wear is already recognised as contributing to bringing 'wearables' to market, for example, in motorbike clothing and corporate wear for hostile environments (interactive-wear.de). As in the case of Fibretronic's ConnectedWear, electronic components are developed ready to insert into garments and textile-based products. These electronic providers collaborate with garment producers to produce garment-ready components with their cabling, powering devices (where one size fits all) to be bonded, welded or snapped into garment shells during, or after, final assembly (fibretronic.com) (Figure 12.2).

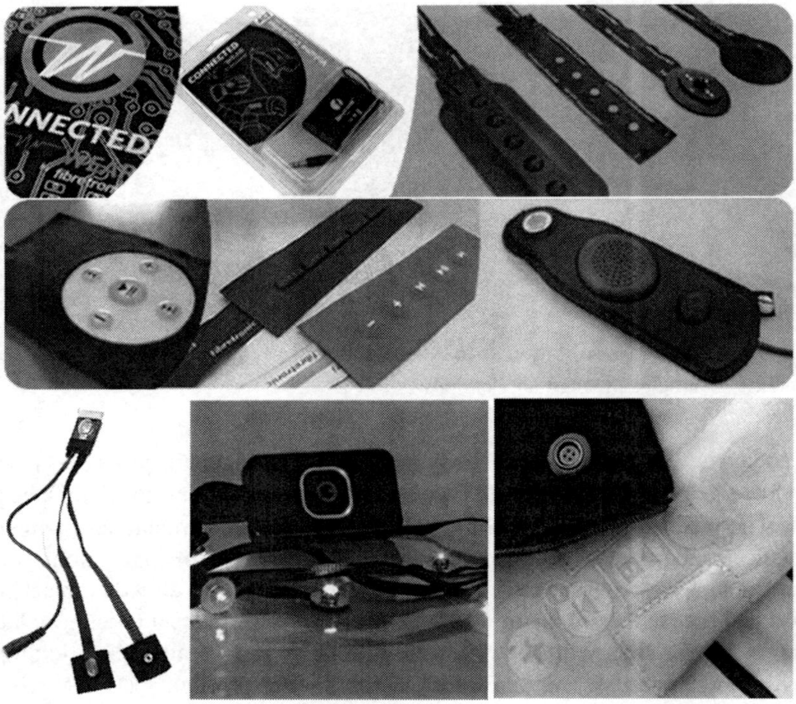

Figure 12.2 Textile-based electronic components.

The integration of wearable electronics into functional outdoor clothing is complex by comparison with fashion-oriented wearable electronics where the agenda may be predominantly aesthetic. Catwalk fashion has comparatively less demands in terms of abrasion resistance and exposure to hostile conditions as well as washing and aftercare. Garments for more extreme use that accommodate electronics for functionalities such as vital signs sensing, positioning and warming require design-led garment engineering to address requirements such as the size, shape and fit, and predominant posture and movement of the wearer for the optimal positioning of the wearable technologies. The skill of cutting for extreme movement, in conjunction with novel garment joining techniques, currently adopted for performance sportswear fabrics, is quite different from mainstream fashion garment engineering (McCann, 2012).

The complexity of the technical textile and functional clothing domains is seldom understood within the electronics community. Growing emphasis on the multifunctionality of the smartphone has tended to increase the divide between clothing and technology sectors. Technologists often concentrate their attention on phone applications, developed in collaboration with private health providers or social networks, with little reference to clothing. This was recently observed at both a trade-oriented event (WT Conference, ISPO, 2012) and at an academic conference (ISWC, 2012). Technology providers promote the smartphone as being universally beneficial to young and old, with scant reference to the integration of wearable technologies in clothing.

The Design for Ageing Well cross-disciplinary, collaborative research has been focused on the design-led application of functional textiles in a 'clothing layering system' for the active ageing, to promote healthy exercise, for those who do not perceive themselves as ill. In the context of this research, the smartphone acts as the computing hub to be worn within the clothing system. The research study, promoting engagement in healthy exercise in a range of weather conditions and environments, acknowledges that users require suitable clothing to encourage them to walk, which now may have the opportunity for enhanced functionality provided by appropriate wearable electronics. Going out in poor conditions, with a smartphone to be taken out of one's pocket, and reading glasses from another, while wearing gloves, is impractical! The technology user interface must be easy to use in hostile environments. As a result of co-design engagement, where soft textile-based exterior controls may be required to operate the phone that is located within the comfortable clothing system, Fibretronic recognises that, for older users, the spacing of their soft controls must be modified to address reduced dexterity.

12.2.5 Definition in the context of Design for Ageing Well research

The aim of the Design for Ageing Well research has been the design engineering of a comfortable clothing 'layering system' suitable for the active ageing body (age range 60–75 years), and adaptable for a range of environments, which constitutes a wearable platform for smart textile-based technologies. A generic clothing layering system that is attractive and fit-for-purpose may provide a basis for rapidly changing wearable electronics. In this context, a working definition of smart clothes and wearable technology is proposed that embraces both the value of nonelectronic smart textile properties that contribute to human comfort, and user acceptability as well as the application of wearable electronics. A proposed working definition, in this context, is as follows (Table 12.1).

12.3 An introduction to the clothing layering system

The comfort attributes of textiles normally developed for the performance sport 'layering system' may be adapted for the benefit of older wearers, in clothing for their everyday lives (Figure 12.3). Understanding the physiological changes of the older body is key to achieving the desired garment cut and fit in order to maintain the comfort of the microclimate within the clothing system. The placement, or 'body mapping', of technical and smart textiles may enhance moisture management, thermal regulation, support, and protection. Aesthetic concerns such as colour selection, fabric handle and overall style will all contribute to the wearer's sense of well-being. Pattern block generation and technical design development is needed with an emphasis on 'cutting in the round' to address predominant postures. The particular demands of the end use will impact on decisions to do with materials sourcing and selection. User-led design process is key to understanding and prioritising issues to be addressed with regard to the nature of the activity, or sequence of events, as well as the range of likely conditions.

Table 12.1 Smart textile terms and definitions

Smart clothes and wearable technology: apparel that constitutes a 'clothing layering system' incorporating differing levels of 'smart textile' and/or 'wearable technology' with electronic functionality

Functionality may embrace

- Nonelectronic *smart textile* and clothing properties including fibres, fabrics, textile and garment assembly, finishing and whole garment engineering
- Interactive 'wearable technology' including electronic functionality that is sewn, bonded, knitted or woven into clothing and operated/controlled by an integrated control panel or operation button (descriptor based on Euratex definition, 2009)

Levels of smart textile functionality

- Passive

Textile structure that is capable of responding to external stimulation (i.e. from pressure, temperature, light, low-voltage current etc.) (Van Langenhove, 2007; Euratex, 2009)

- Active

Textile structure that can sense and react to the environmental condition or stimuli (requires a sensor and actuator) (Van Langenhove, 2007)

- Very smart

Textile structure that can sense, react and adapt their behaviour to the circumstances (requires a sensor, an actuator and a controlling unit) (Van Langenhove, 2007)

Wearable technology

Any electronic device that is small enough and light enough to be worn or carried on the body (as defined by Euratex, 2009)

A wearable textile system is basically comprised of the following components

- Sensors to detect body or environmental parameters
- A data processing unit to collect and process the obtained data
- An actuator that can give a signal to the wearer
- An energy supply that enables working of the entire system
- Interconnections that connect the different components
- A communication device that establishes a wireless communication link with a nearby base station

The challenge for the textile community is to manufacture as many components as possible out of textile material (http://www.ugent.be/ea/textiles/en/research/research-themes/smart-textiles/smart-textiles.htm)

Smart clothing layering system

• Body/skin layer	Vital signs monitoring
• Base layer(s)	Moisture management
• Mid layer(s)	Insulation: the trapping of still air
• Outer layer(s)	Protection of the system

(McCann, Morsky and Dong, 2009; McCann, 2012)

Figure 12.3 Examples of the clothing layering system. (From left to right: Santoni seamless knit base layer; fleece mid insulation layer; outer protective shell.)

Is the activity seasonal, carried out indoors or outdoors, in a known and predictable environment or in unknown conditions (McCann et al., 2009)?

The sports 'layering system', originally a response to military preparedness, has been driven by textile innovation in increasingly light-weight, protective, compact, and easy-care fabrics. The system typically comprises a moisture management base layer, mid insulation layer(s), and a protective outer layer. In the case of the ageing wearer, elements of personal protection (i.e. hip protectors) may be incorporated. Textile innovation now embraces the concept of the 'soft shell' as a hybrid mix of the protective attributes of the outer shell garments with the comfort and insulation of mid layers. The function of the system relies on coordinated cutting, in terms of overall silhouette, and sleeve pitch and articulation, with suitable ease to enable movement and to avoid friction, throughout the layers. Mainstream 'fashion' pattern cutting is normally adapted from pattern blocks that cater for erect figure types, driven by the catwalk, with a limited degree of articulation. Clothing for extreme sport is becoming increasingly tailored to the body due to the enhanced breathability of fabrics, incorporating articulation for arm lift as well as elbow and knee joints. This type of cutting is seldom used for everyday wear where, in particular, it could enable greater mobility in clothing to address predominant postures of the ageing body. With appropriate design, the functionality of a multipurpose 'clothing layering system' has the potential to benefit the well-being of older people (McCann, 2013).

Smart textiles and wearable technologies have emerged predominantly in the areas of performance sport, corporate work wear, and in medical applications with the potential to enhance health and wellness. To date many of the smart textile developments adopted for older people have been focused on medical end use with devices that have little concern for design aesthetics and for applications in relatively static postures in health and hospital environments. Design-led garments, for sports and fitness, have been developed predominantly for young athletic figure types with little consideration of less predictable older figure types, with relatively restricted movement and agility

(McCann, 2009). Catering for the size and shape of the older body is critical to the placement of smart fabrics and emerging wearable electronics, with textile-based sensors, for the recording of vital signs, movement and positioning. Recent developments have included base layer garments with embedded vital signs monitoring (Adidas, miCoach), mid layer heated garments and outer layer garments with soft switches, flexible displays, illumination, communication devices and safety alerts (Fibretronic, ConnectedWear). With the introduction of wearable technologies, including vital signs monitoring, the 'layering system' requires a close fitting 'skin' or 'body' layer for men and suitable intimate apparel for women. The design, positioning, and the usability of the technology user interface are key to user acceptability.

12.3.1 Skin/body intimate apparel and base layer

Close-to-body intimate apparel and base layers are normally of knitted construction with varied structures placed around the body to aid moisture wicking and offer increased support and protection. Key to maintaining the comfort of the clothing microclimate is effective moisture management and thermal regulation, enabled through both the selection and application of special fibres and fabric constructions, and through the provision of ventilation within the garment(s). Clothing can impede evaporative heat transfer by disallowing the movement of water vapour across the various layers, creating a microenvironment next to the skin surface. The surface area coverage (and permeability) can determine the extent and nature of this microenvironment. As perspiration increases there is a reduction in the capacity of evaporation to remove heat energy; this means that most of the sweat from the body becomes trapped within the fibres and textiles. Clothes saturated by sweat can affect the thermal characteristics and influence further heat transfer rates. Older people may be at risk in relation to survival, in extreme cold, if a high moisture level is absorbed within the garment base layer and the temperature drops. Textiles, in applications predominantly for performance sport, may be selected to reduce the effects of sweating through 'wicking away' moisture. There are also obvious advantages in the application of high moisture wicking – quick-drying textiles for the onset of incontinence (McCann, 2009, 2012).

The attributes of stretch fibres have revolutionised the technical and aesthetic design of clothing affecting comfort factors such as fit, support, ease of movement as well as shape retention. In 'underwear' and intimate apparel, stretch may be incorporated, in varying percentages, in warp and weft knitted structures, laces and nets, woven constructions, and narrow fabrics. Elastomeric and mechanical stretch properties are of particular relevance to garment design for older figure types. A varied percentage and direction of stretch provides comfort and fit, through engineered areas of support, or enhanced movement to enable the wearer to put on and take off the garment with greater ease. Stretch also contributes directly to garment fit and to the embedding of textile sensors and wearable electronics (which should be held in appropriate locations within garments, currently found in close-fitting base layer garments and bras) for vital sign monitoring in sport and medical applications. Zip manufacturer YKK has recently launched their stretch zip, with two types of tape available, vertical stretch and horizontal stretch (ykkfastening.com), which further contributes to comfort and ease of movement.

Variable stretch engineered knitted textiles may offer targeted support in maintaining muscle alignment and a reduction in the loss of energy in athletic performance. Compression garments assist blood circulation and are commonly recommended for athletes, surgical patients, oedema patients, and those with a sedentary lifestyle. Body scanning may directly inform the design and fit of customised compression hosiery, as demonstrated in the Bespoke Scan2Knit project, where medical garments were developed for people with severe ulcers. The research of three dimensionally shaped and seamless fibre assemblies for technical textiles and apparel applications was led by Dr Tilak Dias (formerly of the William Lee Centre, Manchester). The aim was to improve the effectiveness of compression bands that apply pressure, which forces lymphatic fluid toward a functioning lymph node. The specialized Shima Seiki computer-assisted manufacturing process involved scanning the leg and foot with a 3D limb scanner to construct a 3D model from the scans, with the information then transmitted to a dedicated computerized flat-bed knitting machine. A bespoke stocking was accurately engineered to apply a prescribed 3D pressure profile to the leg, with the process subsequently adapted for the manufacture of compression sleeves for the treatment of lymph oedema. This work has since been developed further (Moghaddassian, 2010).

12.3.2 Mid layer

The mid insulation layer may be varied in thickness, or bulk, in its ability to trap 'still air' for heat retention. It may constitute more than one garment with examples such as jackets, smocks, gilets, traditional knitwear, and felted garments. Typically, the insulation is made up of lightweight man-made down, sliver knit constructions, known as fibre pile or fake fur, and increasingly sophisticated developments in knit structure fleece fabrics, primarily in weft knit, and predominantly of polyester fibre. Fleece pile garments may have 'body mapped' engineered knit in varying patterning, depths of pile and finishes to provide variable protection around the body (McCann, 2013; McCann et al., 2009). Older people are susceptible to cold and may benefit from adaptations of existing and emerging technologies initially adopted in performance sportswear, such as heated panels (Fibretronic.com). Mid layer garment design should be compatible, in terms of cut and styling, with that of the base and outer layer garments, to avoid impeding movement and friction between the layers.

12.3.3 Soft shell

The more recent hybrid 'soft shell' layer incorporates many of the protective elements of outer shell garments combined with the comfort and insulation of mid layers. Soft-shell fabric technology is normally a laminated assembly of two fabrics with a membrane in the middle. Typically the layers are composed of a combination of polyester or nylon woven face fabrics backed by lightweight, to high loft, knit constructions with an inner membrane intended to allow airflow through the fabric but prevent water from penetrating from the outside. These garments are designed to be water repellent rather than waterproof and do not require seam taping. These fabrics are softer and less noisy than more protective 'hard shell' assemblies and often have a degree of stretch to enhance comfort and fit. As extreme protection is not required on a daily basis, this category of garment, if designed to suit older people's style requirements and figure types, may offer an attractive solution to adaptable and comfortable lightweight protection for everyday use (McCann, 2013).

12.3.4 Outer layer

The outer layer provides protection for the clothing microclimate from the ambient conditions. A range of variations on woven or knit structure protective textile assemblies have become increasingly lightweight, contributing to more stylish garments with an increasingly tailored fit, with comfort enhanced through the moisture management properties of hydrophilic or microporous waterproof 'breathable' membranes, with their stretch properties. The concept of 'body mapping' identifies areas or 'zones' for the placement of textiles in relation to comfort factors such as ease of movement and articulation, moisture management, thermal regulation, abrasion resistance, impact protection, environmental protection, and enhanced visibility. Novel joining methods, such as the heat bonding of seams, zip insertions and laminated panels, have revolutionised appearance, cut down on bulk, and contributed to the encapsulation of wearable electronics, as safety and communication devices, with their textile-based switches and controls. Such garments have seldom been designed with consideration of the changing size, shape and style requirements of older wearers (McCann, 2009; McCann, 2013; McCann et al., 2009) (Figure 12.4).

Figure 12.4 Examples of wearable electronics that may be put into the layering system.

12.4 The identification of user needs: design fit for purpose

Emerging smart textiles and wearable technologies, integrated within the clothing system, have the potential to enhance clothing comfort and functionality and promote a sense of well-being. However, to be acceptable, as a vehicle for self-expression, clothing should look stylish and attractive and function reliably in relation to the technical, aesthetic concerns of the wearer as well as from social, cultural and health perspectives. The aim of the Design for Ageing Well project has been to identify and address the design requirements of active ageing, in relation to healthy exercise (walking), through the application of smart technical textiles and wearable electronics in a 'layering system' that is comfortable, attractive and fit for purpose. This did not involve the development of new technologies but rather design solutions in the application of existing materials and electronics to be wearable and easily usable for the growing active ageing community.

A co-design approach has used designers as facilitators, translators, and visualizers to aid the elicitation of older user needs and aspirations in balancing technical, aesthetic and cultural concerns for the development of working prototypes. A hierarchy of design requirements, shown below, developed by the author in previous work in the context of performance sportswear design, provides an overview of generic topics for consideration (McCann, 1999). The author proposes that, to be worn willingly, for a chosen end use, clothing must both function and look good from the perspective of the target audience. Her extensive tree of design requirements, to guide product development of functional clothing, served as a useful starting point. The author now revisits this hierarchy of requirements beginning with 'function' (Figure 12.5).

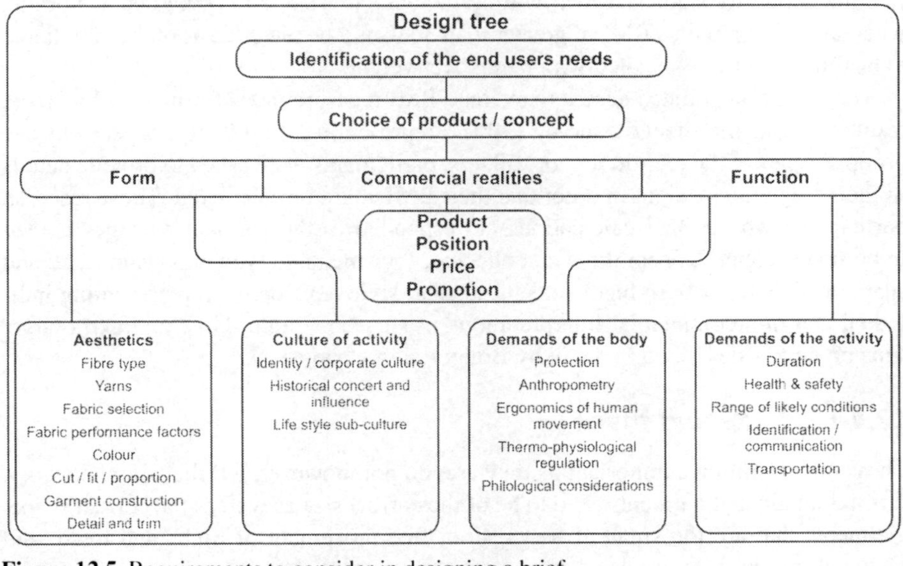

Figure 12.5 Requirements to consider in designing a brief.

Creative design solutions, in product fit for purpose, for an identified category of end user, require the following topics to be considered (McCann 2000, 2009).

12.4.1 Function: the demands of the ageing body

As the human body ages, changes in body shape take place. Increasing body fat, reduction in muscle volume, posture changes, and even height follow a largely predictable progression. These changes, and other age-related phenomena not particularly related to the Design for Ageing Well research, are considered in many publications, for example, in Smith and Gove (Physical Changes of Ageing), Hughes et al. (Longitudinal changes in body composition in older men and women) and Shepherd, 1997 (Ageing, physical activity, and health). It is therefore clear that if well-fitting and comfortable garments are required, a different garment cut is required, taking into account the different body shapes (and movement requirements) of the 60- to 75-year-old target group. The project team, which embraced an expert in size and shape who had made a key contribution to the SizeUK survey, adopted the body scanning of research participants, selected to represent the average body mass index (BMI) classifications for men and women in the 60–75 age range, to inform the development of garment pattern blocks (www.size.org).

12.4.1.1 Body mass index to represent 60–75 year age group

Body mass index was originally developed in the mid to late 1800s, and BMI is usually used in clinical studies of obesity. The function is defined as:

> BMI of a particular person = body mass in kilogrammes divided by the square of the person's height in metres.

In clinical obesity studies, a BMI within the range of 18.5–25 is regarded as normal, whereas a person with a BMI of greater than 30 would be regarded as obese, and liable to health problems associated with being overweight.

The project negotiated access to average BMI measurements for men and women, resulting from the SizeUK survey (2002), representing the 60- to 75-year-old age group. As part of the recruitment of walkers, participants were asked to provide details of their height and weight in order that their BMI could be calculated. Those research participants whose BMI calculations corresponded to the national averages agreed to be body scanned. From the data collected, four men and women became size and shape models for pattern block and subsequent style development, representing individual and non-commercial interpretations of small, medium, large and extra large. This process is described in detail by Bougourd in Chapter 10.

12.4.1.2 Size and shape

These considerations, important as they are, do not, however, tell the complete story. It is axiomatic that garments need to be of the correct size as well as the correct shape. Garments that are too small at best restrict free movement of limbs and torso, and may well not be able to be put on at all. Garments that are too large can also restrict comfort, movement being inhibited by limbs being caught in loose folds of fabric. Ill-fitting garments of the wrong size are also often rejected on the grounds that the wearer is made to look foolish. The phenomenon is well known in the clothing industry, and is catered for by manufacturers producing the same garment in a range of sizes, all, however, of the same shape.

For body size, however, the same pattern of relatively predictable change in body shape with age is by no means the case. Size can vary over both the short term and medium- to long-term periods. There can be many causes, including illness (where increase or decrease in size can occur), voluntary calorie intake reduction in the form of dieting, or traumatic life experiences such as bereavement. These size variations can be accommodated, of course, by buying new clothes of the correct size, but with a complex garment system such as that being produced by the DfAW project, replacement becomes a less attractive prospect. Short-term weight change of one of the co-design user groups even affected the SCWT programme, resulting in body layer garment redesign.

In a recent survey into what women really want from fashion, Rebecca Willis reported that, above all, they would like clothes to fit. A questionnaire was sent to more than 40 women ages 18 to 84, who represented different shapes, sizes and nationalities, living in a variety of places across the Westernized world and with a wide range of incomes. The aim was to hear opinions from those outside the fashion 'bubble' (Willis, 2013). Both male and female consumers are confused by the lack of a universal sizing system, with 50 percent returns reported by some mail order companies (Bougourd). Willis refers to the concept of 'vanity sizing', calculated to flatter you into buying because you can fit into a smaller-than-usual size. She quotes a 49-year-old respondent who stated, 'I used to be a size 12 and now I'm a size 10, but I haven't got any thinner'. Women want clothes to flatter their bodies and not their minds, with the resulting message clear that female consumers want the following: more petite and tall sizes, half sizes in clothes and shoes, legs and sleeves in several lengths and, ideally, facilities for garment alteration in-store. The market needs 'a middle way between couture and off-the-peg, a bit of pin and tuck' with many willing to pay to have their clothes altered to fit properly (Willis, 2013).

The nearest existing clothing types to those required by the project are found in performance sportswear ranges. However, these are predominantly based on the body size and shape of a younger target group, quite reasonably seen as the principal customer by sports clothing manufacturers. As intimated, both intuitively and by individual personal experience, well-fitting and comfortable garments are more likely to be used than others, and so the design of the garments in the case study had to reflect this need. As discussed, older body shapes are different from those of younger people, and so age-appropriate shape and size was required for the project garment layering system, quite apart from any demands made by electronic system incorporation.

12.4.1.3 Full body scanning

Laser scanning captures the complete size/shape data for an individual subject and the data resulting from such a scan can have many uses, including biomedical studies, ergonomics and obesity study and control. One attraction of body scanning is that with modern equipment, the data produced can be made to drive CAD/CAM garment cutters, warp knitting equipment or advanced weft knitting machines. Thus, combined with the designer's style interpretations of user requirements, accurately fitting garments can be made.

A body scanner was made available to the Design for Ageing Well project, and selected members of the user group, based at the University of Wales, Newport, were scanned using this equipment to capture their individual size and shape. The representative group was selected to be size and shape models for the project as described by Bougourd in Chapter 10. User engagement was key to the development of individual basic blocks developed from the body scans that, in turn, were adapted to articulated blocks that were cut for enhanced movement. These were further developed into style patterns to accommodate the major design requirements based on the co-design workshop findings. It should be noted, however, that these attempts to be rigorous in defining body size to ensure accurate and comfortable garments were adversely affected by marked changes in older user body size during the project (see Chapter 10).

12.4.2 Demands of walking as an activity

12.4.2.1 Why do older people walk?

Apart from the trivial answer that walking is a natural human function, there is a range of answers to this question. Within the context of the New Dynamics of Ageing research project, there has been emphasis on the health and wellness of the ageing population with a key focus on *autonomy* and *independence*. In Design for Ageing Well, walking was chosen as an example of healthy exercise, being considered as a less vigorous 'sport' that might be acceptable to a large part of the ageing population. Sometimes walking is associated with pleasure – enjoyment of fresh air, admiring the scenery and, for walkers in groups, companionship. In other cases, walking is a necessity associated with shopping, pet walking, taking children to school, or as a preliminary stage to a longer journey by, for example, bus or train (Table 12.2).

Table 12.2 **Reasons for walking**

Reasons for walking	Percentage
Going shopping	21
Education (including school escort)	20
Leisure/social	20
'Just to walk' (including dog walking)	17

Source: Ramblers (previously the Ramblers Association).

12.4.2.2 Health and wellness

One important reason for walking in later life, however, is often not appreciated, and that is that there are health benefits to walking. In 2006, the National Health Service (NHS) in Scotland defined five reasons why older people could benefit from walking, ranging from reducing liability to osteoporosis to reduction in certain cancer risks. Also reported was that those regularly walking found improvements in some seven factors, including sleeping better and reduced stress levels. In a study in 2008 reported in the Journal of Geriatric Physical Therapy, Moore-Harrison and Cress found in a control-based experiment that older low-income walkers (60+) increased their physical function by 25%, and their aerobic capacity by 19%. (Aerobic capacity is defined as the maximum amount of oxygen the body can use during a specified period, usually during intense exercise.) The health benefits of walking are also recognised by the UK 'Walking for Health' (WFH) initiative, sponsored by the Ramblers Association (now known simply as 'ramblers') and Macmillan Cancer Support. Such is the importance attached by WFH of walking as exercise that the WFH website has an informative short video showing the benefits of walking for older people.

12.4.2.3 Social engagement

As a social activity, walking is popular with all age groups, but there may be special considerations for older walkers. Many older people maintain excellent health and possess active minds, but others suffer from factors that limit their walking abilities. In these latter cases, the companionship afforded by walking with others can lead to increased confidence that if the worst happens, there is someone close at hand to deal with the situation. Even for those with fewer limitations, the companionship of the walking group may help to offset losses in other life aspects, for example, after bereavement. Although not easy to quantify, there are strong indications of the popularity of walking as a social activity. In some cases individuals form an informal group that can be based on such factors as belonging to a village, an urban district or even a workplace. These informal groups form and reform, for varying lengths of time. More formal walking groups also exist, sometimes with formal constitutions and regular memberships, some of which are affiliated with the UK national association 'ramblers', mentioned above. What can be quantified, however, is the membership of this organisation: it has a membership of some 500 formally constituted walking groups throughout the UK, with special emphasis on the social aspects of walking. The social aspects of walking are preeminent in walking holidays, a principal example of which in the UK is HF Holidays. With many centres in the UK and other parts of the world, this organisation arranges specialist and more general walking holidays, with a high social content as well as a walking one (Figure 12.6).

12.4.3 Aesthetic design requirements of the clothing system

Active ageing represents a diverse but neglected community that is little understood by designers and not easily segmented by marketers. The image of ageing is often dreary, with many unattractive clothing catalogues and virtually no upbeat design trend

Figure 12.6 Active ageing target market.

information for this rapidly growing sector. Individuals communicate and express themselves through clothing, and often older people have developed strong personal statements in the way they dress that are not conditioned by transient fashion. The design of smart textile-based products, that are attractive and pleasurable to wear, demands an appropriate blend of form and function, embracing aspects of size, shape, proportion and fit, balanced with appropriate aesthetic style considerations. Colour should be flattering to older complexions and hair colour.

Again, with reference to the recent findings from questionnaires, reported by Willis, not all women as they get older relish the prospect of dressing in baggy cardigans, comfort waistbands and fluffy slippers even though comfort does become more of a consideration. Willis expresses surprise that, despite all that we hear about the West's ageing baby boomer population bulge, how poorly served older women still feel by the fashion industry. Willis quotes an 84-year-old former architect in writing that 'Fashion is pleasant until the age of 60' and 'after that we are forgotten' while a 78-year-old retired solicitor states that the industry 'should accept that women over 60 don't all want to dress in black bombazine' but that 'whatever the effects and restrictions that come with it, age doesn't necessarily wither our pride in our appearance'. Perhaps when today's young designers get their bus passes, they will at last address this (Willis, 2013).

From a functional clothing design perspective, design practitioners within the case study, knowledgeable in technical design considerations, introduced certain given generic recommendations. Design requirements related to garment performance include the avoidance of seam lines over the shoulder that might allow ingress of moisture or that would contribute to friction and abrasion when wearing a rucksack. Cutting for different postures, associated with performance sportswear, demands a degree of arm lift and elbow and knee articulation, to provide enhanced comfort and ease movement in donning and doffing clothing layers of particular benefit to older wearers with restricted agility. In combination with the overall clothing comfort and appearance, a key factor for consideration in relation to wearable technologies, is the design and usability of the technology user interface. Designers and technologists must therefore engage in co-design with older participants, their peers and family (and, in some situations, carers) to capture and understand real user needs and limitations.

12.4.4 Cultural design concerns

Some attitudes remain largely unchanged from youth – hence the common feeling that what is odd about growing old is that you do not feel any different (Metz & Underwood, 2005). However, from a design point of view, the active ageing have a strong sense of self-actualisation with opinions that should be captured. Designers must respect the culture of the end user in that what is fashionable to one target age group will not be attractive to another. Even if the garment cut, fit and fabrication is appropriate, the garment detail, colour and branding may be unacceptable for a different end user and their peer group. In addition there may be design issues to address the corporate image of a team or club.

Smart textiles, in appropriately designed applications, have the potential to provide a range of protective attributes in clothing applications that may enhance the autonomy, independence and well-being of older people in their everyday lives. To date many of the developments in textile-based wearable devices have focused on medical end use with little concern for design aesthetics. In order to encourage the uptake of clothing with emerging technologies, with a view to improving the quality of life of older people, the design and functionality of wearable technology should be attractive and fit for purpose as well as easily understood, easy to use and easy to service. Best Ager findings inform us that the motivation for older people engaging in healthy exercise will be increased if, at the same time, the activity provides joint or individual pleasurable experiences, an enjoyment of nature, a sense of adventure, education and entertainment, as well as increased fitness and the prevention of ill health rather than enforced rehabilitation (Best Ager Gallery, ISPO, 2010).

12.4.5 Commercial realities

The active ageing, from diverse backgrounds, with their lifetime of experiences, have developed strong opinions that are not easy for marketers to categorise and address.

Findings from Best Ager market research carried out in Germany, led by Claudia Boesl of GenerationSport, confirm that the active ageing community is frustrated with what is on offer in the retail environment. This market information has been elaborated over recent seasons in advising the outdoor market of the potential size and value of the ageing market. Findings indicate that many older users engage in active outdoor pursuits but are often unable to find suitable clothing products, in terms of style, shape and fit in the retail market. They were also unhappy with the nature of the retail environment, oriented to the youth market, where it is difficult to try garments on and the service provided is not in tune with older user needs. May older potential consumers are looking for quality and appropriate design. GenerationSport now promotes health

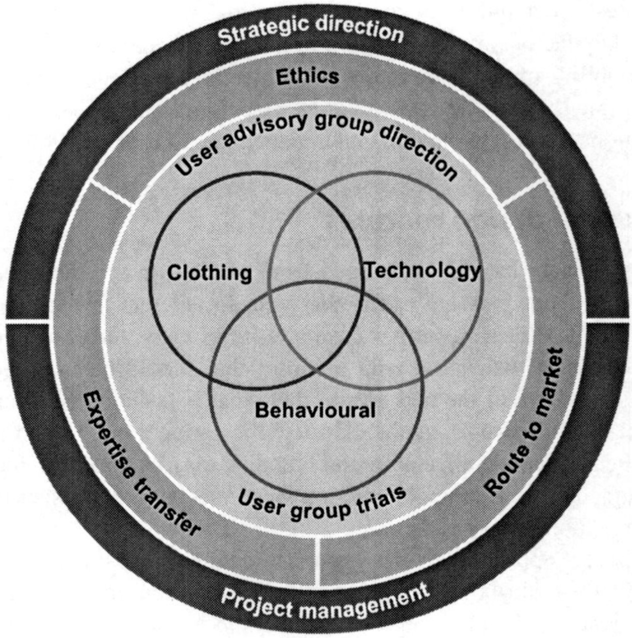

Figure 12.7 Design for Ageing Well collaborative research work packages.

and wellness with continuing relevance to sport and the adoption of assistive wearable technologies (www.generation-sport) (Figure 12.7).

12.5 Co-design approach to smart clothing development

12.5.1 The co-design network

Co-design methodology (new to clothing design), is defined in the context of collaborative research case study: Design for Ageing Well. This design-led collaborative research began with embracing a representative network of cross-disciplinary academic competencies from across the partner universities. Specialist Work Packages (WPs) were established in three areas, which were interdisciplinary in their own right. WP1, *Behaviour*, linked the psychology of behaviour (University of Westminster) with sociogerontology (University of Salford), WP2, *Clothing*, led design and technical development in the application of technical and smart textiles in functional clothing (University of Wales, Newport – now merged into the new University of South Wales), while WP3, *Technology*, comprised the team that addressed wearable electronics (University of Ulster). The teams were led by expert co-investigators with a research associate and a PhD student contributing to each WP and forging links across the disparate disciplines. In addition, the stakeholder network comprised selected research participants (active agers) and other key stakeholders (e.g. industry

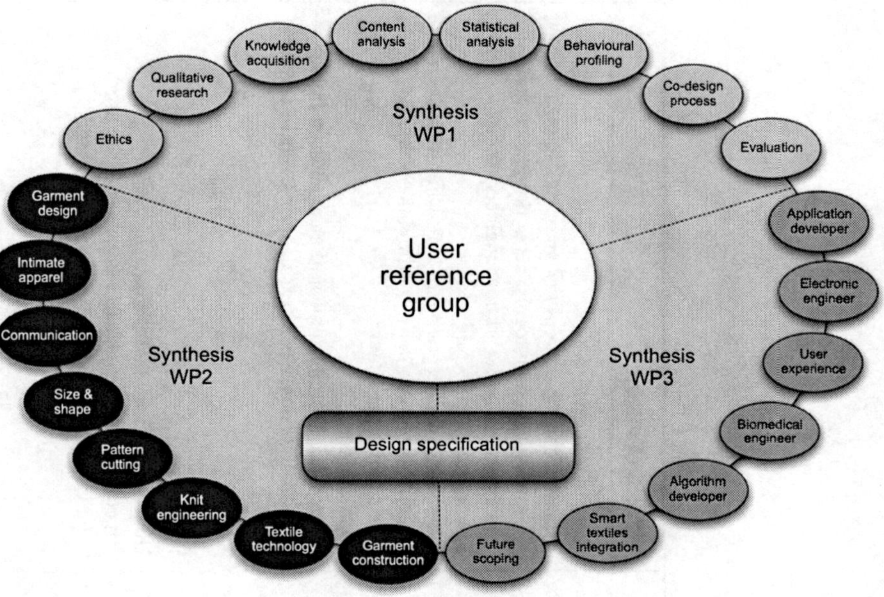

Figure 12.8 Roles of researchers representing skills and expertise needed for the development of wearable technology. (See greater detail in Table 12.3.)

representing the fibre, textile, and clothing product development chain and the emerging wearable electronics sector). The collaborative approach gave equal importance to the contribution of all stakeholders in helping to provide an initial understanding of complex terminologies and practises in breaking down barriers between the disparate disciplines (Figure 12.8 and Table 12.3).

12.5.2 User selection through participation in walking

The process began with the recruitment of the User Reference Group, older research participants within a 60–75 year age range, from within a 30-mile radius of the lead university, and with an interest in walking. Initial recruitment of the research user group involved a range of approaches by telephone, email and word of mouth to local walking groups, such as the University of the Third Age (U3A) Walkers, the 'ramblers', to other local community groups, as well contacts with potential individual recruits. Approaches were made to those within a 30-mile radius of the university to maintain a realistic level of time commitment and costs involved in travel. Word of mouth enlarged the level of response.

Despite a research focus on walking, the research team was conscious of attracting those who were less committed walkers in order to consider any barriers to healthy exercise. The age range was partly defined by the remit of the new Dynamics of Ageing programme that distinguishes research topics between the *Active Ageing*, associated with autonomy and independence, by comparison with the *Oldest Old*, where agers are dependent on care. As this case study involved capturing the size and shape

Table 12.3 **Roles of work package researchers**

WP1: Behaviour	WP2: Clothing	WP3: Technology
Ethics	*Design* Fashion: Performance sportswear design: theory and practice Fibre/fabrics/colour: sourcing and selection	Computer Science Experience in the systems design of ICT-based systems, communication protocols, networking, data flow management and workflow engineering
Qualitative research	Intimate apparel: Body/skin layer/bra: design engineering for the placement of textile-based electrodes	Biomedical Engineering Expertise in the integration of hardware, software and textile-based components within the wearable computing paradigm
User engagement	Communication: Graphics/web Product Architecture Design interpretation and presentation: initial concepts to detailed working drawings and promotional material	Sports Science Governance in establishment of an evaluation framework. Experience is assessing user levels of activities
Knowledge acquisition	*Clothing technology* Body size and shape: capturing ageing size and shape through body scanning and block development and testing	Human/Computer Interaction Expertise in the design and evaluation of the user interaction model and user interface design

Content analysis	Pattern cutting: Block adaptation and cutting for movement/ articulation throughout design interpretation of garment layers	Intelligent Data Analysis Experience in the processing of the underlying raw signals recorded from activity data and vital signs
Statistical analysis	Knitted garment engineering: Expertise in the size and shape of larger figure types in relation to whole garment construction	Wearable Sensors Assessment of appropriate technologies to assess activity and vital signs
Behavioural profiling	Garment construction: Sequence of manufacture with knowledge of novel garment joining methods and the embedding of technologies (team)	Evaluation Framework Design and construction of appropriate structured and semistructured approaches to assess user feedback
Evaluation User Advisory Group (UAG)	*Textile technology:* Lab and field textile testing: Knowledge of textile fibres and materials assemblies that constitute the garment layering system	Technology Watch On-going enquiry about smart innovations in textile-based electronics and clothing concerning materials and components, their applications, interfaces, powering etc.
Synthesis Design specification: Working drawings of garment layers in accurate scale/proportion to link to digitized patterns, with placement of technology and exploded garment detail to communicate integration of technology.		

of the participants, recruitment was aimed at men and women in the 60- to 75-year-old age range, to correspond to the segmentation of recent SizeUK data into 5-year periods. Participant recruits completed a brief questionnaire to establish their relationship to the SizeUK average, with a view to selecting certain representative participants to be scanned as potential size models for garment development.

12.5.3 Developing a shared language

Design for Ageing Well demanded the initial sharing of the different terminology and methodologies across the disparate study backgrounds and cultural practices of the academic stakeholders. Preparatory workshops were planned for the development of a shared cross-disciplinary language. The predominantly male Technology team, based in a smart home lab environment, had little knowledge of user engagement, or of clothing and textile design, while the Clothing and Behaviour teams had little understanding of the concept of the smart home and wearable technologies. The Technology experts introduced a 'Technology Watch' in plain language to other work package researchers (December 2009). Likewise the Behaviour team introduced the concept of public involvement in research and best practice in user engagement (Spring 2010).

In order that the concept of smart clothing, at the core of the research, might be understood by all stakeholders (academic partners, industry collaborators, independent experts and research participants), a 2-day preparatory workshop took place (May 2010) with presentations to develop shared language around the project stakeholders. The recruited research participants, forming the User Reference Group (URG), became involved in the preparatory workshop. Specialist speakers introduced their topics in plain language in order to give the older research participants, and emerging researchers, the initial understanding and confidence to engage in the co-design process (Figure 12.9). Audio/visual presentations and the demonstration of existing examples of products and materials clarified issues and demonstrated possibilities. 'The more that the event locale and schedule can be configured to maximize an equal input from different voices, the better the outputs will be' (Fuad-Luke, 2009, p. 180).

Presentations from experts, within academia and the international sportswear industry, provided a comprehensive overview of technical textiles and novel garment manufacturing techniques with relevance to the sports-type functional clothing 'layering system'. The author introduced her 'Requirements to consider in designing a brief' with relevance to the development of functional clothing. This hierarchical 'tree' of design requirements, central to her previous work, relates to the user needs–driven application of technical and smart textile attributes within the performance sportswear clothing layering system (McCann, 1999, 2009). This design process, based on a balance of 'form' and 'function', provided an initial overview of generic user needs, as guidance, to be further amended and elaborated with respect to the needs of the active ageing community.

The iterative stages of the co-design process were then progressed. As described in detail in subsequent chapters, the ongoing identification of end user needs was to

Figure 12.9 Design for Ageing Well co-design workshops.

inform the design and technical development process through to the wearer testing of near market prototypes. It is important to note that the 'co-design' process adopted was entirely relevant to the design of product for older people carrying out all types of walking – and indeed to those of any age.

12.6 The way forward

12.6.1 User-led practical prototype development

In terms of 'intelligent garment design', or smart clothing with wearable electronics, 'the research phenomenon may be considered as a complex and hybrid field where educators, researchers and practitioners have turned to interdisciplinary work to solve problems that are beyond the scope of any one discipline, and to achieve unity of knowledge' (Uotila, Mattila, & Hanninen, 2006, p. 5). These Finnish authors reflect on the fact that 'an examination of the research on clothing and fashion reveals that in many respects the concept of an intelligent garment system has yet to be analysed. The discussion in the field in recent years has revolved around technological research and knowhow involved, and few if any references can be found in the literature to the conceptual points of departure used in the research on clothing and dress and in fashion design' (Uotila et al., 2006, p. 6). In particular, DfAW recognises that there has been little prior concern for a wearable technology system to enhance the everyday lives of the rapidly growing active ageing community.

The interdisciplinary research project Methods and Models for Intelligent Garment Design (MeMoGa), undertaken in Finland (2003–2005), focused on strategies for research 'on the usability and acceptability of wearable intelligence and in developing new design models' (Uotila et al., 2006, p. 7). 'The idea was to view wearable intelligence or intelligent garments more as a social situation and cultural events than as individual objects' (Uotila et al., 2006, p. 12). The research identified and analysed the protective garment needs of selected workers in heavy industry in order 'to guide the vision stage of the product design process'. The main interest was 'in developing methods that could assist in the early stage of design and could support open communication between users and designers through the concept design and interactive virtual prototypes' (Uotila et al., 2006, p. 8). The output of this team, dispersed across Finland, was to create a multimedia presentation to show the concept garment in a virtual working environment to enable designers to engage with the user experience for the design of a virtual prototype.

Projects concerned with wearable intelligence may have similar user-centred philosophies but quite different working methods and outputs with, for example, virtual models or tangible products that may be willingly worn by 'real' end users. The DfAW research project has been concerned with the needs of the end user with the focus on the practical, in situ, design development of a 'real' prototype garment layering system offering support mechanisms for older people (60–75). Part of that support mechanism is based on incorporating information and communication technology (ICT) systems within garments worn whilst walking, to carry out such processes as vital function measurements, recording routes taken and communication with fellow walkers. The intention has been to incorporate technologies into comfortable, flexible and protective clothing to fit 'real' users and to be suitable for use in everyday life.

Iterative co-design methodology has involved the regular input of stakeholders working alongside active ageing participants to respond to their design requirements. Design-led development has been situated primarily within a studio setting or, occasionally, a sports lab or outdoor environment, to test the comfort and functionality of the integrated technology system and the usability of the technology user interface. In DfAW, the 'hands-on' approach has been seen to be essential for effective engagement with older users, dependent on the input of designers as interpreters. The principal thrust of the DfAW project was in meeting the real wearable technology needs of real (older) people.

12.6.2 Collaborative approach toward technology adoption

In order to meet these needs, new approaches to both the clothing and electronics systems have had to be made and technologies from both these topics designed to overlap. Rapid changes in these technologies need to be accommodated, particularly in the electronic system field. According to Moore's 'law', processing power of computers doubles every two years, and although less startling advances are made in the textile field, steady improvements in fibre, yarn and fabric properties are made. Using a basic 'input–output' model, inputs to the DfAW project arise from four main sources:

12.6.2.1 Inputs

Inputs involved the contribution of system specialists designing electronic system-based support mechanisms related to the concept of 'ambient assisted living'. Often based around the needs of elderly and infirm people being supported in their needs when living in their home, this aspect of the project allowed exploration of fully portable systems, designed around the needs of fitter older people. Such garment-incorporated systems have their own limitations, such as processing power and garment energy supplies, but such is the rate of progress in portable ICT systems that it was found that a mobile phone could be pressed into service as the central processor.

In addition, technically oriented designers developed a layered clothing system where the layers worn are integrated rather than chosen at whim. More frequently encountered in the serious sportswear field, the concept of each layer of clothing (skin to outer layer) working together was used in the DfAW project. Body size and shape of the project's target clientele are very different from active, younger sportsmen and sportswomen, and this set an interesting challenge to the clothing design team.

Members of a formal co-design team linked the garment system design with the needs of real people, involving members of the public as a user reference group. A team of individuals within the age range (the user reference group) supported and were supported by the specialist topic designers at all stages of the project. Advisors from industry and other specialist sources liaised throughout the project.

12.6.2.2 Outputs

The principal outputs of the DfAW project represent, of course, the example working prototype garment systems produced for and tested by the co-design user reference group, and the design parameters for electronically enabled layered garment systems. It is these aspects that will help to drive the product to market and will be discussed further in Section 12.6.3. These outputs, one a physical entity and the other a process description, are by no means the only possible outputs.

The DfAW project is aimed at people who are generally fit, but the principles could be of interest to people who have a limiting disability. Here the interest may lie in measuring pulse, determining physical location and seeking access to emergency services. It is worth noting that the electronically enabled garments have not been subject to clinical trials, and more work on, for instance, skin layer sensors would be needed.

As far as processes are concerned, outputs include specific factors such as methods of measuring body shapes for older people, and developing cross-boundary ICT system/fabric common language. Less specific but of considerable interest across a range of fields of study are the processes of co-design and general cross-topic communication and understanding.

12.6.3 Driving the product toward market

'How technology Innovations will become culturally accepted final products, artefacts and natural parts of our everyday life' (Uotila et al., 2006, p. 16).

This research has confirmed that users who have engaged in the co-design process, over time, are interested in materials and technologies that were previously unfamiliar. However, routes to market for such hybrid products as an electronically enabled enhanced layered garment system are not immediately apparent. Where will it be stocked – in a sports goods shop, an electronics retail outlet or a shop within a shop? There are few similar products on the market that have such a level of complexity, although some examples of simple heating elements enclosed in fabrics do exist. One example of this would be the common electric blanket – hardly a sophisticated product, but one sold through bedding suppliers. With closer relevance to DfAW items, heated garments are quite familiar to motorcyclists, and these are promoted largely through accessories sections in motor cycle retail outlets. Again these outlets are unlikely to be much visited by older customers. One retailer of electrical and electronics goods (Maplin) retails a range of heated gloves, socks and gilets, but these products are a very small part of their range. School blazers, with incorporated wiring for controlling an MP3 player, were produced some years ago by Marks and Spencer but found to be unpopular (particularly with teachers) and were discontinued from sale.

How then can this new product be taken to market? The user is of course the 'market' in the final analysis, and although the project target user is quite tightly specified (aged 60–75, no limiting health problems, interested in walking) it may be difficult to reach them directly. In terms of the industry, Claudia Bieker of GenerationSport, Germany, having researched the gap in the market for 'Best Ager' products, cannot effectively promote solutions to brands without exemplars of proof of concept prototypes. Once the garment system resulting form the research exists, as a saleable product, strategies for routes to market might be better placed in promoting a product through a continuation of co-design engagement with key intermediaries. Bringing together all those with an interest in new routes to market, for this neglected sector, would involve, for example, product developers, designers, sales and marketing teams, clothing buyers, those involved in promotional editorial and advertising and those involved in the retail experience, point-of-sale information, and the training of the point-of-sale team. An intervention needs to be made 'to business as usual' through those that have the power to change company policy.

In addition, the involvement of organisations with concerns about older people and/or walking, such as Age UK and/or 'ramblers' is also required. Finally, the medical profession could have an interest particularly in those parts of the NHS promoting walking as healthy exercise. More difficult to reach, but with some considerable potential, are sons and daughters buying enabled garments for parents living some distance away, wanting the reassurance that emergency situations can be effectively handled. Bringing a new product to market is a difficult process, even where, as here, the size of that market is known. The likely take-up is, however, an unknown statistic. Great efforts have been made, through methods including using external advisers and the co-design process, to make sure the product is right, and only manufacture and selling garments can confirm the market. This project has sought to generate ideas through end user research in order to develop products that come near to market. A further stage is needed to bring the research to a wider audience.

References

Euratex. (2009). http://www.systex.eu/content/definition-smart-textiles

Fuad-Luke, A. (2009). *Design activism: Beautiful strangeness for a sustainable world*. Oxford: Earthscan from Routledge.

Kotrotsios, G., & Luprano, J. (2011). The commercialization of smart fabrics: Intelligent textiles. In A. Bonfiglio, & D. De Rossi (Eds.), *Wearable monitoring systems* (pp. 277–294). New York: Springer.

McCann, J. (1999). *Establishing the requirements for the design development of performance sportswear (M Phil thesis)*. University of Derby.

McCann, J. (2009). Smart clothing for the ageing population. In J. McCann, & D. Bryson (Eds.), *Smart clothes and wearable technology*. Cambridge, UK: Woodhead Publishing Ltd. In association with the Textile Institute.

McCann, J. (2012). Translating the hybrid methodologies and practical outputs of smart textile-oriented research, in clothing for the growing ageing market, for the benefit of all stakeholders. In *The textile institute 88th world conference 2012: Conference proceedings (book)*. Textile Institute.

McCann, J. (2013). Smart protective textiles for older people. In R. Chapman (Ed.), *A smart textiles for protection* (pp. 224–275). Cambridge, UK: Woodhead.

McCann, J., Morsky, S., & Dong, Z. (2009). Production technologies for smart clothing. Garment construction: Cutting and placing of materials. In J. McCann, & D. Bryson (Eds.), *Smart clothes and wearable technology*. Cambridge, UK: Woodhead Publishing Ltd. In association with the Textile Institute.

Metz, D., & Underwood, M. (2005). *Older richer fitter: Identifying the customer needs of Britain's ageing population*. UK: Age Concern Books.

Moghaddassian, S. A. (2010). *The development of test procedures for controlling the quality of the manufacture of engineered compression stockings (thesis)*. The University of Manchester.

Poslad, S. (2009). *Ubiquitous computing: Smart devices, environments and interactions*. UK: John Wiley and Sons.

Shephard, R. J. (1997). *Ageing, physical activity, and health*.

Uotila, M., Mattila, H., & Hanninen, O. (2006). Methods and models for intelligent garment design. In H. Mattila (Ed.), *Intelligent textiles and clothing*. Cambridge: Woodhead Publishing Ltd.

Van Langenhove, L. (2007). *Smart textiles for medicine and healthcare*. Cambridge: Woodhead.

Van Langenhove, L. (2012). *Smart textiles at a glance*. Available: http://www.ita.rwth-aachen.de/1-aktuell/Flyer%20Medical%20Smart%20Textile.pdf. Last accessed 25.06.13.

Van Langenhove, L., & Hertleer, C. (2004). Smart clothing: a new life. *International Journal of Clothing Science and Technology*, *16*(1/2), 63–72.

Websites

Adidas MICOACH: http://www.adidas.co.uk/micoach?cm_mmc=AdieSEM_Google-_-miCoach-B-Exact-_-General-miCoach-X-General (Last accessed 25.06.13).

Baltex: Technical Textiles: www.baltex.co.uk.

Best Ager Gallery (ISPO, 2010): http://media.nmm.de/47/ispo10_bestager_presentation_eng_24372647.pdf (Last accessed 25.06.13).

D'Apuzzo, N. Hometrica Consulting. 3D scanning of the human body, http://www.igp.ethz.ch/photogrammetry/education/lehrveranstaltungen/MachineVisionFS2011/coursematerial/body_scanning (Last accessed 30.06.13).

D3O Shock Absorbing Protection: http://www.d3o.com (Last accessed 25.06.13).

Fibretronic ConnectedWear: http://fibretronic.com (Last accessed 25.06.13).

Flexible & Stretchable Electronics: www.stella-project.de (Last accessed 25.06.13).

GenerationSport: http://www.generation-sport.de/en/home.html (Last accessed 25.06.13).

Grado Zero Espace: http://www.gzespace.com/gzenew/index.php?pg=nitilon&lang=en (Last accessed 25.06.13).

Walking Holidays, hfholidays: http://www.hfholidays.co.uk/ (Last accessed 30.06.13).

Hughes et al. American Journal of Clinical Nutrition: http://ajcn.nutrition.org/content/76/2/473.short (Last accessed 30.06.13).

Interactive Wear: iLightX – Safety and fashion effects with integrated LEDs: http://interactive-wear.de/cms/front_content.php?idcat=72 (Last accessed 25.06.13).

Lerch et al. Journal of Textile and Apparel, Technology and Management: http://www.tx.ncsu.edu/jtatm/volume5issue4/Articles/Lerch/Lerch_full_221_07.pdf (Last accessed 30.06.13).

Moore-Harrison and Cress University of Georgia, http://news.uga.edu/releases/article/study-regular-walking-nearly-halves-elderly-disability-risk/ (Last accessed 30.06.13).

Moore's law: http://www.mooreslaw.org/ (Last accessed 30.06.13).

Philips Blue Touch: http://www.design.philips.com/philips/sites/philipsdesign/about/design/design-news/newvaluebydesign/january2012/blue_touch_pain_relief.page (Last accessed 25.06.13).

Philips Lumalive: http://www.research.philips.com/password/download/password_28.pdf (Last accessed 25.06.13).

Scan2Knit: https://www.escholar.manchester.ac.uk/uk-ac-man-scw:92193 (Last accessed 25.06.13).

SizeUK – results from the UK National Sizing Survey: http://www.fashion.arts.ac.uk/research/projects-collaborations/archive/size-uk/ (Last accessed 25.06.13).

SizeUK: http://www.size.org (Last accessed 25.06.13).

Smart Textiles: http://www.ugent.be/ea/textiles/en/research/research-themes/smart-textiles/smart-textiles.htm (Last accessed 25.06.13).

Smith and Gove, University of Florida: http://edis.ifas.ufl.edu/pdffiles/HE/HE01900.pdf (Last accessed 30.06.13).

Systex: http://www.systex.org/content/definition-smart-textiles (Last accessed 18.09.12).

Tumin and Qian, The Ohio State University: http://paa2011.princeton.edu/papers/111109 (Last accessed 30.06.13).

Walking for Health: http://www.walkingforhealth.org.uk/ (Last accessed 30.06.13).

Willis, R. (2013). Clothes: A manifesto. *Intelligent Life, The Economist*. Available: http://moreintelligentlife.com/content/lifestyle/rebecca-willis/clothes-a-manifesto?page=full (Last accessed 25.06.13).

YKK Elastic tape: http://www.ykkfastening.com/products/zipper/coil_zipper/elastic_tape.html (Last accessed 25.06.13).

http://www.ramblers.org.uk/Resources/Ramblers%20Association/Website/Get%20Walking%20Keep%20Walking/Documents/factsandfigures-2-participation-0510.pdf.

Co-design principles and practice: working with the active ageing

13

J. McCann
University of South Wales, Newport, UK

13.1 Introduction

The Design for Ageing Well project (DfAW), introduced in the previous chapter, has begun to raise the profile of design for the neglected but rapidly growing active ageing community. Ten million people in the UK are over 65 years old with latest projections indicating for 5.5 million more elderly people in 20 years' time and the number nearly doubling to around 19 million by 2050 (http://www.parliament.uk/business/publications/research/key-issues-for-the-new-parliament/value-for-money-in-public-services/the-ageing-population/). This chapter reports on collaborative, cross-disciplinary research that has begun to address the lack of clear design direction for the development of functional clothing, for active agers, to promote healthy exercise, autonomy and well-being. Attention is drawn to the gap in design trend information, that is virtually non-existent, to guide both product development within the emerging wearable electronics sector and, in particular, for the neglected active ageing community. Through an iterative co-design process, stakeholders have worked together to identify wearable technologies that may be incorporated in a functional clothing layering system for active ageing walkers. Academic experts, with older research participants, have benefited through industry collaboration that has enabled access to novel materials and processes. The ultimate aim has been to provide new design direction, in plain language, in both text and clear visual format, to inform the range planning and subsequent practical development of near market prototypes.

By means of a co-design process, adopted within the DfAW case study, older user functional clothing and wearable technology needs have been identified and prioritised to create a hierarchy of design requirements to inform the subsequent practical design development stage. This chapter identifies and collates the smart clothing requirements, within the context of a clothing 'layering system', for active agers to engage in healthy outdoor exercise, with a focus on walking. The co-design process has covered aspects of 'form' (aesthetic design appropriate to the culture of the user/peer group) and 'function' (technical design that is deemed suitable for the intended end use). Co-design workshops have systematically introduced users to the concept of the clothing layering system, and the scope for the application of smart materials, with cross-reference to the demands of different levels of participation in walking. Using plain language, rehearsed between the user reference group (URG), industry stakeholders and academics, co-design process has contributed to identifying and addressing the real world needs of the user group to inform the development of attractive and age-appropriate functional clothing (Taylor, 2011).

Textile-led Design for the Active Ageing Population. http://dx.doi.org/10.1016/B978-0-85709-538-1.00013-4

Designers have acted as facilitators, interpreters and visualizers in helping the older research participants express their needs and aspirations. With their tacit understandings, designers have been able to identify and prioritise the key technical user requirements to be addressed, to inform the iterative co-design prototype development. The resulting hierarchy of requirements, that covers generic concerns of relevance to clothing design for older wearers, may subsequently inform design development for other functional clothing end uses. The resulting hybrid design specification methods developed, to clarify the complex integration of wearable electronics into the clothing layering system, may also be transferrable to other garment sectors. The methodologies developed in the course of this practical research have the potential to help merge the quite different cultures of the clothing and wearable electronics sectors.

13.2 Capturing user experiences: clothing and technology

An initial focus group was staged by the Clothing Work Package (WP) to introduce older research participants to the aims of the project and to representatives of the academic team. A facilitator from the Behaviour WP, as an expert in user engagement, helped to ensure that all participants were able to make a contribution, to sum up and to indicate the way forward. This event was planned to gather initial views from 12 users (men and women) on: What type of walking do active agers participate in, what do they currently wear, and what they might currently use in terms of wearable technology? Three design researchers had individual roles in capturing information: on a flip chart, in making a video recording of the event, and in contributing to sharing specialist terminology. This half-day session identified initial motivations for walking, levels of participation, environments for walking and the degree of social engagement. This ranged from individual dog walking to walking as couples, with friends and family, and walking groups, either informal or recognised, such as the Ramblers and the University of the Third Age (U3A). There was debate between those who considered themselves serious walkers and those who wanted to take a picnic and stop to look at the view!.

At a second workshop, older users were encouraged to bring in personal garments of interest to a 'show and tell' discussion. This initial clothing discussion uncovered preferences, and problems, demonstrating differing levels of user understanding in relation to available products. This discussion provided initial insights into the low level of understanding of: the technical clothing layering system, modern fibres, fabrics and finishes, and novel manufacturing techniques. It showed confusion with textile and clothing terminology and quality differences, mistrust of the claims of brands, and the sourcing of garments from a breadth of retail outlets. A perceived preference for natural fibres (with cotton and wool considered 'good') and a mistrust of synthetics (considered 'bad') was attributed to this generation's early introduction to the poor image of the sweaty nylon shirts and Crimplene dresses of their youth. As older users, they have become accustomed to cotton base layers, often wool insulation, and

Figure 13.1 Design for Ageing Well 'show and tell' workshop.

a variety of waterproof and other outer layer garments including corduroy trousers, a heavy tweed hunting jacket and an army poncho! Examples of walking clothing along with opinions, and stories, related to the pros and cons of the disparate items, included several references to 'clothing as an old friend' (Figure 13.1).

There was virtually no understanding of, or perceived need for, the merging of electronics into textiles. The concept of wearable technology was new to the group, although, when briefly explained, some of the user group acknowledged the use of mobile phones, cameras and pedometers, but health monitoring, in clothing, was virtually unknown to them. Taylor discusses their confusion with the profusion of complex point-of-sale material, prevalent in the retail environment, in more depth (Taylor, 2011). The cost of clothing reoccurred throughout the discussion, with reference made to bargains found at a range of non-specialist outlets. It was generally expected that clothing should be cheap, with the participants retaining little memory of the relative high cost of clothing when they were young.

As with the general public, the users had a lack of appreciation of the sophisticated nature of fibres and fabric developments, and finishes adopted within performance clothing. They were also unaware of the restrictions in colour choice associated with mass produced transient fashion and problems associated with short production runs of specialist colours. There was little awareness of the problems attached to global sourcing and the degree of waste within the textile chain. There was general distrust of environmentally conscious clothing attributes. User engagement within these initial workshops began to raise awareness and showed that preparatory training, in shared lay language, was necessary. As the project progressed, it became increasingly evident that other work package researchers were equally unaware of the languages and cultures of the textile, clothing and electronics trades and would benefit from joint training across the whole stakeholder group. The initial show and tell workshops had provided valuable insights in helping the academic team to plan the detail of the co-design process to embrace both research and industry stakeholders, with their disparate disciplines, cultures and terminology, in communicating with older research participants.

13.3 Explaining the attributes of the 'layering system' to older users

A series of co-design workshops took place over a one-year period that systematically introduced and discussed what was currently available or emerging in terms of textile innovation and wearable technologies in functional clothing that, through design, could be adapted to the needs of older wearers. The concept of the clothing 'layering system', originally driven by military preparedness, is the principle adopted in the performance sportswear industry, a sector that has also been an early adopter of wearable technologies. However, until now, product developers have neglected the design requirements of older wearers. The focus of the co-design process was to consider the attributes of the performance sports layering system with the potential, through age-appropriate design, to offer enhanced overall comfort to older wearers. Through co-design, a layering system was to be developed as a textile-based 'platform' for the application of smart textiles and wearable electronics.

The iterative co-design process proceeded, with all stakeholders, in a series of practical workshops that took place over the academic year November 2009–July 2011 (Table 13.1). Both Technology and Clothing academic participants, with industry stakeholders, benefited from direct user engagement, with best practice ensured through collaboration with the Behaviour team, as preconceptions about ageing were altered as the team began to identify the 'real' everyday needs, aspirations and challenges of the active ageing. The full-day workshops took place adopting a sequence of show and tell discussion in the morning followed by investigation into, and feedback on, commercial product development in the afternoon. Older participants were encouraged to bring in personal garments of interest, related to walking or other functional end use, to the show and tell sessions. In order to prompt discussion and capture design preferences, existing products and brand catalogues were introduced for reference during the afternoon co-design workshops. As the iterative stages progressed, the users gained confidence and became actively involved in hands-on practical work. For example, users became involved in drawing, in working on the mannequins, and in colour and fabric selection. Users, selected from the body scanning process, acted as average size and fit models in making design decisions.

The layering system typically comprises a moisture-management base layer, mid insulation layer(s), and a protective outer layer (Figure 13.2). Textile innovation has now embraced the concept of the 'soft shell', which incorporates a hybrid mix of the protective attributes of the outer shell garments combined with the comfort and insulation of mid layers, offering a softer smooth handle with reduced but adequate water repellence for moderate, multifunctional end use. With the introduction of textile-based sensors, there is the need for a close-fitting skin or 'body layer' for men and suitable intimate apparel for women. Heat bonding, which is now prevalent in sportswear for seam joining, may also be used for the encapsulation of wearable electronic cables with their textile-based switches and controls. With appropriate styling, the system has the potential to enhance the overall comfort of the active ageing in their busy everyday lives. The concept of the layering system is described in more detail

Table 13.1 **Design for Ageing Well table of co-design training and workshops**

Date	Type of workshop	Industry liaison: fibres, fabrics clothing
Nov 2009	Technology researcher training for development of shared language	University of Ulster (UU)
Feb 2010	Behaviour researcher training for development of shared language	University of Salford (US)
	User recruitment	URG-University of Wales Newport (UWN)
	Initial focus group	Led by US at UWN
	'Show and tell' focus group	UWN with US & University of Westminster (UW)
May 2010	Clothing researcher & user shared training re: textiles & clothing	Sprayway/Keela/Lenzing/ Rohan/
	Behaviour research training	University of Westminster
Sept 2010	Base layer	Lenzing/Smart wool/Coolmax etc
Oct 2010	Mid layer insulation	Sprayway/Polartec/Rohan etc.
Nov 2010	Outer layer	Sprayway/Keela/W.L.Gore/ Youngone etc
Dec 2010	Wearable technology	Textronics (Adidas)/ Fibretronic/Shimmer
Jan 2011	Bra workshop with female URG	In-house consultant: Juliette Smith
Feb 2011	Leg wear	Led by Laura Seppala, PhD, US.
March 2011	Colour trend forecasting and selection from trade perspective	Clare Johnstone (RCA) Juliette Mann (Consultant) Julie Greengrass (Sprayway)
May 2011	Santoni knit practical workshop	Users worked with academic design team
June 2011	Fabric selection and colour	Practical workshop to develop age-appropriate colour pallet suitable for prototype layering system
July 2011	Resume of style direction	Final style range plan for men and women

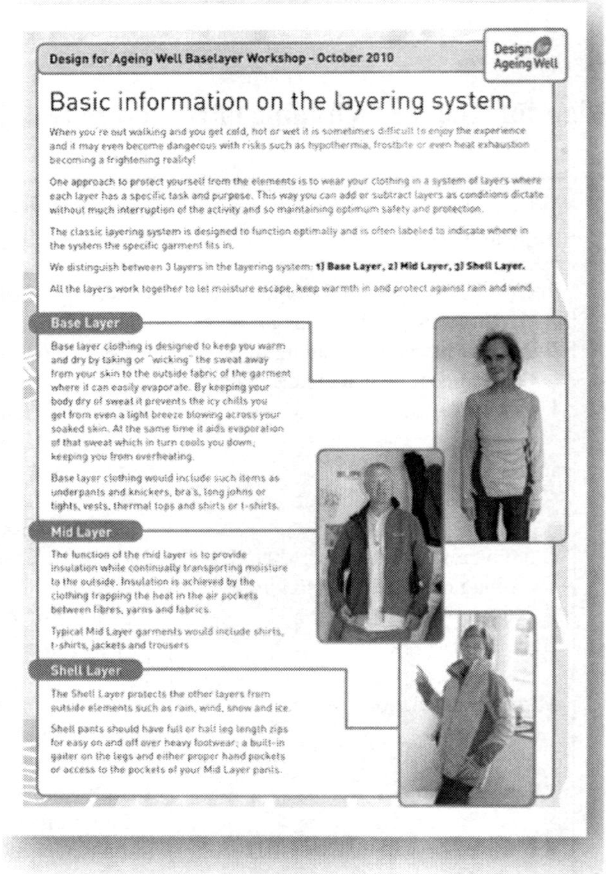

Figure 13.2 Poster to communicate attributes of the layering system to users.
Source: David Taylor.

by the author in "Smart Protective Textiles for Older People", in Woodhead's *Smart Textiles for Protection*, (McCann, 2013).

13.3.1 Base layer co-design workshop (September 2010)

The full day co-design workshop, involving approximately 12–15 older research participants, began with a 'show and tell' session in the morning. The research participants, who had benefited from stakeholder training workshops in the spring, on the concept of the layering system, were asked to bring in examples of base layer items that they would normally wear for various types of walking. Emerging researchers, from all the project academic partners, were welcome to attend these events. Within the semistructured session, each participant, in turn around a table in a studio setting, presented one or two garments drawing attention to positive and negative aspects with respect to attributes

Figure 13.3 Design for Ageing Well co-design workshop: comparative reporting on commercial garments.

such as comfort, appearance, handle, function etc. The academic team asked questions, as did other users, with a view to eliciting sufficient understanding to inform design development. Aspects of interest were to do with user preferences with regard to fibre content, fabric construction and handle, styling, fit, proportion, armhole cut, necklines and neck openings, garment finish, colour selection, and garment branding. At this stage in the project, there was very little evidence of the use of modern fibres and fabrics, for men or women, with a predominant use of cotton, in T-shirts for women, with both long and short sleeves, and with different necklines. Some of the men wore rugby style shirts or woven shirts over simple T-shirts. Female garments were slightly fitted, while male styles were loose and boxy. The session was video recorded.

In the afternoon, industry stakeholders were invited to attend with, or to send, examples of base layer products that they considered appropriate, or wanted to gain feedback on, from users representing this potential growing market. The selection included both long- and short-sleeve styles, for men and women, with a selection of lengths, degrees of fit, and varied necklines and openings. The garments that had been made available were coded, with numbered labels, and photographed by the academic team to enable them to prepare report sheets, with small photographic icons, to request brief written comments by the users, see Figure 13.3. This session provided an opportunity for older research participants to see unfamiliar fibres and blends in performance sportswear garments, often more technical than those that they would have previously been aware of. It enabled users to handle new materials, sometimes already discussed in the earlier training sessions, and, where appropriate, they were able to try on garments. The users were encouraged to be forthcoming in commenting on details, such as necklines, giving reasons in terms of those they liked and disliked. 'Clothing' researchers, representing experts in textile fibres, size and fit, functional clothing design, ergonomic cutting, with graphic and product designers, acted as facilitators and interpreters, ensuring that a comprehensive record was made to inform subsequent practical development. The workshop was recorded in video, and with still photography, by researchers and postgraduate students.

Some companies that were unable to attend, such as Lenzing (2014) (Austria), sent multiples of styles for wearers to trial. The representatives of prominent brands, who did attend the workshops, enjoyed lively debates with the participants, adding real value in terms of stakeholder engagement in the co-design process. Industry stakeholders found the process highly informative while the views of the users were already seen to be altering. An original lack of importance had been associated with what might have been deemed a simple T-shirt, in a basic cut, in cotton, while user emphasis was given more to the choice of outer protective jackets. Users had begun to alter their opinion in embracing synthetic and man-made fibres in better fitting garments. They gradually became much more opinionated on fit, proportion, handle, ease of movement, and collar and neck shape and finish. Women were unanimous in their appreciation of Tencel, with respect to its cotton-like handle and overall comfort. Men were impressed with technical polyester constructions and sport wools and their blends. All the researchers became aware of the potential for 'body mapping' enabling the combination of different fabric constructions and design lines offering enhanced functionality in different 'zones' around the body, with meshes for breathability, more stable constructions for support and varied stretch for movement.

The design of Santoni seamless knit garments was addressed in a hands-on practical co-design workshop that enabled the academic team to gather valuable information as selected users, wearing simple fitted, long-sleeved T-shirts collaborated in drawing the positioning of the desired knit structures onto the selected user fit model.

With particular relevance to the placement of vital signs monitoring technologies, separate workshops took place on the co-design of close-fitting body layers to incorporate knitted sensors based in men's vests and women's sports-type bras. The intimate apparel bra workshop took place exclusively with female users and specialist clothing researchers. These body layer workshops were led by a specialist intimate apparel designer, Smith, working in close collaboration with the size and fit expert and in liaison with the electronics team see chapter 20.

13.3.2 Mid insulation layer co-design workshop (October 2010)

The pattern of the co-design workshops was repeated. The morning show and tell session involved 12–15 older walkers, sharing with peers and researchers examples of what they wore to regulate temperature when walking. Again, the male and female participants spoke in turn, around the room, about clothing items of particular interest, both in terms of attractive and disappointing features. Yet again there was little evidence of the typical categories of garments that make up the mid insulation layer of the performance sportswear 'layering system'. Some male and female users wore varying combinations of traditional knitwear, selected to personal colour preference, with layers added, or taken off, dependent on the weather conditions. Some male users had loose fleece styles, either zip through or pulled over the head, either in plain classic colours (black, navy, forest green) or with a natural textured appearance. Invariably, styling shown was relatively loose and casual for men and slightly more fitted for women. Some women wore padded gilets over lightweight knitwear, and one man wore a simple fleece-lined Pertex gilet over a rugby shirt.

In the afternoon, users were introduced to a variety of more fitted fleece jackets and more tailored soft shell styles. Users were unaware of the variety and sophistication of fleece fabrics available, in varying weights and textures, as predominantly basic fleece fabrics have become integrated into everyday fashion, corporate wear and home wear. There was general preference for a natural 'look' with marl yarn effects. Further discussion on the attributes of soft shell technology follows in reporting on the outer layer workshop.

13.3.3 Outer layer protective layer co-design workshop (November 2010)

Again, users brought in existing garments in the morning, and current commercial products were reviewed in the afternoon. Despite the recruits being dedicated walkers, it was already known that their knowledge of sophisticated outwear garment technology was limited to that discussed in the preparatory sessions. Only one female participant was fully committed to a full-performance clothing system, previously aware of the particular technology of her preferred UK brand (Páramo Directional Clothing Systems, 2014) prior to involvement in the project. However, being relatively petite, she was unable to obtain this clothing in a good fit, having become resigned to 'making do' with the extra length and width, in navy blue. Another couple was introduced to waterproof systems while going into an outdoor shop in the Lake District, to buy a map, and have since become loyal to the UK brand (Berghaus, 2014). A comment from older participants was on the hood volume of certain styles, perceived as unnecessary, being unaware that this caters for more extreme use in the wearing of climbing helmets. Users admitted to buying shell garments because of colour choice, styling and price level.

In questioning the predictable sameness of outdoor clothing, a design-aware female participant brought in a stylish lime padded knee-length coat, with a distinctive asymmetric fastening, that she wore for urban walking. The same user showed a long bright yellow, lightweight, showerproof coat, a knee-length white PVC trench coat, and a lightweight calf-length pale blue trench coat as a 'mac-in-a-sac', all of which were intended to introduce a range of proportions and cheerful colour. Other women within the group agreed that they would wear knee- or calf-length padded coats for everyday use in winter and for urban walking. Despite these garments lacking articulated 'arm lift' and technical textile attributes, they demonstrated styles that could be adapted and modified to give outdoor 'life wear' ranges, a non-transient, classic twist in styling with a sense of colour and fun for travel and walking in urban environments.

The users were interested in trying on a selection of garments provided by the brands, with a range of clothing technology that embraced three-layer outer shells, soft shells and 'dual protection' systems for jackets and protective trousers. The outer layer, or 'shell' garment, provides protection for the clothing microclimate from the ambient conditions. This layer may constitute waterproof breathable textiles, in a range of variations on woven or knit structure protective textile assemblies, with additional properties and finishes, such as abrasion resistance for more extreme environments (McCann, 2009). Outer garments have become increasingly lightweight, tailored and stylish, with stretch properties to enhance comfort and fit in combination with improved moisture management, in hydrophilic or microporous waterproof 'breathable' membranes. Novel joining methods, such as the heat bonding of seams and zip insertions, and laminated panels, have revolutionised garment appearance and reduced bulk. The heat bonding of seams also contributes to the encapsulation of wearable electronic devices with their textile-based caballing, interconnections, soft switches and controls.

The 'soft-shell' concept incorporates many of the protective elements of outer shell garments combined with the comfort and insulation of mid layer fleeces. Soft-shell fabric technology normally constitutes a laminated assembly of two fabrics with a membrane in a 'sandwich' construction. These garments are designed to be water repellent rather than waterproof and do not require seam taping. These fabrics are softer and less noisy than more protective 'hard-shell' garments and often have a degree of stretch to enhance comfort and fit. As extreme protection is not required on a daily basis, this category of garment, if designed to suit older people's style requirements and figure types, may offer an attractive solution to adaptable and comfortable lightweight protection for everyday use.

System Dual Protection (SDP by Keela, Scotland) is claimed to be 100% waterproof, breathable and warm, reducing inner condensation to keep the wearer protected. Garments are made up of an outer layer fabric, with a breathable membrane, and an inner mesh lining, with a laminated high-wicking hydrophilic (water-loving) membrane (Keela International Ltd., 2014). The two layers create an 'air gap', providing a natural thermal barrier. The water-loving membrane rapidly transports moisture away from the body, reducing condensation and maintaining comfort. Outer layer and soft-shell technologies are discussed in greater detail by the author in Woodhead's 'Smart Protective Textiles for Older People' in *Smart Textiles for Protection* (McCann, 2013).

13.3.4 Introduction to the smartphone

One of the early considerations in the Design for Ageing Well case study was a decision about the extent that electronic systems should be integrated into the structure of the garment. In theory the necessary data collection, processing and display could be closely integrated, or embedded, within the garment itself, and be invisible to the user, apart from inputs and outputs. The idea of embedding electronic devices has its attractions but also considerable drawbacks. These advantages and disadvantages are outlined in the concept map (Figure 13.4) 'Integrating Electronics into Clothing'. In the early stages of the project, the design team was agnostic about the degree of integration that would be appropriate for the DfAW-layered garment system.

As the project progressed, however, it became apparent that major and rapid advances were being made in mobile phone technology to such an extent that designing a fully integrated customised processing unit, built into the garments, would represent longer development times and increased expense. Garments were still to contain sensors and some displays, but the central data processing would be carried out by a (more loosely integrated) smartphone. Smartphones can easily link with computers, and so an additional feature was to be added, for the recording of data such as route travelled that could later be downloaded from the smartphone to a laptop. This feature

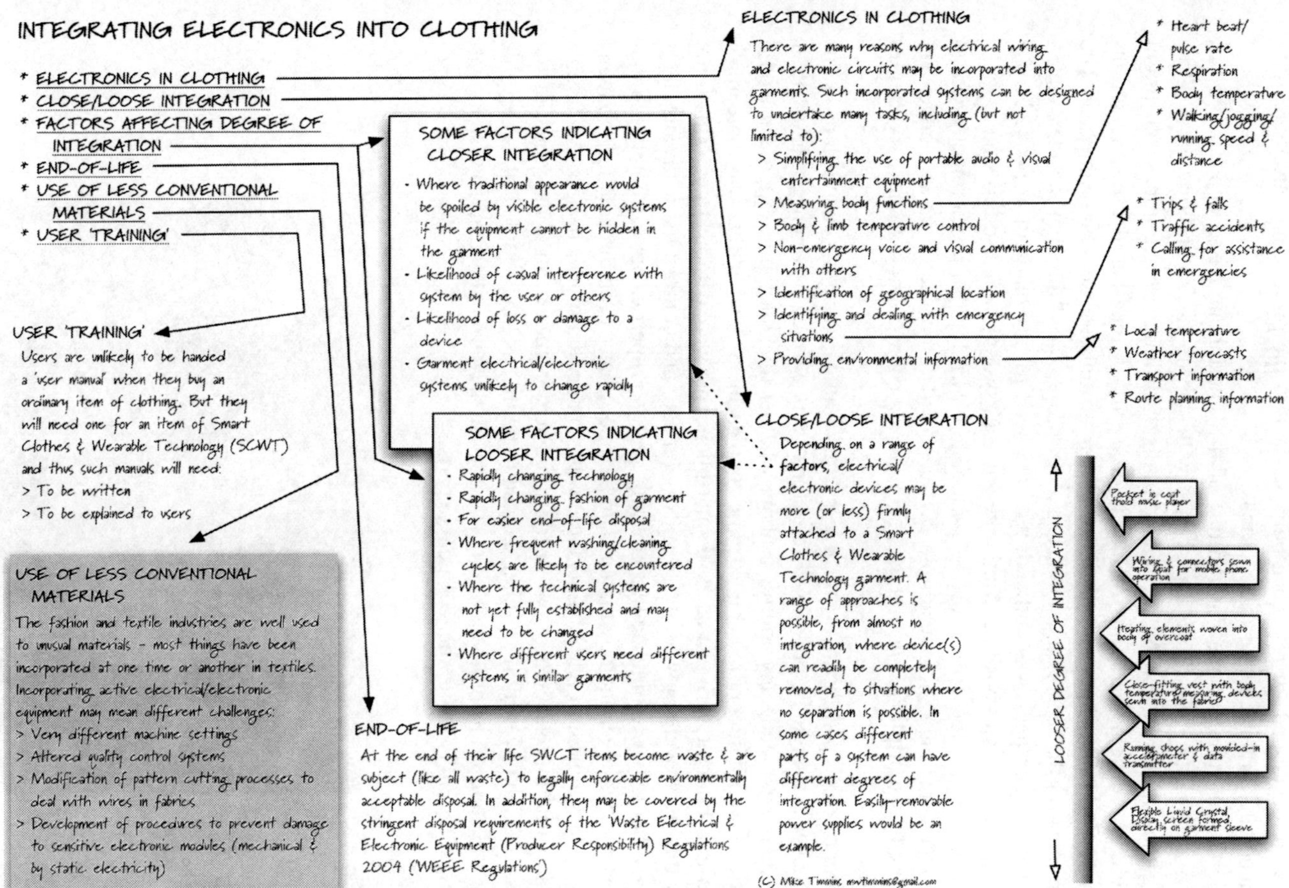

Figure 13.4 Design for Ageing Well concept map 'Integrating Electronics into Clothing'.
Source: Mike Timmins.

is perhaps more likely to be of interest to enthusiastic walkers undertaking longer walks, defined within the DFAW project as 'Type 4', or possibly 'Type 3'. These walk types were developed for reference purposes within the project, and are more fully described later in this chapter in Section 4.1.

In order to introduce the possibilities of the smartphone-based approach, a full co-design workshop was held. This began with one of the Technology team introducing the topic followed by a demonstration of different smartphones that might be used. These demonstrations included some early screen-based user interactions (Figure 13.5). Following these demonstrations, a discussion session took place. The requirement for the sensors to include use of the proprietary 'Shimmer' accelerometer, to provide certain data, was outlined, with the need for provision of a 'pocket' to be placed in a convenient position within the garment system, to house this small, unobtrusive device, was described. (The smartphone's own built-in accelerometer would not be able to distinguish, for example, between a fall of the wearer or swift movement in taking the smartphone out of a pocket, requiring each situation to have a different response.)

Older research participants seemed less comfortable in reaching conclusions in this technical area than in the previous discussions about garments. No one, users or other group members, made any comment about whether the smartphone approach would be better than full integration of electronics into garments, and neither was there much discussion about the various makes of smartphone that were demonstrated. This is not to say that users were unwilling to discuss, compare and contrast the ideas and items on show, but rather to indicate that clothing and its properties are extremely familiar to everyone, while technology matters can be unfamiliar – even threatening – to many people. Among the user participants there was a range of attitudes to new technology, but this variability seemed to have little influence on the eventual outcomes of the discussions. The overriding opinion seemed to be that as long as good fit, acceptable colour and appropriate protection were provided in the garment, technical system matters could be safely left to the technical experts.

Figure 13.5 Design for Ageing Well co-design workshop to consider usability of wearable electronics and user interfaces.

13.3.5 Wearable technology co-design workshop (December 2010)

Now that the user group was becoming more familiar with the sports-type layering system, in terms of both garment engineering and the body mapping of technical and smart textiles, it was appropriate to stage a workshop dedicated to wearable electronics. This co-design workshop brought together technology providers with academic researchers and the older research participants. The various technologies demonstrated were of particular relevance within the different garment layers. In looking at the body layer, Textronics (2014) (now part of Adidas) brought in their sports bras and a range of tight-fitting engineered knit developments with embedded knitted electrodes or sensors. There were both single two-layer vests for men and camisoles with 'secret support' for women. The electronics expert and garment knit technician explained what was possible within the scope of the Santoni seam-free knit technology, with sizing limitations in terms of the dimensions of the barrel of the machine and the degree of potential support available, within the knit structure, for larger figure types (Santoni, 2014).

After discussion on the degree of fit and pressure required to capture vital signs signals, the level of bust support required, and ease required for putting on and taking off garments, it was decided not to proceed with seam-free knit (Santoni). The revised plan was to incorporate the knitted sensors, supplied by Adidas, into 'cut-and-sew' sports bras in technical jersey knit fabric that could be seam bonded. The design engineering of the sports-type bras was made possible through collaboration with a specialist intimate apparel designer, working with the user-fit models, and in liaison with the Technology team in respect of the placement of the technology. Walking tests were set up in the sports lab, with users wearing harnesses with sensors attached, to monitor the optimal positioning of the sensors on the body to gather reliable data.

Fibretronic, an early developer of soft switches and controls, brought both garments and separate electronic components to the technology workshop. They demonstrated their textile switches, flexible keypads, smartphone and MP3 player controls, lighting, and garment heating systems, all designed specifically for garment producers to integrate into clothing. Their CONNECTED*wear*™ system has a keypad or joystick to be integrated in the garment and a controller module that connects the garment to a range of personal electronics such as an iPod or iPhone. The function interfaces are supplied in standard cable lengths and designs that normally would suit most apparel products and applications. The keypads and joystick cables can be applied by conventional sewing or bonding techniques terminating in a standard connector plug that is compatible with the full range of CONNECTED*wear*™ (http://fibretronic.com/connectedwear). The female members of the user group were most impressed with HEATwear technology in a heated vest with a flexible fabric heating panel for body warming.

Co-design engagement prompted discussion on the usability of the proposed wearable devices, and the challenges in the integration of the powering and communication components, with particular emphasis on the intuitive use of the technology user interface. Following the workshop, the academic Technology team had more detailed discussions with the technology providers with a view to resolving the compatibility of the technologies. It was agreed that Fibretronic would supply CONNECTED-*wear*™ components, to be adapted for use in outer layer garments to be connected to the smartphone, and HEATwear panels, with their interconnects and batteries, to be incorporated into female gilets. Fibretronic found the feedback on their products from this new potential market sector useful. It was acknowledged that, ideally, buttons on a keypad for older users should be further apart for users with limited dexterity but that this would require further development if it were to be taken to commercial scale.

13.3.6 Colour workshop (March 2011)

The colour session began with three visiting speakers who explained colour forecasting on a national and international scale and how buyers and retailers may interpret the trend information, as discussed within this publication (Chapter 4). Greengrass drew attention to the commercial challenges for manufacturing companies in making colour selection many months before the product is to be delivered to the retail environment (Chapter 18). It was observed that the fashion industry has subjected consumers to more than a decade of black as the predominant colour on the high street having been adopted by Punk, couture fashion, and by innovative Japanese designers. In recent years, black has replaced navy blue as a classic colour within the outdoor sports sector. This has resulted in older users being confused as to where to find interesting and flattering colours while not wishing to resort to black or beige.

In preparation for a practical co-design colour workshop, research participants had been asked to bring in examples of garments and accessories in colours and textures that they liked to wear and that they considered sympathetic to ageing complexions and hair colour. The group was unanimous in that colour is a major initial factor when selecting clothing, balanced by the importance of fit and proportion. There was a preference for marl effects and textures, to introduce a natural 'feel' to man-made fibres, with reference to colours of the countryside, of fruits, flowers, berries and lichens and natural stone tones. These were not transient choices but levels of colour that the older participants had become familiar and comfortable with and enjoyed wearing. They were keen to see such colours brought through to outdoor clothing.

Within this workshop the topic of branding arose, prompting a lively debate with the industry representatives. The discussion confirmed a major concern for the brand-led sportswear sector in that the active ageing do not wish to wear logos on the exterior of garments. When young, this generation was accustomed to labels being either in the back neck of a garment or inside the front facing. They are not happy to advertise even discreet tone-on-tone branding.

13.4 Segmenting types of walking

In order to provide clear design direction, a starting point has been the segmentation of walkers, based on types and levels of participation, through individual user stories, surveys and workshops to clarify and elaborate categories of walking, as there

Table 13.2 **Types of walking**

Type	Characteristics
1	Day-to-day mobility, walking to shops, visiting friends
2	Externally time controlled: work, doctor's appointment, school run, first stage of travel, pet walking
3	Woods walking, beach walking, sightseeing, pet walking
4	Hiking, fell walking, mountain walking, urban power walking

are many different ways of describing walking types, with, for example, the walking grades described by the Ramblers association (www.ramblers.org.uk/go-walking/find-a-walk/walks-grading.aspx). It was early in the project that a common set of walk-type descriptors were devised, with the intention of contributing to the cross-project common language. These were based on consideration of levels and degrees of participation, the likely range of environments and whether participants walked as individuals or were involved in social engagement.

13.4.1 Types of walking

These categories are as described in Table 13.2 and restated briefly as follows:

The first two walking types are those associated with ordinary everyday activities and broadly tend to be closer to home, and on well-known routes. As described above, these might be of short length and duration, purposeful, and may be the first stage in a longer journey. For these walks that are precursors to using other forms of transport, the average British adult makes 78 journeys per year that combine walking with public transport (The Ramblers 2014). It should be noted that these descriptors, although complete in themselves, do not provide a full picture of the 'walking experience', whether using technology-enabled garments or not. Other factors come in to play, including terrain/location and seasonal weather conditions.

13.4.2 Location/terrain and seasonal clothing requirements

In all types of walking, factors regarding location, terrain, and seasonal weather conditions can play an important role in a walking programme and need to be taken into account in the selection of appropriate smart textiles within the layered clothing system design (Figure 13.6). Weather conditions can make great differences to a particular walk. Atrocious weather can turn the shortest of a type 1 walk into a real ordeal, and change walking types 3 and 4 into life-threatening situations. Activity that demands multifunctional clothing, for a range of conditions, and varied locations, is more difficult to cater for in terms of design detail and textile selection, than more predictable activity in a known, anticipated environment. Garment design requirements for hot, dry conditions, or very cold climates, will benefit from specific fibre types and fabric constructions. Contrasting seasonal changes and unpredictable daily weather conditions demand an adaptable clothing system. More robust garments are needed

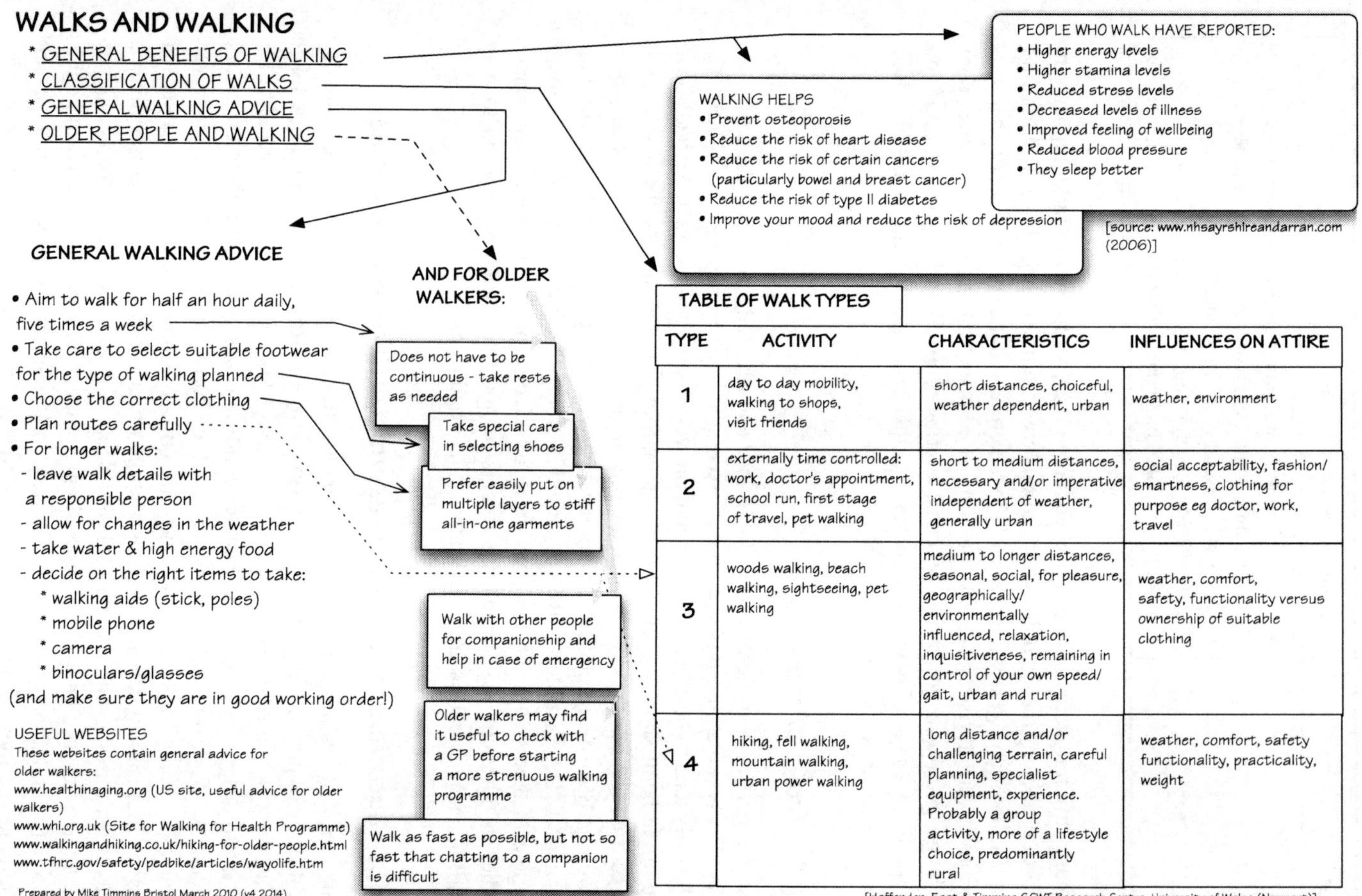

TYPE	ACTIVITY	CHARACTERISTICS	INFLUENCES ON ATTIRE
1	day to day mobility, walking to shops, visit friends	short distances, choiceful, weather dependent, urban	weather, environment
2	externally time controlled: work, doctor's appointment, school run, first stage of travel, pet walking	short to medium distances, necessary and/or imperative independent of weather, generally urban	social acceptability, fashion/ smartness, clothing for purpose eg doctor, work, travel
3	woods walking, beach walking, sightseeing, pet walking	medium to longer distances, seasonal, social, for pleasure, geographically/ environmentally influenced, relaxation, inquisitiveness, remaining in control of your own speed/ gait, urban and rural	weather, comfort, safety, functionality versus ownership of suitable clothing
4	hiking, fell walking, mountain walking, urban power walking	long distance and/or challenging terrain, careful planning, specialist equipment, experience. Probably a group activity, more of a lifestyle choice, predominantly rural	weather, comfort, safety functionality, practicality, weight

Figure 13.6 Design for Ageing Well 'Walks and Walking' concept map.
Source: Mike Timmins.

for difficult terrains over longer distance walks than those on well-marked smooth paths. The duration of the activity, and means of reaching the point of departure, will affect the weight and compactness of the clothing system; requirements will vary if using transport to arrive at the starting point as opposed to carrying personal clothing and equipment over longer distances (McCann, 1999).

13.4.3 Relationship of walking types to clothing to wearable technology requirements

The concept of the layering system has been proposed as a basis to offer comfortable and adaptable clothing solutions for a range of environments and situations. Through co-design engagement, the team endeavoured to designate clothing types to match the demands of the identified walking types, addressing multifunctionality where possible. The design of the smart garment systems must address the added complication that, as well as protecting the wearer in foul weather, the electronic system must also be protected to preserve operational capability. Conversely, in very hot weather, the usual method of cooling the body is to remove a layer of clothing, but within the integrated technology layering system, a method has to be devised to enable this to take place without destroying the integrity of the electronic system.

In considering the initial identified categories, type 1 and 2 walks are less likely to need the confidence-building attributes of a full multi-layer electronically enabled garment system. It is in walk types 3 and 4 where the smart garments come into their own, as experienced by walkers in the project case study. The needs of those carrying out type 1 and 2 have not been neglected, however, since users of type 3 and 4 walks could reasonably be assumed to take 1- and 2-type walks, albeit in different circumstances. In fact, during the lifetime of this case study, the outdoor industry has increasingly recognised the need for lifestyle clothing, with less extreme protection, and more inclusive styling, to cross over from 'Hill' or 'Trek' categories to cater to walking in urban environments (e.g. Sprayway, 2013 and Rohan, 2014).

For all end uses, robust electronic systems for smart garments, particularly interlayer connections, need to work faultlessly and power supplies for them made reliable. If global positioning system (GPS) is a feature of the layered system, as in the Design for Ageing Well case study, then terrain can markedly affect signal recognition and transmission, particularly as more remote locations are less well served by earth satellite systems. The design of the technology user interface, with easy access to controls and feedback, should be useable even in hostile conditions.

13.5 Creating personas to guide the design process

13.5.1 Personas

One of the main pillars of the Design for Ageing Well project was that the garment layering system, with the enhanced functionality of wearable electronics, should address the real needs of real people. The challenge was to develop garment designs with attributes to correspond to the demands of users within the context of the walking types identified. To bring this to a wider audience, the research team considered the concept of 'personas' as a way of developing a clear description of the type of person likely to use the garments to guide the design development process. Originally used, along with a companion concept 'scenario', as a method of describing the desired attributes of a character in a theatrical performance, the persona approach has become used in a number of widely different fields.

'All the world's a stage, and all the men and women merely players…' Shakespeare, *As You Like It,* Act II, Scene VII.

Personas may be used as 'scenarios or profiles created to inspire and guide design. They are typically visual and textural descriptions, but, ideally they are results of studying real people'. They 'are less effective if the audience is diverse – by definition personas are narrow descriptions. They're best suited to homogenous audiences or niche markets' (Ireland, 2003, p. 28). This approach was widely used in advertising in the 1980s and 1990s and subsequently adopted in technology design. 'Personas are archetypal users with specific goals and needs based on real market and design research' (Don & Petrick, 2003, p. 75). These authors suggest that persona descriptors often include: a name, a photograph, demographic characteristics, specific goals and needs, technographic characteristics, behavioural characteristics, barriers and/or challenges.

Specific user requirements can be aligned with business needs and technical goals to create a prioritised list of features and functions. 'To this end, the personas provide a common understanding of whom the product is being designed for' to 'prevent the project team from making decisions based on personal preferences and biases, thus helping to curtail "feature creep" and feature debates' (Don & Petrick, 2003, p.76). For each product category, more than one persona is usually created, but one persona should always be regarded as the primary focus for a specific design. The feeding in of real data from ethnographic research allows design teams to avoid generating personas as stereotypical users that may bear no relation to the actual user's reality. The number of personas to be created may be determined by developing a target market segment matrix.

13.5.2 Sorting and prioritising persona needs

'Although personas have unique and distinct goals and come from unique demographic or even technographic situations, they may share common needs. Some needs may be shared by all personas while some may be unique to a single persona'. (Don & Petrick, 2003, p. 78).

For example, for all walking types, users require comfort that embraces common aspects such as protection from the elements, moisture management, thermal regulation and ease of movement. In terms of product development, lightweight articulated shell garments with hoods, for extreme conditions, would represent one end of the matrix, while calf-length quilted coats would be more suited to the casual/leisurewear end of the scale. In terms of multifunctional end use, for example, soft-shell garments could be worn across the walking types and cross over into everyday dress. Aesthetic choices in the styling, fabrication and use of colour are more personal. In looking at

Table 13.3 **Persona needs**

	Persona 1	Persona 2	Persona 3	Persona 4	Persona 5
Need 1	X	X	X	X	X
Need 2	X		X		X
Need 3		X	X	X	
Need 4	X		X	X	
Need 5	X				
Need 6		X	X		
Need 7				X	
Need 8			X	X	
Need 9	X	X			
Need 10	X	X	X	X	

Source: Based on Don and Petrick's guidance, 2003, p. 78.

Table 13.4 **Persona needs: sorted by shared needs**

	Persona 1	Persona 2	Persona 3	Persona 4	Persona 5
Need 1	X	X	X	X	X
Need 10	X	X	X	X	
Need 2	X		X		X
Need 4	X		X	X	
Need 3		X	X	X	
Need 8			X	X	X
Need 9	X	X			
Need 6		X	X		
Need 5	X				
Need 7				X	

Source: Based on Don and Petrick's guidance, 2003, p. 78.

the technology requirements of the user reference group, a breadth of responses varied between a serious walker, being adventurous with technology, another with little interest in electronics who preferred a traditional map, a more sedate user being relatively technology oriented to another who was only interested if the 'device' was entirely reliable. Don and Petrick suggest two steps: one in creating a Persona Needs chart, to identify which personas share each need (Table 13.3), and a second chart to prioritise the 'Persona needs sorted by shared needs' (Table 13.4).

13.5.3 Design for ageing well persona development

It was recognised that the range of older research participants, both men and women, within the specified age bracket (60–75), represented differing degrees of fitness. The project aim was to develop clothing, and wearable technologies, to enable varying levels of participation in healthy exercise, in different types of walking, in contrasting terrains, and in a range of seasonal environments. At an early focus group, the active ageing user reference group had offered general agreement in specifying four different walking types, given that participants might identify with more than one category. For example, a serious hill walker could also be an urban walker, with quite different levels of protection needed for the different activities, and with obvious contrasts in terms of the requirements of style and appearance. It was also evident that there were quite varied levels of interest and competence within the group in the functions and usability of wearable technologies. It is worth noting here that although 'fitness level' is associated with health, with the project aimed at healthy and fit 60- to 75-year-olds, it would be surprising if this age group was not subject to some health problems. The 'fitness' factor was included so that medical issues could be toned down but not forgotten, given that the research remit has also been to bring comfortable and functional clothing to a wider community, beyond dedicated walkers, to encourage participation in healthy exercise.

An initial persona workshop took place, with representatives of the cross-disciplinary academic research team, where persona profiles were fleshed out based on older participant feedback but with a degree of creative imagination. On reflection, however, it was agreed that, in the context of the research, the personas should be more closely representative of the characteristics of the project user reference group in terms of behaviours related to their clothing and technology needs, wants and aspirations. The scope of the findings from real data collected from the co-design workshops, through video transcripts, still photography, and the observation of users, would be synthesized into persona profiles mediated through the tacit understandings of the academics as experienced designers and technicians. The technical and aesthetic garment style needs captured, in relation to the breadth of walking requirements, would be prioritized and balanced with findings from technology-oriented user questionnaires to inform the development of the persona profiles. The resulting personas, representing typical users, would influence technical and aesthetic style decision making in combination with prescribing the positioning and functionality of the wearable electronics.

Four male and four female personas, in the 60–75 year age range, were created to correspond to the different types of walking that, in turn, linked to the different levels of functionality within the clothing and technology that would constitute the 'layering system'. The initial overview was based on generic aspects of the personas related to the walking types, without the characters being fleshed out in detail. Clothing attributes were designated in relation to the demands of the walking types, from casual to extreme requirement (Table 13.5). Photographs were then reworked to represent anonymised persona characters. The persona descriptors were designed to relate to user situations and needs designated in relation to: Life stage, Living situation, Family/Peer group, Fitness level, Interests. The overview of needs and activities was to be associated primarily with clothing design requirements in terms of style and level of functionality in relation to the demands of the walking types. The persona profiles were also to reflect differing levels of potential acceptance and uptake of technologies.

Table 13.5 **Design for Ageing Well persona planning**

Persona Male 1	**Persona Male 2**	**Persona Male 3**	**Persona Male 4**	**Persona Female 1**	**Persona Female 2**	**Persona Female 3**	**Persona Female 4**
Anonymised photographs of 4 men and 4 women with their obvious contrasts in style from clothing designed for casual to more extreme requirement.							
Walking Type 1	Walking Type 2	Walking Type 3	Walking Type 4	Walking Type 1	Walking Type 2	Walking Type 3	Walking Type 4
Context to be given for each persona regarding: Life stage, living situation, family, fitness level, interests, etc.							
Clothing: Smart/ Casual	Clothing: Multifunction	Clothing: Moderate End use	Clothing: Extreme End use	Clothing: Smart/ Casual	Clothing: Multifunction	Clothing: Moderate End use	Clothing: Extreme End use
Technology: Highlighting differing levels of demand for smart textile attributes to cater to the different walking types while allowing for varied wearable electronics functionality in terms of the differing persona characteristics and their interests in, and acceptance of, electronic devices.							

13.5.4 Persona profiles

Personas were created with behavioural characteristics synthesised into four male and four female profiles. Personas were attributed with varied design requirements with respect to certain generic needs of relevance to overall human comfort, in relation to the demands of the ageing body, and to the project-specific requirements of different types of walking (Figure 13.7). The personas were created to reflect particular cultural and social habits with peer group influences that would impact on their aesthetic design considerations. In response to insights gained in observation and in the feedback captured from the user reference group, there was little need for correlation between the walking type and the demand for technology. This reflected the fact that real users might be extreme walkers either committed to using traditional maps or conversant with the latest communication devices, while quite sedate users could also be confident technology users! The personas were devised with respect to both generic shared needs and more individual needs sorted through findings in co-design engagement with requirements that would directly affect the selection of clothing product types. Shared needs, in terms of clothing, were predominantly to do with size, shape and proportion, age-appropriate styling, colour and fabrication within the context of the different walking types. Persona size and shape was predetermined relative to SizeUK designations of small, medium, large and extra large men and women typical

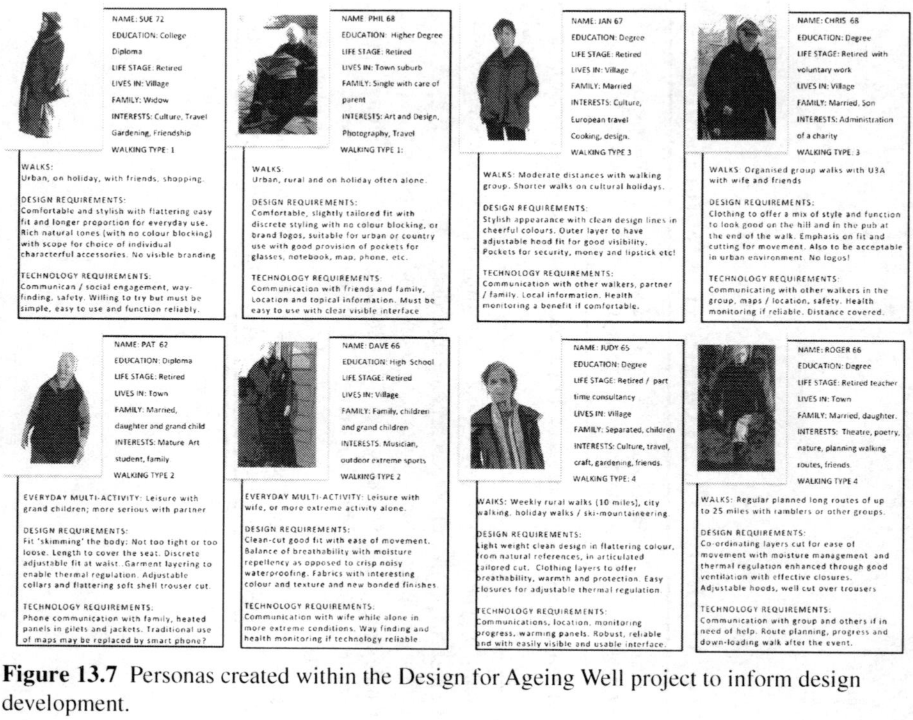

Figure 13.7 Personas created within the Design for Ageing Well project to inform design development.

in relation to their age group. As noted, persona technology uptake did not correlate with levels of interest in clothing technologies or walking types.

13.6 Creating a range plan to cater for different walking requirements

The persona profiles, based on real user needs, contributed to informing the design engineering and technology provision in collating a 'range plan' for a clothing layering system to cater for the different walking categories with varied levels of user participation.

Desired styling for the active ageing was found to be less to do with transient fashion and more to do with comfort and style lines and proportions that are flattering to a range of figure types. Both male and female research participants were conscious of their ageing body shapes. They had favourite categories of clothing, with strong feelings about fit and proportion and length, with some personal items that were brought to workshops described as 'an old friend'. Some users admitted to buying more than one garment if they found a suitable style. They communicated anxiety at the speed of fast fashion that tends to prompt decisions that have to be made quickly and with the inability to replace a favourite garment at end of use. They wanted more enduring styling in a greater choice of colours and fabrics. An important consideration for bringing design to near market is that women often make the purchasing decisions for men. One male user didn't know the origin or price of his clothing!

The following selection of garment categories was prioritised during the co-design workshops as users became familiar with the principles of the functional layering system, having tried on and often having tested garments (Table 13.6). As the participants communicated with each other and with industry stakeholders, they gained confidence in contributing to specifying a range of garment types to cover the designated walking types to encompass casual, urban, and more extreme exercise and travel. A general preference for organic, rounded, style lines was agreed upon throughout the layering system. Style lines were planned for seaming in strategic positions, such as shapely side panels, to provide varying degrees of fit across the four sizes, and following the contours of the arm, to provide cleaner lines while enhancing movement. A range plan of approximately 12 styles for men and women provided the starting point for the subsequent iterative co-design development phase (Figures 13.8 and 13.9). This is described in the following chapter where users continue to contribute to the experimentation and refinement of working prototypes.

The co-design style development was carried out over the following year (July 2011–July 2012) in close collaboration with the academic and industry stakeholder network. Designers continued in their key role as translators, interpreters and visualisers, helping older users to communicate their practical design requirements. Prototype development demanded specialist academic knowledge and skills, supported by industry expertise, and with access to novel manufacturing processes. The

Table 13.6 Design for Ageing Well range plan

12 styles for Women Sizes S, M, L & XL	*F1: Body Layer* Sports bra with sensors	*F2: Base Layer* Santoni body mapped knit	*F3: Base Layer* Short sleeve wicking 'T'	*F4: Base Layer* Long sleeve wicking 'T'	*F5: Base Layer* Camisole with secret support	*F6: Mid Insulation Layer* Gilet with heated panel
	F7: Mid Insulation Layer Textured fleece jacket	*F8: Soft Shell* Multi function Jacket	*F9: Outer Shell* Hooded waterproof jacket	*F10: Outer Shell* Calf length urban coat	*F11: Soft Shell* Trouser with stretch for comfort	*F12: Outer Shell* Waterproof over trouser
12 styles for Men Sizes S, M, L & XL	*M1: Body Layer* Close fitting vest with sensors	*M2: Base Layer* Santoni body mapped knit	*M3: Base Layer* Short sleeve wicking 'T'	*M4: Base Layer* Long sleeve wicking 'T'	*M5: Mid Layer* Long Sleeve shirt	*M6: Mid Insulation Layer* Gilet with Cargo pockets
	M7: Mid Insulation Layer Textured fleece jacket	*M8: Soft Shell* Jacket with bonded zips	*M9: Outer Shell* Hooded waterproof jacket	*M10: Outer Shell* Calf length urban coat	*M11: Trouser* Water repellant with cargo pockets	*M12: Outer Shell* Waterproof over trouser

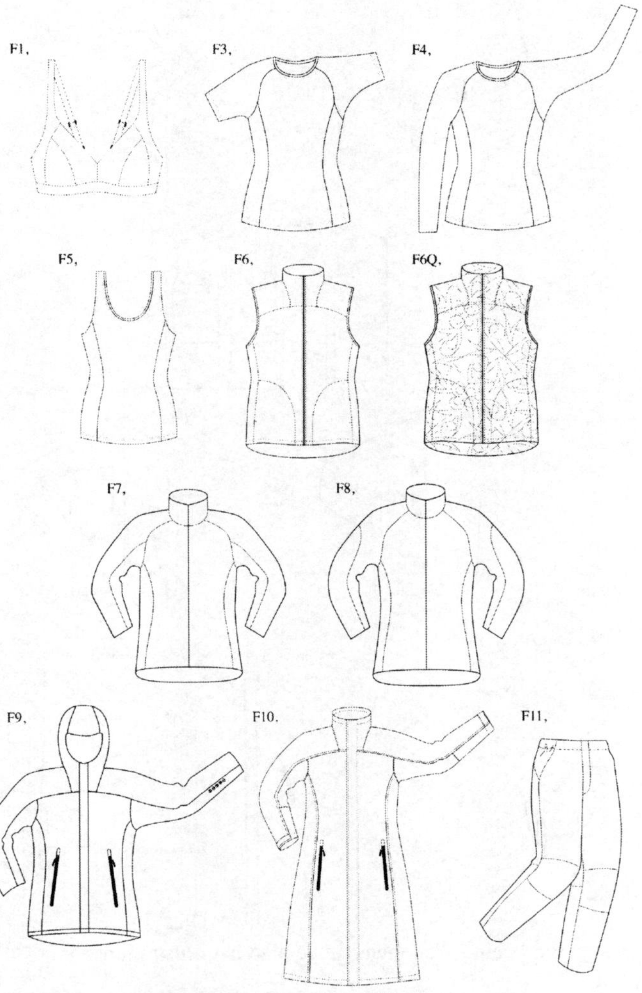

Figure 13.8 Design for Ageing Well ladies range plan flat illustrations.
Source: Chris Lewis.

development focused on close collaboration with the user reference group size and shape models for regular fittings with further design adjustments normally made in discussion with the wider group of participants. The overall shapes of the co-ordinating layering system were developed and the positioning of the design detail was confirmed, including the incorporation of the electronic components. Emphasis gradually moved towards industry collaboration in the production of prototypes. Fabric selection, directly relevant to garment cut and fit, represented a realistic compromise between end user and designer aspirations and what was commercially available within the time span and scope of the project. This phase is discussed in greater detail by the author in the following chapter.

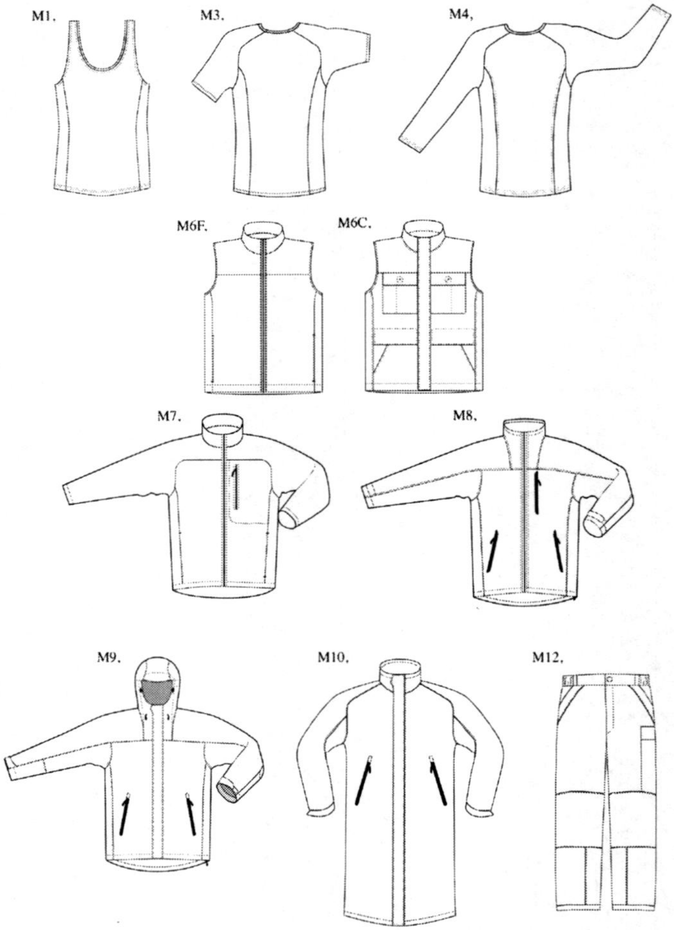

Figure 13.9 Design for Ageing Well mens range plan flat illustrations.
Source: Chris Lewis.

13.7 Future trends

Co-design has initiated understanding between older users and the outdoor clothing industry. While researchers have continued to progress the co-design development of working prototypes, described in chapter 21, supporting activity has been on going in bringing an awareness of this potential market to the attention of textile and clothing manufacturers at trade events. Along side the in-house co-design workshops, a lead user (Chapter 15), also a member of the older Persons Reference Group (OPRG) within the New Dynamics of Ageing (2014) research programme, attended both project workshops and the major international trade event, ISPO, Munich in 2010. Her challenging discussions with sales and marketing representatives made leading brands more aware of problems within existing product ranges (Figure 13.10). The

Figure 13.10 Lead user at ISPO trade event in 2010.

challenges of difficult closures, hood volume, and size and fit issues for older consumers had not been brought to their attention before. In addition, selected representatives of the DfAW User Reference Group attended ISPO in 2012 to gain an overview of the essence and scale of the performance sports industry and the outdoor market in particular. Building on co-design stakeholder engagement has begun to alert industry collaborators to the importance of the potential size and value of the active ageing market with the realisation that innovative approaches are needed to address new market needs. The lead user also took part in a comparative review of the retail environment (in London), from an active ager perspective, with project researchers observing. Older user confusion with complex point-of-sale information is the topic of research papers associated with the case study (Taylor, 2010) while the training of retail staff to address the ageing consumer has been the topic of a conference presentation by Gladstone (Mischke, 2012) (Figure 13.11).

Current research is timely in contributing to the growing recognition of demographic change, within the international sports trade. ISPO, 2012, Munich, introduced their lead article, 'Live Healthy, Live Longer', on the front cover of 'ISPO News', 2012, with an upbeat image of a happy, active ageing couple in the outdoor environment. Within this technical textile and smart materials–driven industry, seminar sessions and master classes are taking place alongside trade exhibitions, raising awareness of the link between healthy exercise and wellness. In addition a WT: Wearable Technology conference runs concurrently alongside the ISPO trade fair. It follows that there will be more of a demand for appropriate clothing to cater for the particular design requirements of this growing active ageing sector and, hence, the need for clear design guidance. In the last four years, the Design for Ageing Well project has had regular exposure at the ISPO under the banners of 'Best Ager' in 2010 and 2011 (Bieker 2010), 'ISPO Generations' in 2012 and in 2013 'ISPO Healthstyle' (2013) (Figure 13.11). At this final event, within the case study time span, the working proof-of-concept prototypes, resulting from the co-design process, were exhibited. Early research findings, shared with collaborating companies, have resulted in initial trade acceptance and uptake of modified product detail in current ranges as described by Greengrass in this publication (Chapter 18).

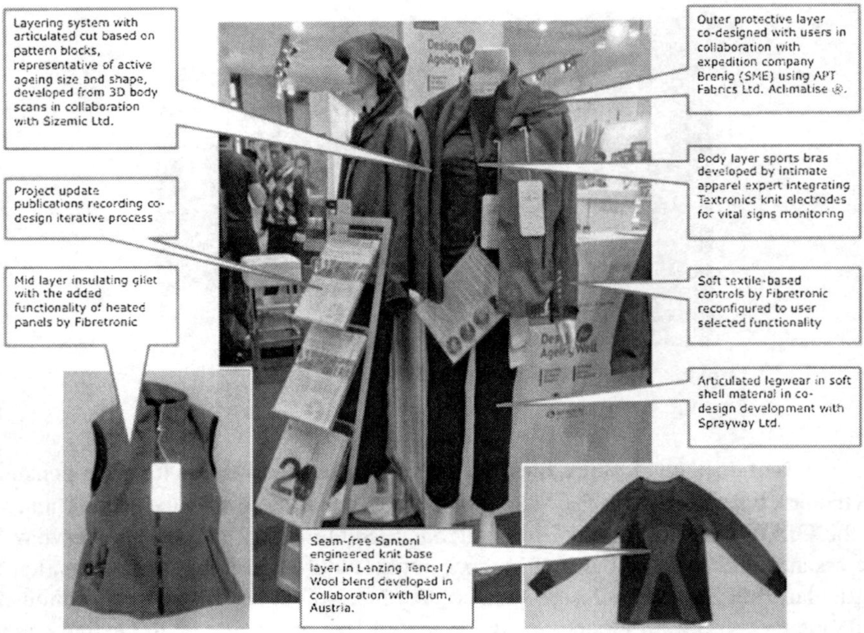

Figure 13.11 DfAW exhibit at ISPO HealthStyle, Munich (2013).

13.8 Conclusion

This chapter has reflected on a project that brings a new co-design approach to functional clothing design for the benefit of a rapidly growing and currently neglected generation that, since teenage years, has had design throughout their lives but now feel abandoned. It has shown how co-design engagement can uncover their real needs, wants and aspirations, as well as the limitations and perceived barriers to be addressed in designing for this diverse community. It hopes to raise the profile of active ageing and help to break down preconceptions to do with younger practitioners' gloomy view of ageing. In particular, representatives of the outdoor trade have learnt that older users, and even the general public, do not understand the concept of the sports-type layering system. Terminology such as 'wicking', the attributes of lightweight insulation, and 'breathable' outer protection are undervalued in comparison with the fixation on waterproofness. Co-design process, involving academic and industry technology experts meeting with older research participants, has drawn attention to the 'technology push' of predominantly youth-oriented electronic products that are unusable by the growing ageing consumer with failing eyesight and reduced dexterity. The active ageing should not be perceived as old and infirm and, as Dame Vivienne Westwood has warned us, in criticising the rise of disposable fashion, 'it makes us all look the same and only 70-year-old women display any sense of style these days' (Whitehorn, 2012).

References

Berghaus. (2014). *Berghaus - Outdoor clothing, waterproof jackets & rucksacks*. http://store.berghaus.com. Last accessed 17.05.14.

Bieker, C. (2010). *ispo Best Ager market research*. http://media.nmm.de/47/ispo10_bestager_presentation_eng_24372647.pdf. Last accessed 17.05.14.

Don, A., & Petrick, J. (2003). User requirements: By any means necessary. In B. Laurel (Ed.), *Design research: Methods and perspectives*. Cambridge, Mass, USA: The MIT Press.

Ireland, C. (2003). Qualitative methods: From boring to brilliant. In B. Laurel (Ed.), *Design research: Methods and perspectives*. Cambridge, Mass, USA: The MIT Press.

ISPO. (2013). *ISPO Healthstyle Guide*. http://media.trainer-magazine.com/epaper/2013/ispo_en/index.html. Last accessed 11.07.14.

http://www.ispo-mediaservices.com/prj_153/_inc002/epaper/Magazin/index.htm (Last accessed 26.06.13)

Keela International Ltd. (2014). *System dual protection*. http://www.keela.co.uk/system-dual-protection. Last accessed 17.05.14.

Lenzing. (2014). *Tencel The new age fibre*. http://www.lenzing.com/en/fibers/tencel.html. Last accessed 17.05.14.

McCann, J. (1999). *Establishing the requirements for the design development of performance sportswear* (M Phil Thesis). University of Derby.

McCann, J. (2009). Smart clothing for the ageing population. In J. McCann, & D. Bryson (Eds.), *Smart clothes and wearable technology*. Cambridge, UK: Woodhead Publishing Ltd. (in association with the Textile Institute).

McCann, J. (2013). Smart protective textiles for older people. In R. A. Chapman (Ed.), *Smart textiles for protection* (pp. 224–275). Cambridge, UK: Woodhead.

Mischke, J. (2012). *A review of wearable futures: Inclusive design conference*. http://www.wearable-technologies.com/2012/05/268/. Last accessed 17.05.14.

New Dynamics of Ageing. (2014). *New dynamics of ageing: A cross-council research programme*. http://www.newdynamics.group.shef.ac.uk/. Last accessed 17.05.14.

Páramo Directional Clothing Systems. (2014). *Påramo waterproof jackets, trousers and outdoor clothing*. http://www.paramo.co.uk/en-gb/index.php. Last accessed 17.05.14.

Rohan. (2014). *Outdoor & travel clothing*. http://rohan.co.uk. Last accessed 17.05.14.

Santoni. (2014). *Santoni - Seamless story*. http://www.santoni.com/en-azienda-seamlessstory.asp. Last accessed 17.05.14.

Sprayway. (2013). *Sprayway - Outdoor clothing specialists*. www.sprayway.com/. Last accessed 17.05.14.

Taylor, D. (2010). *Developing a Visual Language to Enhance Knowledge Transfer in the Design of Smart Clothes and Wearable Technology for the Active Ageing*. Conference, Design Principles and Practices, Rome.

Taylor, D. (2011). Developing a visual language to enhance knowledge transfer in the design of smart clothes and wearable technology for the active ageing. *Design Principles and Practices: An International Journal*, *5*(6), 369–387.

Textronics Inc. (2014). *Bio-monitoring, wearable monitoring, electronic Textiles & wearables by textronics*. http://www.textronicsinc.com/. Last accessed 17.05.14.

The Ramblers (2014). *Making the case for walking*. http://www.ramblers.org.uk/what-we-do/making-the-case-for-walking.aspx. Last accessed 17.05.14.

Whitehorn, K. (2012). *Vivienne Westwood may think women are stylish at 70 – but where's the help?* http://www.guardian.co.uk/commentisfree/2012/feb/20/vivienne-westwood-stylish-7. Last accessed 26.06.13.

Public involvement in garment design research

14

T. Williamson[1], S. Hinder[2]
[1]University of Salford, Salford, UK; [2]RaFT Research, Clitheroe, UK

14.1 Introduction

INVOLVE (2009) describes public involvement in research as an activity where the public is actively involved in one or more stages of the research design or undertaking process, e.g. as research team members or study advisers but not as research participants or subjects. Policy drivers for public engagement in health research are increasingly evident in a variety of global settings including developing countries (Tindana et al., 2007). Yet there exists reluctance amongst many health and social care researchers to engage with the public in their research endeavours. Researchers are increasingly being influenced by others who expect demonstrable input of the public in the design and conduct of research, including ethical review bodies and research funders (Hanley, 2005). Evidence is growing concerning the benefits and challenges of public involvement in health and social care research. Whereas in the past available evidence has not been well collated, this has recently been rectified, and an overwhelmingly positive impact of involvement has been reported on across the different elements of the research process. For example, positive impact has been found in relation to qualitative research and clinical trials where involvement has helped elicit participants' views and aided trial design and determination of outcome measures. Furthermore, participant recruitment improved through involvement (Viswanathan et al., 2004) and the likelihood of getting research funded as a result of involvement increased (Lindenmeyer, Hearnshaw, Sturt, Ormerod, & Aitchison, 2007). Less evidenced impact has been found in relation to commissioning and research funding.

Policy moves have been towards promotion of public involvement in research (Department of Health, 2006) yet the research workforce has not been in a state of full readiness to embrace its principles. Training in public involvement has previously been lacking (Lockey et al., 2004). Some health care professions such as nursing and midwifery have been identified as having particular need of capacity building to develop their research activity in general (Green, Segrott, & Hewitt, 2006) although these professions have made good headway with regard to public involvement in research (Smith et al., 2005). To address the need for greater public involvement in research, significant effort is being invested by a number of providers, including higher education institutions, in associated training and conference provision. Additionally, in England, researchers can benefit from a number of regional and national support and advisory services such as National Institute for Health Research (NIHR)

Textile-led Design for the Active Ageing Population. http://dx.doi.org/10.1016/B978-0-85709-538-1.00014-6

Design Services and INVOLVE. The authors' experiences in roles within these groups suggest researchers often struggle with the application of principles of public involvement in their research. Consequently, the quality of research funding applications with regard to public involvement is often lacking.

The authors' position is that they advocate involvement as the morally correct thing to do (Williamson, 2006). This view is based on the belief that not only do people have the right to be involved in issues and decisions that affect them, but also that research outputs stand to be of a better quality (relevance, utility) than in the absence of public involvement. An increased likelihood of achieving research funding could well be viewed as an incentive to some researchers to embrace public involvement in their research activities. For others it is the potential for research of greater quality and potential impact that appeals.

Whilst within the literature many terms including *user*, *service user*, *public and patient* and *end user* are used interchangeably with the associated terms of *involvement* or *engagement*, they may have very different meanings for some people. This chapter adopts the INVOLVE explanation of involvement to mean active involvement in the research itself (INVOLVE, 2009), perhaps as an advisor or a co-researcher, and not merely a role as a subject or participant in research.

14.2 Background to public involvement in design research

Whilst policy drivers exist around promoting public involvement in research (Hanley, 2005), the filtering down of these into the day-to-day business of doing research presents challenges. There exists substantive academic literature (Faulkner, 2005; Warburton, Bartlett, Carroll, & Kendig, 2009) and information materials (INVOLVE, 2012) concerning the principles of involvement in research. What is lacking is published material aimed at facilitating the translation of those principles into practice. Commonly researchers appreciate the 'what' aspects of involvement but not the 'how' elements. Whilst the authors argue that these materials and insights are of direct value to those concerned with garment design research, they have been developed with a particular audience in mind. Therefore, this chapter seeks to address the dearth of guidance for garment design researchers and translate the 'how to' of public involvement for this audience. It is then up to others to evaluate the application and impact of these principles in this field and develop an evidence base for the garment design industry. The principles of involvement explored here are widely relevant to research practitioners, research managers and commissioners of research alike.

This chapter begins with issues concerning the planning, design and undertaking of research and latterly considers late stages of the research process including dissemination and life after the study ends with regard to public involvement. The additional processes of study management and evaluation of involvement are also presented.

14.3 Planning for public involvement

In determining your public involvement approach, there are several issues to consider at the outset. These include who to involve and how to locate them, payment issues, training and support needs and how to end the involvement aspects of a research study. Each issue will now be considered in turn.

14.3.1 Whom to involve

An often forgotten about or ill-addressed aspect of involvement comes prior to the start of any study. It is the preparation element whereby those planning a research proposal often do not think to gain the views of those who the research outcomes are for. It is always a good idea to seek the views of any stakeholders even informally, but this is essential with regard to the public or users of products or devices being developed and possibly their informal carers (spouse, family etc.) depending on the research topic. These people can be available to you very easily, for example a neighbour, family member or community group member. Approaching even one or two people informally from the population the study intends to target, to give their thoughts on a research idea, is permissible. They may help you think through whether you need to have an opportunity to get a few informants together in a more structured way such as a listening event (facilitated group discussions) or by the research team attending an occasion where these people meet already, such as a local luncheon club, walking group, Carers' Forum meeting etc. These people would need to be as close in situation and personal characteristics as possible to those who the research is to focus on. You may also wish to gain insights from people who do not have the characteristics you want but hope they will develop them, e.g. would-be keep-fit enthusiasts rather than those already committed to regular exercise. Such involvement may help prevent you perceiving issues with 'tunnel vision' and open your eyes to alternative viewpoints or a greater understanding of people's behaviour. In this way you can help prevent research from being professionally dominated and so more likely be focused on the needs and wishes of the population and target audience it affects.

People to involve can be located and approached through existing local, national or international organisations and groups, both those that are formally constituted as well as informal social groups or clubs. Be aware that some may charge for access to members for research, but in the authors' experience, most are happy to help with a good quality study of relevance to its members. Make use of a relationship you have or can forge with a worker or community leader relevant to your field of interest who has access to the population you are interested in. Work with the public to locate others like them especially if they are difficult to locate, such as marginalised or seldom heard groups.

14.3.2 Participant remuneration

When thinking about who to involve, it is also a good time and opportunity to think about payment, reward and recognition for involvement. Everybody's time is valuable and you may want to seek the public's own views regarding payment. Some people may be insulted by this as they want to 'give something voluntarily'. Others will want payment for their time and/or reimbursement for travel. These costs can then be built into any subsequent research funding application; however, if the involvement is to help design a study or funding application, funds need to be found. The key is to be non-tokenistic, although it is acknowledged that some funders do not yet support the appropriate funding of public involvement in the pre-application or funded stages of a study. The authors suggestion is to ask people what they would ideally want early on and make clear what is possible. By way of example rather than as a guide, the authors generally try to offer the public £75 per half-day attendance at a meeting or other research activity. Sometimes we can only provide refreshments and lunch but always offer travel reimbursement.

It is also the authors' belief that the public should be signposted to support agencies to check that any payment and reimbursement will not affect any state benefits that they may receive. Individuals should always be advised to check with their local benefits office as to their individual circumstances regarding payment issues. It may be helpful to establish early on what policies, mechanisms and constraints your organisation has regarding payments; a conversation with a finance department may save time later. For example, the authors advocate that cash is paid on the day of involvement (if it is what the individuals prefer), yet you need to be sure you can do this before promising it. Wherever possible, liaise with payments departments in your organisation as some people are not eligible to pay tax, such as retired people, and the mechanism for claiming tax deductions back can be very onerous.

Travel is usually paid in full rather than an allowance, which can be inequitable depending on individual's travel requirements. No one should profit from excessive travel expenses and nobody should be out of pocket for expenses. Pay travel costs on the day of involvement and ideally arrange transport and book travel tickets so people do not have a large financial outlay to then claim back. Remember that a small amount of money can be a lot to some people. Also consider whether people need overnight accommodation if travelling far or if they are attending early or late in the day.

14.3.3 Training and support needs

Often the past experiences, skills and aptitudes of the public are not appraised during a study, which is a missed opportunity to draw and build upon their individual qualities. If the public are to receive any training for their role in a study, it may help to ascertain their preferred learning styles rather than a 'one-size-fits-all' approach. Similarly, it would be a shame to not make use of skills gained in previous employment and which may be useful within your study, such as creative drawing (for recruitment posters) or recruitment interviewing techniques (for qualitative interviews). People often have transferable skills, and through discussion it can be established whether they would like to make use of them or simply prefer to develop new ones. Members of the public might like to keep a personal record of their learning and experiences, and it may be worthwhile to provide them with a letter or certificate confirming their involvement and contribution.

Having identified training and or support needs (such as accessibility issues, e.g. mobility, visual impairment, dietary needs), you need to identify someone within your team to address them. Plan with the public the best way to meet these issues, and then implement the necessary actions.

14.3.4 How to end a study

The authors advocate thinking about the end of a study at its outset. Researchers need to think early about how to plan withdrawal of the public from what can be a lengthy study where close working relationships and even friendships may have formed. People do not want to feel 'dropped' or abandoned after having 'served their purpose', and so a staged and/or supported withdrawal needs planning. A discussion about this early on is advisable, although for short-term involvement such as a discussion group or consultation meeting, a warm thank you may well be sufficient and possibly a payment for their contribution or a shopping voucher.

14.4 Designing research studies

Several opportunities for public involvement are possible in relation to many research studies' design processes and are described here.

14.4.1 What topic to research

It makes sense to pursue studies of topics that are of particular concern and importance to people affected by them, making sure that there is a particular population you wish to reach or who form a particular market for whatever you are producing (new knowledge or new/adapted products). This could help avoid studies that are based on the whims or 'pet' interests of the research team who may or may not be misinformed. This is arguably the morally correct thing to do for research that is funded by public money. The authors believe this may lead to research that is more relevant to your field of work and more likely to be implemented. Findings are more likely to be credible and thus believable. Sometimes there may be several feasible topics to research, and public involvement may help identify more critical topics amongst others that appear similarly important.

At times the priorities for research will have been determined by the research funder, and funding applications may be in response to a call for a defined study contract to be competed for. For others such as responsive mode funding, research teams gather themselves together and agree on a topic focus prior to formulating a proposal that they then use to apply for funding. There is scope for meaningful public involvement in both types of research, funded research and indeed unfunded or partially funded studies (such as those undertaken for Masters level study or as part of service development) to determine any subthemes or issues of particular importance. Depending on the individual study, stakeholders (such as potential or returning customers) or professionals (such as buyers or designers) can be invited to a forum to compare and agree priorities for research. Venues for such meetings need careful consideration to ensure that they are suitable for the intended purpose. For example, they need to be accessible (from a mobility, visual and audio angle and so on) and suitable for the intended audience, e.g. young people, people with a disability. Other considerations include appropriate time of day, duration, diet/refreshments, frequency of breaks, expenses, the needs of carers, seating, toilets, map and parking and so on.

Support mechanisms need to be inbuilt, such as someone to meet and greet people, a 'buddy' system whereby someone is appointed to support a person, mentoring and issuing of large print documents (that are not unwieldy) at least a week prior to any meetings where they may be discussed. If a meeting or event is held to gain views in this way, then they should always be followed up with a thank you, some means of feedback about what was discussed and an opportunity to hear about the study findings at a later date when available. Researchers need to be aware that people will be from different social, professional and educational backgrounds, and have varying levels of confidence, understanding and experience of working in groups in this way. This may necessitate bringing in the services of a skilled facilitator to lead or support the proceedings and to marry up the needs of the research team with those of the public involved.

Involvement at this stage may be about agreeing on what is the most relevant research question. Whilst a particular question may be important, it may not be the most critical issue for the different stakeholders, hence the need for discussion. Whilst some kind of meeting will likely work well for gaining the perspective of the public, there are other means to consider. For example, depending on the population or patient group, other media may also be used such as questionnaires, online discussion forums, attending existing community meetings, telephone discussions, face-to-face visits to the person's home, or by holding fringe events or workshops at conferences. Sometimes it may be appropriate to seek views from public representative organisations such as AgeUK or sporting organisations such as health clubs to complement rather than replace the direct view of members of the public. Ideally, gaining the views of the public directly is desirable, yet where this is difficult, groups or organisations can provide a general view based on considerable experience of the views of its members or a particular population.

14.4.2 Informing design

Whilst there are no guarantees, public involvement can significantly enhance the quality of planned research. In particular, the public perspective can be of help when applied to development of research questions to ensure that they are the right questions and are asked in the right way so as to gain meaningful answers. The public as informants can help ensure that methodologies are deemed appropriate by the population they are to be used with, such as being fit for purpose and ethically sound. Sampling strategies and recruitment approaches for participants can be made realistic and workable in the real world. Study methods can be selected based on what will be well received by participants. Study tools and documents such as participant information leaflets, questionnaires and interview guides can be reworked in order to be more readable and appealing. Misconceptions can be avoided about the study population such as the belief that older adults do not carry mobile phones.

Study design stage is also the time to consider the possible roles for the public in a study, for example advisor, co-researcher, reference group member, critical reader or simply as an initial informant with no long-term role. It is up to researchers to decide what they want these roles to comprise, but in the authors research, we tend to have advisory groups for projects of 8–60 months rather than short studies. Advisory groups of 6–10 allow for attrition over time. The public cannot always commit to sustained involvement and should be supported to be only involved in aspects of a research study if it suits them and fits with the study. They may not be able to contribute during holiday periods or simply may become limited by ill health or family commitments. However, some study types may need the public to endeavour to attend meetings throughout and this should be made clear when recruiting the public into such roles. Some studies engage the public as full co-researchers if they want to be very 'hands on' and be part of the research in the fullest sense, perhaps being involved in data collection and analysis. It can be helpful to have someone familiar with the research topic and population available on a day-to-day basis rather than being asked for occasional advice. Reference groups tend to have more transient involvement concerning the sharing of members' opinions or commenting on research design and findings and typically have a larger membership than an advisory group. Critical readers tend to comment on draft funding applications, draft reports and other written study outputs. In some studies, the public may have substantive roles whereby they lead the research or indeed control it fully (Turner & Beresford, 2005), but this is less common even in health research.

At a minimum within many kinds of studies, the public can assist in the design of participant information sheets to ensure they are understandable to a lay audience. Furthermore, they can often verify emerging analytical interpretations and findings prior to finalisation. One-off engagement in the form of a consultation, often as an event, is also valuable, but there are limitations should you want to gain advice on further issues or wish to clarify insights after the event.

Researchers need to explore the degree of involvement that is needed for the study and wanted by the individuals concerned with them. If they meet the study requirements, the initial informants you work with in the pre-application for funding stage can stay engaged with the study in later roles should it be funded. It is a good idea to seek continuity in this way. This is not always preferred by the public, whereas others then find they want to stay involved once they have had some input. Whilst it can be useful to have continuity from the outset, this is not always possible, and, for whatever reason, people's commitments may change. The key is to ensure that involvement is meaningful and people are genuinely engaged to add something that may otherwise not be achieved in their absence.

There is a need to be realistic and balance what is desirable with what is possible within confines presented by any ethics committee requirements, funding and time available. Maximise opportunities for involvement and provide training where needed both for the public (such as committee skills, interviewing techniques) and the wider research team (such as public involvement in research, participative enquiry). There is a need to ensure time and funding is built in to undertake these exploratory discussions, plan and provide training and support, and then gain people's involvement as the study is underway. It is important not to underestimate the time and energy these activities will take and to expect the unexpected. Members of the public may have disagreements, wavering confidence, difficulty in working in ways preferred by the research team and so on. During the design stage, mechanisms to support, reward and recognise the public should be in-built. It may be an idea to bring in an outside facilitator if the skills to manage these training and support needs are not present internally amongst research team members. Funding panels will likely be looking to see who has this responsibility on a research team.

14.5 Conducting the research

14.5.1 Roles and activities

Whilst in some studies you might be very clear about what roles you want the public for, in others you can work it out with them during the design stages. Whichever roles are required by the study and whatever the preferences are of those interested members of the public, some careful negotiation may be needed. There will then need to be adequate training for the public and other members of the research team to be able to embrace these roles that may be quite different from previous experience and ways of working.

When planning training for the public, it is important to remember to establish their previous work roles, life experiences and so on from which they may have transferable skills and knowledge. It is important to realise that the public may not know at the outset what kind of involvement they would like. Where possible build in a degree of flexibility that will enable decisions to be made at a later date when people are more familiar with what a study is about and when they have gained a better insight about their ability to contribute. Undertake any preparation and training in a suitable environment. University or similar settings can be quite daunting to some people, and a more relaxed community venue may be preferable.

14.5.2 Drawing out findings

There are a number of opportunities for the public to help in maximising the accuracy and relevance of study findings. They may potentially add to the rigour of analyses and validation of emerging and final findings. The public can examine anonymised chunks of data – perhaps the more problematic chunks that the research team is unsure about – or read one or several anonymised transcripts of interviews, for example. Others may play a valuable role in reviewing full-draft findings chapters depending on their interest and abilities. Some may need to be taught skills to do this meaningfully and skilfully. Suitable ethical body approval will need to be sought for such activities, yet use of anonymised data should meet their requirements. A workshop approach may work best for these activities building upon some individual pre-reading at home. Particular insights that stand to be gained include the public's views as to gaps in the data and suitable questions to fill these. Similarly, study recommendations can be critically commented on or devised with the public to be realistic and accurate. They can help make sure that recommendations do indeed arise from the study findings and not the professional concerns of the research team. Future research ideas can also be generated as the public identify and prioritise what to address next after the particular research has been completed.

14.5.3 Ending the study

Following sustained involvement, this should be designed with those affected and may include lessening the frequency of meetings towards the study end, planning a debriefing session or a celebration to mark the end of the study, and perhaps with a review meeting after several months of dissemination to see how findings were received. It is good practice in the authors' opinion to consider where the public go next. Ideally researchers could help to identify further opportunities for involvement in research or other activities that make use of the public's experiences and skills should they want to. This is not always possible, and people should be warned at the outset that the original study is a finite project so their expectations are realistic.

14.5.4 Evaluating the involvement

It is also important to plan evaluation of the public involvement at the study outset. This is so the notion of evaluation does not lose attention at the study end when researchers' focus may have moved on to finding funding for the next study. Evaluation in itself is important as a means of learning from what worked or did not work and in what context. This is helpful to the study itself (especially if it comprises formative and summative evaluations) and also to add to the evidence base for what works in public involvement in research. Again, the authors' advise discussing evaluation with the public early on so that any measures or evidence of any impact resulting from their involvement are meaningful to them. Both quantitative and/or qualitative approaches should be considered. Foci for questions may include whether expectations were met, impact of involvement on the person and research process, reasons why processes and outcomes were positive/negative and suggestions for improvement. Once evaluation is complete, share what has been learnt with others through presentations and publications and especially with the public you involved.

14.6 Beyond the study

14.6.1 Dissemination

The role of the public need not be limited to contributing to plain language findings summaries or findings posters, although these can be very useful activities particularly in ensuring that they use accessible language. Remember that plain language makes your report more readable to professionals outside your specialism too. Some members of the public may want to be involved in informal presentations and dissemination events and activities to local groups or study participants such as through patient forums. They can help make sure that such audiences are not forgotten about in favour of academic presentations and publication in professional journals. Some appropriately skilled, supported and confident members of the public may wish to contribute to academic conferences and writing for publications and this should be considered and supported whenever possible. Involvement can significantly add to the resonance of study findings when relayed by public representatives of the population with which the study has been concerned.

The dissemination element of a study should always be considered at the outset of a study in the authors' view, and this is the time to mention to members of the public that there will be opportunities for them. It is important to gain their preferences for involvement in dissemination and train and support them accordingly, e.g. presentation skills, poster development. Often the public has really good ideas as to where dissemination is best, what formats will work best and can helpfully arrange access.

14.6.2 Uptake of findings

In many studies, the public involvement may have a large degree of influence when it comes to the uptake of findings by commissioners, policymakers and others affected by the outcomes of research. Research can be viewed as more trustworthy if it has strong public involvement, and if done well involvement can add to the validity and resonance of findings. In the authors' experience, this likelihood is further enhanced if the public is involved in any face-to-face presentation of findings as audiences receive their views first hand.

14.7 Additional processes

14.7.1 Study management

Whilst study management may lend itself less to public involvement than other aspects of the research process, there are possibilities worth exploring particularly in larger projects. A role in the management of a study can be helpful when agreeing on support mechanisms, payments, venues and communication methods with research team members, to highlight just a few. If not present in these discussions, the risk is run that others will make decisions for the public that may not be the most appropriate. The public can often ask the awkward questions and challenge study direction and progress. For it to work, the chair or facilitator of management meetings needs to be aware of the needs of the public and manage the meetings accordingly – that is, ensuring that the public voice gets heard, which includes having sufficient numbers of the public present in the first place and supporting those who are less vocal. Another opportunity is for the public to be involved in the appointment of researchers to a study particularly if they are to work with the public members' peer group, as they will know what personal characteristics will be best received.

A buddy to prepare users prior to any meetings may be helpful. Think about the nature of support the public might need for what may seem obvious to a professional research team member, such as how to ask a question/raise a point at a meeting or how to feed back on a document that they have been asked to review.

14.8 Conclusion

Overall, this chapter has illuminated many of the different opportunities for meaningful public involvement in research studies. Its pragmatic content is drawn directly from the extensive research experiences of the authors in their own studies and consultancy and advisory work regarding public involvement in research. The principles discussed, if incorporated by researchers, stand to improve the quality of the research design and undertaking. Furthermore, such involvement can promote a study's moral standing and commercial appeal in terms of engaging people in matters that concern them and in promoting development of research outputs that are fit for purpose and meet user requirements. Researchers in the garment design field can develop an advantage through consideration and application of the public involvement principles described here. High quality public involvement in research funding applications may well increase chances of gaining funding if it is combined with a well-designed study with a suitable research team. The key is to demonstrate authenticity in involvement and listen to what the public has to say and acting on it wherever possible. Engage the public early and fully and do not be afraid to not have all of the answers. Simply work with the public to realise the best possible way forward within the situation and constraints that you have. Clearly demonstrate that you appreciate the key principles of involvement and how you have operationalised them in your research proposal. Finally, if your research goes ahead, deliver the public engagement element to the best of your ability and evaluate it.

References

Department of Health. (2006). *Best research for best health*. London: Department of Health.

Faulkner, A. (2005). *Guidance for good practice*. Service user involvement in the UK Mental Health Research Network. Service User Research Group England (SURGE).

Green, B., Segrott, J., & Hewitt, J. (2006). Developing nursing and midwifery research capacity in a university department: a case study. *Journal of Advanced Nursing*, *56*, 302–313.

Hanley, B. (2005). *Research as empowerment?* York: Joseph Rowntree Foundation.

INVOLVE. (2009). *Good practice in active public involvement in research*. Eastleigh: INVOLVE.

INVOLVE. (2012). *Briefing notes for researchers: Involving the public in NHS, public health and social care research*. Eastleigh: INVOLVE.

Lindenmeyer, A., Hearnshaw, H., Sturt, J., Ormerod, R., & Aitchison, G. (2007). Assessment of benefit of user involvement in health research from the Warwick Diabetes Care Research User Group: A qualitative case study. *Health Expectations*, *10*, 268–277.

Lockey, R., Sitzia, J., Gillingham, T., Millyard, J., Miller, C., Ahmed, S., et al. (2004). *Training for service user involvement in health and social care research: A study of training provision and participant's experiences (The TRUE Project)*. Worthing: Worthing and Southlands Hospitals NHS Trust.

Smith, E., Ross, F., Donovan, S., Manthorpe, J., Brearley, S., Sitzia, J., et al. (2005). *User involvement in the design and undertaking of nursing, midwifery and health visiting research*. London: NCCSDO.

Tindana, P. O., Singh, J. A., Tracy, C. S., Upshur, R. E. G., Daar, A. S., Singer, P. A., et al. (2007). Grand challenges in global health: Community engagement in research in developing countries. *Public Library of Science Medical*, 1451–1455.

Turner, M., & Beresford, P. (2005). *User controlled research: Its meaning and potential*. Eastleigh: INVOLVE.

Viswanathan, M., Ammerman, A., Eng, E., Garlehner, G., Lohr, K., Griffith, D., et al. (2004). *Community-based participatory research (CBPR): Assessing the evidence*. Evidence Report/Technology Assessment 99. Rockville MD: Agency for Healthcare Research and Quality.

Warburton, J., Bartlett, H., Carroll, M., & Kendig, H. (2009). Involving older people in community-based research: Developing a guiding framework for researchers and community organisations. *Australasian Journal on Ageing*, *28*, 41–45.

Williamson, T. (2006). The individual in research. In T. Long, & M. Johnson (Eds.), *Research ethics in the real world: Issues for health and social care* (pp. 9–28). London: Elsevier.

The co-design process for apparel for the active ageing population: the participant experience

15

M. Sinfield
Older People's Reference Group (OPRG), Sheffield, UK

15.1 Introduction

Mary is a graduate and had been a civil servant in several government departments, with a career break to work in the private sector. A self-maintained divorcée and mother, she had also had 20 years' experience as informal carer to her blind-and-disabled mother and to her elderly former mother-in-law. Post-retirement, Mary became heavily involved with Older People's campaigning groups, through research she had undertaken relating to local use of assistive technology. That research provided knowledge of the Economic and Research Council 'Growing Older' (ERSC – GO) programme, of the Engineering and Physical Sciences Research Council 'Extending Quality Life' (EPRSC – Equal) Initiative, and of Roger Coleman's writings on inclusive design (Coleman, 2002).

Mary had also volunteered as a social and consumer rights advocate for disabled and older people, and joined several London community and interest groups. Through membership of one of the latter groups, she became a volunteer subject for a well-run university medical research project concerning older people's health and well-being. The consequence of these many and varied activities was contact with a range of older people – locally, London-wide, and nationally.

A product of the emergent post-WW II youth culture herself, Mary had become increasingly irritated by the fashion and leisure industries' lack of interest in any older person's clothing requirements, other than office wear. Thinking appeared to have stagnated, being obsessed with, and market-focused on, the late teenage/early 20s body type for leisure wear. The *Design for Ageing Well* project appeared to give an opportunity to change such thinking.

Although a serious feminist since a teenager, politicised by the early 1950s campaigns for equal pay for women, at the same time Mary had maintained an interest in fashion developments and nuances, and learned at evening classes to make her own clothes. She is not one to condemn young people's seemingly outrageous dress styles, being herself notorious for her miniskirt and 'Joan-of-Arc' hairstyle in 1957; moving on to multi-coloured beehive hair a little later, and industrial-type footwear and clothing from the 1960s.

Textile-led Design for the Active Ageing Population. http://dx.doi.org/10.1016/B978-0-85709-538-1.00015-8

Whilst career progression and later marriage and motherhood brought dress sobriety and conformity, experience had given Mary the basics for appreciating developments in materials, styles and cut, colour and pattern, and body shapes, the last with particular note to changes over time in her own and those of her peer group and of her elderly charges.

15.2 The New Dynamics of Ageing (NDA) programme and the Older People's Reference Group (OPRG)

Through links with Help the Aged and the Better Government for Older People initiative, Mary was made aware of the Joint Research Councils' New Dynamics of Ageing (NDA) eight-year multi-disciplinary research programme with 35 collaborative projects. Innovatively, the programme was seeking to create a reference group of older people (OPRG) to complement that of the advisory group of professionals: scientists, researchers, and representatives of organisations focused on services to older people. From their 'insider' perspective, OPRG members were to help to monitor outputs from the programme and to consider their implications for older people.

The key issue here was user engagement. Members were required to act as a resource on issues concerning the involvement of older people in all stages of the research process, and, where requested, to give advice to individual research projects within the programme. In due course, the group was expected both to write a report on their combined NDA experiences (now published, see reference below) and to encourage discussion of the findings, to disseminate the outputs from the programme – especially to older people.

Given her life experiences, Mary was encouraged to apply to join the OPRG. She volunteered because she had discovered that, despite her close interrelationship with her mother and former mother-in-law, it is only by being an older person that one truly understands the problems and issues of ageing in later life. Further, Mary wanted to be involved with a programme that included older people actively working from the outset with other older people, and not merely with their self-styled representatives.

With hindsight, it is unfortunate that the OPRG was not established in advance of the initial selection of NDA projects. Thus it was not possible to include an OPRG representative member on panels determining either the focus of the overall programme or much of the subsequent commissioning.

OPRG meetings were held three times yearly, from October 2007. The first few were concerned with developing both the interrelationships of members and to clarify the aims of the group. Members were also provided with details of each project and the contact details of the principal investigator (PI) and asked to choose with which three (or more) projects they wished to become involved.

It was left to each OPRG member to determine when and how to make the initial contact, and – again unfortunately – by the time choices had been made, the initial project steering groups had usually been established, and the OPRG members found themselves 'parachuted in' to PIs who were uncertain of the expected, or possibly predetermined, OPRG role.

Mary chose to become involved in five NDA projects, and to become a volunteer research subject in two others. Exceptionally, she had volunteered to help without cost to project budgets and was willing to consider involvement at all levels. Initially, Mary attempted to make telephone contact with PIs, following up by e-mail with providing details of her background and particular interest in the project topic. Frustratingly, often contacts were elusive and responses were slow and sometimes discouraging.

15.3 Engaging in the Design for Ageing Well project

Contact with the project began with long phone calls made by Mary in November 2009 to the PI and her postgraduate research assistant to introduce herself, and to arrange for her to make a personal visit to the Caerleon Campus of University of Wales Newport to discuss in depth the requirement for, and the extent of, her role as an external volunteer.

The first meeting was held on December 8, 2009. Detailed explanations were given of the roles of the multi-disciplinary team and the levels of expertise and experience of its advisory group members, researchers, and consultants, as well as of the expected contributions from volunteer older people. Mary was briefed on where and when it was intended to hold meetings, research workshops, and conferences, and she was also provided with a copy of the initial project plan.

For her part, Mary offered advice on potential sources of older people as volunteer recruits and their subsequent support needs, and to provide a critique of intended communications to targeted groups. She made clear that, although NDA funding assumed her expenses would be met from the project's budget, to enhance the flexibility of her input, her participation would be self-funded.

As an introduction to the current offerings in the outdoor leisure wear and fashion industry, Mary was pleased to accept the invitation to accompany the team to Munich on February 8–9, 2010 to attend the Winter Internationale Fachmesse für Sportartikel und Sportmode (ISPO Munich, 2010. The 4-day International Sporting Goods Trade Fair at which more than 2400 international exhibitors presented their latest products to some 80,000+ visitors). This was to be followed by a browsing expedition to sportswear/leisure clothing shops in central Munich, and subsequently by a shopping expedition to Covent Garden in June 2010 (when Mary intended to buy clothes for a planned cross-Europe train journey). On both occasions, Mary's activities were observed by the team, photographed where possible, and her comments noted – in Covent Garden, on tape. To help the team, Mary even volunteered to transcribe the recordings afterwards – a task she vowed after the work never to repeat!

15.4 Training of volunteers in user engagement

Supporting a training event at Salford University in July 2010, Mary was greatly impressed by the wide focus on the principles of service user involvement in research projects. Whilst the presentation had been based on research within NHS fields, the principles were clearly shown to cover volunteer involvement in research generally.

Issues ranged through justification, topic selection, management, payments and support of volunteers (input, ethical concerns, monitoring, output, and evaluation), and impact on the researcher. Throughout the event, Mary was encouraged to participate in the workshop, whether supporting or challenging the speaker.

Impressive though the arrangements and output of the entire day had been, Mary was disappointed that this training had not been available at the beginning of the *Design for Ageing Well* project, and also to PIs and their research assistants more generally. From the literature provided at the end of the training event (encapsulated in Williamson & Hinder, 2009), there is obviously much information available on user engagement to prospective researchers. Mary wondered whether academics now give preference to studies clearly relevant only to their own particular field, overlooking those from other disciplines that might act as catalyst to their thinking on user engagement.

Of the Newport and Salford older people volunteer groups, some were committed walkers, others content with rambling or urban walking. Several invitations to research workshops had to be declined because of Mary's other commitments, but she was able to attend three of those held at Newport and one in Salford. Whilst the early meetings were informative, learning jargon and sharing experiences, Mary became the interested observer (and occasional contributor) at later ones that involved fittings and technical demonstrations of the prototype garments.

15.5 Getting to know terminology in the clothing for active ageing sector

There is available technology that provides very lightweight, shaped, bonded materials; reports on what new fabrics provide in terms of warmth, insulation, and/or protection and on how they are sourced, sustainable or recyclable; and information on how all these textiles are or can be used. Very little of all this is communicated to the public by the media – and Mary subscribes to popular science magazines, as well as reading fashion reviews.

Mary attended over 25 meetings, workshops, conferences, and research events connected with the project, including meetings with individual researchers. She spent considerable time additionally on background reading, phone calls, 'admin' concerning research matters, and talking about the project at external events.

As every field has a jargon, it was necessary to learn those of the textile, communications, and computing industries. Language sharing was essential if volunteer consumers were to make realistic input to the development of innovative, functional, leisure wear related to older people's wants and needs.

A lead user is always concerned that they may be totally unrepresentative of their peer group. In November 2010, Mary joined members of the volunteer representative body shape group to learn the 'languages', at University of Wales Newport's Caerleon Campus. Together, they learned of layering, wicking, shell garments, seamless construction, laser welding, moulding, textile structures with inserts, spacers, bonding, and components. 'Body mapping' recalled human anatomy lessons from long ago.

During the training sessions, Mary was relieved to find so much commonality of views and experiences with the other volunteers.

At a later workshop (April 2011), all volunteers, Mary included, took along favourite items of lower body clothing that to them satisfied needs such as fit for purpose and comfort, cost, and care. They discussed styling, appearance, sizing, weight, quality, accommodating movement and posture, body-to-limb proportions, warmth and heat regulation, fastenings, washability, costs, shopping outlets and experiences, environmental/sustainability concerns, and motivation to buy.

At the last of the Newport workshops Mary attended (October 2011), she observed the fitting of the several garments developed through consultations during the project. True to considerations of layering, these ranged from brassieres to jackets and trousers. Into all of these could be fitted aspects of the technology requirements (identified during the co-design consultations), without giving discomfort to the wearer. A Salford workshop in May 2012 provided Mary and the many other older volunteers with insights into the technology developments themselves, plus the opportunity to 'test-drive' prototype technology-enabled garments by a walk around the Salford campus.

15.6 Getting to know the textile industry: the International Sporting Goods Trade Fair 2010 (ISPO 2010)

The size of the Munich exhibition space is staggering: 14 huge halls, with conference/meeting rooms, restaurants and other services on the upper floors. Configured in three rows (A–C), as two sets of six, and one of four, each hall had five rows of around 30 exhibitor stands. Given its size, Mary limited her visits to each of Adventure, Indoor and Outdoor Sports, Sports Equipments & Accessories, Sportswear & Accessories, and Electronics & Technologies.

The project team were based on ISPO's own 'Best Ager' stand, tucked away along one side of Hall B. Mary expressed disappointment about how the stand presented its intended message. It was not hosted by a representative older person, but by young women in tight-fitting leisure sportswear. Greeted by 'slim young men' store mannequins, the significance of the over-50s market was lost on the visitor from the outset. The three rooms did not invite interaction, nor did the exhibition space give real information on why older people were different, in terms of body size/shape, required design features and garment functionality, and needful approaches to retailing.

The exhibitors were truly international, indicating the global reach of the industry. On her own, Mary toured the stands, asking exhibitors if and how their products were designed with the over-50 market in mind; inspecting materials, looking for examples of applications of materials technology; scrutinising designs and trying on clothing; inspecting different types and uses of fastenings; taking an in-depth look at what might be offered by adaptations of 'bionic' garments; and viewing available technologies and training equipment. With the team, Mary discussed the following with exhibiting companies' representatives: developing garments with a medical application, and product design of socks, running sportswear, and jackets.

Despite the size of the exhibition, and the number of exhibitors, there was very little variation in basic design:

- Styling variations were centred around colours, orientation, and placement of pockets, and types of fastening. There was no standardisation of sizing, and few symbols to indicate the function of a particular aspect of the garment (such as different materials or pockets), or to address sustainability concerns.
- Versatile and stylish caps and hats for men were on show, but it seemed to be assumed that in cold weather, women only wore ear-muffs or Peruvian-style caps and fur-framed jacket hoods. (The last are comfortable only in *dry* cold!)

Mary expressed disappointment that so much of the clothing took no account of

- The similarity of shape of older women and men.
- Putting on, sitting down in, and taking off a garment.
- Problems with pocket placement and design, two-way zips, small zip toggles, press studs or buttons, pieces of Velcro (rather than a continuous strip) for older fingers and projecting stomachs – *particularly* when fumbling in poor light or cold.

Questioned about selling to the older market, very few exhibitors knew of the existence of the 'Best Ager' stand, and knowledge of population and wealth statistics about that market was almost non-existent. Responses were generally along the lines of: 'There's no point in designing for the over-50s as a group… We target the 30–40 age range on the basis that those younger aspire to the quality and brands, and those older are youth-seeking, and would discard the garment and buy another if the first did not suit'.

Exhibitors from the Far East indicated that it was more profitable to follow the design lead of the top market brands and sell at lower cost, than to spend on their own research and design. However, they did have interesting ranges of materials based on natural products such as milk, hemp, and bamboo.

Unsurprisingly perhaps, the vast majority of exhibitors and visitors were under 50. Those who confessed to being designers appeared to be in the 20s/early 30s age range; company sales and marketing representatives were in their late 30s to 40s. There is obviously a huge lacuna in marketing training, given the ageing of the populations in the Northern Hemisphere and Australasia.

15.7 Getting to know the textile retail sector

Had Mary wanted to go skiing, cross-country walking, or on safari, the clothing on offer in shops in both cities differed little in styling, sizing, or fastenings from that on show at ISPO 2010. Equipment was similarly focused. Colour ranges were predominantly at the sombre end of the spectrum, with attempts to brighten made in terms of trimmings, zips, piping, and seam sealing strips. Clothing seemed to take no account of the flexibility of clothing needed for city/town walking, or family or group rambles across local common land or parks.

Racking in shops was often crowded, with clothes displayed by gender, type, and/or colour, and sometimes on price-reduced/sale rails. Too often, racks were close together, making movement difficult for examination of detail, including sewn-in manufacturers' care labels. There was no standardisation of displaying sizing ranges; some shops had clear plastic flags, some displayed size on the coat hanger, some had to be sought on the garment itself. Lighting was generally poor, and shop assistants not immediately evident.

No point in being in denial: the five senses decline gradually with age, though fortunately not at the same rate! Not all older people can afford or use varifocals; often they have different glasses for long-distance and reading, trying not to keep changing from one pair to another. Yet shop information signs were *not* always large with clear font contrasted against its background, and garment labelling was in small print and sometimes an overly fancy design.

Few older people bother to have their hearing checked until difficulties are only too obvious. Over-sensitivity to certain ranges of noise seems to come with hearing loss. Ambient noise is a problem, yet background music is constantly – and distractingly – played, interfering with concentration and communication. Whilst being allowed to browse was welcome, it was often difficult to attract the attention of an assistant for supplementary information. Neglected or ignored (the more so when not wearing a hat, and grey hairs are evident), Mary would either have had to seek help, or leave the shop to seek a purchase elsewhere.

Mary was surprised to find how, on entering a shop, she had automatically taken note of a shop's lighting level, ambient noise, crowding (racks and/or people), sale rail, availability of assistants – and its smell (rubberised products, mustiness, artificial scents.) Browsing, she did not take note of the floor information display boards.

She did visit both male and female sections, and looked for the manufacturer's care label, rather than focus on the price tag. Mary's reasoning was that a male garment was often better suited to her shape and had more of the required functions, and the cost of the garment was secondary to whether it met all needs, including maintenance.

15.8 Getting volunteer participants

To assist Westminster University's research assistant, Mary agreed to seek the participation in the *Design for Ageing Well* project of the largest UK Ramblers Association group. She received permission from its management committee to give a talk about the project at their annual general meeting (AGM), with a view to encouraging their members to respond to survey forms which she would distribute following the talk. The visit provided an unexpected additional learning experience.

The talk appeared to have been well received at the very well-attended meeting. At a later stage, Mary passed on further survey forms that were intended to seek information to supplement the previous research. In events, it was merely presumed without discussion that Mary would take action regarding the latter, and she was never informed of the outcome of each of her efforts, either in terms of number of completed forms submitted or of the usefulness of the exercise.

Aside from the excellent food with which the Ramblers group were provided at their AGM, they also enjoyed a fashion show presented by the local *Cotswold* shop.

The shop was the preferred partner of the Ramblers group, as it gave a 10% discount on purchases and a personal shoe-fitting service.

Four male and five female members of the OPRG, obviously well-briefed, modelled a selection of the latest offerings by North Face, Berghaus, Craghoppers, Mountain, Jack Wolfskin, RAB, and Icebreaker. The jackets included hiking waterproofs, soft shell, and insulated (quilted and down-filled). The intermediate layers were thermal fleeces, microfleece sweaters; baselayers were merino or synthetic T-shirts, but no shirts. Trousers were stretch fabric, waterproof over-trousers, or zipped short and long convertibles (described as 'Kiwis').

At each couple of tables, the model stopped to exhibit a particular feature described by the commentator, such as lightness, pack-ability, convertibility, and colour range. None of the garments' useful or beneficial design features were mentioned, other than zips (length or contrast colour).

From Mary's strategically placed seat amongst the audience, although there was much approval of the range of bright-but-muted colours, the most common comments were not laudatory:

- Nearly all jackets were too short: 'They didn't even cover the bum in most cases, let alone the upper legs. Presumably that's to sell the trousers'.
- Double-ended zips were far too fiddly for cold and arthritic fingers.
- Quilted jackets were too bulky to store and 'made you look like a Michelin man'.
- No mention of the essential 'long johns'.
- Too much choice without real explanation (on labels too): 'It gets bewildering'.
- The clothes were 'quite pricey, and no-one is into buying the latest fashionable colour each year'.

15.9 The co-design process and outcomes

Caerleon Campus is 3½ miles (5.6 km) from central Newport, so the decision to take part required a lengthy journey by bus, taxi, car, or perhaps the offer of a lift in another member's car. Mary's journey required a train journey from London to Newport, and then a taxi to the campus. Journeys to Salford (by train, and then tram) could usually be combined with a visit to another NDA project at Manchester Metropolitan University.

Co-design required cooperation with the project's volunteer group. Had Mary's views been wildly different from those that emerged subsequently from the group as a whole, her usefulness in pilot studies would have been very limited.

That said, Mary and members of the Newport and Salford older people volunteer groups had similar profiles – white, educated, middle class, with no inhibitions or problems about articulating their views, whether this was on fit, length, colour, or fastenings. Considerations on how tactile, breathable, sustainable a fabric might be, or how it or the garment was structured, were relatively new concepts, but were taken on board, to emerge at later workshops.

Maintaining body shape was important, because volunteers had been selected as examples of the representative average, but generally this did not inhibit them enjoying the excellent and generous lunches provided at workshops – always a major factor in obtaining and maintaining older volunteer support. Whilst Mary was not involved in consultations with volunteer older people regarding positioning technologies within the garment, she was able to see the outcome of such at the workshops at which the prototype garments were fitted and demonstrated.

Sourcing garments was a consideration at one workshop. Many older people are brand conscious, equating brand with good quality, for which they are often willing to pay the premium. Seeking lower prices resolved the conflict between sustainability through buying new clothing (labelling specifically checked and approved for such details), and buying recycled garments cheaply, regardless of environmentally friendly attributes, perhaps through charity shops.

Over time, Mary learned of the 'Hawthorne effect' on being involved in research (Draper, 2002) Being asked for help and advice as an older and retired consumer inflated self-esteem, stimulated cooperation with researchers and the other older people volunteering, and generated the belief that older people might – at last – be gaining some control over clothing products on offer. But new factors have been brought into play. Thus textile engineering has become a much more interesting subject, whilst Mary is more appreciative of the quality of natural fibres such as silk and wool. Already conscious of 'layering', she is now more aware of the types and interactions of materials that make layering garments more efficient. And she *listens* to and *smells* the fabrics; try being a bird watcher, observer of wild animals, or a fisherman, in some of the rustling new materials. The sound negates their other qualities here, and one needs to ensure laundering with perfume-free detergents.

Now further disillusioned by the shopping experience, yet still looking for more suitably styled and sized clothes, Mary has begun to take more interest in offerings from the many home shopping catalogues and via the Internet. She continues to be irritated by the lack of both standardisation of garment sizing and information about fabrics and their care. Another problem is that the minimal data given on the latter is often difficult to relate to laundry detergents and washing machine programmes. More industry co-ordination required here, please.

Mary remains sceptical of the *need of* – as against *want for* – wearable electronics. The industry may see 'toys for the boys' in terms of tracking, displays, or 'garments as battery' for the phones, music, and GPS equipment of committed countryside walkers – which may well be 'wants' satisfied by spin-offs from defence industry developments. Except perhaps for medical monitoring, 'need' for electronic components in clothing has yet to be demonstrated.

Because she gives presentations to London older people's groups on 'current UK age-related non-medical research', Mary is able to publicise the project's work. Audience feedback inevitably includes complaints about lack of choice, shape and sizing difficulties, and fiddliness of fastenings and openings. However, it also includes strong approval of the objectives of the project, and much interest in the new features, such as articulated sleeves, new materials, and incorporated technologies.

Mary finds it very difficult to gauge the contribution she might have made to the thinking of academics, retailers, and industry, much of it being through oblique or second-hand comment. She hopes (but it can be no more than that) that she has helped each realise that there are many well-off older people (they have weathered past recessions, and will cope with the future) who have the pride in their appearance, the very strong desire to remain active and independent for as long as possible, and the money to buy the products that will meet all those requirements. Those with less money will benefit through the buying garments recycled via charity shops.

At ISPO 2010, Mary visited the Bridgedale trade stand, and in discussions, learned a great deal about the textile engineering within socks to support various parts of the foot and leg. Given a pair each of ankle- and knee-length socks, she agreed to give these a thorough testing and report back to Bridgedale.

The test took the form of several 3-mile (4.8-km) fast walks to and from the southern end of Kennington Oval to the eastern side of St Katharine's Docks, Tower Bridge. The walk was occasioned by the weekday need to push Mary's grandson from his parents to the home of his great-aunt and great-uncle, with the evening return when his parents' were back from work. The size of the buggy (one of the new 'tanks'!), the locations, and the timing meant that bus, tube, or rail journeys were not really feasible. The child was also taken for a longish buggy ride in the afternoon, as a means of inducing sleep. (Readers with children will appreciate that a small, active child takes a lot of looking after, especially when you are older!)

The short socks gave support around the instep and the ankle, and performed excellently, becoming quite favourite for much urban walking afterwards, in Mary's subsequent studies to become a Green Badge Guide. The knee-length socks had an attractive heather pattern, which disguised the integrated structuring. However, Mary had to report to Bridgedale that they gave insufficient support to a female's less well-developed calves, and were constantly slipping down – a great source of irritation.

At the Wellcome Trust's London building in late March 2012, Mary was asked to give an afternoon workshop presentation at a design conference held for academics, industry and the media. She talked about her involvement with the *Design for Ageing Well* project, and the case for user engagement when developing design and older consumer projects.

Whilst long experienced with PowerPoint, Mary's preference is rather to face her audience without technological aids, to ensure that they are looking fully at her as an individual – but older, and retired-from-employment; one of the many, usually overlooked, volunteers on which age-related academic research is so dependent, and about the treatment of which it is often so careless. This style always brings rewards, focusing attention on the message, and generating thoughts brought out in the question-and-answer (Q&A) session at the end of the presentation.

The image of older persons as grumpy old men or women is one of which the media is all too fond. Yet older people (Mary amongst them) have justifiable reasons to make complaints, and occasional compliments, to manufacturers' or retailers' customer services. Talk with older people, and often it will be found they have presumed that such correspondence will not only rectify their problem (or send praise to where it is due), but should serve as useful input to future design or marketing considerations. However, Mary doubts whether this source of feedback ever passes internally to those it hoped to influence.

Mary understands – although not from the manufacturers – that Sprayway not only has already acted on her feedback about zips, hood fastenings, etc., but also that major change is in view in their garment ranges. Thanks to the contacts made, and publications by the project team, fabric suppliers and manufacturers are far more aware of the needs of older consumers.

15.10 Conclusions

Mary's commitment to NDA projects generally has been affected by:

- Being the one having to make contact initially, or to obtain a subsequent progress report, or to chase up an academic who has promised to provide information and failed to do so.
- Presumptions being made about what unpaid effort will be committed to aid a project; not being given enough advance notice of required input or presence; no choice of meeting dates – all as though Mary's (indeed, any older person's) time was at the project's disposal, and there were no opportunity costs of cancelling some other engagement or commitment.
- Having to wait for papers, or a missed meeting's minutes (at times, several months), rather than being contacted with an apology or a quick resumé, preferably by phone.
- Receiving e-mails with attachments, rather than phone calls and hard copy information by mail.
- Printing documents sent via e-mail, and making lengthy long-distance daytime phone calls – which are not a pensioner's usual household costs.
- Not being asked to scrutinise papers that are intended for volunteer research subjects or for non-academic reading. (Do academics ever seek 'Plain English' awards?)

The *Design for Ageing Well* project ranks highly in that it has largely (but not completely) avoided these discourtesies.

Mary has found her OPRG engagement with projects exhilarating, frustrating, socially enjoyable, time-consuming, boring, self-satisfying, and at times, really annoying. All these reactions are mirrored by her OPRG colleagues, as their report on the NDA Programme illustrates. And as well as journeying with her pensioner's 'Freedom Pass' to various London venues, Mary has self-financed travel to Cardiff, Glasgow, Manchester, Milton Keynes, Munich, Newport, and Salford. Yet the depth of her involvement was only possible because Mary is London-based and fortunate enough to have sufficient income not to be a burden to any project's budget other than, occasionally, to be fed and watered.

Mary had contributed this chapter as both a report on and warning about her experiences of participation in the best-managed project with which she was involved within the NDA programme. She would remind readers that – daily – *they* are growing older, and one day (hopefully) will retire and volunteer to participate in some future research project. Bear in mind the maxim 'Do as you would be done by' and be an example to others. You are most likely to find you will get better cooperation, more positive results, and older people willing to volunteer for your colleagues' projects.

Mary – and fellow OPRG members – hope that this case study shows how useful an experienced older person can be as volunteer. Consider the involvement of such at the outset, when you are framing your research proposal for funding; question not only whether the outcome will *really* be useful to older people, but is it likely to be realised?

A non-academic older person on your steering or advisory group will bring practicalities to the table to counter lack of experience of interacting with older people generally. An older person can give advice on sources of other older volunteers and how to maintain their interest and support; can act as 'guinea pig' in a pilot study; and can scrutinise and give sound advice on wording and presentation of documents intended for a lay audience. And at the end of the project, an older person can act as your project's continuing ambassador to help your colleagues recruit for their age-related studies.

Mary has attended several conferences, at which she has brought the attention of participants to the *Design for Ageing Well* project. Given the positive feedback, her presentations seem to have been successful also at workshops for 'early career' post-graduates intending to undertake age-related research, and at the International Federation of Ageing in Prague in May 2012.

On the other hand, presentations to older people's groups need visual aids. Q&A sessions following have always been lively, with a queue afterwards of older people with questions that were prevented by meeting's time considerations. Mary strongly urges using an older person as publicist of your project, both whilst it is ongoing and to disperse news of your findings.

A closing conference was held for the NDA Programme as a whole in October 2013, at which the *Design for Ageing Well* project made presentations and exhibited findings. Unfortunately, it was not possible to give a fashion show for delegates, to illustrate and contrast the previous and the now improved clothing styles for older people.

The Hawthorne effect (Draper, 2002) is also evident in Mary's extensive knowledge and experience of being involved in so many research projects in the 10 years from retirement in 2002. Whilst she has thoroughly enjoyed the diversity of learning experiences, and making so many new friends and contacts, as a matter of principle she feels that researchers should seek more widely for volunteers (ESRC, 2002).

Too often approaches are made to the 'easy pick' social, community, interest, or health-based sources from national or federated older people-focused organisations. It takes time, commitment, legwork, and courage to find a representative group of older people. It takes effort, real communication, and an adequate budget to maintain their interest and commitment. It takes all the above and a courteous and memorable thanks to ensure good public relations for the sake of your own and colleagues' future projects. Good luck!

References

Coleman, R. (2002). *Living longer: The new context for design*. London: Design Council Policy Papers Series, Available from http://www.designcouncil.org.uk/Documents/Documents/Publications/Health/Living_Longer_Design_Council.pdf. Accessed 09.11.12.

Draper, S. W. (2002,May 14). The Hawthorne, Pygmalion, placebo and other expectancy effects: some notes [WWW document]. Available from http://www.psy.gla.ac.uk/~steve/hawth.html. Accessed 01.05.14.

ESRC. (2002). *The fifth newsletter of the growing older programme: Recruitment and retention of research participants*. London: Economic & Social Research Council. Available from http://www.growingolder.group.shef.ac.uk/Newslet_5.pdf. Accessed 09.11.12.

ISPO Munich. (2010). The annual international trade exhibition of sporting goods, fashions, and athletic footwear. Available from http://www.munich.ispo.com/en/TRADE-SHOW-VISITORS/About-ISPO-MUNICH/.

New Dynamics of Ageing Programme, Older people's reference group end of programme report. Available from http://www.newdynamics.group.shef.ac.uk/assets/files/OPRG%20End%20of%20Programme%20Report%20FINAL%202013.pdf. Accessed 24.01.14.

Williamson, T., & Hinder, S. (2009). *Online 'How to' guide for public involvement in research*. NHS National Institute for Health Research. Available from http://www.invo.org.uk/resource-centre/training-resource/. Accessed 09.11.12.

Key choices in developing sustainable apparel for the active ageing population

16

V.L. Knowles[1], *C.J. Hussey*[2]
[1]Pebble International, Manchester, UK; [2]Consultant, UK

16.1 Introduction

For those who may read this chapter to gain insight into sustainability, it not only gives an overview of the important concepts but it has also been written to put the subject in context with the active ager as with the rest of the book. It will address the many facets of sustainability to demonstrate how as consumers active agers have an awareness of sustainability issues but may not realise that they are being sustainable.

16.2 Ageing market

As this book focuses on the active ageing population, in writing about sustainability it is important to understand where as consumers they are positioned and how they value their clothing in a way that has been influenced by life experiences. This topic is covered at length in the chapter 'Baby boomer interactions and expectations with regard to apparel needs', but in this chapter consideration of the ageing market is important to contextualise their attitudes to garments as well as sustainability.

The active agers have varying opinions on how as consumers they feel about clothing. There are those that fall into the bracket of being more considered purchasers, who will retain garments for longer periods of time, maximising the use and functionality of the items. On the other hand, there are those that seem to thrive on diversity and choice, purchasing many garments from many sources: new, discounted and second-hand.

16.2.1 Influences on the active ageing consumer

All wearers have their own historical experiences of garments. As children in the 1940s and 1950s, they grew up at a time of great austerity, with the global economy impacted by world wars. Most children of that time would have worn hand-me-down clothes, garments their mothers or family members had made for them, all these items would be carefully looked after, be resized, passed down to the next child or recut to suit another person or purpose. By the 1960s and 1970s, the wearers were in their teens and twenties, experiencing first-hand the introduction of the newly developed

Textile-led Design for the Active Ageing Population. http://dx.doi.org/10.1016/B978-0-85709-538-1.00016-X

'synthetic miracle fibres' such as nylon and polyester. These new fibres, although they had revolutionary easy care features, show that actual wearer experiences mean that they are remembered as harsh to the touch, sweaty, static, and would stain and smell through wear. During the 1980s and 1990s as consumerism was on the rise, in-store choices of clothing began to increase, and socialising and a workplace society provided a 'catwalk' with a need to parade with multiple outfits. Into the 2000s and 2010s, an already full market became saturated with discounted and fast-fashion retailers, changing the dynamics of the high street, and therefore consumers' shopping habits – buying garments that have little inherent physical value, low value, giving the consumer the perception that the garment is dispensable and something they can easily dispose of (throw away).

This knowledge and experience of garments has informed wearers throughout their purchasing lives. Those who grew up during the war period were predisposed to avoid waste where possible, as household phrases devised by the British government, such as 'waste not, want not' and 'it's a sin and a crime to be so wasteful' (Hinton & Redclift, 2009) would have been used. These phrases were widely publicised, ensuring that valuable resources were not wasted. Although the messages have been retained and passed down to younger generations, the meanings have been lost, and in a faster, more commercial, all-consuming environment they no longer have the relevance they once did. A generational shift has produced a market segment that is very complacent about the clothing they wear.

For the active ager the influence of change through knowledge, combined with strong government messaging, articulated into a make do and mend philosophy for their parents, which was adopted across all levels of society. In relation to garments, this meant items that were not considered to be worn out could be kept for many years, or garments that were damaged would be adapted or repaired, extending the life of that item. But this new generation (now the active agers) was also the first to have a disposable income, to grow up surrounded by 'design', to want what they had not previously had. For the first time fashion became something for the average person, no longer only available for the rich. So we have a generation who know what they like, understand design, and have money to spend!

During the last 30 or so years, the clothing retail sector has mushroomed in size, with many more types of garments available in stores, in wide-ranging sizes, styles and price points. Within the active outdoor sector, a specialist area that has also grown significantly, the market has expanded from being worn only by the mountaineer and specialist hiker, climber or walker to everyday wear, worn by many purely as fashion, never actually seeing a hill let alone a mountain.

16.2.2 Purchasing decisions influenced by personal priorities and preferences

For the active ager group the purchasing of outdoor garments generally occurs for practical reasons rather than being a fashion-driven decision. Hinton and Redclift (2009) identified common stages that occur when purchasing; they refer to a six-stage decision model (see Figure 16.1) that was devised by Blackwell, Miniard, and Engel (2006).

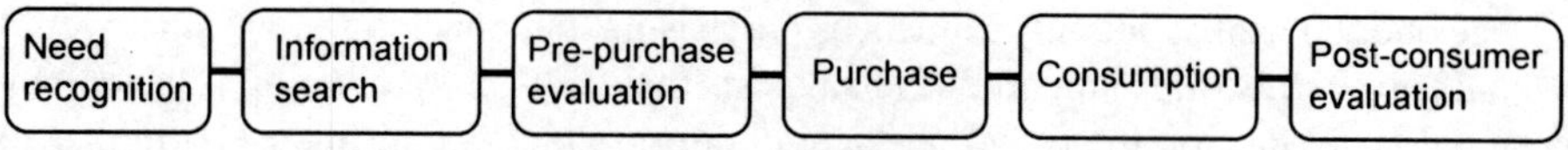

Figure 16.1 Visualisation of the six stages of consumer decision making.

The model identifies that the consumer: *recognises need*, an example could be when a consumer realises they need a new performance jacket, for walking; then an *information search* would be undertaken, this can be when the consumer considers what they already have, so that any new purchases will not unintentionally replicate what they already own; this will be followed by a *pre-purchase evaluation* of alternatives, consumers will often consider the options available through different retailers or brands, decisions can be influenced by many aspects including availability, price, style, size, fit and colour; then *purchase*, once the selection has been made the consumer will buy the item, this can be in-store, online or through catalogues (remote retailing); finally *consumption*, in terms of garments this is the in-use or wear phase; and *post-consumption evaluation*, with clothing the wearer will have different triggers that influence when they want to dispose of garments, which could be when the garment no longer fits, it becomes damaged, or even when it goes out of style or in the outdoor sector when a new technology is developed and made available. Clothing is considered a form of consumable, where items are bought, used and replaced, however, unlike other consumable items like pens or toner cartridges that run out of ink making them unusable, garments are often still wearable when it comes to the replacement stage.

Levels of consumer awareness regarding sustainability issues can vary from no interest or understanding to advocates who prioritise sustainability when purchasing. These consumers will seek out items and garments that have environmental or ecological credentials, and are generally willing to pay a premium for products. This type of consumer ideology and ethics puts them in the position where they can apply their priorities to influence market transformation (Sebastiani, Dalli, & Montagnini, 2011).

16.2.3 Fit as an attribute that can influence the purchasing decision

The fit of a garment influences the purchasing decision, but it is also considered important when judging the satisfaction of a garment (Rahman, 2011).

When a consumer identifies a garment they intend to buy, Suh, Carroll, & Cassill (2012) suggest that the appeal of the garment can be characterised 80% by aesthetics and 20% function, in relation to the active ageing sector it would be presumed that the function would supersede the aesthetic appeal, but one element still seems to be missing for this sector and that is the importance of fit!

When considering the types of garments that are sold, the author Belk (1988) identified that contemporary consumption shows that the feeling of identity invested in material objects can be extraordinarily high. When related to clothing this suggests that it would be important for many consumers to achieve the desired 'look' and that this can be as important as other more functional aspects. That desired 'look' will undoubtedly include the fit of the garment.

The one aspect that not only influences purchasing, but how long a consumer would keep a garment, is fit. Rahman (2011) suggests that well-fitting clothing is shaped by personal taste, fashion trend and physical comfort, which are attributes particular to the individual. In some situations at point of purchase, if a garment fits particularly well, and especially with the active ageing consumer (remembering that disposable income), they may opt to buy multiple items in different colour options, and once having found a brand providing a good fit may become a favourite, securing future purchases.

16.2.4 Attachment to garments

Individuals purchase garments for different reasons, influenced by different aspects including fashion and function. For example, if a person needs an outfit for a black tie event or wedding, style and fashion might be prioritised, for lounge or nightwear, comfort would be important, and for outdoor pursuits functionality and performance are likely to be considered. Chosen garments portray an individual's style choice and taste, and according to Belk (1988) define their sense of self. The experience of wearing the garment, for a particular event, or with outdoor gear when an individual achieves a significant goal (e.g. first mountain ascent), provides the individual with an association with the garment, connecting it with the event or achievement, and this can lead to forming a type of emotional bond to the garment that stimulates a desire to hold onto the item and therefore the associated memory. This personal attachment can make it difficult for someone to give a garment to a charity or even a friend, and will in many cases cause it to sit in the wardrobe for many years.

16.3 Understanding of sustainability

Sustainability is a term that can be used to refer to many different aspects and processes, but in this instance we are choosing the understanding of sustainability that 'meets the needs of the present without negative impact on the environment and present and future generations' (The Textile Institute Sustainability Special Interest Group, 2011; adapted from the Brundtland Report, 1987) rather than in relation to practices that are 'capable of being sustained' (Collins, 2012).

To establish a standpoint, the question should be considered: What does the active ageing consumer understand as sustainability? At a global level, most nations have been reusing unconsciously for many years (Dunne, 2007), through hand-me-downs, jumble sales, the 'rag and bone' man (this is a British term used to describe a man who would walk around the streets buying and selling discarded items such as clothing, textiles, furniture or metal goods, the forerunner of the current recycling company), and donating goods to a charity shop. Most readers will be familiar with these different routes and aware that even though reuse does occur, there is a proportion of discarded clothing that will be disposed of into the refuse, which will inevitably end up in landfill (Figure 16.2).

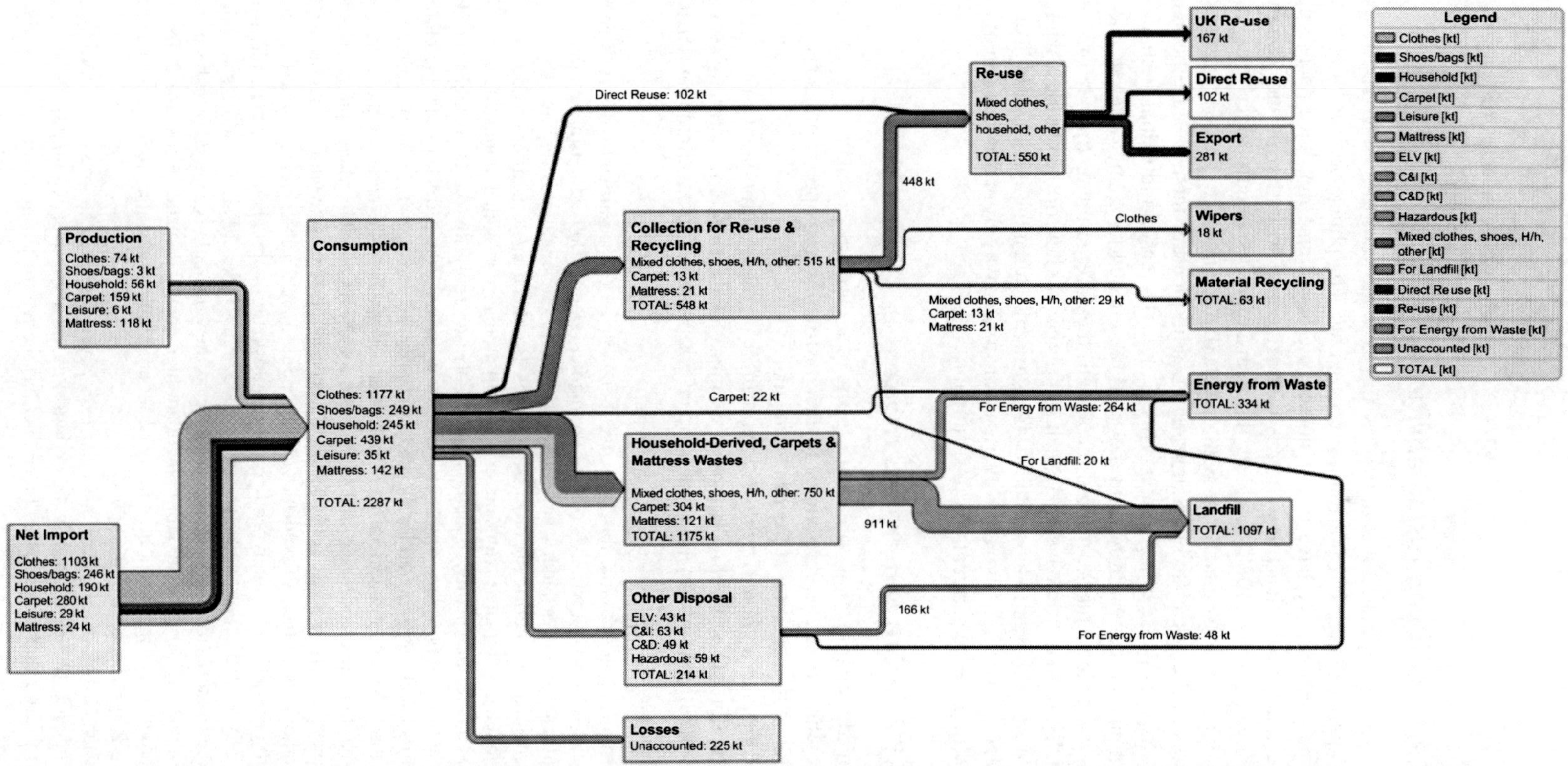

Figure 16.2 Mechanisms for disposal showing the split between different routes within the UK market (Oakdene Hollins for WRAP, 2012).

16.3.1 Clothing sustainability

It has been estimated that in the UK the national wardrobe (WRAP, 2013) accounts for approximately 232,000 tonnes of garments. The phrase *national wardrobe* relates to the garments we have in our homes, clothing items that have been accumulated over many years that the wearer is often reluctant to part with and also may never have worn. Belk (1988) suggests that consumers, especially older ones, often have memory-evoking connections with objects that are linked to past experiences. He also writes of the associated nostalgic regret (with pleasant memories of the item) felt at the disposal phase of a garment. This is often the reason individuals will amass many garments in the wardrobe, and they will be retained for many different reasons; they like the style or fabric, it fitted them well, they are garments that are only worn seasonally, or in extreme weather, such as holiday clothes, or the warmest winter coat.

We now appear to have a 'need' for garments, and although it would take a massive systemic change to move the average consumer away from this, we can at least learn how to produce garments more sustainably.

16.3.2 Fibre and process choice

Consideration of many different aspects of a garment can make it more sustainable, everything from the fibre choice to the construction and opportunities for disposal can be taken into account at the design stage.

Considered choice of fabric composition – a major component part of all garments is the fibre. Given below is a brief summary of the different materials available and, highlighting a few key issues associated with some of the available raw materials and fabrics.

- Natural fibres: cotton, flax/linen, wool and silk, etc. The garment industry is heavily dominated by cotton with 35–40% of the market (Defra, 2009); this is a crop that requires significant levels of land, water and pesticides to grow, and also requires water, energy and chemical use during the processing stages. But this usage can differ greatly from country to country, and many options are available to produce cotton in a more sustainable way.
- Synthetic: acrylic, nylon and polyester. Polyester (polyethylene terephthalate) or PET, is by far the most prevalent synthetic. The fabric is derived from petrochemicals, therefore oil-based fossil fuels are required for its production, requiring energy-intensive processes to create the fibres. A further consideration of polyester is that due to the nature of the fibre composition, fabrics do not biodegrade, and recycling is not always possible. Many options do now exist for recycling and if a way could be found to continuously recycle without loss of integrity of the original fibre, then all synthetics could become more feasible, and sustainable alternatives.
- Regenerated: viscose, lyocell. These cellulosic fibres are regenerated generally from chemically treated cellulose, derived from wood pulp (Burke, 2008). Cellulosics are often considered a more sustainable choice of fabric, as the wood pulp is sourced from beech forests, but in order to produce the fabric, there are high energy demands across the different areas of production, significant amounts of water are required and also toxic gasses can be released as a by-product (Defra, 2009).
- Blends: blends of fibres are used excessively in fabrics across the clothing sector, to increase the performance, fit or longevity of a garment. Common blends are: cotton/polyester that will create a fabric that has the feel of cotton, with the added durability of

the polyester; cotton/elastane blends allow knitted and woven cotton fabrics to incorporate inherent fit, stretch and recovery properties. Although blends do offer obvious benefits during the use phase of the garment, it has become apparent that once garments enter the end-of-life stage, the way that the fibres are tightly bound together makes them potentially more difficult to recycle.

For more detailed information regarding fibre and materials choices, the Material Sustainability Index (MSI), originally developed by Nike and now incorporated into the Higg Index, helps in understanding about the potential environmental impacts of materials choices in the product creation process. Developed over an eight-year period, MSI is a cradle-to-gate index informed by life cycle assessment (LCA)–derived inventory data to engage designers and the global supply chain of apparel and footwear products in environmental sustainability.

Choice of colour – there are a number of ways that the colour of a garment can improve its sustainability credentials; fabric could be dyed with natural dye stuffs; or brands could look at reducing the colour options within their ranges, as has been undertaken by the adidas Group; or more research could take place in creating synthetic dyes that copy the chemical makeup of natural dyes, therefore producing a stable dye, less likely to fade and with less liability to break down at both the dye house and consumer phase.

Fabric colour – achieving the desired fabric colour that works for the garment, within a wider range, is another complex process. Colour continuity within the clothing sector is notoriously problematic, due to the many options of colours, hues and shades a designer can choose from to create a garment. When a fabric or garment is dyed, very precise recipes are used to ensure that the correct shade is continuously achieved; when this does not happen, and shades across garment ranges do not match, redying may be required, even order cancellations that can result in significant wastage. Understanding that the dying process is often heavily reliant on chemicals, to prepare, dye and finish a fabric, it is also inevitable that the effluent formed as a result of these processes can contain chemicals that remain in the expelled water.

Chemical usage and stewardship – the issue of toxic chemicals became the focus of a large-scale Greenpeace campaign in 2011; this high profile international non-governmental organisation (NGO) has published a series of exposé reports entitled Dirty Laundry (Greenpeace, 2011a,b, 2012), which focus on issues of water scarcity and pollution connected to the textile industry in countries such as China and Mexico. Although the industry has for many years been working to improve its practices, these reports have resulted in major reactions from some very high profile clothing brands that have come together to form an industry group, the ZDHC's, Roadmap to Zero Discharge of Hazardous Chemicals (2013). The Roadmap has seven work streams looking at different aspects, including developing a generic audit protocol and the development of alternative (greener) chemistries. Brands involved in the Roadmap include Nike, the adidas Group, PUMA, C&A, Li-Ning, H&M, G-Star Raw, Levi Strauss, Marks and Spencer, and Jack Wolfskin.

Another reaction to the reports has been from practitioners specialising in the area of colouration (Patterson, 2012), who acknowledge this topic is something the industry has been aware of for many years, and that in the long term these exposés will no doubt change wider industry practice in a positive way. However, Patterson also emphasises that this should be seen as an opportunity for industry to make changes in all areas of the sector that use questionable chemicals, processes and practices.

On an international level, legislation is being introduced to ensure that organisations producing goods are more considerate of sustainability issues. In Europe REACh, a European Union regulation concerning the Registration, Evaluation, Authorisation and restriction of Chemicals has and continues to have considerable impact upon the list of restricted substances being continually updated. Individual country government bodies have also been introducing many restrictions regarding the use and handling of chemicals in the textile supply chain, and many apparel and footwear brands now have their own restricted substances lists, used to advise their suppliers of what is acceptable and also to advise the consumer of what level of commitment the company has to ensuring environmental stewardship.

Considered choice of fabric finishing – most consumers are not aware of the chemical and mechanical finishes that can be applied to fabrics to enhance or improve the appearance, texture, drape or performance. Having even a basic awareness of fabric finishes not only informs the wearer of what to expect from the garment but can also provide an understanding of the likely chemical footprint of the garment.

- Fabric finish: mechanical and/or chemical treatments can be applied to the surface of fabrics to enhance the look and feel of a fabric; techniques include brushing to give texture, embossing to apply patterns, or shrinkage to avoid dramatic shrinkage during the use phase of the garment.
- Fabrics with performance finishes: when considering outerwear garments, finishes are often necessary to meet performance needs, for example, to make a fabric water repellent or to help with crease resistance or oil/stain repellency. Some of these chemical finishes have been found to bioaccumulate, causing environmental problems.

16.3.3 Ethos of the brand or retailer

Companies are in the position to use their business approach to influence the products they create. More sustainable companies often make conscious decisions to source sustainable fabrics, ensure that garments are produced ethically, even providing opportunity for consumers to return goods for repair, therefore extending the life of the garment. According to Black, & Shaw (2009), consumers are becoming increasingly aware of how their own purchasing decisions impact on global issues; this awareness also demonstrates agreement with associated practices. In relation, consumers who buy into fast-fashion goods need to be aware of the questionable production processes that may occur, also that the garment they purchase may be of inferior quality and less durable. On the other hand, there are the consumers who will willingly pay a premium for goods that are more sustainable, for companies targeting this market; Hinton and Redclift (2009) suggest that it is possible to present sustainability as an achievable goal in existing approaches to consumer purchasing.

Brand visibility – although some brands have a very strong marketing profile that shows their sustainability ethos (Marks and Spencer with Plan A (2012); Patagonia (2012a,b); Hiut Denim Company (2012) etc.) many are choosing to do the 'right thing' but not to communicate this to the consumer due to a fear of being perceived as 'green washing'.

16.4 Achieving sustainability through considered design

The design process of any garment is important to the success of a product, style and range. When thinking of the products aimed at consumers of the active ageing market, the design process is vital to ensure stylish and functional garments. Elsewhere in this book it has been identified that the ageing consumer can have a very discerning eye, looking for particular attributes within a product.

Once the consumer has decided to purchase a garment, Rahman (2011) identified six intrinsic cues that could influence the final purchasing decision: quality, style, colour, comfort, fit and fabric; and there are three further extrinsic cues – price, brand name, and the country of origin.

Away from the constant fluctuations in fashion trends, according to Grossman (2011), changes in consumer tastes are also being influenced by warmer temperatures and the physical effects of climate change. In one respect this can work well for outdoor garments, as the layering systems provide wearers with multiple options and the flexibility to adapt when weather conditions change. Functionality also has noticeable impact, when we think of other functional garments such as denim jeans, which, initially associated with utilitarian purposes and physical labour (Rahman, 2011), now are entirely mainstream, with many consumers owning multiple pairs. With outdoor gear, the original concept was to create garments for protection against extreme weather conditions; in the current market the fashion and trend aspects, as well as the function, influence the development of styles and ranges, and clothing is just as likely to be worn going to the shops or walking the dog as it is to be worn scaling a mountain or hiking in a national park.

16.4.1 Role of the designer

In different companies the approach to design can vary, but the design phase is critical to clothing production, and according to Dickson, Loker, and Eckman (2009), the use of a scalable, flexible working method can challenge the designer, enabling and encouraging them to include dimensions of sustainable development in the design process.

Jane McCann, editor of this book, has identified the importance of the co-design process when developing garments for the active ageing, Suh, Carroll, & Cassill (2012) explain the process as being where designers and consumers share ideas on a product during the initial development phase. Going one step further to where a company integrates supply chain management with NGO and consumer expectations which will lead to a more sustainable business practice, Wolf (2011) explains the process and findings around the importance of understanding in more detail the environmental impacts of each tier of the supply chain, and concludes that several factors will need to be addressed to ensure impact.

The chapter in this book 'Developing a strategy for the effective specification of functional clothing with integrated wearable electronics', co-authored by M. Price and C. Lewis, also gives a clear indication of the importance of the designer, showing how better specification cuts down iterations and therefore waste in time and resources.

16.4.2 Importance of smart design decisions

In the outdoor sector, the design concepts of different brands are normally targeted at particular consumers for an intended function or range of similar end uses. The garment types can be casual/fashion, fashion/performance, and extreme/survival, and as consumers generally look for multiple attributes to satisfy their divergent needs (Rahman, 2011), the designer will use their knowledge and skill to achieve this. They will consider the themes, colours and level of functionality for each range, the choices will influence the fabric selection, which in turn impacts the cost, as a higher level of technical specification and functionality generally means higher cost.

16.4.3 Proactive design practices

Progressive design processes and techniques can be utilised to alleviate future sustainability issues associated with garments. These approaches include but are by no means limited to:

Design for longevity – generally at the design concept stage, a decision is made to create a garment that has the potential to have a longer life. This prolonged existence can be achieved by considering: the intended function of the garment; if appropriate, more durable fabrics; increased consideration of proportion and fit; and where possible incorporating in-built adjustability. Good examples already exist in the outdoor sector, including Swedish company Fjällräven's trekking trousers, supplied with unfinished leg hems to allow for customisation of fit (Seppällä, 2013).

Design for reuse – this is an increasingly important aspect, especially within companies that prioritise sustainable practice. When designing for reuse, it is important that during the initial stages the choice of fibre is considered, prioritising non-blended fibres and avoiding fabrics that have a finish applied, as both of these can affect the recyclability when the garment reaches its end of life.

Once discarded by its original user, there are different routes that a garment can take. The preferred option is for the garment to be reused as it is, but alternatives would be, if the fibre used in the garment has suitable properties, it can be recycled using a closed loop system, or it could be recycled into lower grade fibre; in countries where the relevant systems are in place, the garment can be thermally recycled; and the worst case scenario would be the garment entering landfill. A great example of a company thinking about the ability to reuse garments is the Finnish company Halti. They have printed information about the jacket on their jacket's wind flaps (see Figure 16.3). Hang tags are very rarely saved, but print enables the second or third user to have key information regarding the garment (Seppällä, 2013).

Design for deconstruction – this is an approach where fibre, fabric, components and construction methods are considered to ensure that where possible the

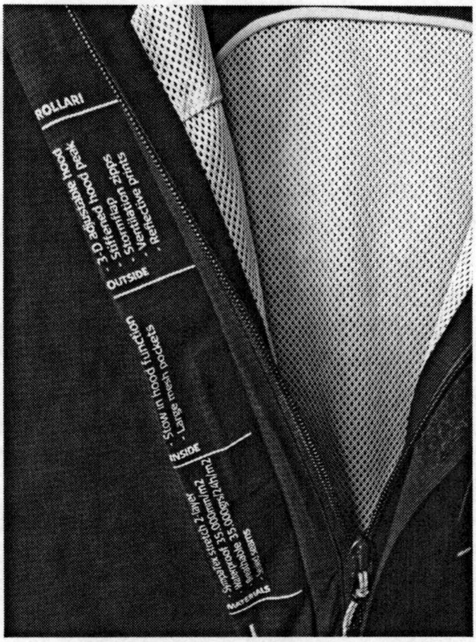

Figure 16.3 Halti wind flap printed with jacket information (Seppällä, 2013).

garment can be easily taken apart and the resulting waste can easily be reused or recycled.

Design for recycling – this could allow for complete garment recycling, with no deconstruction necessary. This is easiest when a garment has been produced using a single fibre, in the way that Patagonia have for their Common Threads Garment Recycling Program, where the company collects base layer garments made from Capilene, and Teijin (2012) through its Ecocircle™ system and recycles them into new fibres.

16.5 Conclusion

It is important for companies developing garments to generate appropriate messaging to ensure that consumers understand company priorities (Grossman, 2011). Companies providing high-end, outdoor gear will need to rigorously test garments to see whether they perform as required whilst also ensuring the least harm is done to the environment and the people producing them. Wearer trials will form part of the testing, with comments received used to inform design choices, including fit and fibre selection.

As a sector, the outdoor industry has been active in the pursuit of sustainable industry practice for a number of years, with international companies proactively working in collaboration to develop solutions that can be utilised by all brands and retailers. The Sustainable Apparel Coalitions, Higg Index, should prove to be an invaluable tool to help international businesses and their suppliers to understand and work toward a cleaner supply chain. Its long-term aims to improve water-use efficiency, minimise water and energy usage in garment production and care, develop effective uses for textile waste, creating a second life for materials and therefore minimising waste throughout the supply chain, as well as reducing the use of chemicals and other potentially hazardous materials should provide industry with a mechanism for self-regulation.

It is important to remember that sustainability may be implemented throughout the supply chain, and the consumer can influence (by demand) this action but cannot control it. Our current fast-fashion acceptance raises the question, has this come about because of consumer demand or has it been forced on us by industry, marketed well and the consumer manipulated to think they 'need' more clothes 'now'? Could the consumer stop buying, or should the industry stop selling, or sell at a higher price, therefore reintroducing perceived value into textiles and clothing? The outdoor sector has shown that many brands are committed to influencing change and continue to collaboratively work toward a more sustainable business environment. A new business model and a major directional shift need to happen within the wider clothing sector. Clothing is too cheap – both ethically and for the world to survive environmentally – and we need to look at paying more for garments. Then, perhaps, we will start to appreciate them more!

References

Belk, R. W. (1988). *Possessions and the extended self [Online]*. Available at http://homepages.abdn.ac.uk/david.j.turk/pages/dept/Lab%20Papers/BELK(1988).pdf. Accessed 26.06.12.

Black, I., & Shaw, D. (2009). *Consumption as voting: A path to sustainability? [Online]*. Available at http://sydney.edu.au/business/__data/assets/pdf_file/0017/56600/Consumption_as_Voting.pdf. Accessed 26.06.12.

Blackwell, R. D., Miniard, P. W., & Engel, J. F. (2006). *Consumer behaviour* (10th ed.). (Mason, OH, USA; South-Western).

Burke, M. (2008). *Green couture*. Available at http://www.rsc.org/chemistryworld/Issues/2008/March/GreenCouture.asp. Accessed 29.08.12.

Collins. (2012). *Definition of sustainability. [Online]*. Available at http://www.collinsdictionary.com/dictionary/english/sustainable. Accessed 12.09.12.

Defra. (2009). *The role and business case for existing and emerging fibres in sustainable clothing*. Available at http://www2.wrap.org.uk/downloads/Emerging_fibres_summary.84922001.10840.pdf. Accessed 01.09.12.

Dickson, M., Loker, S., & Eckman, M. (2009). *Social responsibility in the global apparel industry*. New York: Fairchild Books.

Dunne, S. (2007). *Retro-philia: Fashion and identity in postmodern society [Online]*. Available at http://www.thedunnething.com/retrophilia2.pdf. Accessed 26.06.12.

Greenpeace. (2011a). *Dirty laundry*. Available at http://www.greenpeace.org/international/en/publications/reports/Dirty-Laundry/. Accessed 12.09.12.

Greenpeace. (2011b). *Dirty laundry 2: Hung out to dry*. Available at http://www.greenpeace.org/international/en/publications/reports/Dirty-Laundry-2/. Accessed 12.09.12.

Greenpeace. (2012). *Dirty laundry: Reloaded*. Available at http://www.greenpeace.org/international/en/publications/Campaign-reports/Toxics-reports/Dirty-Laundry-Reloaded/. Accessed 12.09.12.

Grossman, D. (2011). *Physical risks from climate change [Online]*. Available at http://www.ceres.org/resources/reports/physical-risks-from-climate-change. Accessed 26.06.12.

Hinton, E., & Redclift, M. (2009). *Austerity and sufficiency: The changing polotics of sustainable consumption [Online]*. Available at http://www.kcl.ac.uk/content/1/c6/03/95/42/HIntonRedcliftWP17.pdf. Accessed 26.06.12.

Hiut Denim Company. (2012). *Homepage*. Available at http://hiutdenim.co.uk/. Accessed 01.12.12.

Marks & Spencer. (2012). *Report – How we do business*. Available at http://plana.marksandspencer.com/media/pdf/how_we-do_business_report_2011.pdf. Accessed 01.12.12.

Patagonia. (2012a). *Company information page*. Available at http://www.patagonia.com/eu/enGB/patagonia.go?assetid=2047&ln=140. Accessed 01.12.12.

Patagonia. (2012b). *Common threads garment recycling program with teijin [Online]*. http://www.patagonia.com/pdf/en_US/common_threads_whitepaper.pdf. Accessed 05.12.12.

Patterson, P. (2012). Report gives brands toxic shock, *Textiles* (1), 18 (Manchester, UK; The Textile Institute) previously published in *Ecotextile News*, Issue October/November, 2011 (Normanton, UK; MCL Global).

Rahman, O. (2011). *Understanding consumers' perceptions and behaviours: Implications for denim jeans design [Online]*. Available at http://ojs.cnr.ncsu.edu/index.php/JTATM/article/viewFile/845/909. Accessed 26.06.12.

Sebastiani, R., Dalli, D., & Montagnini, F. (2011). *How consumers shape the market: Ethics and value co-creation [Online]*. Available at http://www.naplesforumonservice.it/uploads//files/Sebastiani,%20Dalli,%20Montagnini.pdf. Accessed 26.06.12.

Seppällä, L. (2013). *Understanding sustainable design, textiles*. Manchester, UK: The Textile Institute, Issue No. 3, pp. 19.

Suh, M., Carroll, K., & Cassill, N. (2012). *Critical review on SMART clothing product development [Online]*. Available at http://ojs.cnr.ncsu.edu/index.php/JTATM/article/viewFile/702/728&embedded=true. Accessed 26.06.12.

Tejin. (2012). *Homepage*. Available at https://www.teijin.co.jp/english/index.html. Accessed 01.12.12.

The Textile Institute. (2011). *Sustainability special interest group – Agreed definition of sustainability*. Available at http://sustainabletextiles.org/AboutUs. Accessed 03.01.14.

Wolf, J. (2011). Sustainable supply chain management integration: A qualitative analysis of the german manufacturing industry. *Journal of Business Ethics*, *2011*(102), 221–235.

World Commission on Environment and Development (WCED). (1987). *Our common future (aka. Bruntland report)*. Oxford: Oxford University Press, pp. 43. [Online]. Available at http://conspect.nl/pdf/Our_Common_Future-Brundtland_Report_1987.pdf. Accessed 05.12.12.

WRAP. (2012). *Valuing our clothes*. Available at http://www.wrap.org.uk/sites/files/wrap/VoC%20FINAL%20online%202012%2007%2011.pdf. Accessed 01.09.12.

WRAP. (2013). *Textiles flow and market development opportunities in the UK*. Available at http://www.wrap.org.uk/system/files/private/Textile%20Flows%20and%20Market%20Development%20Opportunities%20Updated%20June%202013v2.pdf. Accessed 14.11.13.

Zero Discharge of Hazardous Chemicals. (2013). *Homepage*. Available at http://www.roadmaptozero.com/. Accessed 01.09.12.

Issues and techniques in the inclusive design of apparel for the active ageing population

17

J. Cassim
Royal College of Art, London, UK

17.1 Background

As we age, we incrementally acquire disability to differing degrees and require increasing levels of embedded functionality in the products and services we buy, in the environments we use, and the communication modes we employ. Thus, for the purposes of discussion, the focus of this chapter will be on the disabled body. My argument centres on the idea that by understanding and responding to such extreme scenarios, one can understand and innovate for mainstream ones too, but this cannot necessarily be achieved the other way around. In addressing the extreme scenarios generated by disability in relation to clothing, one can highlight uncompromisingly the greater issues relating to aesthetics and stigma alongside functional concerns that are shared by the ageing population. One can then demonstrate how the adoption of an inclusive methodology allows designers to innovate solutions that are mainstream in nature, achieve the right balance between aesthetics and functional need, and in so doing capture a market that is waiting to be taken by storm.

Over a decade ago, I presented a paper at Include 2003, the biannual conference on inclusive design organised by the Helen Hamlyn Centre for Design at the Royal College of Art (HHCD) entitled *Smart Wearables: A New Frontier for Inclusive Design Innovation*. The paper outlined the inclusive possibilities inherent in the congruence of separate developments in international fashion, technical textiles, CAD/CAM/CIM design and production techniques, performance sportswear, extreme environment clothing, and the emergence of embedded technology (Cassim, 2003a).

This set of converging scenarios seemed to offer the exciting prospect for clothing that could 'combine mainstream fashion aesthetics with functional properties and act as interfaces in their own right, integrating the separate assistive devices currently used by disabled people for communication and mobility'. Conceptually, I felt that much as the mobile phone and computers have had a decisive impact on the ability of disabled people to enter the workforce and live independently, these developments augured the possibility of a brave new world of innovative, desirable inclusive clothing for disabled and older people alike – two groups largely ignored or misunderstood by the clothing industry and ill-served even by manufacturers with a specific focus on that market.

Textile-led Design for the Active Ageing Population. http://dx.doi.org/10.1016/B978-0-85709-538-1.00017-1

17.2 Mechanisms of engagement

The paper aimed to achieve two things. The first was to articulate in detail a nascent arena of creative possibility to design professionals and RCA students that would interest them and to which they could contribute their creativity and expertise – the first group through the medium of the annual *DBA Inclusive Design Challenge,* which I organised from 2000 until 2010 in collaboration with the Design Business Association (DBA), and the second through the Helen Hamlyn Center's annual *Design for Our Future Selves* competition. Both had been devised as a means of engaging designers at different stages of their careers with the subject of inclusive design.

17.2.1 *The power of the prototype*

Five years after the Disability Discrimination Act of 1995 (http://www.legislation.gov.uk/ukpga/1995/50/contents Accessed 25.07.12) and before the official definition provided by the 2005 British Standard (British Standards Institute, 2005), inclusive design was still a new subject on the design agenda with few concrete exemplars of inclusively designed products, services, or communication initiatives in evidence on the market, which could illustrate what it embodied in terms of process, result, or business and creative advantage. Thus, it was important not only to tell but also to show designers how inclusive design methodologies could be applied across the span of design disciplines if they were to arrive at better, more inclusive design results. For this, realized design concepts for the mainstream market were essential. Thus, the second underlying aim of my paper was to generate actual exemplars of 'smart wearables' and inclusive clothing from the participating DBA Challenge firms that would create a virtuous circle of interest and activity.

17.2.2 *Disguised tutorials*

The DBA Inclusive Design Challenge was framed as a five-month long tutorial in inclusive design (Cassim & Dong, 2007) aimed at the professional design community. The 500 designers or more from 41 DBA member consultancies, who took part over its decade of activity, worked closely with disabled and older people, brainstorming with them from the start in user forums and individual ethnographic sessions where they documented the context of their interaction with the world of products, services, and environments. They were mentored throughout the design ideation and development process by the author. Put simply, it was an intensive tutorial in inclusive design practice.

17.2.3 *Countering indifference*

However, the take-up for the subject of inclusive clothing by DBA member firms mirrored our experience with students entering the *Design for Our Future Selves* competition since its first iteration in 1998 as part of *DesignAge*, the action research programme led by Roger Coleman, which led to the creation of the Helen Hamlyn Research Centre in 1999. In comparison to other design disciplines, particularly those where ergonomic considerations are a routine part of the design process such as product design or architecture, students from fashion and textiles were particularly hard to engage and remained resistant to our efforts to interest them in the subject.

17.2.4 The DBA Inclusive Design Challenge and Design for Our Future Selves competition

Each year the *DBA Inclusive Design Challenge* and *Design for Our Future Selves competition* saw many entries for wearable digital products, which boasted a range of inclusive features. Increasingly they were conceived as a product that was the primary interface to the provision of a new service of some kind with the wearable element the least realised element of the design – an afterthought rather than a core concept. The device would be worn around the neck or wrist but not integrated into any item of clothing, eventually evolving over time into yet another smart phone application as mobile phones became ubiquitous devices.

Few competition entries related to clothing were deliberately designed with the functional needs and aspirations of older and disabled people in mind and, which involved no technology element. Where they did, it was often a design response to the personal situation of a family member or friend, such as the clothing collection designed by RCA student Rebecca Paterson for her sister Susana, who is a wheelchair user (Paterson, 1996). Only one project by the design firm Pearlfisher in 2002, of the total of 53 entries to the DBA Challenge, centred on a clothing collection for young disabled people. This exploited the temperature-regulating qualities of the microfibre Outlast and had other deliberate design features co-developed with the young disabled people who collaborated with the design team (Cassim, 2003b, pp. 18–21).

Karen Welman of Pearlfisher, the creative director behind the project, and textile designer Suzy Holland then went on to develop the multiple award-winning 37° brand of temperature-regulating clothing for babies based on the same principles and using the same materials. It was launched in 2004 and has since become the basis for a successful business (http://www.37degrees.co.uk/). Conceptually, the project was framed as a product idea rather than as a fashion one, since its core aim was the reduction of sudden infant death syndrome (SIDS). Since then, the *Design Bugs Out* (http://www.designcouncil.org.uk/our-work/challenges/Health/Design-Bugs-Out/Case-studies-equipment/ Accessed 28.07.12) and *Design for Dignity* (Design Council, 2010), both projects led by the UK Design Council in collaboration with the Department of Health, have generated clothing or wearable concepts related to health care. However, the overarching focus has been on the issues of disease control in the first and in the latter, a range of concerns. Differences of patient size, cultural and religious preferences, laundry, disposal, the garment's lifetime costs, and the specific context of use in an ICU unit were among the factors that the designers had to consider. The eventual style was determined by functional considerations in a classic utilitarian clothing scenario. Dignity was preserved by the reduction of exposure through considerate design rather than styling. As Ben di Lisi, the lead designer, noted:

> *This project isn't about glamour it's about well being. This gown has to be hardworking and user-friendly and help clinicians to do their job—without costing NHS Trusts more money.*
>
> *Design Council, 2010*

17.2.5 Reframing clothing as product

Therein, I feel, lies the crux of the problem of inclusive clothing. Use of the word 'fashion' in relation to it seems to blur many of the issues that have to be addressed if it is to be realised in concrete design terms – ones that will satisfy clients, customers, manufacturers, and marketing experts alike. Perhaps it is time to reframe clothing as a product like any other – to deconstruct the qualities that are necessary for older and disabled people, including the all-important one of aesthetics and style. Then one could reconstruct the design development process following the more rational, effective, and participatory methods of inclusive design and not the more discursive and spontaneous ones used by fashion designers in the first concept development stage. Only then, perhaps, will the subject lose the intense emotive power that it currently has for those who must suffer the indignities that the fashion industry serves up to them, with disabled and older people leading the list of victims. As Michael Shamash, the disability activist who has restricted growth, noted: 'Clothes are a major component in our self-definition and self-esteem. Disabled people want the profile to be a positive one. We need glad rags, not sad rags' (Shamash, 2004, p. 11).

It is a sentiment with which the swelling global ranks of older people (Coleman, 2003), whose bodies are undergoing the metamorphosis of the ageing process, will agree. For ageing changes the size, proportions, and posture of the body, and these changes affect the fit and sizing of clothes relatively early (Goldsberry, Shim, & Reich., 1996). As the late Marion Bieber wrote: 'Clothing is probably the area of the greatest frustration for the older consumer. The attention to fashion instead of style and the constantly changing requirements of the youth market have put all quality and style design on the back burner' (Bieber, 2003).

17.3 Inclusive design: origins, definitions, and the limits of terminology

The terminology concerning user-centred design in all its manifestations is many and varied. The three most commonly used terms are 'Design for All' (Scandinavia, Western Europe), 'universal design' (USA, Asia, etc.), and 'inclusive design' (UK, Europe). All are expressions of a design philosophy based on the desire to understand, engage, and respond to what the poet Louis MacNeice called the 'incorrigible plurality' of the world (MacNeice, 1935). Their shared aim is simple: to avoid the marginalisation or active exclusion of those groups most impacted by design's failure to anticipate their needs in service, communication, product, or environmental terms.

17.3.1 Universal design

All three movements evolved in their respective contexts against a common background of population ageing and the demands for a socially just society in which everyone regardless of age or ability could be an active member. In the United States, the universal design movement found its inspiration and modelled some of its tactics on the civil rights movement of the 1960s. Its major focus on the creation of 'barrier free' and 'accessible design' of the built environment was propelled by the reality of large numbers of returning disabled veterans from the Vietnam War and the realisation that social and financial independence was highly dependent on an individual's ability to physically access goods, employment, education, and public services via the environments that housed them.

Of the three, only universal design has seven core principles to uphold its definition and a host of legislated guidelines to steer its practice, particularly in relation to the built environment. These were developed between 1995 and 1997 at the Center for Universal Design at North Carolina State University by a team led by the late Ron Mace, an architect, educator, and wheelchair user (Ostroff & Preiser, 2001). Mace used the term 'universal' to express the aspiration that the design of products and environments should be with the aim of ensuring that they are 'usable by all people, to the greatest extent possible, without adaptation or specialised design'.

17.3.2 Limitations of universal design principles

One could argue that these seven principles have limited universal design's ability to evolve and embrace other disciplines or be seen within the design profession as something more than relating to physical and sensory accessibility alone. None deal with the all-important issue of aesthetics – they concern ergonomics, accessibility, and function, not desirability. Too often, alas, design solutions presented as universally acceptable are bland, functional perhaps, but entirely lacking in that all-important design quality – distinctiveness. As a result they are not viewed as aspirational to designers (Cassim, 2005; Pullin, 2009). Neither did the principles address the greater societal changes that were then taking place particularly in Western European societies with the expansion of the European Union and immigration and need for the social integration of people of different cultural and religious backgrounds.

17.3.3 Other scenarios of design exclusion

As a result, new scenarios of design exclusion unrelated to ergonomics were emerging for which designers, particularly those working in the areas of service design and visual communications, had to find solutions. The accessibility that needed to be addressed related to more subtle issues of cognition, language, and culture. In an era of respect for cultural diversity, the term 'universal' seemed not only unattainable but also anachronistic since it seemed to deny this key aspiration. As Coleman noted, 'The European approach has been to think in terms of "Design for All" and "inclusive design", which reflects concerns to integrate diverse groups of people without enforcing conformity, and suggest a degree of flexibility in how that goal is achieved, whereas universal design tends to imply that each individual should be able to access a building or service, or use a product. The ultimate aim is the same, but the approaches reflect different historic and social conditions' (Coleman, Bendixen, & Tahkokallio, 2003).

17.3.4 The broader church

'Inclusive design' and 'Design for All' have been left open-ended in both definition and methodology, with no prescriptive guidelines to support their practice. BS7000-6 2005 (BSI, 2005; Coleman et al., 2003) defines inclusive design as 'the design of mainstream products and/or services that are accessible to, and usable by, as many people as reasonably possible … without the need for special adaptation or specialised design', while the EIDD Stockholm Declaration of 2004 (http://www.designforalleurope.org/Design-for-All/EIDD-Documents/Stockholm-Declaration/ Accessed 25.07.12) similarly states that Design for All is 'design for human diversity, social inclusion and equality'. Neither specify a modus operandi or a fixed set of principles about how to achieve those goals beyond the general philosophy expressed in their respective definitions and the emphasis on a participatory design process.

As a result, this broad church with open doors has allowed Design for All and inclusive design practicioners the freedom to experiment across and within different design disciplines, including the area of clothing, combining and adapting tried and true design research methodologies such as task analysis and ethnography with experimental new ways of responding to that core understanding of the purpose of design – to include rather than exclude in terms of process and designed result. And practitioners have been allowed to do so in ways that go beyond accessibility and embrace the other key qualities of successful design such as desirability, cost-effectiveness, sustainability, simplicity, and so on.

17.3.5 Inclusive design as process

Since the emergence of inclusive design as a discrete subject, design researchers have mapped the multiplicity of methods used by individual designers and design firms to capture user needs and aspirations in order to present them back to designers themselves. This has been in recognition of their need for information that is practical and experiential rather than theoretical (Cassim, 2005), given the high rates of dyslexia within the creative industries – Steiffert, for example, estimates that between 15% and 25% of art and design students are dyslexic and/or dyspraxic (cited in Rankin, Davies, & Riley, 2005).

The Methods Lab (Aldersey-Williams, Coleman, & Bound, 1999) was the first such designer-centric compendium of 53 user research methods, which were positioned within a diamond-shaped matrix in relation to two axes. The horizontal axis ran from *designer-centred* to *user-centred*, while the vertical axis ran from *visual qualities* to *functional ones*. The methods were then positioned according to their effectiveness in eliciting information relating to these concerns (Figure 17.1). This highly practical information set has since been followed by two

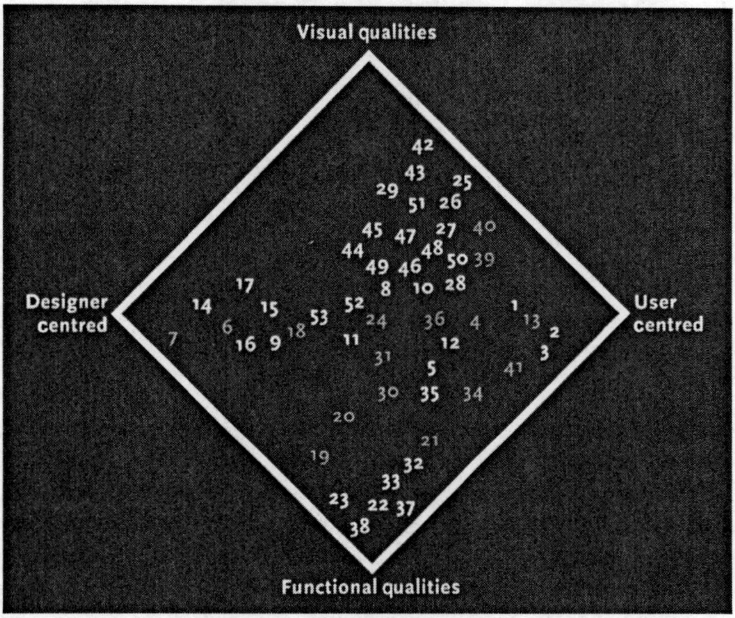

Figure 17.1 The Methods Map. The map positions each method at a point along two axes that reflect the designers' concerns.
Source: Helen Hamlyn Research Centre, RCA, 1999.

web-based resources – the UK Design Council's 24 Methods (UK Design Council) (http://www.designcouncil.org.uk/about-design/How-designers-work/Design-methods/ Accessed 26.07.12) and the 20 described on the Helen Hamlyn Center for Design's *Designing with People* website (HHCD, 2011) (http://designingwithpeople.rca.ac.uk/methods Accessed 26.07.12).

17.3.6 Design process models

Similarly, different organisations and design firms have attempted to define the key stages that make up their own user-centred design development process. As a result of a survey of 11 leading companies, the UK's Design Council summarised the process in four distinct stages: *discover, define,develop and deliver*, which were represented graphically in the form of a double diamond joined at the center by the reality of the brief to indicate the divergent and convergent nature of this complex process (UK Design Council, 2005) (http://www.designcouncil.org.uk/about-design/How-designers-work/The-design-process/ Accessed 26.07.12). In contrast, the *Inclusive Design Toolkit* published by British Telecom and the Engineering Design Centre (EDC) at Cambridge University describes it as a 'waterfall model' with four similar stages of *discover, translate, create and develop* cascading away from the starting point of understanding user needs (Clarkson, Coleman, Hosking, & Waller, 2007).

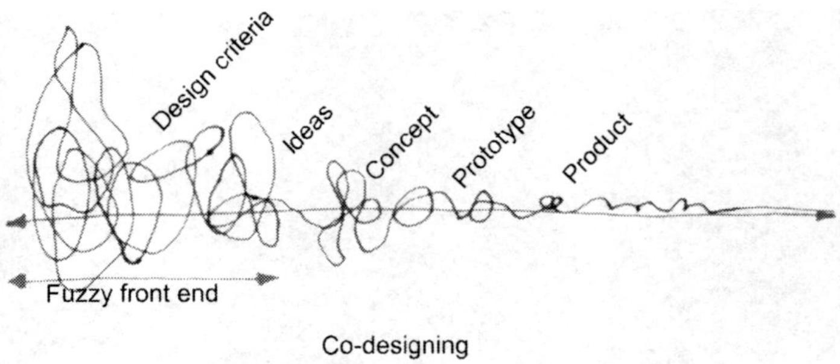

Figure 17.2 Diagram of front end of participatory design process.
Source: Sanders and Stappers (2008).

The realities of any participatory design process are probably closer to Sanders and Stapper's far messier diagram of the co-creation process[1] (Sanders & Stappers, 2008) (Figure 17.2), in which users become actual members of the design team and their participation is more substantive and active than in conventional user research, which places emphasis on quantitative data collection based on the ergonomic or functional assessment of user needs and capabilities. The intensity of creative activity described in the diagram is weighted heavily towards the front end of the design process and the brainstorming, information gathering, and delivery of the all-important creative stimuli that precede the development of design ideas. While there is always a danger of a loss of objectivity given the closeness of the relationship between designer and subject or subjects, the co-creation model allows for the more emotive issues relating to clothing to come to the fore and for the eventual design concepts to be developed, calibrated and refined to achieve a balance of function and aesthetics.

17.3.7 Inclusive exemplars

It is particularly suitable for clothing design and wearables such as shoes (Figure 17.3), as with those by shoe designer Helen Richards co-developed with a young ambulant graphic designer with cerebral palsy, in which her specific functional needs have been addressed through clever detailing, transforming them into powerful visual features:

- The leather was perforated to provide ventilation, given her poor temperature control in her feet.
- Support for her ankles was provided by padded curves around them and to the whole foot via a rigid metal plate embedded in the sole.
- The shoe was laced to the point of the toe to allow for maximum ease of access and this aspect emphasised aesthetically to give the shoes a distinct style.

[1] Sanders and Stappers define co-creation as 'any act of collective creativity that is shared by two or more people, which can take place in fields other than design while co-design is seen as an instance of collective creativity as it is applied across the whole span of a design process. Thus, co-design is a specific instance of co-creation'.

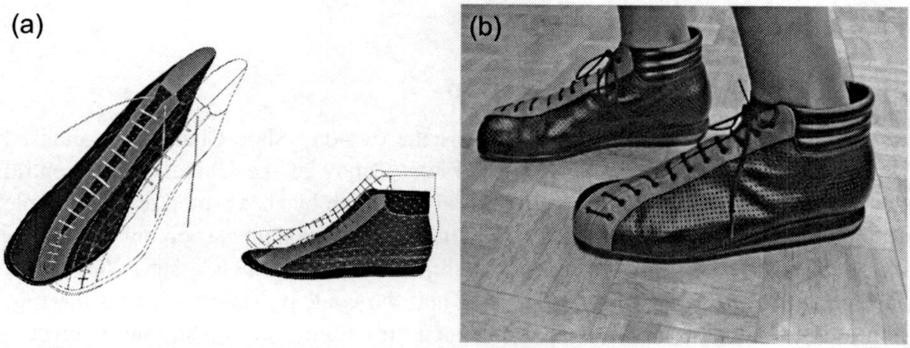

Figure 17.3 (a) Design drawing *(Courtesy of the designer)*. (b) Prototype of bespoke orthopaedic shoes by Helen Richards.

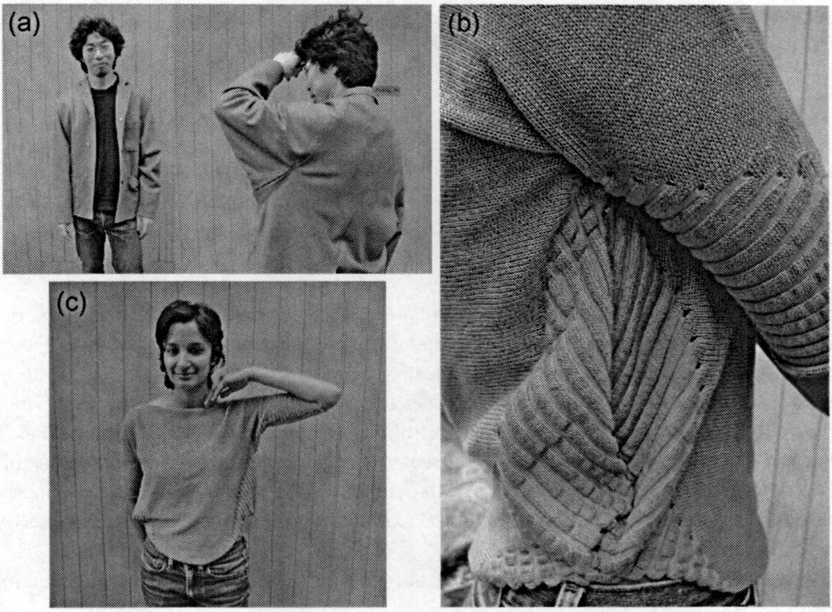

Figure 17.4 (a) Daijiro Mizuno wearing own design showing underarm gussets. (b) & (c) Underarm detail and front view of Fatima Saifee's sweater worn by the designer.

Similarly in Daijiro Mizuno and Fatima Saifee's joint 2003 project 'Garment', simple gussets have been inserted into the underarm of a man's tailored jacket and woman's knitted sweater to allow for ease of putting on but which also act as signature design details (Figure 17.4).

17.4 Immersive workshops

17.4.1 Shoe challenge workshop

Another example of this methodology was the two-day Shoe Challenge organised by the author, Dr Lynne Goodacre, and Fiona Candy of the University of Central Lancashire in 2010 and funded by ArthritisUK, which has been cited as an example of good practice (Arthritis Research UK, 2012). The workshop was organised following the Challenge Workshops model developed by the author (Cassim, 2007a) and adapted as is customary to both the brief and the context. This telescopes the four stages of a design process into an abbreviated time frame culminating in the presentation of initial detailed design concepts and/or simple prototypes. In other Challenge workshops involving textile-based products such as clothing, iterative prototyping leading to a finalised product has been possible within the relatively short period of five days since they are easy to make in comparison to other products that may require detailed CAD/CIM data in order to prototype via 3D printing (HHCD, 2011) (http://www.hhc.rca.ac.uk/4264/all/1/Zagreb-2011.aspx Accessed 27.07.12).

17.4.2 Workshop structure

The aim of the Shoe Challenge was to develop formal footwear concepts for women of different ages with rheumatoid arthritis (RA). This had been identified as the area of greatest concern and difficulty in relation to their wardrobe (Candy & Goodacre, 2007). The key workshop elements were:

- The three teams were multidisciplinary in nature, with product and shoe designers, orthotists, and podiatrists sharing their skills and expertise, allowing aesthetics to be balanced with clinical and functional need at all times.
- Each team comprised a senior shoe designer and two junior designers with experience of designing footwear as part of a broader portfolio or who specialised in footwear alone. This ensured that there was sufficient specialist knowledge of shoe design and manufacturing. The rationale for including designers with different skills was to broaden the conceptual base of the participants, and encourage the exploration of new approaches to shoe design beyond conventional generic types.
- The design partner was present throughout, allowing iterative development and evaluation of the designs to take place.
- The three design partners were a former knitwear designer in her 30s, an image consultant and mother of four in her 40s, and a business executive in her 50s, with differing degrees of severity and duration of living with RA. This allowed three distinct scenarios to be explored in terms of lifestyle and stylistic taste alongside the core theme of RA and footwear.
- Detailed briefings were given on the clinical, emotional, and design issues relating to RA and footwear from a participating orthotist and podiatrist and on inclusive design from the author prior to the start of any design work.
- The workshop environment was equipped to allow simple iterative prototyping to take place.

This type of workshop represents the fullest expression of the user-centred co-creation cycle possible within the structure of a two-day workshop. However, shorter, more intense formats can deliver similarly rapid design results where they are structured carefully to ensure delivery of the core stimuli and information and an actual design result rather than a written wish list alone.

17.4.3 One-day clothing design workshop

The 40 attendees at *Inclusive Clothing and Materials for Different Needs: Alternative Solutions for the High Street* , a one-day Challenge Workshop held at the College of Occupational Therapists in London (COT) in 2007 and organised by the author (HHCD, 2007) (http://www.hhc.rca.ac.uk/1276/all/1/preview.aspx Accessed 27.07.12), were a mix of occupational therapists, clothing manufacturers, textile specialists, designers, design educators and disabled consumers – a cross-section of the inclusive clothing constituency. The aim of the event was to explore the innovative possibilities of new materials in the design of clothing for extreme needs, and come up with new design directions. Delegates came from all over the UK as well as Finland and Sweden – an indicator of the high level of interest in the subject and the dearth of such opportunities.

The morning was devoted to detailed briefings from experts on textile innovations and applications, clothing adaptation, and inclusive design in order to provide an overall context of the subject. Delegates were then divided into four teams and each assigned a design partner with a disability from a professional creative background – a mouth and foot painting artist, a graphic designer who is a wheelchair user, and two former textile designers one of whom is partially sighted and the other with rheumatoid arthritis. The teams were led by RCA product designers with a background in inclusive design methodologies, with RCA textiles students acting as visualisers for the ideation process. The deliberate choice underlying this expressed three strategic choices in the creative direction of the workshop.

In such design workshops, which are cross-disciplinary in nature with few or no participants with drawing skills, the importance of visualisers cannot be underestimated. The brainstorming process throws up many ideas and issues expressed often in the terminology of the participant's respective discipline, which may not be familiar to other participants or may have a specific and different meaning to someone from another discipline. The visualiser can capture, define, and synthesise these in a visual shorthand that is understandable to all and will promote further discussion. Where the team leader has a background in product design, they can structure the process within the framework of product development rather than fashion and ensure that a visualised design result is possible within the given time. Lastly, where the design partners have a professional understanding of design and personal experience of its failure, they can simultaneously articulate the issues objectively and subjectively. As designers themselves, they will also have an understanding of the type of information that is necessary to stimulate and progress a successful design. These three factors can accelerate the development of innovative, visualised, and feasible design concepts.

17.4.4 The design concepts

The design results of the COT workshop were diverse, including menswear with ingenious magnetic fastenings for Tom Yelland, president of the Mouth and Foot Painting Artists, which he could put on unaided and an attractive but durable evening glove for the wheelchair user who complained of the damage caused to the sleeves of garments by the dirt and friction that results from propelling her chair. For the designer with rheumatoid arthritis, her team designed a two-piece wraparound swimsuit with simplified fastenings. For the partially sighted textile designer, the team devised a set of modular solutions to address her need for security when outside and for stylistic cues on garments to ensure that her clothes are correctly worn (Figure 17.5).

All four concepts have a direct relevance to ageing consumers who are likely to develop dexterity, mobility, vision, and cognition problems of varying degrees, which will impact on their relationship with clothing. For example, a person living with dementia may no longer be able to remember the learned patterns of dressing – how to put on an item of clothing correctly so that it is not inside-out or how to button a cardigan correctly with the dexterity they once had. Thus, the cues required by a visually impaired person will benefit other groups and other contexts too where cognition and vision are compromised.

The format of combining contextual study with actual design work jointly facilitated by designers and design partners proved popular with delegates. A comment in the survey taken at the conclusion of the workshop stated: 'It has made me realise that thinking about inclusive design can be quite simple but make a big difference'.

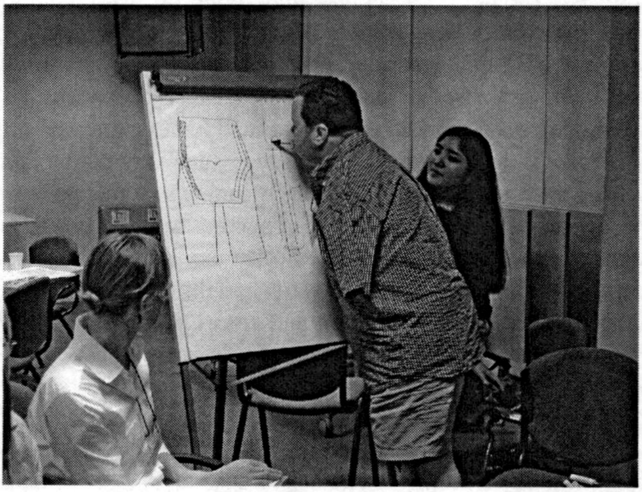

Figure 17.5 Tom Yelland draws his suggestions for clothing with magnetic fastenings.

17.5 User forums and interviews

The user forum or individual user interview is the most common form of user input to the design process. While individual interviews offer in-depth insight into the emotional, aesthetic, and functional issues of one person, a sufficient sampling may be required for a broader picture to emerge for design that transcends the bespoke or individually specific. In contrast, the biggest advantage of the facilitated user forum is that designers are presented with a range of information allowing them to identify and evaluate common issues, concerns, and strategies, assign a hierarchy of importance, and conceive a range of different scenarios.

Importantly they can provide the greater context of the whole user journey – for example, that clothing may not be the only issue; instead, the retail environment in which it is sold may be the greater hurdle. Changing rooms are often cited as problematic, with dimensions too small to allow older shoppers to change comfortably, no seating for those whose balance may be poor. And in the retail space itself, there may be an insufficient supply of larger sizes in view of the fact that we tend to expand as we age.

This was noted as far back as 1993 by John Harrison, Managing Director of the TMS partnership at *Fashion and Textiles for Our Future Selves*, a seminar organised by *DesignAge* at the RCA as part of the 1993 EC *Year of Older People and Solidarity Between Generations*: 'Women get larger as they get older, with the result that 60% of the female population aged sixty and over takes size 16, with a considerable proportion requiring size 20 or above…. Do retailers offer real choice on their rails, and in sizes 16, 18 and 20? Do we ghettoise larger women in more expensive parts of department stores and specialist shops that make them feel unwanted and embarrassed about shopping?' (Harrison, 1993) Almost two decades on and despite the competition from internet shopping, has the situation in the High Street really changed?

17.5.1 The structured user forum

As with the two immersive workshops described, user forums can similarly be structured to deliver specific results as were those organised for the DBA Design Challenge teams, which were tailored to the demands of the brief.

For product concepts, task analysis and existing product evaluation were included alongside a more general discussion, with participants chosen to highlight the issues that their respective disability raised. Care was taken to ensure that the designers were presented with a range of issues and not the viewpoint of a single disability alone so as to avoid any 'special needs' scenario. Through these discussions, the disabled participants were able to mediate a common inclusive ground, which worked better for all whilst ensuring that their individual concerns were highlighted. The designers in turn were presented with a general view and a detailed individual one, which informed the design details of their final idea. Sometimes too a 'wild card user' was introduced to steer the designers away from conventional cosmetic improvements alone towards a more radical 'back to basics' design approach in order to answer the design question embodied by the user. In a project centred on the redesign of a sticking plaster, a participant with no arms was introduced to challenge the designers with the fundamental question 'How can you put on a Band-Aid when you have no arms?' (Cassim & Dong, 2007; Harrison, 1993)

Initially there was a focus on young disabled people rather than older people as in the *DesignAge* programme, although this changed over time to allow a generational balance, depending on the project. This was a deliberate strategy to help the designers to understand the importance of aesthetics and the impact of the stigma of poor design on everyone, but particularly on the young. The fact that the message was conveyed through the medium of their disabled contemporaries made the message more powerful.

The only other clothing-related project to emerge from the DBA Challenge was the *C system* by the visual communications and branding firm Coley Porter Bell (Figure 17.6) – a 'smart' clothing tag system consisting of a tactile code of 16 shapes

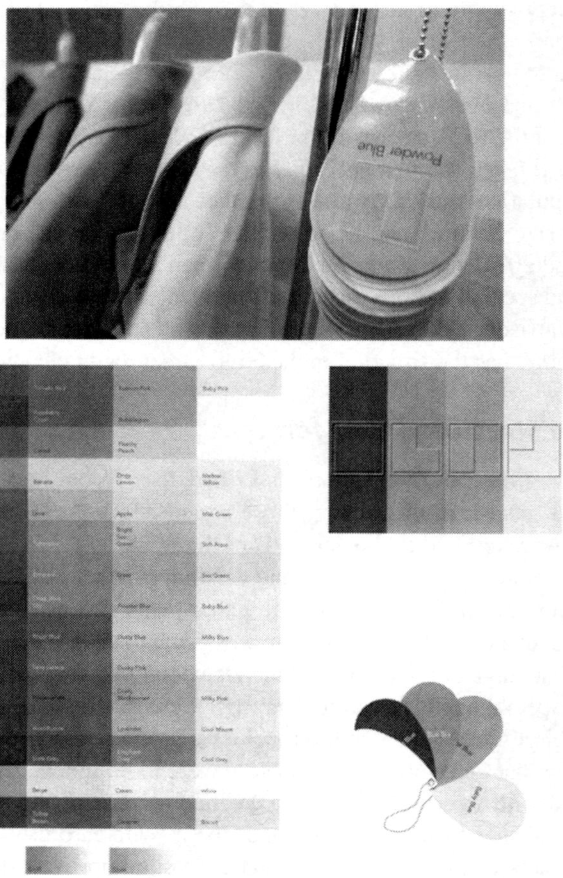

Figure 17.6 'C' system: smart clothing tag system by Coley Porter Bell.
Source: HHCD and courtesy of Coley Porter Bell.

representing colour and size, allowing 60 shades to be identified with a bar code reader for additional product information (Cassim & Dong, 2003).

This was the direct response to three of the young visually impaired users who took part in the initial user forum at the start of the project. All relied on the opinion of others to select their clothing. All shared a fear of generational mismatch in which clothing chosen by a parent or friend was not expressive of the image they wished to convey, a reverse take on the 'mutton dressed as lamb' fear of older consumers. One, for example, refused to shop anywhere except at Topshop, knowing that what he chose would be 'of the moment' and shunned larger retailers such as Marks and Spencer because he could not negotiate with confidence the range of choices available.

17.5.2 Individual interviews

The *Methods Lab* (Aldersey-Williams et al., 1999) demonstrated that designers adapt existing methods and develop their own creative ways to gain user-centred information and stimuli. Cultural probes are a design-led approach to understanding users that stress empathy and engagement and consist of 'collections of evocative tasks meant to elicit inspirational responses from people – not comprehensive information about them, but fragmentary clues about their lives and thoughts' (Gaver et al., 1999). Self-reporting through user diaries, contextual photography, and audio recording are a large element. As a mechanism, they are particularly suited to the subject of clothing, since designers rely heavily on visual and material information of the kind that cannot be captured by conventional interview or survey techniques alone.

17.5.3 The emotional context

While people may happily describe functional issues in user forums, talking about more deep-seated emotional ones in relation to clothing may be more difficult – they may not wish their concerns to appear trivial or foolish particularly in a public forum, and the Gok Wan technique of getting people to shed their inhibitions via the shedding of their clothes, which represents the climax of his *How to Look Good Naked* TV show, is not universally acceptable.

Thus, having a method that combines discretion and an ethical understanding and approach to this sensitive area is important, as epitomised by Candy and Goodacre's interviews of women with RA. These were conducted in their homes in close proximity to their wardrobes, allowing them to 'show and tell' spontaneously and for the researcher to photographically document in detail the wardrobe items discussed. As they note: 'This draws on the fashion and textile design practice of making mood or storyboards as a means to record permutations of aesthetic qualities that cannot be adequately described using words' (Candy & Goodacre, 2007; Gaver et al., 1999).

Similarly, using physical artifacts in user forums as the basis for discussion in the form of beloved clothes or clothes that are problematic to the subject can prompt detailed practical discussions on the issues of fit, style, and manufacture particularly in the area of fixings and closures as well as the range of material, sensory, and cultural qualities embodied in a garment.

17.6 Making the case for inclusive design

If inclusive design is to be embedded within the core constituencies that make up the clothing industry, it seems important that it is articulated to each in turn in a way that makes undeniable sense and is couched in their own terms.

17.6.1 *The business case*

Speaking at the DesignAge 1993 seminar Fashion and Textiles for Our Future Selves, John Harrison was blunt in his assessment of the potential take-up of the inclusive clothing agenda by the clothing industry: 'On the basis of our research, I can say that the over 50s present a real opportunity for the textile, fashion and retail industries, but, at the expense of sounding mercenary, I would suggest that whatever the social values, and social benefits that might come from the facts…we will only see real change in the High Street if there is money in it for somebody. Without that economic imperative there will not be change' (Harrison, 1993; Gaver et al., 1999).

Since then, with the ageing of the baby boomer generation, the business case for inclusive design has become ever more glaringly obvious even to those areas of the industry that have been the most reluctant to look into the demographic mirror. A headline in the *Daily Mail* of April 17, 2012, declared 'The Rise and Rise of the Boomer Models: As 50-Year-Olds Spend $47 billion on Fashion, Agencies Cash in on a Lucrative Market' (Wellman, 2012); but it has taken more than two decades for the idea to sink in.

Undeniably, the business case will remain the most powerful trigger for uptake of the idea given the sheer numbers involved. The UN Population Division has calculated that 'In the more developed regions of the world, almost one fifth of the population was aged 60 or older in the year 2000; by 2050, this proportion is expected to reach one third. In the less developed regions, only 8 percent of the population is currently over the age of 60; however, by 2050 older persons will make up nearly 20 per cent of the population' (UN, 2002).

Given the extended arc describing the life expectancy of the ageing population, in comparison to previous generations, we will be older longer than we will be young. 'The fastest growing age group in the world is the oldest-old, those aged 80 years or older. They are currently increasing at 3.8 percent per year and comprise more than one tenth of the total number of older persons. By the middle of the century, one fifth of older persons will be 80 years or older', which means that the proportion of disabled people within the population will also rise to unprecedented levels. And given the drop in fertility rates there will be fewer young around since the total fertility rate is below the replacement level in most of the industrialized countries (UN, 2002).

The Engineering Design Centre at Cambridge (EDC) bases its arguments to industry on the importance of inclusive design on the cost of design exclusion by calculating how much businesses will lose through their failure to recognise the needs of their older and disabled customers (Keates & Clarkson, 2004). However, while economic arguments may convince businesses that this is a market worth considering, how they address it in product terms and how they speak to that potential market may be way off the mark.

17.6.2 The marketing case: repositioning age and disability

George Davis at *Next* was one of the first mainstream manufacturers and retailers to use demographics as a strong sales tool; but his colleague Brenda Polan, speaking at the 1993 *DesignAge* seminar, explained how he initially got it wrong. 'Although George was trying to cater for 25–40 year olds and actually belonged to that generation, he had rather set ideas about what they would want to wear…when the PR and stylist looked at the clothes their reaction was "but these aren't fashionable, they are safe, dull and dowdy!" So on the fashion shoot, the stylist took scissors to the back of the jacket, made it look bigger – everything was taken apart and put back together on the session…they showed the pictures to George and he said: "Great – aren't they wonderful!" and the reply came back "Yes! But these aren't your clothes George – and you've got to make them like this!" And George is clever, George did' (Polan, 1993).

Here Polan identifies an enduring problem: the demographic of the design industry is overwhelmingly young and male dominated. The Design Council's 2005 survey of the design industry based on official Labour Force Survey statistics shows that 62% of designers are under 40, 61% are male, and only 6% are from minority ethnic groups (Design Council, 2005, p. 10).

Like George Davis, their ability and perhaps that of their female colleagues to interpret the aspirations of older consumers may be based on received ideas or outright stereotypes. In recent years, the fashion pages of *The Guardian* newspaper have shown a welcome shift in their choice of models. Each week, four to five models pose together in a selection of different clothes. The line-up has always included a rake-thin older woman wearing clothes that are indistinguishable aesthetically or generationally from her younger and ethnically diverse counterparts. The message while welcome is mixed – it is OK to wear the clothes but only if you conform to the same aesthetic codes and have no trouble putting them on.

As Julia Twigg has noted, 'Fashion and age do not fit easily, or happily, together. There is a discordant quality in the mix. Fashion is assumed to be all about youth and beauty – so far removed from the world of age'(Twigg, 2010). So the question remains: How does one describe the older consumer in terms that make sense conceptually and stylistically without ignoring their key functional concerns? And how can this be conveyed in a way that engages rather than stigmatises?

17.6.3 Dove campaign for real beauty

The Dove Campaign for *Real Beauty* for Unilever showed that when it is done well, consumers respond. Unilever reported a 12.5% increase in sales in 2005 and a 10.1%

increase the following year (O'Donnell, 2008), but this did not lead to the development of age-friendly packaging for the product range. It was a triumph of marketing over ergonomics but a powerful lesson also in the importance of understanding how we perceive ourselves.

17.6.4 Ageing and context

This power of mind over matter in relation to ageing was demonstrated conclusively in a famous 1981 experiment by the Professor of Psychology at Harvard, Ellen Langer, who had long suspected that the fixed ideas internalised in childhood can affect the way that we age. Two control groups of men in their 70s and 80s were invited to stay for a week at a retreat in the New Hampshire countryside. The first group was asked to pretend that they were young men reliving their youth in the 1950s in an environment that reflected the period and via the events of the period, which were transmitted through publications and recordings on a vintage radio and black-and-white television. The second group, who followed a week later, were told to stay in the present and only reminisce using the artifacts and materials provided. Before and after the experiment, the men were subjected to a battery of cognitive and physical tests and the researchers recorded dramatic positive changes in each group – both were stronger and more flexible – their height, weight, gait, posture, hearing, vision, and performance on memory tests improved, with the group who had acted as if they were actually back in the 1950s showing the most significant changes (Langer, 1989).

17.6.5 YOYOs: expressing the duality of age

In 2002, Jo Taylor, then a Helen Hamlyn Centre research associate, worked with Fitch, the design firm, on *See You Later*, a communication project that explored new ways for young designers to connect more creatively with older people without marginalising or patronising them (Taylor, 2002). She came up with the term YOYO to describe someone who is both old and young simultaneously and whose undeniable physical age may bear no relationship to the age that they consider themselves to be – the functional and emotional duality with which we all live. As a term, YOYO has been particularly effective in helping designers to understand the dichotomy of the ageing consumer: that while they may share the same taste in products as their children or even grandchildren, they require the extra levels of functionality demanded by their physical age; these are achievable through a balance of considerate ergonomics and aesthetic design. However, there must also be an understanding of the messages that they wish to transmit through their belongings.

One project, which epitomises this trinity of aesthetic, functional, and symbolic issues, was Wunder, which won the 48-Hour Inclusive Design Challenge held in Singapore in 2007. The female design partner working with the team from the international advertising agency Grey was a wheelchair user and amputee. Her greatest desire was attractive underwear, which, although unseen, was crucial to her sense of self-esteem. The team designed underpants that fasten at the side, open flat at the crotch, and can be slipped on easily for carer-assisted or independent dressing (Cassim, 2007b, p. 41) (Figure 17.7).

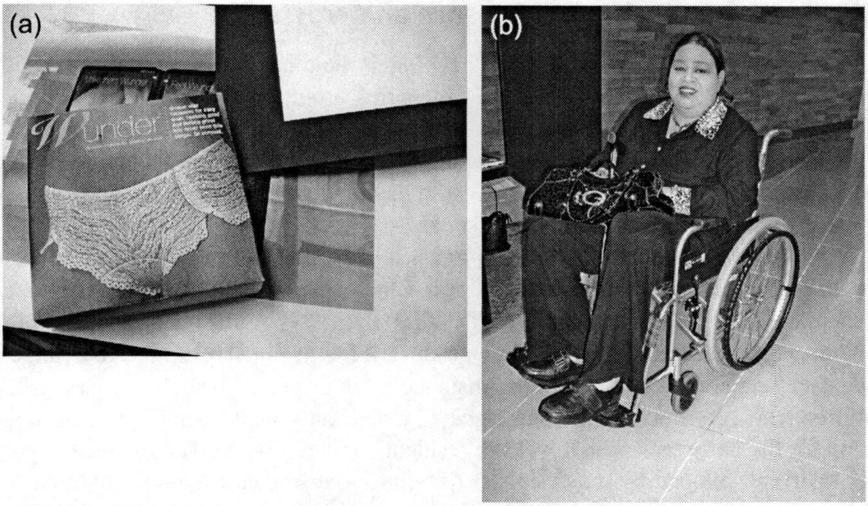

Figure 17.7 (a) Wunder panties by Grey Design *(Source: image courtesy of design team)*. (b) Their design partner.

But perhaps the most eloquent plea for understanding of clothing's important symbolic role was by Brenda Polan at the 1993 *DesignAge* seminar: 'Fashion is very much about sexual allure and sexual allure is not just about body display, it is about an attitude of mind that people who believe that they are at the cutting edge of all that is important – of ideas, trends, philosophies – consider clothes to be a symbol of. Those of us who are getting older will not want to sacrifice that element of youthfulness, which makes us think of ourselves as still sexy, still powerful, and in control; we are also strong enough to insist that what we demand be given to us. As a generation we are reluctant to relinquish centre stage, and interestingly…this is the generation of women who have had careers – women with the confidence given by those careers and the habit of spending on themselves which that confidence and income give' (Polan, 1993).

17.6.6 The creative case for inclusive design

The creative case for inclusive design is one that has to be made to designers since ultimately they are the key to good design. Simultaneously, inspiring and informing young designers so that they consider the needs of older or disabled people and see them within a continuum of which they themselves are part has been a major thrust of work of the Helen Hamlyn Centre for Design. It was the rationale behind the choice of the term 'Design for Our Future Selves'. The inspirational aspects and stimuli have been delivered overwhelmingly by the extreme and highly lateral scenarios provided by the design partners, who have introduced the young able-bodied designers to a world that they have difficulty imagining. But beyond that, a more formal creative case has had to be made that ties in with a design innovation process that they recognise within the framework of their own design discipline.

17.6.7 Scenarios for convergence and crossover

A core strategy of inclusive design teaching is helping designers understand the importance of creating multiple design scenarios around a product or service idea. Identifying congruence between groups of people who superficially appear to have little in common, and then melding this information into a concept that has relevance to them all, can be said to be the acid test in deciding whether something is inclusive or merely 'special needs' or niche market alone.

A product that exemplifies this approach is 'Mo' the radically redesigned cushion that won the DBA Inclusive Design Challenge in 2009 in response to the 'Sedentary Lives' brief that year (HHCD, 2009) (http://www.hhc.rca.ac.uk/2395/all/1/%27mo.aspx Accessed 30.07.12). As a product, it has as much relevance to a range of sedentary scenarios – including long-distance truck drivers, wheelchair users, school children, call centre workers, concertgoers, and so on – as it does to the core target group for the designers, which was the residents of care homes. The innovative polymer matrix developed by the designers to replace the standard foam cushion has the advantage of being able to adjust to the micro-movements of fidgeting, an important aspect of comfort for those who are seated for long periods. Its mesh structure allows the cushion to be hosed down and sterilised in response to the very real problem of incontinence and the high costs that this entails for care homes in terms of furniture damage (Figure 17.8).

'Mo' of course is not an item of clothing, but identifying scenarios of convergence and crossover has the same relevance to the creation of new inclusive clothing typologies. Working collaboratively, as in the two workshops described, is the foundation stone of this approach to the ideation process, since it allows designers to collaborate with specialists in such fields as technical textiles, who may be actively searching for applications for technology they have developed for a different purpose. As Geoffrey Beene observed, 'The future of fashion will come from the laboratory not the sketchpad. (Yusuf, 1994).

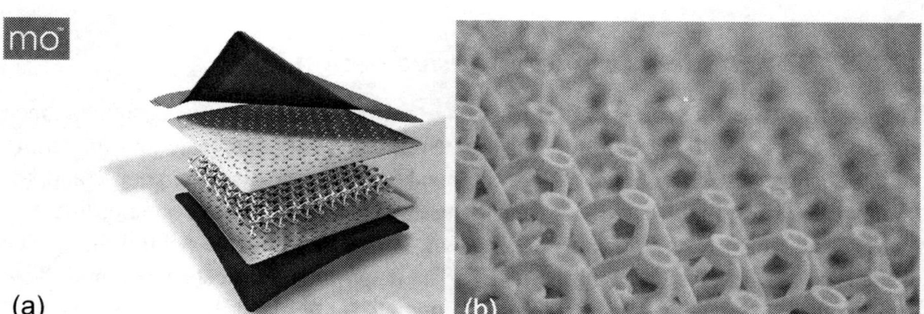

Figure 17.8 (a) 'Mo' cushion structure. (b) Close-up of polymer matrix.
Source: Images courtesy of Matter.

17.6.8 New clothing typologies

Designers such as Issey Miyake, Yohji Yamamoto, and Hussein Chalayan have all used this collaborative, experimental approach to re-examine clothing conceptually as a product and by allying it to modern production techniques, produced garments, which in Miyake's case are supremely wearable by people of all ages and body shapes. As he noted: 'I do not create a fashionable aesthetic, I create a style based on life, not a style for its own sake' (Benaim, 1998).

Working with textile designers Makiko Minagawa, Jurgen Lehl, and Junichiro Arai, Miyake has drawn on Japan's rich textile and clothing tradition, allying it to modern production techniques to produce clothing that is modular, does not follow or display the contours of the body, but can be draped or layered from a single straight piece of cloth – a concept based on the way the *kimono*, *haori*, and *monpe* have traditionally been constructed. It means that anyone can wear them irrespective of their body shape. The kimono is traditionally passed from mother to daughter, and when too old is deconstructed and made into other textile items. Inclusivity meets sustainability.

Similarly, the designer Caterina Radvan has brilliantly exploited advances in seamless knitting technology to produce high-end asymmetric knitwear for unconventional body shapes that is also supremely cost-effective (Radvan, 2011).

17.7 Conclusion

It is undeniable that the clothing choices offered on the High Street or the Internet are more varied and comfortable in comparison to the past – 'Comfort is the prerequisite of clothing today. Clothes which stretch and mould with the natural form rather than imposing a rigid shell, have become the norm' (Radvan, 2013; Yusuf, 1994). In this regard, the impact of Lycra and Velcro on the greater comfort of clothing that stretches to fit the body and can be easily fastened and undone cannot be underestimated. Neither can the cross-pollination of styles and materials from previously separate clothing cultures such as performance sportswear, skateboarding, aerobics, and dance cultures, where ease of movement is sine qua non. These have resulted predominantly in leisure clothing rather than formal clothing typologies that can be worn confidently by a far broader spectrum of the population. Similarly, the shift from traditional pattern-cutting methods and clothing adaptation to CAD manufacture allied to body scanning seems to offer the potential for customisation at source. All have been immensely positive factors in affording choice to the two groups in the population whose voices are still muffled or misheard. It seems that the greatest challenge that remains is to help the fashion end of the industry understand that there is nothing to fear.

As Michael Shamash noted: 'When a relatively non-radical campaign such as Dove soap, which uses supposedly average women, causes a major debate, then you realise the journey the fashion industry must embark on before it addresses the fashion needs of disabled people. There is a status quo that does not want to be challenged' (Shamash, 2008). But challenge it we must!

References

Aldersey-Williams, H., Coleman, R., & Bound, J. (1999). *Methods lab*. London: Royal College of Art.

Arthritis Research UK. (2012). *A call to action – Providing better footwear and foot orthoses for people with rheumatoid arthritis*. London: Arthritis ResearchUK.

Benaim, L. (1998). *Issey Miyake*. London, UK: Thames and Hudson.

Bieber, M. (2003). The struggle for independence. In J. Clarkson, S. Keates, R. Coleman, C. Lebbon (Eds.), *Inclusive design – Design for the whole population* (p. 56). Springer Verlag(2003).

British Standards Institute. (2005). *British Standard 7000-6:2005*. Design management systems – Managing inclusive design – Guide.

Candy, F., & Goodacre, L. (2007). *The wardrobe and well being: Exploring relationships between women living with rheumatoid arthritis and their clothing*. Include 2007 Proceedings. London: Royal College of Art.

Cassim, J. (2003a). *Smart wearables: A new frontier for inclusive design innovation*. Proceedings Include 2003. London: Helen Hamlyn Research Centre, Royal College of Art.

Cassim, J. (Ed.). (2003b). *Innovate 5, Can insights on disability help to design better products?* London: Helen Hamlyn Research Centre.

Cassim, J. (2005). *Designers are users too! – Attitudinal and information barriers to inclusive design within the design community*. Include 2005, Proceedings. London, UK: Royal College of Art.

Cassim, J. (July 2007a). *It's not what you do it's the way that you do it, the Challenge Workshop a designer-centred inclusive design knowledge transfer mechanism for different contexts*. HCI International Proceedings, Beijing.

Cassim, J. (2007b). *Challenge 2007*. London: Helen Hamlyn Research Centre, Royal College of Art.

Cassim, J., & Dong, H. (2003). Critical users in design innovation. In J. Clarkson, R. Coleman, S. Keates, C. Lebbon, et al. (Eds.), *Inclusive design – Design for the whole population* (p. 551). London: Springer Verlag.

Cassim, J., & Dong, H. (2007). Empowering designers and users: Case studies from the DBA inclusive design challenge. In R. Coleman, J. Clarkson, H. Dong, & Cassim, J (Eds.), *Design for inclusivity: A practical guide to accessible, innovative and user-centered design* (pp. 89–109). UK: Gower.

Clarkson, J., Coleman, R., Hosking, I., & Waller, S. (Eds.). (2007). *Inclusive design toolkit* (p. 24). Engineering Design Centre, Cambridge University.

Coleman, R. (2003). Living longer. In J. Clarkson, R. Coleman, S. Keates, C. S. Lebbon (Eds.), *Inclusive design: Design for the whole population* (p. 121). London: Springer Verlag.

Coleman, R., Bendixen, K., & Tahkokallio, P. (2003). A European perspective in inclusive design. In J. Clarkson, R. Coleman, S. Keates, C. Lebbon (Eds.), *Inclusive design – Design for the whole population* (p. 290). London: Springer Verlag.

Design Council. (2005). *The business of design*. London: Design Council.

Design Council. (2010). *Design for patient dignity*. London: Design Council.

Gaver, W., Dunne, A., & Pacenti, E. (1999). Cultural probes. *Interactions*, *vi*(1), 21–29.

Goldsberry, E., Shim, S., & Reich, N. (1996). Women 55 years and older: Part I. Current body measurements as contrasted to the PS 42-70 data. *Clothing and Textiles Research Journal*, *14*(2), 108–120.

Harrison, J. (1993). *A changing marketplace: A new opportunity in designing for our future selves*. London: Royal College of Art, ISBN: 1 874175 45 4.

Keates, S., & Clarkson, J. (2004). *Countering design exclusion: An introduction to inclusive design*. London: Springer Verlag, 125–139.

Langer, E. J. (1989). *Mindfulness*. Reading, MA: Addison Wesley, ISBN: 0-201-52341-8.

MacNeice, L. (1935). *Snow in poems*. London: Faber & Faber; New York: Random House, 1937.

O'Donnell, D. (2008). Unilever's dove and axe: examples of hypocrisy or good marketing? In *Case Study Competition Journal 2008*. New York: Arthur W. Page Society.

Ostroff, E., & Preiser, W. (2001). *Universal design handbook*. New York: McGraw-Hill.

Paterson, R. (1996). *Fashion and disability* (MA thesis). London: Royal College of Art.

Polan, B. (1993). Where next? In R. Coleman (Ed.), *Designing for our future selves-fashion and textiles for our future selves* (p. 94). London: Royal College of Art.

Pullin, G. (2009). *Design meets disability*. Cambridge, Mass & London: MIT Press, ISBN: 978-0-262-16255-5.

Radvan, C. (2011). *Inclusive design solutions for womenswear through industrial seamless knitting technology*. Fashion Practice: Design, Creative Process and the Fashion Industry. ISSN 1756-9370, 1756-9389.

Rankin, Q., Davies, M., & Riley, H. (2005). *The effect of dyslexia and dyspraxia on the cognitive modelling process within the context of art and design. Part one: Drawing on drawing*. Proceedings, Include 2005. London: Royal College of Art.

Sanders, E., & Stappers, P. J. (March 2008). Co-creation and the new landscapes of design. *Co-Design, 4*(1).

Shamash, M. (2004). *The disabled also like to be fashionable*. The Guardian. September 9, 2005.

Shamash, M. (January 2008). A dedicated (disabled) follower of fashion. *Occupational Therapy News*.

Taylor, J. (2002). *See you later: Bringing originality and delight into communication with older people internal report*. Helen Hamlyn Research Centre, Royal College of Art.

Twigg, J. (2010). *How does Vogue negotiate age?* Forthcoming in Fashion Theory, 2010. Available online: http://www.kent.ac.uk/sspssr/staff/academic/twigg/paper9.pdf. Accessed 31.07.2012.

United Nations Population Division.2002). *World population ageing: 1950–2050*. New York: Department of Economic and Social affairs, Population Division.

Wellman, V. (April 17, 2012). The rise and rise of the boomer models: as 50-year-olds spend $47 billion on fashion, agencies cash in on a lucrative market. *Daily Mail*.

Yusuf, N. (1994). Fabric in fashion. In M. Mahoney, & S. Braddock (Eds.), *Textiles and new technology*. London, UK: Artemis and Crafts Council.

Part Four

From design to apparel for the active ageing population

From co-design to design specifications and manufacture of apparel for the active ageing population

18

J. Greengrass
Greengrass Design Studio Ltd, Glossop, UK

18.1 Introduction

Sprayway is one of the UK's leading walking clothing and accessory brands. The company tag line is 'for all walks of life' and this is reflected in the design brief: to design a range that is suitable for all age groups from zero to ninety years old. Often we considered the more interesting parts of the collection to be those aimed at the under-60 age group, and before starting the project Sprayway had not carried out any specific research into whether there might be other needs or wants for the over-60 age group, e.g. fit, features, marketing and wearable technology that were not being addressed by the current collection.

When the company was approached about being involved in the Design for Ageing Well project, we saw it as a really exciting opportunity. Not only would it be able to aid the design process in producing products that were better suited to our active, older consumers but also it would allow us to try again with the introduction of wearable technology. We had introduced push-to-talk modules and iPod-compatible systems before, but not had great success with them.

We realised from the beginning that we have synergy with the project and it could really improve current products as well as guide our choices for new products. In a struggling retail market new ideas and a new niche customer with such great credentials was worth further research.

As research has shown, the active ager is a growing market with more time than their younger counterparts for leisure activities such as walking. With less financial restrictions and dependents that mostly support themselves, the older consumer is in a better position to choose quality over price, understanding that very often value is not always based on the cheapest product, but the product most suited to the consumer. Confident in their own identity, they are seldom concerned with social acceptance underlined by their dress, unlike the less confident, younger consumer. Elaborate marketing campaigns often have the reverse effect with a consumer more concerned about attributes such as being fit for purpose, flattering fit, comfort, adjustment, colour and texture etc. Independence and comfort are paramount, and clothing that allows the wearer to do more, or go further and fulfil their passions or dreams,

Textile-led Design for the Active Ageing Population. http://dx.doi.org/10.1016/B978-0-85709-538-1.00018-3

may be bought regardless of expense. Ultimately, they also want to look good and stylish with it.

18.2 Design brief to point of sale (POS), the current process

18.2.1 Research

Co-design is a different way of working for Sprayway. Our current design and development process is 18 months long from the initial research to product in store. The process starts with gathering information. This includes feedback from the sales team and retailers and data from the sales office regarding orders in and out; this gives us current market information.

We then add to this information with trend research carried out at trade shows such as ISPO, Outdoor and Bread and Butter. Here we gather information on competitors as well as new fabrics and technology. Other sources are trend books and reports for colours, styling and detail such as Pantone View, Imouv, View 2 and Sport and Street.

When we started the research for the Autumn/Winter 2013 collection we found the trend research for this area very difficult. Publications for menswear, womenswear or sportswear did not cover this age group; colour prediction books were aimed at the more general sportswear consumer and were not age specific; and at the trade shows there were few brands catering for this consumer specifically. There are many trend books for menswear, womenswear, childrenswear, young fashion, sportswear, bridalwear, underwear, footwear, accessories etc., but there are no publications or online trend websites that are forecasting trends for this part of the market specifically.

18.2.2 Testing

During the development process we send prototypes to our team of wear testers for their feedback: generally retailers, shop staff, outdoor activists that have contacted us or we have met at events such as the Ullswater walking festival, Sprayway team members or their family friends and children, outdoor pursuit centres etc. Each wear tester submits their measurements, location, activities and activity frequency to us, we then choose whom we send clothing to test based on this information to check performance and detailing. All feedback is collated and reviewed by the designers.

Another area that we look at is product that has been returned. Our customer service department sends the design team a returns report every month – firstly, to check that there are not any problems with one particular style, which could suggest a manufacturing or component fault; and secondly, to see which areas of the garments are failing and wearing quicker than others.

We also use laboratory testing for fabrics, components and whole garments to ensure that our products are fit for purpose. All Gore-Tex fabrics have a guarantee to keep the wearer dry, but we have our garments tested by W.L. Gore to ensure that they will not leak due to the design of the garment. Baby wear is tested for safety and that it meets EU standards. Fabrics and components are also sent to the lab for testing to check performance characteristics, for example abrasion resistance, hydrostatic head, breathability, tear strength and pilling resistance.

Finally, we look at customer reviews submitted to our website in addition to magazine and website reviews from consumer and professional gear testers such as Trail magazine, Country Walking, Outdoor Enthusiast, Outdoorsmagic.com, Livefortheoutdoors.com and TGO.

18.2.3 Range brief

Once our initial research has been carried out, we then use that information to create a range brief. This will include reviewing the previous season and deciding which products need an update to make them more current in terms of colour, detail and fabric; which product is needed to complete the range based on price point and function; and which products are not for that season or have not sold and need to be removed from the collection. Here we also look at projects, which could include a new collection, e.g. active agers and wearable technology, or a new treatment such as UV, anti-mosquito or new trim packs.

18.2.4 Design and development

On completion of the design brief draft (see Figure 18.1), design and development can begin. The design brief draft is a plan of products that will be developed for that season, including the products that will carry forward. Often the brief will include a description of the new product, perhaps an existing or past style that we will use as a starting point for fit, shape and detail, a target price point, ideas of which fabric might be used, possible colours and the factory that it will be developed and made in.

Products are carefully thought out and specified in detail to the factories using Skype or email to send diagrams, photographs, size charts and written notes, to give the factory as much information as we can for them to create the prototype as closely to our idea as possible (see Figures 18.2 and 18.3). The merchandisers that work with us at the factories translate our notes into the relevant language, as generally pattern makers and machinists do not understand English. Sometimes a limitation is that the merchandiser has not made garments or patterns before but has very good English. Their translation will be literal as they have only a moderate understanding of what we are asking the pattern maker or machinist to do. The pattern maker or machinist can only take information from the translation and visuals, so sometimes the details in the diagrams are more important than the words.

Years ago, Sprayway made all of their patterns and prototypes in-house. Currently, they are all made at factories in the Far East. This has some limitations on time, creativity and also on fit as the pattern maker has to be shown using diagrams, measurements and photographs what needs to be changed rather than having them be involved in the fitting session with the fit model (a model with body measurements chosen to best represent the customer).

Prototypes are received and feedback sent back to the factory via email or Skype in the same or similar format as the initial specification. This may include comments to

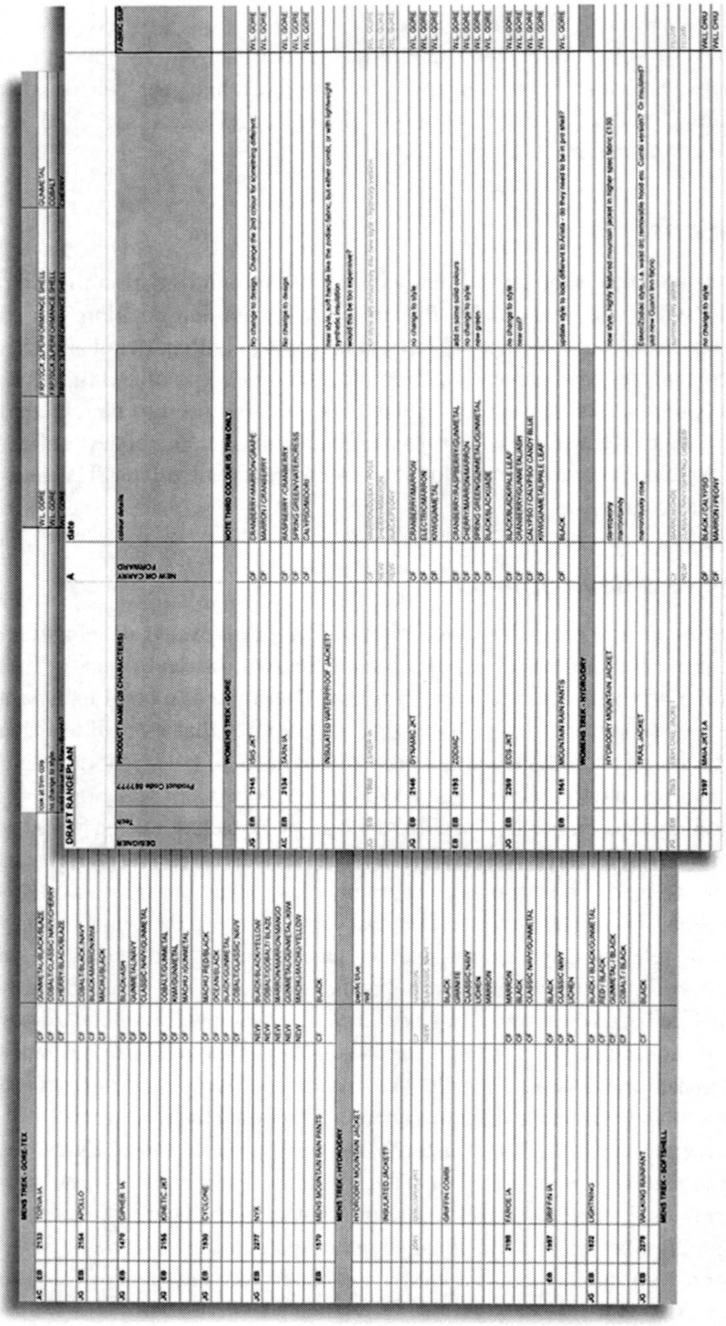

Figure 18.1 Example of a draft range plan.

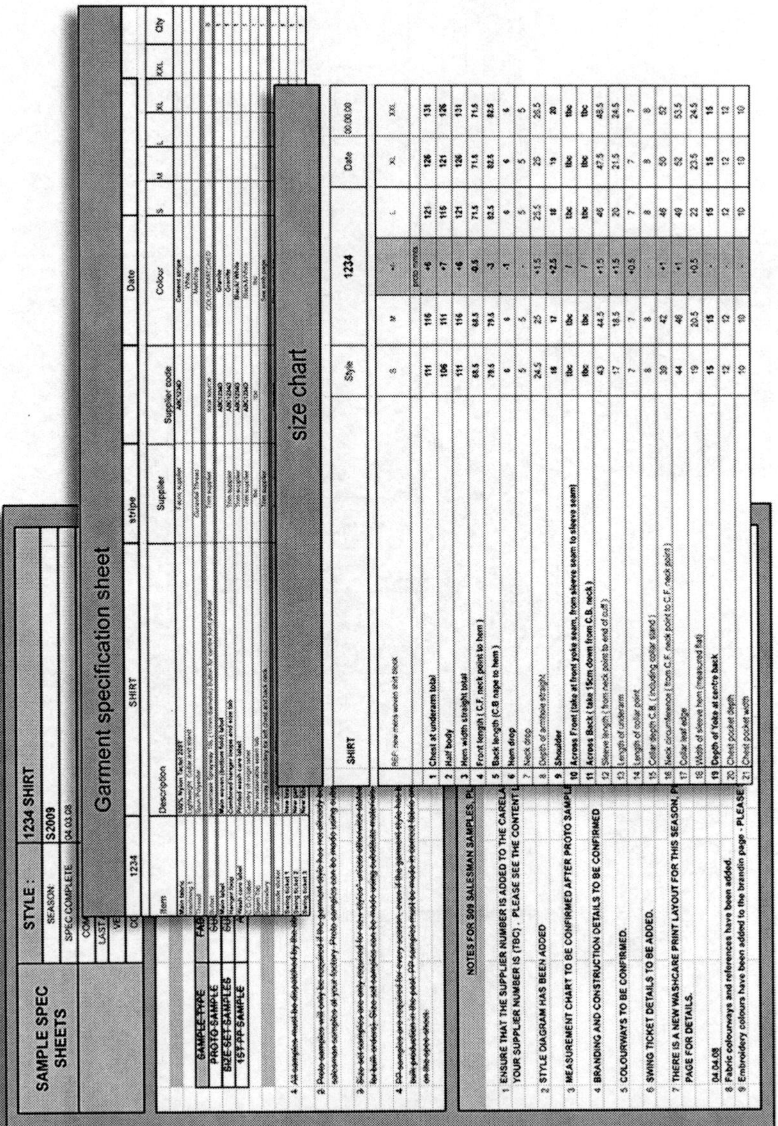

Figure 18.2 Examples of size chart, bill of materials and notes that are just some of the documents sent to the factory to produce a prototype, salesman's sample or bulk production.

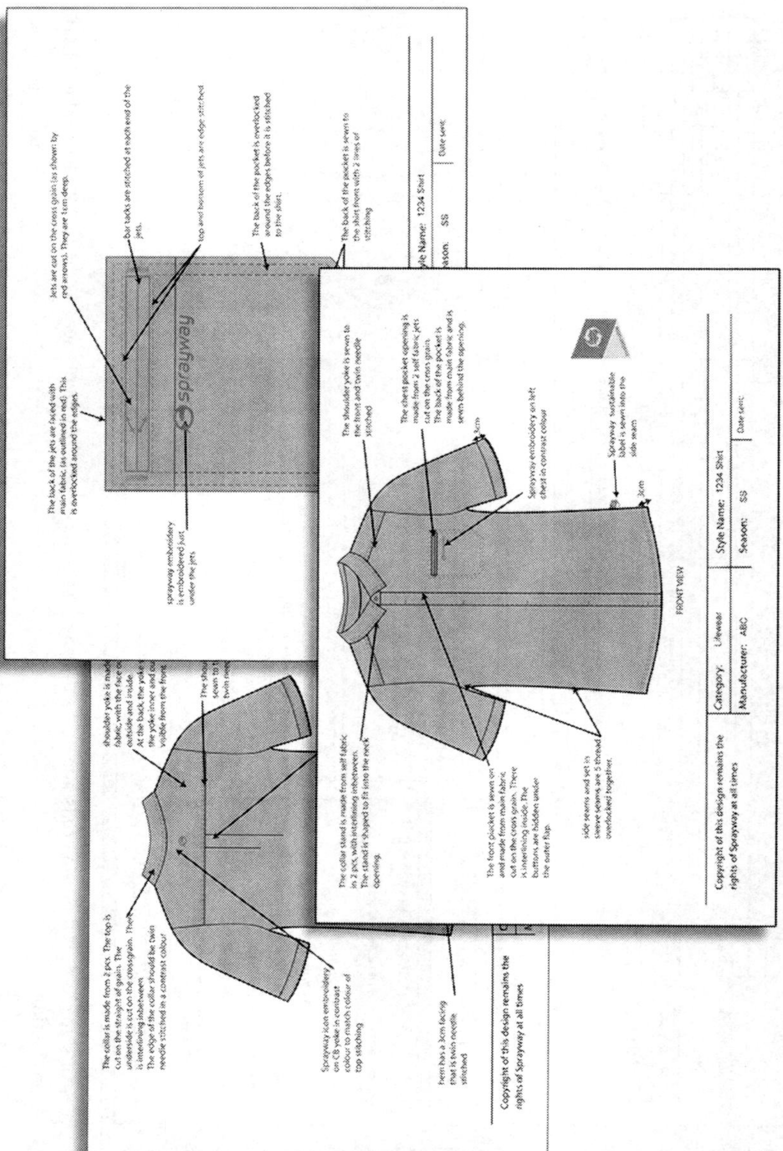

Figure 18.3 Examples of annotated diagrams sent to the factory.

details that are not correct or have not worked and new details included; fit and sizing and what needs to change and by how much; new trim and how it should be attached to the garment; and changes to the fabric. During this process, the design team continues to gather feedback from sales, the sales team and retailers, wearer trials and lab tests.

Before final product sign off for salesman's sample production, members of the design team will visit the factories to feedback directly on final details and ensure that salesman's samples will be as correct as possible. This is a great opportunity to have contact with the pattern makers and machinists and to see what the factory is capable of, e.g. manufacturing techniques such as bonding, laser cutting, embroidery, printing as well as different stitch types. These can all be experimented with on this visit. Sometimes an interim trip may be organised depending on the level of detail of the new products, which allows for development to be carried out at the factory.

18.2.5 Salesman's samples

Salesman's samples are the tools that our salesmen use to show retailers what new products will look like for that season. This allows them to place a forward order and decide on when they would like it delivered to their store (we tell them which delivery dates will be available for them to choose from). The samples are generally 95–99% correct to the final product that will appear in store as sometimes trim components may be substituted, or the fit may need to be altered following feedback. When the salesman's samples arrive, this is the first time that we see products in the correct colours, with correct trims. They are sent to be photographed for our workbook and website.

The workbook is another tool used by our sales team to sell products to retailers (see Figure 18.4). Each retailer is given a copy and generally notes are made during the sales presentation next to relevant product. Often a retailer will not place an order at the presentation as they will first need to see the ranges from all of their main suppliers to make sure that they pick the best product for their customer. The workbook is very important for them, as it is their aide-mémoire for our collection. This is also

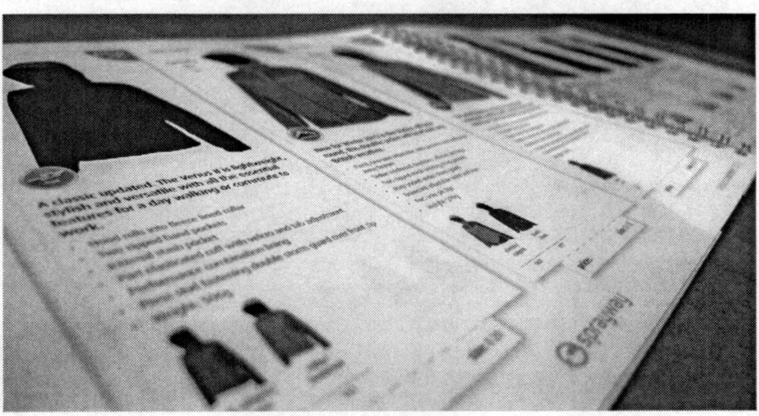

Figure 18.4 Sprayway workbook showing information given to retailers.

why we choose to use photographs over line drawings, as it is better to see an actual garment rather than a drawing of one to base an order on.

The workbooks are used by the stores to choose products to top up their orders in season and also to identify specific products requested by consumers. They are used during staff training by our sales team during their visits to stores where they present the products to the retail staff and teach them all of the details, fabrics and the customers or activities that the products are aimed at.

18.2.6 Product launch

The design team presents the new products to the sales team, distributors and agents at a two-to-three day workshop, where all new details and concepts are introduced.

The sales team sells our new product to the retailers. Often the retailers' openness to new products depends on several factors: changes in the weather, the economy, past season's sales and how much stock they have left from the previous season. If market conditions are poor, yet you have a great collection, it can often be disheartening if sales are down.

18.2.7 Limitations

We are always limited with time and we aim to stick to a predetermined critical path or timeline (see Figure 18.5). The start-to-finish process is around 18 months long, e.g. from gathering info to product arriving in stores. Retailers' confidence in the brand or on risking new products literally changes with the weather; this also affects the sales team's confidence. Fabric and factory minimum order quantities (MOQ) and lead times can change what is made and what it is made from, and cheaper-price

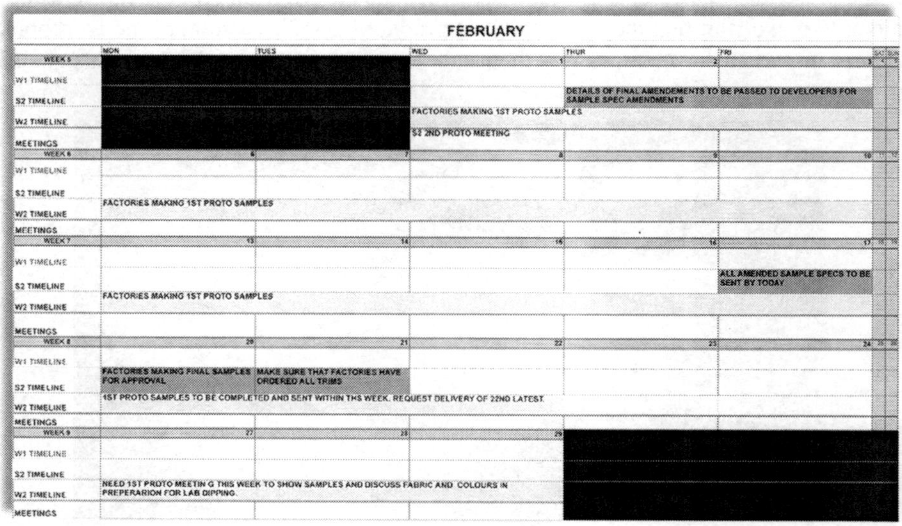

Figure 18.5 Example of one month taken from the critical path.

factories generally can limit what is designed as they tend to be less skilled than the more expensive factories or have older machinery.

18.3 Growing awareness of the ageing market

The preconceived idea of what someone in their sixties or seventies is capable of or should look like is often wrong. The stereotypical image of the ageing population suggests that they are typically inactive, reliant on other people and not concerned with their appearance. Working on the project, we met a group of individuals that were not only more active than the average 30-year-old but were also stylish and incredibly keen to be heard – and importantly very proud of their independence.

It is rare to discover a new market with such enticing credentials. Once your eyes are open, it is easy to see this growing market. Many of our retailers are very aware that their key customers are from the older age group and are often the greatest part of their footfall during the week. Their stores are welcoming, with helpful, attentive staff and unintimidating; from the research this is what the active ageing consumer is looking for.

Following meetings with YKK, leaders in their field of zips and trims, they are very keen to develop products for the older age group to facilitate fastening clothing in an easier way – zips that are simpler to fasten and snap fasteners that have a positive lock, yet are easy to pull apart for a user with limited dexterity.

Waterproof fabric giant W.L. Gore was also keen to work with the project, as they had identified that their core customer base is the walker and general outdoor user. Experts in lightweight fabrics, Pertex was also aware that this is a growing area of the market and has contributed to the project.

18.4 Co-design – listen, learn, develop, repeat, refine and repeat

It is easy to sit inside your bubble of experience and come up with ideas being fed by the same sources. It is arguably better to take the time to move outside of that bubble and get feedback about those ideas and have new ideas that come from new sources. When the new source is the consumer, this can spark a whole new line of thinking that you may never have discovered within that bubble. Co-design is a means of escaping the bubble and is often exciting and eye opening.

As a designer it is relatively easy to understand the needs of users in sports that you are involved in and can participate in. The one thing that it is almost impossible for a designer to do is to get inside the body and mind of a person nearly twice their age and understand and empathise with their needs, wants and experiences.

Communication with the target consumer is key to gaining insights into what they do not want, what they do not understand, what they do not like and what they cannot find in stores. In contrast, it is also important to discover what they do want, what they do understand and what they have learnt that they need.

On this project those involved in the focus groups are the experts and the real soul of the project. They are real people with very real opinions and some of the things we have learnt a lot by working with them include respecting their culture, functional requirements, body demands, aesthetic wants and independence. The more that you listen and see, the more knowledge is transferred. Sometimes the hardest part is to just let them do the talking.

18.4.1 The layering system

At an early stage of the project it was key to find out what clothing was already being worn for activity. Within the groups were people that had been participating in walking and other outdoor activities for many years yet did not know how to make their clothing more comfortable.

We spent an afternoon looking at their current clothing and then at the layering system, which is one of the most versatile systems used to create comfort during activities – other systems can be used, but this is the most well known and easiest to understand. The layering system, because it comprises a series of functional layers, allows the wearer to create their own personal, adaptable system that works to protect and maintain comfort across seasons and activities (see Figure 18.6).

The main aims of the layering system are to move water, provide protection and keep the wearer comfortable. When water ends up in the wrong part of the system at the wrong point in the wearer's activity, one result could be that they can become too sweaty, with difficulty cooling. At worst the moisture inhibits their insulation from trapping air, and conducts heat away from the body. Lack of adequate protection can affect the wearer in the heat as well as the cold, and depending on the environment, the consequences of either could be fatal.

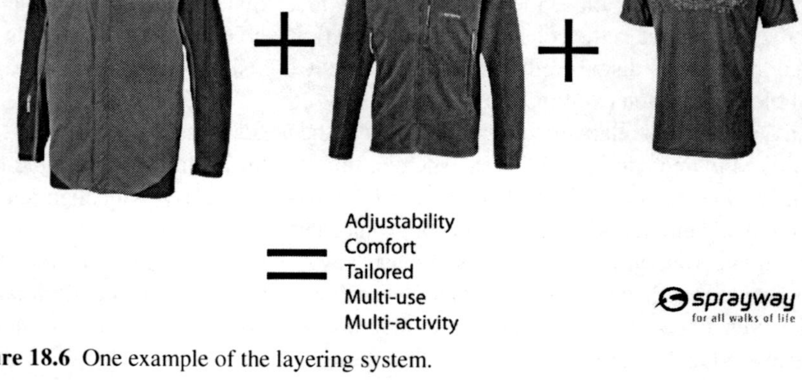

Figure 18.6 One example of the layering system.

There are three basic parts to the layering system:

18.4.1.1 Next to skin

Clothing that moves moisture away from the skin to enhance cooling and prevent chills. There are several ways to do this; in simplistic terms, the fabric can either hold moisture away from the skin, of which merino wool is one example, or the fabric wicks the moisture from the skin as found generally with synthetic fabrics with or without a wicking treatment. With the latter, the fabric moves moisture over a large surface area through wicking action. Body heat then evaporates the moisture from the fabric. This action of removing moisture from the skin helps to cool the wearer. The rapid removal of moisture from the next-to-skin layer also prevents this layer from becoming a conductive layer drawing heat from the body when the wearer stops.

Fabrics such as cotton become extremely good conductors when saturated. This can be lethal in cold conditions if the wearer stops or is injured, as when worn next to skin a sweat-laden cotton layer will continue to cool the body through conduction even though the wearer is adding extra layers in an attempt to warm themselves. We found in this session that the majority of the focus group wore cotton next to skin for all activities due to the perceived comfort level that it would provide versus a synthetic layer. This may also be due to the lack of awareness of the benefits of a layer that moves moisture away from the skin and how it performs in relation to cotton.

18.4.1.2 Insulation

Clothing that traps still air to be warmed by the body. There are different levels of insulation from the very lightweight option of wearing two next-to-skin layers, trapping air in between to the far warmer option of a down-filled jacket. With both, the warmed, trapped air insulates the body against the cooler air and also slows the escape of body warmth by radiation.

18.4.1.3 Protection

Protects the wearer and the other layers against elements from the outside and moves liquid or moisture vapour from the inside to the outside of the garment. This layer can be simple like a wind-shirt, or complicated like a full mountaineering waterproof shell. Designed to keep wind or rain or both out of your system: preventing wind from disturbing your warmed air in your insulation with a lightweight packable windproof, wind-resistant, or soft-shell garment; or to keep you dry in a heavy downpour yet let moisture vapour out like a waterproof shell.

This system is very flexible and allows you to tailor it to your activity and the weather.

We found that this was a good starting point for us as we could determine the level of understanding along with some demands that were made on their clothing. For the users within the group, the system along with new fabrics opened up a whole world of possibility for them, which fed in nicely to some other groups that we were involved with for colour, styling and fabric groups.

18.5 Co-design – industry involvement

Had Sprayway been approached to find a solution to clothing for the active ager it would have been difficult to find information about this consumer that already exists in relation to sportswear. As a relatively small company, we do not have the resources, expertise and specialist knowledge in-house to be able to cover all aspects of the research in such detail. Co-design has been the best route, not only working with the target consumer but also working alongside a group of professionals from around the country and from different disciplines to get to a conclusion.

18.5.1 Co-design: what i means to the company

Co-design can be defined in many ways; these are a few words that we now associate with it: *collaborate*, *community*, *co-operate*, *communicate*.

18.5.1.1 Collaborate

This is a partnership that hopefully gives a better quality outcome for *all* involved that you cannot find in the bubble. It doesn't remove the risk associated with launching new ideas, but it helps us make the right decisions about those ideas and how to put them into practice.

18.5.1.2 Community

In terms of innovation, it is perhaps not the best route to totally new concepts, but what it can achieve is solutions to problems that we were unaware of via collective thinking: this is where we develop and explore those solutions together.

18.5.1.3 Co-operate

With this project, I see two different sides to co-design with a high level of knowledge transferred. We are working alongside professionals in their fields from all over the country including graphic designers, clothing designers, fabric technologists, pattern makers, technology experts, colour, sizing and fit specialists, research fellows, lecturers, doctoral research students and the project leader and principal investigator of cross-disciplinary collaborative research Jane McCann, an experienced performance sportswear designer and director of the Smart Clothes and Wearable Technology Research Centre at the University of Wales, Newport. Through co-operation we gain ideas and insight into how to make our products right for this area of the market via users and professionals involved in the project. In return, we are the commercial voice and realising the project whilst also helping practically with fabric and manufacturing resources.

18.5.1.4 Communicate – user group engagement

As a whole it is about: asking the right questions and listening to the answers without leading; learning the needs and wants of the target consumer and developing the products; referring back to the consumer and then refining the products and then keep repeating until they are right.

From working with the User Reference Group (URG) based in Wales, the User Advisory Group (UAG) based in and around Salford and Mary Sinfield OBE, lead user for the New Dynamics of Ageing Older Peoples Reference Group (NDA OPRG) we have learnt that it is the small details that are really important and can make a difference to their comfort and independence. Mostly, these key details are missing but are easy to change and will improve the usage of the garment for all, even including younger consumers.

18.5.2 Key points to change

From working with the user groups, we learnt that there were some very obvious things that we could change in our existing products and add to new products that would make our garments more appealing to this age group. Our first encounter was with Mary Sinfield OBE at the 2010 ISPO outdoor trade fair in Munich, Germany. When presented with our full Autumn/Winter 2011 collection, Mary was quick to find garments that she liked. On trying on her first garment, a 3-in-1 jacket (a jacket designed with a separate zip-in fleece), she found that she could not get out of it. The friction between her jumper and the inner fleece was so great that it would not let her take off the jacket. By trying on several garments, other key points to change became apparent. These points and others were also reaffirmed later by the User Reference Group (see Figure 18.7).

18.5.2.1 Fastenings

When dexterity is reduced, things that we take for granted, such as grabbing with finger and thumb tips, are less easy to do. Fastenings need to be adapted to suit this. We

Figure 18.7 Mary Sinfield OBE struggles to escape a Sprayway 3-in-1 jacket.
Source: Image courtesy of David Taylor, University of Newport.

found that many people struggle with two-way open-ended zips (see Figure 18.8) – the biggest problem being the struggle to push the bottom pin into the slider whilst pulling the top slider up (see Figure 18.9). Further research revealed that other people struggled with this feature, including children and less regular users of outdoor-specific clothing. Sometimes it would affect the sale, and understandably, a customer that could not fasten a jacket would not buy it.

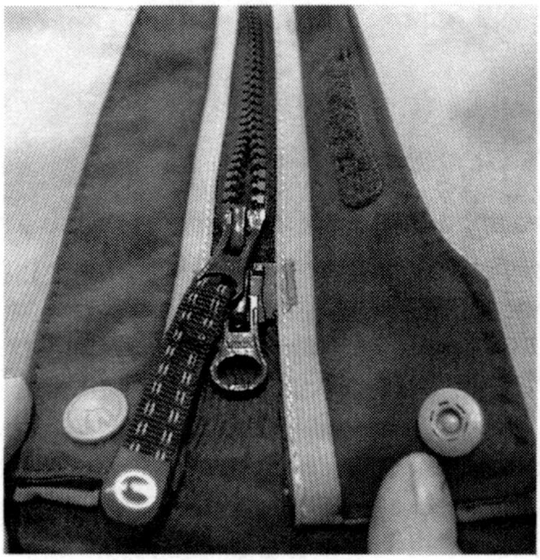

Figure 18.8 Two-way open-ended zip.

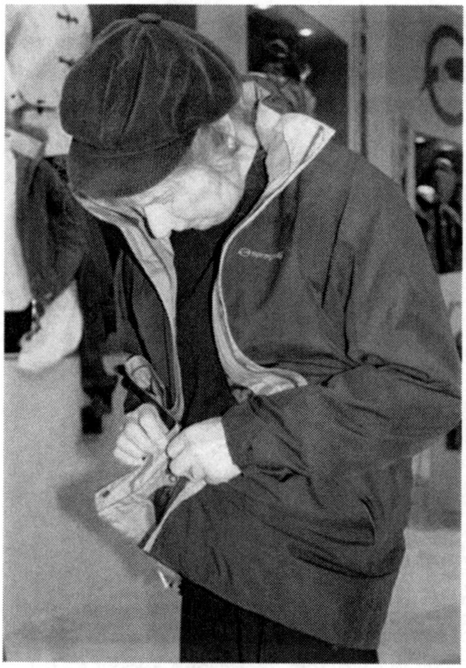

Figure 18.9 Mary trying to fasten a jacket with a 2-way open-ended zip.
Source: Image courtesy of David Taylor, University of Newport.

Starting with our Autumn/Winter 2012 collection, we changed the zippers to one-way open end on shorter jackets, removing the two-way slider to have just a single slider. This was an easy change to make, with no real loss of function and no cost implication. This has ignited our further research into zippers that allow the user to dock the zipper pin easier on even single-slider zippers.

Another part of the zipper that needed to be updated was to add easy-to-grip pullers on all zips again, due in part to the lack of dexterity of the user. We changed all zippers to have an additional puller, e.g. a knotted cord or moulded puller, which adds to the functionality of that garment for all users.

Velcro® or other non-branded hook-and-loop fasteners, were seen to be the most detested part of an outdoor garment. Users from the URG complained that the hook part snags other clothing worn with the offending garment. In addition, many of the women liked to wear scarves with jackets to add colour around the face, and these were also snagged. The snagging problem was not just a problem with this age group; others also would prefer another type of fastening that snagged less. Therefore, we looked at where and how we use hook-and-loop fasteners so they do not snag other garments. We are also looking at different less-snagging forms of hook fasteners, as well as options to replace it altogether.

Snap fasteners, although good replacements for hook-and-loop fasteners, have their own drawbacks. Many in the URG and Mary found that snap fasteners were too strong to undo with their reduced dexterity. Often less-than-adequate amounts of fabric were left around the snap to grip to open the fastener, or it was difficult to dock the male and female parts of the fastener together to close. We created a guideline for a minimum amount of fabric visible at the side of the press fasteners so they can be opened easier. We are now also looking at alternatives with YKK and Duraflex to find a snap that has a softer open and an easy close, yet are still fit for purpose.

18.5.2.2 Garment design

With the 3-in-1 jacket that Mary tried on and subsequently could not get out of, we looked at how we could reduce the friction. From further research we also found that children were finding the same problem with their 3-in-1 jackets, especially when worn with school uniform jumpers. Our solution was to amend the fleece sleeve design to accommodate a taffeta lining, without losing the comfort and the fit of the fleece.

Pocket entry was also a problem for Mary and the URG, as sometimes finding and opening the zipper was awkward. This was mainly due to extra flaps over the zipper specified to keep rain out of pockets. We also found that the zipper snagged easily on the welt, making it difficult to open or close the pocket. Following testing, we found that the extra welt was not always required depending on the construction and angle of the outer flap. Where possible the flaps were removed; where not possible piping was added to the edge of the welt to prevent snagging.

Sizing and fit has been a big part of the project working with Size UK to arrive at correct body shapes for the age group. However, Sprayway already has established fit and sizing across the range, which would be difficult to change. Instead we have started to look at lengths and adjustments in our clothing to improve our fit offering for this age group.

18.5.2.3 Branding

Many in the URG felt very strongly about the manufacturer's branding on garments. It transpired that they felt the branding was added to the garment for the manufacturer to use as an advert, and they did not want to be a walking advertisement or have the look of the garment spoilt by the logos. As detailed earlier, the 60–75 age group has different values than the younger age groups and does not need to feel like they are part of a tribe. Brands mean something different to this age group. Brands are a reference point for something that has been purchased before, stands for good quality, or fits well and functions as required. It is not something that needs to be displayed on the outside of the garment. Indeed, with a men's or lady's suit, a bold branding statement is rarely found on the outside of the garment. Instead, a label to confirm the brand is found inside the garment in a prominent position.

Many of the URG felt so strongly about removing the brand names that they cut off, or unpicked branded labels sewn to the outside of their garments, or overcoloured embroidery with permanent marker. Often within a collection there are branding themes associated with the function of the clothing item or its target consumer. We could not remove the outer branding altogether as it would not sit well with our other products that needed the branding for the younger consumer, especially the male consumer, where the outer branding is often essential as a status symbol and part of the tribal recognition. However, we could make it as inconspicuous as possible. On certain products many fabric logos were made smaller, removed or embroidered in colour-matched thread. Main logos were reduced to a simple logo on the chest or hip in tonal or colour-matched thread.

18.5.2.4 Fabric

We then looked at fabric texture and colour. Outdoor clothing and sportswear clothing in general has sheen to the fabric and a synthetic feel. As with their cotton tees the user group really preferred the feel of natural fabrics, but now understood that to remain comfortable during activity, function was also important. Being from the era of sweaty, smelly Bri-Nylon shirts that was understandable. Fortunately, there is a vast selection of quick-drying, anti-odour, insulating and waterproof synthetic fabric with a natural hand feel as well as natural fabrics such as merino wool, long staple cotton and silk that inherently have their own performance characteristics. Some of the fabric textures such as these were already in the Sprayway collection, so we were starting on the right track.

18.5.2.5 Colour

Colour in clothing is really important when you have lost the colour from your hair and your skin is not as dewy and glowing as it was as a twenty-something. For this reason we found that very bright colours and black are often not chosen by the user groups; this is because such colours tend to draw colour from the skin, rather than enhancing it. The older consumer knows what they like, want and what suits them. Unlike the current market trend for acid-bright colours, many colours that were chosen could be described as 'organic' in that most were colours found in nature: rich berry colours, warm ruby pinks, burnt oranges, mossy greens, warm greys and creams, and Moroccan blues. Colours such as these are classic and rarely look out of date, which suits the customer. The user groups agreed that they would be happier if manufacturers did not change colours every season so that they could return to the store the following season and purchase items that co-ordinate with the clothing they already have. Fortunately, a selection of our colours fitted the bill already, so we just have to ensure that this continues.

18.5.2.6 Wearable technology

We incorporated wearable technology into our clothing in the Autumn/Winter 2007 collection in the form of a push-to-talk module into a Gore-Tex mountaineering jacket and an iPod controller in an everyday waterproof jacket. Disappointingly, we found that the interest was more on a novelty level. As a result, for the following seasons we no longer had wearable technology as a priority for our consumer.

We were surprised to see the amount of interest in this area from the URG, which was mainly from a health and well-being point of view. This was not a focus that we had considered. The advancement of technology now means that electronic systems are less cumbersome, batteries are smaller and many different options are available to enhance everyday life, safety and independence. It is a misconception that all of the older generations are scared of all technology. Generally, existing technology is not designed for or aimed at this age group, so as with the clothing, the user interface is not appealing for someone that has deteriorating eyesight and needs glasses to read the screen of their telephone, less dexterity so pushing small buttons is difficult or impossible or they have no interest or need to follow technology trends. As a result, this is an area that we have started to research, not only for the older consumer but across the collection.

The majority of the above criteria are not age specific, but for this age group, subliminally it makes a huge difference at point of purchase. If the garment fulfils their colour choice, fabric specification, functional needs and is easy to get in and out of or open and close without experiencing aggressive, snagging hook fasteners or being obviously branded, includes technology that is relevant and it fits them, then something like the price may not be as critical as it may be for the younger consumer.

18.6 Getting to store

During one of the focus group sessions with the UAG in the Sprayway showroom, we were asked why Sprayway did not make what they were looking for. We were able to show them that we did have some of those products and colours in our collection and why they did not see them in stores was due to several factors. One factor was that things only arrived in store if retailers bought them, as we do not sell our products direct. The other factor was the quantities that the retailers bought determined whether or not our factories would make them.

In the current poor economic climate, many retailers are suffering. They have excess stock due to low footfall and crazy weather and are not prepared to take huge risks on styles or colours that are unproven. Despite all of the research that brands do into colour, fabric and style, this is wasted sometimes when the retailer chooses colours and styles that sold through in the previous season.

Some of our retailers in the UK have identified that the older consumer is a target customer base that should be catered for in store. During the week, it is the older age group that is free to browse in the stores, and walk in the hills and on trails. Their stores have been tailored to this consumer by creating an environment that is less intimidating and more of an enjoyable experience, much the same as visiting a regular clothing store. Larger stores such as George Fisher Ltd in the Lake District in Keswick, Cumbria, is one such store. The layout is open and inviting with clothing easy to see and reach, shop staff are friendly, attentive, approachable and knowledgeable and a mixture of old and young men and women, and there is an enticing smell of coffee from the in-store cafe. This is very similar to stores that we had visited in Europe where the outdoor lifestyle is a part of everyday life for all age groups, and stores reflect this by making the experience more like a treat than a chore.

The users from all groups generally felt intimidated walking into outdoor stores and often struggled to read swing tickets without glasses, or understand what was written on them; the older age group were sometimes baffled by the things the younger, more initiated take for granted and sometimes found little empathy from the youthful shop staff.

The 60–75 age group is and will continue to be a big part of the market. Nevertheless, clothing developed for this age group cannot be labelled as such. Marketing material for products needs to be as subliminal as the detail in the clothing. Labelling clothing as 'clothing for old people' would not work on many counts, mainly because who would admit to being 'old' in this society where youthfulness is more respected than old age?

Labelling and point of sale (POS) need to change to suit the customer. From the focus groups it was clear that most POS uses a lot of jargon: words and phrases that those working in the industry and the more technically minded customer understand to describe performance. It was clear that jargon simply baffled our focus groups, terms such as waterproof, water resistant, shower proof, breathable, UPF, hydrostatic head, windproof, wind resistant, moisture vapour permeable, RET, double weave, soft shell, moisture vapour transfer rate, Cordura® reinforcements, and grams/m^2/24 hrs etc., meant very little to describe in laymen's terms the actual function and benefit that the garment would provide.

Often POS is not easy to read, nor is it clear where to find the information about the garment's function once the consumer has worked their way through the fabric and finishes swing tickets. Even then it still may not be clear whether the garment is what the consumer is looking for – that is, of course, if they remembered to bring their reading glasses to the store, as many swing tickets are either so crammed with information or specified so small that it is nearly impossible to read them without glasses.

Simplified information and less of it is something that we are already working on for all of our consumers with jargon busters on our website. New POS will look at size of text so that it is easy to read for most customers without the use of reading glasses.

18.7 Conclusion

18.7.1 The co-design process

The co-design process has really opened our eyes to the amount of information it is possible to gather when outside resources all work together. As a small company, our resources and expertise are limited, as well as our time. At this time it would be impossible for us to justify employing pattern cutters, colour specialists, graphics designers, photographers, project managers, designers, developers, researchers etc. for one project. Within the project we have all used each other's knowledge and skill to its best effect.

Working with the users in the focus groups was invaluable. It is too easy to work in a bubble relying on past experience to guide decisions on new products. With the research we can give guided answers to questions, where before we may have worked on hunches. As a brand we are not alone in this process. The speed of turnover of new products does not allow for thorough research for every product much as we would love to. Time is short, and time is expensive.

On the flip side, using focus groups has its limitations. As the project has progressed, the users have learnt a lot. Without trying, we have changed the way that they feel about performance fabrics and details. Those that would only wear cotton next to skin for activity, now swear by their polyester base layers. It is difficult to quantify how this affects the information that we have received. There is also no guarantee that the clothing that will be produced at the end of the project will be something that they will buy. Do we invite a fresh new group of users to test the theories?

18.7.2 Changes made to garments

Many of the changes made to garments to suit the active ager also suit other age groups. Dexterity problems are not unique to the old. There are many users of all ages who struggle with zips and press studs and detest the hook-and-loop fastener. Not surprisingly, children also struggle with many of the same problems, and so we have updated some of our children's clothing too.

18.7.3 Specifications

The current process of sending detailed diagrams and size charts and using existing garments as a starting block has many limitations: adjustments to fit to tailor to activity and body shape and movement is not easy to communicate accurately in an annotated diagram; subtle shaping and articulation in three dimensions can be misinterpreted, or even overlooked; and alterations are difficult to explain fully.

We have seen how creating garment blocks with correct shaping and articulation can remove some of the ambiguity that a factory has on receiving a specification. They can concentrate on the detail of the garment and there is more of a guarantee that the garments will fit as they were intended.

When working with wearable technology, being able to assess component placement in or on a garment is essential before a full prototype is made. This can often remove the need for more prototypes due to the realisation that a component would not fit on that area of the body, e.g. a battery, or heating device. The toile, or mock-up of the block, allows this to be considered from the very beginning.

We are researching into a proposal for in-house prototyping and pattern making to present to management with an aim to improve our fit, detailing and the speed that we can receive prototypes and therefore assess and test them for longer.

18.7.4 Predictions

If the predictions are true, then it is interesting that so many managers within sportswear clothing companies are of an age now that in the next decades they will be included in that large section of the UK population that will be over 60. We are already working on clothing for 2014, so this leaves little time to get ahead of the market. Company direction cannot change overnight, in the same way that the ageing population will not appear all of a sudden. They are out there already, and the market is growing that is guaranteed. At what point should this market lead with colour, design, fit etc. leaving other age groups to wear clothes that are not designed for them specifically, e.g. that do not fit them as well as they could and in colours that they do not really want to buy. Will they notice?

With regard to wearable or inherent technology there are ongoing challenges to be addressed, and we are still limited by what is available, its size and cost. It will be important to ensure that the right technology is used in such a way that it feels second nature and a genuine benefit to the consumer in much the same way that a garment being waterproof or breathable is now.

From this project we have gained confidence in our new products, because of the amount of knowledge backing up the designs. Knowledge is power as they say, and knowledge is pretty limited inside a bubble.

What textile fibres are applicable for the layering system for the active ageing?

19

R. Hibbert
Line Consultants, London, UK

19.1 Introduction

While the various stages along the textile production chain, such as a fabric's construction, weight and finishing, all contribute to the overall final performance and character, it is the fibre choice and the selection of the applicable raw material that remain fundamental and are the essential ingredients in enabling the wearer to feel difference in the product. The main requirements that everyday sports clothing needs to fulfil are chiefly optimum heat and moisture regulation for the activity, rapid drying, durability, easy care, lightness of weight and a pleasant handle – all factors contributing to the wearer's measurement of their 'comfort' in the garment. It is not always possible to achieve all these functions simultaneously or satisfactorily using one fibre. Consequently choice will force some compromise, despite modification by polymer engineering, fibre treatments and finishes to improve the weaker aspects of a particular fibre. Synthetic fabrics have generally been considered the best choice for sportswear as they are able to provide a good combination of moisture management, softness, lightweight, insulation and low cost with fast drying. Stretch fibres play a vital part in allowing freedom of movement, comfort and minimising creases, but also have potential to provide personalised muscle support. Natural materials such as cotton and wool can present a good combination of softness and comfort but have fallen out of favour in active sportswear due in part to poor moisture management. For example, the slow-to-dry and cold-when-wet characteristics of conventional cotton make it unsuitable in conditions in which there are high levels of moisture, either perspiration or precipitation, where the ambient temperature is low.

The performance of sportswear is synonymous with its comfort characteristics. In active wear for outdoor use, clothing should be capable of protecting the wearer from external elements such as wind, sun, rain and snow. At the same time, it should be capable of maintaining the heat balance between the excess heat produced by the wearer due to increased metabolic rate on the one hand and the capacity of the clothing to dissipate body heat and perspiration on the other (Shishoo, 2005). Sports apparel is also often subjected to greater physical abuse and undergoes more frequent wash cycles than other clothing, so aftercare is an important consideration.

Added to these functional elements must be considerations of fit and appearance, both basic and primary factors affecting consumer choice and an area where fibre

Textile-led Design for the Active Ageing Population. http://dx.doi.org/10.1016/B978-0-85709-538-1.00019-5

selection also plays a part. As Bartels (2011, p. 389) suggests, all clothing has to meet the demands of thermo-physiological comfort, such as moisture management, but also skin sensory comfort, ergonomic comfort (ease of movement/fit according to the particular sport or movement) and psychological comfort (image, fashionability, status, brand image). Comfort is one reason for the significant influence of sportswear and the crossover between sports and leisurewear. Consumers who have experienced the superior comfort of sportswear look for the same advantages in leisure garments, while image and fashionability from everyday clothing are also necessary in sportswear. Additionally, fibre selection is directed by the customer's personal affection, knowledge, history and aspiration. As many of those in the 'Third Age' will have grown up wearing natural fibres, they may continue to have a preference for them in sportswear. Requirements of comfort, fit, performance, durability and ease of care are necessary for all consumers, particularly those in the mature Third Age. Clothing comfort and function are governed by many different factors of textile materials, including fibre types, yarn type, fabric structure, number of layers and thickness – but although it can be difficult to separate these elements, this chapter will examine the role played by appropriate fibre choice and the options available in the creation of sports apparel applicable for an older demographic. Paramount issues of sustainability are included, with considerations of innovative developments in intelligent, adaptive and interactive materials as part of the creation of functional active sports clothing.

19.1.1 The layering system

The recognised arrangement for creating practical clothing for active sports uses the layering system, as several thin layers are considered more efficient and versatile than one thick layer. Not only do more layers trap more air for insulation, but layers can be added or removed as required by conditions and the level of activity. The Ramblers Association provides a clear description of the system, whereby the 'base layer' nearest the body should use a comfortable, knitted, soft, hydrophobic fibre capable of wicking moisture away from the skin and drying quickly. The 'midlayer' is for insulation and can take the form of a lofty fleece or open structured fabric. Lastly, the 'shell layer' or outer layer will be lightweight and weatherproof to protect both the wearer and the inner layers, while being breathable, possibly via a permeable membrane system. This arrangement can be reduced to two layers, via a 'soft shell' hybrid combination offering the comfort of a midlayer with the protection of an outer layer. Whether two or three layers, the selected fibres must work in harmony to provide optimal protection and functionality.

19.2 Natural fibres

Prior to the introduction of synthetic materials to the sports market in the late 1960s and early 1970s, natural fibres were the only available choice for functional outdoor clothing. It seems amazing to us today that groundbreaking explorers struggled to the top of Everest and to the extreme Poles clad in heavy layers of wool, cotton and silk. As McCann (2005, p. 46) describes in her history of the evolution of sportswear, fibre choices in the 1960s and 1970s were limited. Closely woven proofed gabardine, cambric and proofed cotton poplin were recommended, with wool knitwear and cotton fleece for midlayers and textured cotton jersey shirts and string vests for base layers. Materials such as moleskin or woollen Bedford cord were preferred, while for cold weather, several thin wool layers were considered more flexible than one heavy sweater. Since this advice was given, naturally derived fibres have dramatically lost market share to synthetic fibres.

Based in agriculture, often in areas of the developing world, natural fibres use either plant or animal sources for the raw material, and have provided humans with hard-wearing textiles that protect, cool, warm and cover for centuries. Cotton still forms the major proportion of natural fibre use, with approximately 31% share (Gherzi, 2011) of the global textile market, having declined steadily from over 50% since the 1960s. The other two natural fibres suitable for sportswear use are wool, with approximately 1.5% market share, and silk, with a negligible stake. Tobler-Rohr (2011, p. 421) states the most common fibre in the United States is cotton, while Europeans also like some wool, linen and silk, mainly for the 50+ generation. Bast fibres such as flax and hemp, in their conventional forms, are too stiff and rigid in character, but recent processing routes, covered later in this chapter, could bring these fibres to the attention of the sports market.

At their inception, sports fabrics from synthetic fibres were a revelation, heralded as the answer to comfort, functionality and performance in textiles. Natural fibres were slow to respond and fight back against the onslaught and highly efficient marketing campaigns waged by the branded synthetic competition. Recognising the growing market for athletic clothing, research was funded by organizations such as Cotton Incorporated in the United States and the Commonwealth Scientific and Industrial Research Organisation (CSIRO) in Australia – countries where fibre production is a major industry. This resulted in a resurgence of natural fibres in advanced forms that combine their traditional and familiar attributes of comfort and versatility with new technical developments. Vital marketing and promotion also became possible, as these innovations were branded to allow their profile to be raised in a competitive market. 'Sportwool' weatherproof technology is one example, where the constituent fibre, yarn, fabric properties and fabric finishes combine to create a drier and cooler microclimate.

19.2.1 Cotton

Cotton, the most universally recognised of all fibres, remains associated with purity, freshness and naturalness. Popularity over the decades has been driven by commercial availability, ease of laundry, comfort, functionality and variety of end uses. Cultivation has been traced back beyond 7000 years, and the fibre's development has been a showcase for human ingenuity and ability, with links to political struggle, conflict and industrial mechanisation (Ferringo, 2012, p. 10). Present-day cultivation takes place in some 80 countries of the world and is enormously important to the global economy and to its population, with some 50–100 million farmers, 250 million jobs and possibly, as Ferringo (2012, p. 21) suggests, as many as a billion people gaining their livelihoods in some way from cotton textiles. For smaller countries, such as Mali, where agriculture contributes 50% of GDP and supports 80% of the population, cotton production from small farm plots is often the only link to the global economy. Despite a dramatic increase in the volume of cotton produced per hectare since the 1950s, without any increase in the acreage, cotton's share of the fibre market, and in particular the active sports market, has declined relatively to synthetic fibres and especially polyester. As an agricultural crop, prices are volatile and fluctuate in accordance with growing conditions, but as an internationally traded commodity, market demand, quotas, government subsidies and global trading patterns are also key factors. Standard cotton fibre, American upland cotton, makes up over 50% of today's use, with more costly grades of long staple fibre available, such as pima cotton, Sea Island cotton and Egyptian, all with superior softness and lustre.

A cotton T shirt is a ubiquitous garment familiar to everyone, and possibly the first garment of choice for recreational sports use, crossing over as it does with everyday fashion. Evolving from undergarments and US Marine uniforms, the crewneck short-sleeved single jersey T shirt was introduced to the general public post–World War II, when military surplus clothing flooded the market. It became the default garment of choice, particularly amongst young men, with its connotations of GI American war hero, and cool macho ruggedness, being worn by icons such as Marlon Brando and James Dean in the 1950s. It was also the perfect choice for displaying a fit, toned body, as attitudes towards sport and fitness changed. Whilst the function of a standard cotton T shirt is adequate for everyday wear, more serious sports enthusiasts need improved function in base layers, particularly with a closer fit, wicking capability and moisture management. Cotton can retain moisture, be slow to dry in cool conditions, cling and feel cold against the skin when wet. The addition of stretch fibres and LYCRA® yarns has ensured a good fit and shape retention, while several recent technical developments with fabric construction and finishing have been aimed at making cotton suitable for active sports. Additional anti-microbial and insect repellent finishes can also be imparted during finishing.

Wicking Windows technology, introduced by Cotton Incorporated, claims that the treated 100% cotton fabric transports moisture vapour more effectively than any other fabric, via a chemical formulation based on repellent fluoro-chemistry, preventing garments from clinging to the skin. The Wicking Windows fabric surface-treatment process is applied in a fine pattern to the inside of a garment, as a 'discontinuous print pattern', so that moisture is more quickly transported to the fabric's exterior via small 'windows' through which moisture can pass, without affecting breathability.

The finish is invisible, is odourless and does not wash away with home laundering, plus it can be 'renewed' by ironing to bring back the original performance of the finish (WSA, 2008).

Similar developments such as the TransDRY™ patented technology, also from Cotton Incorporated, aim to counter accusations that cotton has fallen behind in terms of function and performance, and to increase the fibre's usage in the athletic industry. The moisture management process treats cotton yarns to make then hydrophobic (water repellent), for use in engineered double and single knit structures that reduce absorbency and move moisture from the skin to the fabric's surface (see Figures 19.1 and 19.2). The function can also be applied as alternating hydrophilic and hydrophobic cotton yarns in a knit structure.

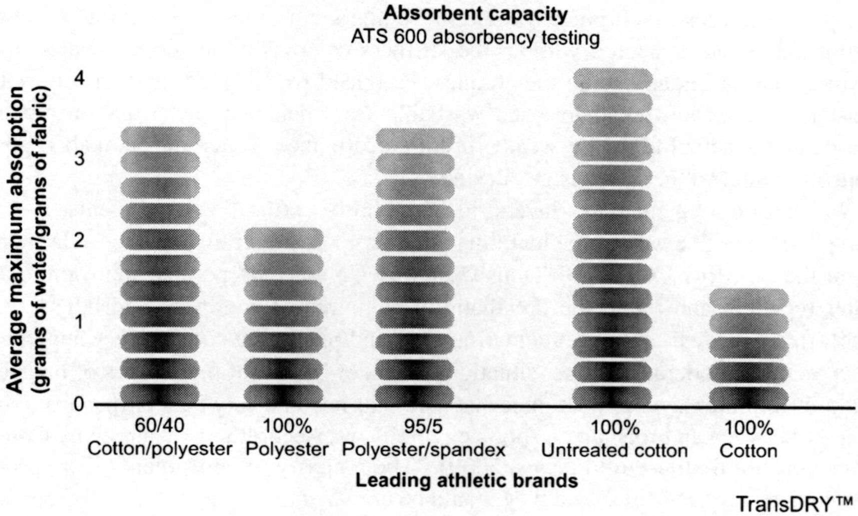

Figure 19.1 Comparative absorbency chart of TransDRY™ cotton with other fibres. Printed with permission from Cotton Incorporated.

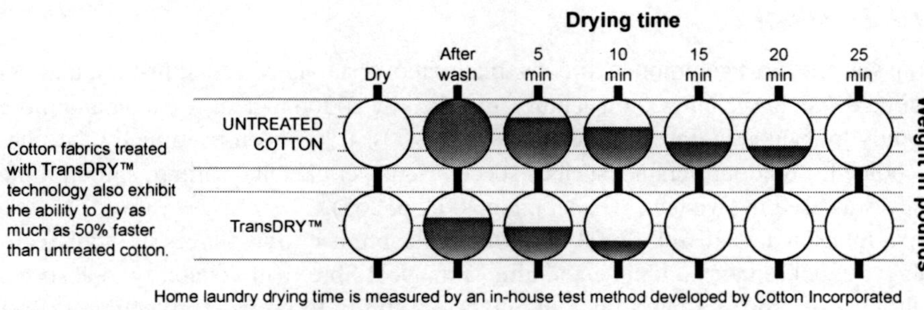

Figure 19.2 Comparative drying time between treated and untreated cotton. Printed with permission from Cotton Incorporated.

Re-examining older technologies can also play a role in the creation of cotton fabrics with inbuilt weatherproof functions for outer protective layers, negating the need for any additional coatings or treatments. As Holmes (2000, Chapter 12) tells us, 'Ventile, probably the first effective waterproof breathable fabric, was developed in the 1940s for military purposes, as a response to keeping air crew safe in very hazardous wartime conditions.' The finest types of long staple cottons are selected, ensuring very small spaces between the fibres. Cotton is processed into combed and plied yarn, to improve regularity and ensure the fibres are as parallel as possible with no large pores where water can penetrate. When Ventile's close Oxford construction is wetted, the cotton fibres swell transversely, reducing the size of the pores in the fabric and requiring very high pressure to cause penetration. In additional to being weatherproof, Ventile fabrics, still in production and use today, are 'quiet', durable and reliable, as the function will not break down or deteriorate with age.

Epic Cotton also incorporates some of Ventile's attributes, in retaining a natural cotton touch and character, without the stiffness or crunchy handle associated with coatings, laminates and some membranes. Produced by Nextec® the fabric is water resistant, windproof, breathable and washable via a patented encapsulation process that treats the actual fibres for weatherproof performance. Water and stain absorption is also minimal so the fabric is fast drying.

Weatherproofing for outer layers, in particularly cotton fleece garments, is now also possible with a water repellent finish that protects from rain and snow. Garments given the STORM COTTON™ finish, another Cotton Incorporated innovation, are water repellent and retain the breathability of untreated cotton, making them very comfortable to wear in different environments and redefining cotton fleece apparel as outerwear for moderate weather situations. A major benefit of these types of invisible methods for outer layer protection is that garments become very dual purpose, moving happily between an urban and a sports environment. According to a survey by Cotton Incorporated (Textile World News, 2009), 'The majority of consumers (77%) prefer cotton in general, and 70% said they would be prepared to pay more for cotton apparel that mimicked many of the performance features more commonly found in synthetic textiles.'

19.2.2 Wool

Wool was the most common textile raw material over a long period of history, until its decline began in the nineteenth century, replaced by cotton to a large extent and more recently by synthetic fibres in the 1950s and 1960s. It has been historically the fibre of choice for outdoor pursuits such as soccer, rugby, cricket and sailing, and until the 1950s was used in swimsuits (Holcombe, 2008, p. 265).

As Johnson and Russell (2008) describe in the preface to Advances in Wool Technology, 'wool remains a unique and almost magical fibre, with its ability to absorb a high level of moisture but with a simultaneous ability to provide protection against showers, plus resilient mechanical properties and natural odour resistance, and can be engineered to be either warm or cool and to function superbly in extreme sporting conditions or in the most elegant of formal occasions'. This versatility in functionality and position as nature's own high performance fibre has evolved from protecting sheep from extremes of heat and cold, and has endeared wool textiles to a generation that grew up with it. In the nineteenth century, woollen flannel was thought to be an effective weapon against colds and rheumatism and was worn as long underwear and vests (Brunel, 2002, p. 22). Military T shirts were made from wool before World War I until being superseded by lighter and faster drying cotton, while knitted wool vests and underwear continued well on into post–World War II.

Although different animal hairs are used for textile production, sheep wool remains the largest resource. Today, wool is produced in around 100 countries worldwide, Australia being the largest production area, with Argentina, the United States and New Zealand also major sheep rearing areas.

Wool fibre has a unique surface structure of overlapping cuticle scales, providing a tough exterior, with a waxy coating for water repellency that still allows absorption of water vapour. The interior of the fibre, the cortex, has a springlike structure, giving the fibre flexibility, elasticity, resilience and wrinkle recovery plus addition benefits of natural fire retardancy, ultraviolet (UV) protection and antistatic properties (Srivastara, 2012). Wool's insulating properties and ability to retain heat even when wet have made it highly valued for sportswear. Wicking is the most desirable property required in a base layer, and wool fibres can be treated to improve this function. The weight of wool garments, particularly when wet, is a major obstacle, but can be improved by the use of finer hair types and mixing with other fibres. The choice of fibre diameter next to the skin is crucial to performance, and the introduction of finer quality more expensive Merino wool fibre was a major breakthrough in active wear use. Merino fibres are finer than traditional wool, feel luxurious next to the skin and can be machine washed. Sportwool™ and MerinoPerform™ are branded examples of lightweight composite double knit fabric forms combining a layer of Merino wool next to the skin with a layer of polyester on the exterior. This separation of the two fibres produces a higher level of wicking than a 100% synthetic equivalent (Holcombe, 2008, p. 41). Blends in the form of layering or yarn structures have the benefit of offering the best of both elements. Optimum wicking and insulating properties can be provided by bicomponent yarns, such as in combinations of polyester/wool or wool/polypropylene. The higher cost of wool garments can also be a drawback, plus the slight tendency to felt and longer drying time in laundry, but durability, low odour retention and good classic appearance, again crossing over between everyday casual wear and sports use, make wool an attractive option for moderate activity.

19.2.3 Silk

Nature's only natural protein filament, silk is produced from the chrysalis of silkworms. Today over 50% is still produced in China, where its origins date back many centuries. This luxurious, comfortable, noble fibre is lightweight yet has good thermal regulating properties. In common with wool, silk was also used extensively for underwear prior to the arrival of synthetic fibres, and it is this functionality that makes it an attractive and somewhat luxurious option for active base layers today. On its own, silk may not be a logical choice for athletic apparel. Although one of nature's strongest fibres, silk loses up to 20% of its strength when wet and can be weakened by too much sun, has poor elasticity and is discoloured by perspiration. On the plus side, silk's good absorbency makes it comfortable to wear in all temperatures, and low conductivity keeps warm air close to the skin during cold weather. Bi-component structures can add additional functions, such as combinations with polypropylene and Merino wool. With stretch fibres, silk is better fitting and more comfortable, and with wool, silk becomes a warmer, heavier material. To increase moisture management, silk can undergo an application to negatively charge the fabric, changing the actual molecular structure of the silk (Shallard, 2008). Consequently, silk makes an ideal fibre for thermal base layers, glove linings and socks, particularly for sensitive skins, as it is naturally hypoallergenic and so will not itch or irritate, and has the added bonus of feeling and looking luxurious. Seracin is the sticky protein that holds the fibre together on the cocoon and has to be removed and sometimes wasted during processing. Containing amino acids, this can be applied as a natural finishing agent to other fibres, imparting anti-microbial, anti-static and moisture absorption benefits.

19.2.4 Natural fibre for insulating fillings

Insulation in clothing is created by trapping still air between fibres and apparel layers. Lofty insulating fibres are required in the fillings and waddings of mid- and outer layer protective garments. Finer fibres make better insulators, while the retention of bulk and resistance to compression are important. Keeping the fibre protected from moisture is vital as all insulating materials lose function when wet. How quickly fibres can shed moisture and dry is also a deciding factor on the choice of filling. Down feather fillings from duck and geese remain the optimum source of insulating fibre in terms of weight to warmth ratio, but can be slow to dry. Hollow polyester or polypropylene staple fibres, which can include recycled elements, are often a lower cost option. They are heavier in weight than down, do not retain loft as long, but shed moisture more rapidly. Wool is unsuitable as it retains too much water, but UK clothing brand Finisterre has experimented with a blend of alpaca and other natural fibres such as hemp, in an attempt to work with more locally sourced and sustainable materials (WSA, 2007).

19.3 Synthetic fibres

Major advances in fibres for textiles have historically gone hand in hand with innovations in technology. With synthetically produced fibres, it was the discovery of a chemical reaction in 1934 by DuPont chemist Dr Wallace Carothers that became the first polyamide. This revolutionary fibre claimed to be 'finer than spider's thread, stronger than steel and more elegant than silk'. First used to produce nylon stockings in 1940, and then parachutes, this discovery led on to the introduction of polypropylene, polyacrylonitrile (acrylic) and polyester or polyethylene terephthalate (PET) in the 1950s. From the late 1960s, everything began to move rapidly and the pace was to accelerate markedly throughout the next decade, with waterproof anoraks and nylon fibre pile 'polar suits' advertised and described clearly in terms of their insulation and moisture management properties (McCann, 2005, p. 47). Today synthetic fibres are not a mere alternative to natural fibres, but are materials of high functionality, with refined and specific properties playing a key role in many textile end uses, including important sports textile applications (Honga, and Philips 1997). These thermoplastic materials can be given additional functions and properties during the liquid stage of manufacture, they can be heat set for pleats and articulated garment sections, plus they facilitate recent laser garment cutting and welding construction methods in the creation of waterproof seams.

Known as the second generation of fibres, after the 'natural' plant- and animal-based family, synthetics caused great excitement in the textile world. Most importantly commercially, as opposed to traditional natural fibres, they were owned and could be branded by the manufacturer, with their use controlled and licenced. A third generation of fibres have followed that includes higher performance and high strength fibres of polyethylene, carbon and polyaramid fibres, providing super lightweight and highly technical fibres initially used in the aerospace industry. Some of these have made the transition to speciality sportswear, where their attributes of lightness in combination with tensile strength and abrasion resistance are valued.

In principle, the manufacturing of all synthetic fibres is the same whereby continuous filaments are produced by extrusion of melt spun polymer chips through fine spinneret holes. Thermoplastic synthetic fibres can be heat-treated to introduce permanent profiles such as coils, crimps and loops. Texturising the yarn will add functionality such as stretch or thermal air-trapping qualities. In addition to continuous filament fibres, short staple fibre production is possible, whereby bunches of crimped filaments are cut into specified lengths, which can be then spun into yarn. Speciality cross-sections can be created to provide added functionality, such as lustre or moisture transport. These cross-sections are available in staple, filament and microfibre forms, and as bicomponent fibres. One example is Thermolite® from Invista, a hollow-core fibre trapping air for greater insulation and lightness of weight, but with a robust performance. Coolmax® is a branded group of moisture-wicking technical fibres also produced by Invista using specially engineered polyester fibres to improve breathability, compared to natural fibres like cotton. Coolmax® fibres are slightly oblong in cross-section with grooves running lengthwise along the threads. The series of closely spaced channels creates a capillary action that wicks moisture through the core and out to a wider area on the surface of the fabric which increases evaporation.

Rather than being a comparatively recent invention, polymeric materials, such as the naturally occurring polymers of amber and horn, have served humans for centuries. Today the primary resource of synthetic materials are fossil fuels, and it is estimated that 8% of the world's annual oil production is used either as feedstock or energy in the production of synthetic plastic materials (O Ecotextiles, 2010).

19.3.1 Polyester

As the trade magazine World Sports Activewear (2007) tells us, polyester now makes up 75% of the synthetic fibre market. Since DuPont launched PET, it has had a huge impact on the textile industry, stealing nylon's leading position in the synthetic fibre market in the late 1970s and opening up vast opportunities in the production of apparel as a result of its strength, ease of care, chemical resistance and wrinkle resistance. To that one must add its durability, dimensional stability, high modulus and competitive volume production costs. The fibre is very hydrophobic in nature – repelling rather than absorbing water, and consequently is easy to launder and quick drying. According to Dow Corning, a manufacturer of polyester, the fibre has a moisture regain of only 0.4%, while cotton has a regain of 7%. The hydrophobic nature of polyester traditionally resulted in a relatively low level of comfort, initially restricting the fibres use in sportswear, but now a distinct advantage. One method of increasing the hydrophilic function of polyester uses a topical chemical finish, but another significant alternative lies in the use of microfibres, which must be mentioned for their importance in improving active wear functionality.

19.3.1.1 Microfibres and nanofibres

Japanese fibre manufacturing companies introduced the first 'micro-denier' products during the 1970s, followed by developments in Europe and the United States during the 1980s and 1990s (Purane, 2007). The initial aim was to produce a more silk-like textile than could be obtained from conventional synthetic yarns, and it is this distinctive silky characteristic, fluidity and superior functionality that has had such a great impact on the use of microfibres in sportswear. Conventional polyester fibres are in the range of 1.5–4 deniers per filament (dpf), while microfibres are about 0.5 dpf. Between 60 and 100 times finer than a human hair and containing up to four times as many filaments as conventional fibres, they have more surface area, which in turn presents a higher level of functionality, including absorbency. The resulting fabric will also have a finer more fluid and silky handle, plus an improved tensile strength-to-weight ratio. Using these extremely fine fibres, it is possible to construct a textile that is wind- and weatherproof while remaining breathable, without the use of coatings or membranes. Fibres are so densely packed that they do not allow the external water drops to enter, but can allow internal smaller perspiration particles to evaporate. The microfibre process is used in all synthetic fibre areas of manufacture and frequently combines materials in high performance bicomponent fibres that merge the optimal functions from each source. One such innovative development is branded as Evolon®. This unique non-woven construction of micropolyester and polyamide filaments, entangled and consolidated by high-pressure water jets, forms a wind and weatherproof breathable lightweight textile ideal for midlayer sportswear. Microfibres are also used to create high-performance filling and insulation materials, such as Primaloft's hollow air-trapping fibres.

Nanofibres, with a diameter of less than 1000 nm, have progressed this technology and moved into the textile arena. They can be added to a textile or a fibre's surface via a coating treatment, to add strength, durability or other performance functions, while minute nanoparticles can be combined with a polymer during the creation of a synthetic fibre. Nanofibres have also allowed the development of innovative spray-on textiles, such as by Fabrican, suggesting potential for the creation, repair or customisation of a garment. US company Nano-Tex has introduced improved wicking properties for polyester using nanotechnology, while odour control, stain resistance and crease resistance can also be improved. Swiss producer Schoeller's NanoSphere© fabrics imitate nature in the manner that dirt is shed from a leaf's surface. This non-stick self-cleaning effect is permanently transferred to the textile surface using nanotechnology finishing. While functionality is greatly enhanced, there is uncertainty as to the potential for negative environmental impact as particles may be released into wastewater during use and laundry, with a potential risk to the environment. This question has been raised particularly with reference to the use of silver nanoparticles to introduce antibacterial functions.

19.3.2 Polyamide

As the first successful fully synthetic fibre, polyamide has continued to perform effectively in today's marketplace for outer protective layers due to its desirable attributes of high strength, durability, wear and tear resistance and dye uptake. Ripstop nylon is one well-known example. In common with polyester, nylon launders well, but is smoother in character, with less bulk, is more hydrophilic and so holds moisture for longer. As Invista, producers of SUPPLEX® nylon, state in their marketing, the fibre 'offers consumers the feel of cotton with the benefits of advanced fibre technology. Cotton's natural texture and aesthetics make it the fabric of choice in many garment categories, but garments tend to crease and shrink easily, and they often fade in colour. Invista scientists wanted to give consumers the benefits of cotton, without the pitfalls. SUPPLEX® fabrics are breathable, hold their shape, dry faster than cotton and retain their colour.'

19.3.3 Polypropylene

Initially considered an industrial fibre, first produced in 1957 in the United States and used for packaging, polypropylene has made the move into knitted sportswear textiles in staple form for base layer products. It remains an underdeveloped polymer in comparison with polyester and polyamide, and has much ground to make up with regard to image. As the synthetic fibre most resembling wool, with a dry touch even when wet, other positive attributes include high strength, low cost, extreme lightness of weight and shrink and crease resistance. The fibre also has an excellent resistance to bacteria (Hibbert, 2006). These factors, coupled with a bulky character and heat retention properties, have made it popular for socks and thermal underwear. For the performance wear developer, its low heat transference, meaning it is demonstrably cooler in the summer and warmer in the winter than any other fibre, and natural hydrophobia, are compelling properties. Unusually for synthetic fibres, it cannot be permanently heat-set, so knitted fabrics require the addition of an elastomeric fibre. In continuous filament form, polypropylene can be used in woven textiles, in conjunction with other synthetics, and has some very good properties for less demanding sports where its natural water repellency, abrasion resistance, lightweight and antisoil properties can deliver some distinct advantages, for instance, as high-performance walking trousers with a percentage of woven polypropylene in the weft (WSA, 2006).

19.3.4 Stretch fibres

The use of stretch fibres has pioneered the engineering of movement in textiles, and must be acknowledged with dramatically altering how people look and feel in the clothes they wear. Revolutionary developments in stretch fibres have made it possible to produce new types of functional clothing, from highly technical and personalised muscle compression garments helping to improve an athlete's performance, to joint supports following injury, and casual sports clothing that retains its shape, does not crease and offers greater freedom of movement. Tight-fitting clothes can, however, suppress ventilation and do not transport or absorb perspiration. Today, ease of movement and the shape retention derived from stretch fibres are expected as the norm, and as in all areas of textiles, various choices are available to create these functions in accordance with the end use. Always used with other fibres, a stretch yarn is defined by Kosa (1999, p. 189) as 'any yarn with the ability to stretch to a significant degree when tension is applied and contract when tension is released, including elastomeric filament yarns (synthetic polymers with stretch and recovery properties that mimic rubber) and manufactured filament yarns treated by various texturising processes to impart bulk and stretch', also known as mechanical stretch yarns.

19.3.4.1 Stretch fibre variations

It is possible to impart a gentle 'comfort' stretch to textiles by adding an elastic fibre created by heat-setting a crimp or by using differential shrinkage in a multicomponent fibre. LYCRA® T400® fibre uses this method and has introduced a new generic description of 'elasterell-p' for fibres produced in this manner. As illustrated in Figure 19.3, this bi-component fibre is made with two distinct polymers, including a wicking polyester, which shrink differently, causing the fibre to form a permanent coil. After being exposed to heat in the finishing process, the coil gets even tighter. To achieve a given level of stretch, the mechanical stretch percentage content of a fabric will normally be greater than that with an elastomeric yarn such as LYCRA®, but the extension will be less. For example:

Cotton with LYCRA®, 98% cotton/2% elastomeric = 25% stretch
Cotton with LYCRA® T400®, 77% cotton/23% T400 = 20% stretch

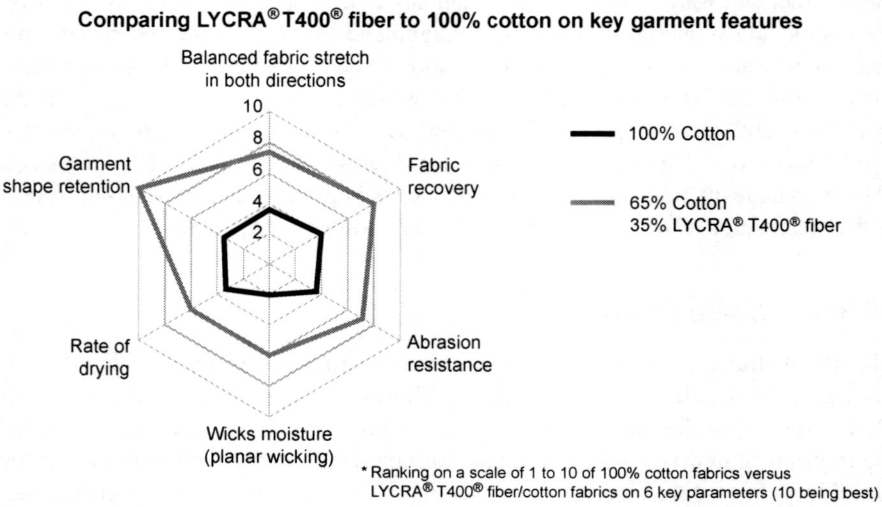

Figure 19.3 LYCRA® T400® fibre.
Printed with permission from INVISTA Technologies S.à r.l.

Figure 19.4 shows the key performance features of LYCRA® T400® compared to standard cotton. Other branded mechanical stretch yarns include Elité PBT and Trevira's Expand, both texturised stretch polyesters. XLAnce is an olefin-based fibre modified to give high stretch. This is different from traditional elastomerics where the yarn develops much of its power in processing and is fed into the machine under tension. This can cause fabric contraction in finishing and the fabric has to be stretched and 'set' to a finished dimension. XLA is a latent stretch product by DOW and is plated into the fabric in a very much more relaxed state, therefore causing much lower fabric contraction. This is ideal when used with its sister polypropylene products as it does not need excessive heat to finish it and cuts down on processing costs. Benefits also include the ability to launder at higher temperatures and the use of harsher finishing techniques such as stone washing. For sport outer layer use, it is a combination with cotton in a woven fabric that would be the most compatible and applicable, where both weft and bi-stretch products can be created.

Spandex is the generic name for the group of stretch fibres produced from polyurethane, the most recognised and well used of which is LYCRA®, now produced by Invista. Invented in 1958, originally as a replacement for rubber in corsetry, it gained popularity in 1968, when the medal-winning French Olympic ski team became the first high-profile sports personalities to wear ski suits with LYCRA® fibre – a trend that soon influenced other sports, particularly swimming. Popularity continued and by the mid-1980s, over half of all women's hosiery and underwear relied on LYCRA® fibre for a close, comfortable fit. During the 1990s, the LYCRA® fibre brand position in the sports market strengthened through the development of hi-tech fibres such as LYCRA® Power® fabric, used in compression shorts to help reduce athletes' muscle fatigue. Other innovations include XFIT LYCRA® for multidimensional fit in both warp and weft, and LYCRA® freshFX® fabric, a moisture wicking and quick-drying combination of nylon and LYCRA®. As a synthetic elastane fibre with unique stretch and recovery properties, it is always blended with other fibres. Any natural or synthetic fibre can be mixed with stretch fibre and core spun yarns are a popular method, combining the properties of two fibres for specific end uses. A central core of one stretch fibre is surrounded by an outer layer of another fibre and is very beneficial in the creation of functional stretch fabrics with the touch and character of natural fibres such as cotton and wool.

Figure 19.4 Comparing the features of LYCRA® T400® fibre to 100% cotton.
Printed with permission from INVISTA Technologies S.à r.l.

Recovery is a key element in stretch fabrics. As is pointed out on the Website of Australian spinners and users of the fibre, Eclipse Textiles, 'the LYCRA® itself doesn't actually stretch the fabric, that's the knit construction. The construction allows the fabric to be pulled out, and the LYCRA® is what brings the fabric back into place'. LYCRA® can be stretched up to seven times its initial length before springing back to the original position once tension is released. The modulus or the power is also an important factor to mention, especially when it comes to compression garment fabrics in the performance sportswear industry.

19.3.4.2 Compression

Textiles containing a high percentage of stretch yarns can apply controlled compression to specific areas of the body, generally the lower torso, in order to assist with regulating blood flow. When the elongation of the fabric is shortened due to the knit construction and a high LYCRA® content is added to the mix, this considerably increases the modulus or the power behind the fabric when it returns to its original position. This assists in reducing the athlete's muscle fatigue through garment support, in turn decreasing muscle strain and improving performance. Currently restricted to high-performance sports such as cycling and sprinting, where a minimal 20% LYCRA® Power® content is used, for the recreational athlete, and particularly for an older demographic, increased external compression could impart great benefits, both actual and perceived. Appropriately designed garments would provide firm yet gentle muscle compression and micro massage, to fit unobtrusively like a second skin, with the potential to support injured joints and muscles, reduce or control oedema, assist with venous conditions, improve skin health and support tired or aching legs. In addition, it has been suggested that these support garments can improve the wearer's ability to sense body movements, and impart a greater awareness of the precision of movement.

19.4 Synthetic cellulosics

So titled as they are manufactured from naturally occurring polymers, mostly from plant-derived cellulose, with the main sources being wood pulps from pine trees and cotton linters – the short, coarse fibres adhere to the cotton boll after removal of the longer textile fibres. Polymers of protein origin can also be used, such as casein from milk, and soya. As the raw materials are of natural origin, these fibres share similar functions and properties, and have more in common with the natural fibre group than the synthetic family.

According to a report by producers Lenzing (Hammerle, 2011), while the price of petroleum increased 21 times over the last 50 years, the price of soybean protein increased only 6.5 times. They also suggest that demand for textile fibres and per capita fibre consumption will continue to grow, with increasing world population, pressure on natural resources, peak oil and pressure on land for arable crops creating a fibre gap that cannot be filled by cotton or oil-based synthetic fibres. This is the so-called 'cellulose gap'. Whilst interest has been shown in novel and unusual sources for fibre, such as soya, it is only the synthetic cellulosic group of viscose, modal and Lyocell that are poised with the volume production capabilities in place to fill this potential supply gap. Lenzing alone plans to increase production by one-third before 2016, and has a broad range of end uses for its fibre to assist in increasing economic feasibility.

19.4.1 Modal

Originally manufactured in France as an inexpensive alternative to silk in the 1880s, rayon is the oldest form of manufactured fibre.

With poor wet strength, high shrinkage and low abrasion resistance, a viscose rayon fabric is considered unsuitable for sportswear use, but other branded forms are available that are more suited to base layer sportswear. Using the trade name of Modal™, High Wet Modulus (HWM) rayon is a modified high wet strength viscose derived from beech wood that will perform much like cotton when laundered. Fibres can also be mercerised for increased strength and lustre. Lenzing's Modal® has an extremely soft hand and lightness of weight, while MicroModal® AIR has a softness factor four times higher than cotton. Additional stretch and recovery can be added via a combination with spandex fibres and it blends well with cotton. The tactility and soft fluidity of Modal™ has focussed it on the underwear and fashion market, and although offering wicking functions suitable for base layer use, costs can be a barrier.

19.4.2 Lyocell

In 1992, Courtaulds developed a new process for the production of regenerated cellulose fibres, the Lyocell process. Encouraged by a growing concern for environmental production coupled with reduction of costs, the new fibre was also engineered to bring additional functions to its performance. Lyocell, branded by Lenzing as Tencel®, the generic name for this cellulose fibre, is obtained by a closed loop organic solvent-spinning process. One key property is the potential to fibrillate, where the wet fibre, through abrasive action, develops microfibrils (or tiny fibres) on its surface. By manipulating or controlling fibrillation, a variety of different base layer fabric characteristics can be achieved that benefit the athlete. According to Lenzing, 'the sub-microscopic canals between the individual Tencel® nanofibrils regulate the absorption and release of moisture which proves to be ideal when performing sports. Tiny fibrils guarantee optimum moisture transportation and excellent thermal regulating properties. The space between the body and the performance textile remains dry.' Offering a pleasant coolness and smooth surface, Lenzing suggest that skin irritations are minimised, and the growth of bacteria is limited as moisture is immediately spirited away to the inside of the fibre. Irritation and abrasion are also limited when a base layer is produced of a seamless construction, a manufacturing process particularly suited to Lyocell fibre. Of potential interest to a mature sports market is Tencel®C, a version of Lenzing's Lyocell fibre which features microparticles of Chitosan. When directly in touch with the skin, the fibre is said to provide a cell-renewing effect, acting as a moisturiser and skin protector, keeping the skin elastic by maintaining the optimal moisture content. When compared to cotton, the moisture loss is considerably lower.

19.4.3 Bamboo

Fibre from bamboo presents the possibility of producing textiles from a sustainable agricultural source, as bamboo grass is a fast-growing renewable crop, grown mainly in China. Two methods are used to process bamboo into a fabric. The mechanical process is similar to that used for flax but is labour intensive and costly. Chemically manufactured bamboo fibre is a regenerated cellulose fibre similar to rayon or modal. The bamboo leaves and woody shoots are 'cooked' in strong chemical solvents to form a viscous pulp solution that is then extruded. The properties of chemically manufactured bamboo rayon are similar to others in the synthetic cellulose family, in being soft against the skin, antibacterial, hypoallergenic, highly absorbent with good wicking properties and quick drying. It is, however, extremely soft and silky, with a luxurious handle, and has proved a popular 'crossover' choice for base layer garments suitable for walking that can also be worn for leisure.

19.4.4 Soya

The choice of soya beans as a raw material seem like a novelty today, but in 1941, Henry Ford experimented with them as an alternative to steel panels for automobiles, when metal was in short supply. One aim of his revolutionary idea, in mixing soybean fibre with resin, was to look for a project that combined the fruits of industry with agriculture, an ideal that would resonate with today's socially aware 'green' customer. The fibres were never commercially produced due to the lack of functional characteristics, and following the end of World War II, petroleum became the major source of synthetic textile fibres, with a lower cost and easier processing. However, since the 2000s there has been a renewed interest in soybean fibre (Dionysios, 2006). Described as 'comfortable as a second skin', soya textiles have a very smooth cashmere feel, with a highly 'feminine' elegant drape and lustre (see Figure 19.5). In common with bamboo fibres, they are ideal for wearing next to the skin. Swicofil, spinners of soya fibres, state that soya possesses many amino acids necessary to the health of the human body. This botanically derived protein fibre can activate the collagen protein in the skin, has good biocompatibility and is beneficial to human health. It also is antibacterial, has a high SPF and UV resistance compared to cotton, silk, viscose and wool, and is smooth to the touch. However, shrinkage can be a problem and laundry must be undertaken at low temperatures.

19.5 Biofibres

The investigation of new plant-based raw materials will be essential for the next generation of textile products. Finite oil stocks are the source of polymers from which the three most widely used synthetic fibres are currently produced – polyester,

Figure 19.5 Soya fibre from Swicofil AG.
Swicofil, image provided by them and printed with their permission.

polypropylene and polyamide. These three also play a highly significant role in the provision of fibre for active sportswear clothing. The drive to find alternatives to these polymers is fuelling research into the innovative area of bioplastics. Usually derived from plant materials, bioplastics use enzymes to break starch from the plants into glucose. This is then fermented, made into lactic acid, polymerized and converted into polylactic acid, or PLA. With PLA textile products already commercially available, the bioplastic market is expanding rapidly and it is estimated that by 2030 'it could account for 10% of the plastics market' (O Ecotextiles, 2012). As with conventionally produced synthetic fibres, PLA fibres can be used alone or in blended form with all other fibre types, both natural and synthetic, to derive benefit from combined functionality. Teijin's ECO CIRCLE™ fibre is a mix of 30% bio-based (sugar cane) plant material and regular oil-derived PET resin, having the physical properties and quality identical to an oil-derived PET, with recycling possible via the ECO CIRCLE™ closed loop recycling system.

The main promoted benefits of these fibres lie in their suggested sustainability credentials – as no fossil fuels are used apart from during processing, they require less energy to produce than conventional plastics, they are recyclable and are made from renewable resources. In order to be commercially successful, other considerations of price, functionality and availability are paramount. Currently the cost of PLA fibres is higher than that of equivalent oil-based products and their performance properties are generally of a lower standard. One problem is the lower melting point of the polymer, which sets restrictions on production, applications and ironing temperatures. However, it should be remembered that the perfection of polyester took many years, and although discovered in the 1930s, work on PLA fibres for textiles has been limited to the last decade – plus ironing of sportswear is generally a very limited activity.

19.5.1 PLA fibres

The two branded fibres currently associated with PLA are Sorona® and Ingeo™. NatureWorks, a joint venture between Cargill of the United States and Teijin of

Japan, launched Ingeo™ in 2003 as the world's first synthetic fibre derived from annually renewable resources, corn in this case. New conversion technologies that use feedstocks such as corn stover (the residue left in the field), grasses, wheat and rice straws and bagasse (the residue of sugar cane production) are being investigated. According to NatureWorks, the biopolymer production process uses 62–68% less fossil fuel resources than traditional plastic materials. No fossil fuel resources are required to build the polymer chain, and electricity for processing is from wind power, all helping to reduce greenhouse gas emissions. Products can be composted in industrial composting facilities, and future chemical recycling back to the polymer may be possible (Burke, 2008). Functionality takes the form of moisture management, wicking properties, low odour retention, quick drying, hypoallergenic and rapid soil release. For woven materials, to combat the problems with low melt temperatures, it is suggested that the fibre is used in the weft, in combination with natural fibre warp faced constructions. NatureWorks also maintain that Ingeo™ outperforms fabrics made from soya and bamboo fibres in shrinkage, after wash appearance, drying time, pilling, wicking and flammability tests. Similar claims are made for Sorona®, used for base or midlayer sports fabrics. Produced by DuPont using corn sugar in a proprietary fermentation process, Sorona® has an impressive list of environmental benefits claimed by the manufacturers. It does not have the famous tenacity of nylon, but producers claim it offers other advantages over both nylon and polyester, being softer, with two to three times more stretch than nylon for ease and freedom of movement. As it is also chlorine and UV-resistant, it holds bright colours well.

19.5.2 Additional biosources

Other plant-based sources of feedstock for polymers are being considered for textile use, to emulate the characteristics of synthetic materials. Toray Industries are producing 100% bio-based fabrics based on the castor plant, which is very robust, growing in dry farming areas and requiring significantly fewer pesticides and herbicides than other crops. Similarly, Japanese manufacturer Unitika's Castlon fibre, also from the castor plant, has more in common with polyamide, being lightweight, durable, and resistant to stains and abrasion. Casein protein from milk is another fibre source, obtained, as described by Swifcofil, the European manufacturer, by 'de-watering and skimming the milk, creating the polymer by a new bio-engineering technique and finally the creation of a high grade textile fiber'. Containing amino acids that are beneficial to human health and nourish the skin, the fibre assists with retaining the skin's moisture, has superior wicking capabilities than synthetic materials and is very soft and comfortable to the touch. First produced in the 1930s and branded with the names of Lanital and Aralac, the staple length fibres had similar characteristics to wool and is still mixed with cashmere today to reduce the cost and improve function of pure cashmere. Other bio-based feedstocks from hairs, such as poultry feathers and wool keratin, are being considered, plus plant sources of mushrooms, flax plants and other waste materials from industry that currently have little value.

19.6 Textiles and fibres for health and well-being

The largest organ in the human body is the skin, presenting therapeutic potential for imparting health and well-being benefits from direct contact with impregnated fibres and textiles. Topical delivery via the dermis can in some cases and for certain medications be a more efficient method than the more customary oral approach. This could be of benefit to a mature generation as the human body is generally in 24 h constant contact with textiles. This area, sometimes entitled 'Cosmotextiles' or bioactive textiles, uniquely links the worlds of cosmetics and medicine with fashion and textiles. Cellulosic Lyocell, with its skin-friendly characteristic, is the favoured substrate carrier for beneficial elements permanently incorporated into the fibre, blends of which are then possible with cotton, polyester and elastic fibres. In order to support the skin's protective function and enhance the wearer's experience, many additional elements can be added, all with the aim of maintaining health. Among those used with Lyocell fibres are vitamin C, zinc and antibacterial functionality from chitosan, obtained from the hard outer skeleton of shellfish, including crab, lobster and shrimp. Whilst the technology is available and some functions are proven, there remains a lack of scientific data in the area, meaning that they are mostly considered as alternative therapies. Communication of benefits also remains a problem, as does the measurement and assessment by the wearer of the attributes, for which they may have been asked to pay a premium.

19.6.1 Microencapsulation

As Nelson (2002) comments in an article for the International Journal of Pharmaceutics: 'the desire for a healthier and more productive lifestyle will continue to generate a market for textiles that promote "well-being"'. Textiles that 'interact' seamlessly with the consumer, reducing stress, promoting comfort and relaxation, are possible through active delivery from microcapsules. They can allow sensing chemicals to be attached to sports clothing and medical products, as seen in Figure 19.6, to warn of damage or hazard to the wearer. Systems can also be developed that deliver measured dosages of chemicals to combat muscle pain or other more serious injuries. The potential applications of microencapsulation in textiles are as wide as the imagination of textile designers and manufacturers (Nelson, 2002). With synthetically derived materials for sportswear, functionality can be embedded in the fibre during its manufacture, ensuring durability and longer life, as opposed to the alternative method of application by external coating or topical treatments for natural fibres, with a shorter life span from wear and laundry erosion. Microencapsulation of liquids and solids is an innovative technology that is opening up new marketing opportunities for performance apparel markets. The process involves the production of microcapsules acting as tiny containers of solids or liquids that release their core contents under controlled conditions to suit a specific purpose. The active substance may be released by friction, by pressure, by temperature change, by diffusion through the polymer wall, by dissolution of the polymer wall coating or by biodegradation. Although microencapsulation methods have been in existence since the 1940s, it was not until the 1980s and 1990s that applications for textiles were developed. The technique has made considerable progress in

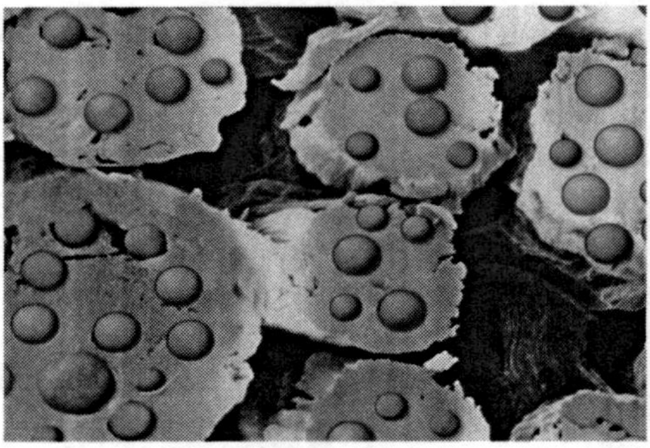

Figure 19.6 Microcapsules embedded in acrylic fibre.
Printed with permission from Outlast Technologies LLC.

the fields of phase-change materials for thermoregulation, in aromatherapy and in the controlled release of fragrances. Anti-microbial and deodorising finishes offer opportunities for freshness retention for garments worn during periods of strenuous activity. Insect-repellent and insect-resist treatments demonstrate higher levels of performance if microencapsulated (Textile Intelligence, 2005).

19.6.2 Alginate

Alginate is used worldwide in the manufacture of wound dressings where gelling and absorbency are key features for improving wound-healing outcomes. The skin-protective and anti-inflammatory properties of this marine plant, rich in vitamins, minerals, trace elements, fats and carbohydrates, offer healing properties that have been well known since ancient times. German spinner Smartfiber AG specializes in the production of active Lyocell-based yarns, including SeaCell® incorporating the benefits of seaweed. The inclusion of North Sea harvested seaweed is said to support skin level blood flow and simulate skin cell regeneration, as well as having a soft, breathable quality as the beneficial seaweed ingredients come in contact with the natural moisture of the wearer's body. Effectiveness remains unchanged even after numerous launderings, an important point with regard to sports base layers. A jersey fabric from a blend of New Zealand Merino wool and seaweed fibre, called Ultrafine SeaMerino™, remains supple and soft, while the embedded SeaCell® fibres add natural minerals and amino acids to the skin.

19.6.3 Silver

The use of silver to kill microbes has been long known. With antibacterial properties said to control unpleasant odours caused by bacteria, silver nanoparticles can be added to fibres such as synthetic cellulosics, at the time of manufacture. Combined with Lyocell, one version is branded as SeaCell® active, also produced by Smartfiber AG, where the active alginate ingredient is further enriched by the addition of silver's long-lasting anti-microbial effect. The fibre can be used in many ways, such as in blends with cotton or polyester. There has, however, been some criticism in this use of silver, which is totally benign to humans but not to aquatic life, prompting questions about environmental safety following laundry.

19.6.4 FIR – far infrared textiles

Far infrared textiles are part of this new category of functional textiles that have putative health and well-being functionality, with the aim of enhancing blood circulation in targeted regions of the body. The function is based on the principle of absorbing energy from the sun and then radiating it back onto the body at specific wavelengths. Thus fabrics with this function achieve thermal retention and are therefore ideal for activities such as mountaineering and hiking and in medical supports or orthotics worn during exercise. Claims are also made that the function could assist to improve increased oxygen in the blood, the rejuvenation of skin and muscle tissue, promotion of regeneration and rapid healing, improved function of the nervous system and relaxing muscles. As Dyer (2011, p. 184) suggests, this could be particularly useful for conditions of arthritis, muscle pain, joint stiffness, as the majority of people over 50 will exhibit some signs of arthritis as joints degenerate. FIR functionality is mainly incorporated via blending the fibre polymer such as polypropylene and a ceramic powder to make the FIR yarn. The functionality can also be applied by impregnation, printing, coating and laminating.

19.6.5 UV protection

UV blocking fabrics not only are an important element in the campaign against skin cancer, but also counter the effects of skin damage by exposure to sunlight, a major factor with ageing. The ultraviolet protection factors of a garment depend on a number of parameters, such as construction, fibre type, colour, weight, thickness, finishing process and UV-absorbing additives such as brightening agents, and consequently can be difficult to measure, but fibres can play an active role. In general terms, wool and polyester fibres inherently absorb UV rays while cotton and nylon allow them to pass through. Dark colours tend to be more absorbent, and the denser, more compactly woven and heavier the fabric the higher the degree of protection. UV protective coatings can also be used on both fibres and fabrics. A functionality can be obtained via the inclusion of activated carbon from coconut shells, such as Cocona by Traptex. Aimed at cycling and sports areas, the function is added to nylon and polyester polymers and can be used with natural fibres in blends. The pores in the coconut structure absorb moisture, odours and in particular UV rays.

19.7 Smart, sensory and adaptive materials

Materials and structures in nature rely on adaptive mechanisms that respond to changes in environmental moisture and temperature for survival (Kapsali, 2009). These bio-mimetic principles are now being applied to textile products, to create fabrics that respond automatically to external stimuli. By adjusting to changes in the surrounding environment, these activities can bring superior protection to the wearer and subliminally combine different functions in one garment, potentially limiting the number of layers or garment types required for an activity. Thus a clothing system is possible that has the ability to sense changes in the microclimate and alter its properties without the intervention of the wearer. Older people are less tolerant of extreme changes in temperature, while feelings of being too cold or too warm play a large part in the personal assessment of comfort in clothing. Phase-change material, known as PCM, and branded under names such as Outlast, can act as a smart thermal regulator (see Figure 19.7). Via direct spinning, paraffin wax is embedded into the fibre, which will absorb excessive body heat as temperatures rise or dispense it as required during exercise. Function can also be applied to the textile by coating or lamination of an active membrane, assisting in regulating the body's temperature by the prevention of heat accumulation and cooling before, during and after exercise. Mondal (2011, p. 163) reports that the temperature range can be variable and is dependent on the required end use of the textile. For example, for underwear and base layers, a near-skin temperature is appropriate, whereas for lining of a ski suit, a much lower phase-change temperature of the PCM is required.

A bio-mimetic concept in thermo-regulating textiles for outer layers is the use of shape memory polyurethane fibres or membranes with temperature-dependent water vapour permeability, such as C-change by Schoeller. Inspired by botanical structures such as the pinecone, able to change shape in response to humidity changes in the microclimate, fabrics use membrane technology that adjusts in high ambient temperatures, allowing sweat to easily evaporate and cool the body. At low temperatures the membrane closes on a molecular scale, water vapour cannot escape but condenses on the membrane and releases its stored energy, keeping the body warm (Bartels, 2011, p. 405).

19.8 Interactive technologies

Fibre choice plays a vital role in wearable electronics, a technology that offers potential for a totally new approach to clothing. It has introduced interactive and digital functions that previously would have been the preserve of rigid devices such as smart

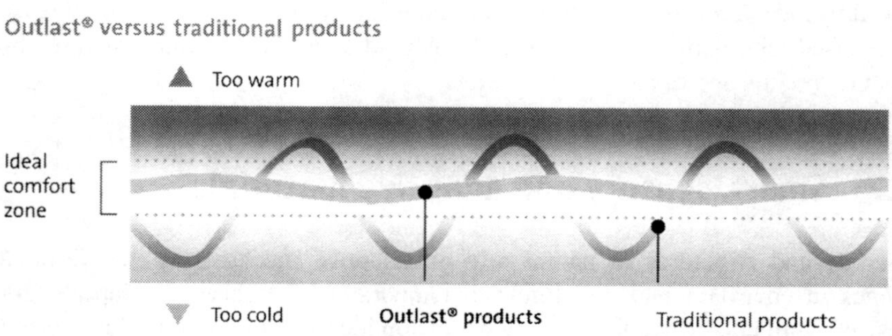

Figure 19.7 Outlast thermal regulation compared to traditional products.
Printed with permission from Outlast Technologies LLC.

phones and computers. By embedding these functions directly in the textile substrate, bulk and rigidity are reduced, assisting with unobtrusive and comfortable use. These wearable interfaces cannot be dropped or lost and are always ready for use. It is fundamental that the 'smart' elements do not detract from a garment's traditional uses and that it can be treated like any normal piece of clothing in everyday situations, particularly regarding laundry and aftercare, to form a piece of clothing that can happily multitask. Despite creating initial excitement at the suggested potential and opportunities for cross-scientific collaboration, only a handful of products have arrived on the market. One of these has been the Smart Shirt program, for monitoring vital signs of the wearer, demonstrating that where a real need can be highlighted, the product outcomes will prove successful.

Characteristics and functions of an intelligent textile are the result of intrinsic combinations of the technicalities of the fibre, the yarn, the fabric's structure and the finishing. While power supply can be an issue, conductivity is paramount, a role played by metal content in fibres, membranes, embroidery threads and printing inks. Intelligent displays and visual read-outs are also possible via the use of ultrathin optical fibres, developed by France Telecom, that emit light along the fibre's length. Intelligent clothing has the potential to introduce new concepts in health care, with a better quality of life for patients and reduced social costs (Solaz, 2006, p. 369). As present-day society is undergoing changes such as ageing of the population, integration of those with disabilities and an increase of chronic diseases, control of disorders such as diabetes and heart disease is vital and may be assisted by textile monitoring. Older active individuals will have been accustomed to being 'digitally connected' to the outside world at all times, and will expect this to continue wherever and whatever the situation. Additionally, any technology should be reprogrammable for different tasks depending on the user's changing requirements. Clothing as a medium for technology is ideal as it is versatile, mobile, universal and adaptable. Initially the preserve of professionals and sports 'geeks' obsessed with gadgets, smart technology is now lower cost and more accessible as big high street brands like Decathlon offer smart products with elements that are permanent and fully washable, such as the CONNECTED*wear*™ range from Fibretronic.

19.9 Environmental and sustainability concerns

In the selection of fibres suitable for active sportswear, it is impossible to ignore the sustainable performance of textiles, from a social and environmental perspective, all along the length of the supply chain. However, it is extremely challenging to make any direct comparisons and there is no single frame approach (Fletcher, 2008). As Dr Meyer (2011) suggests in the preface to the *Handbook of Sustainable Textile Production*, on scientific levels, data and statistics abound. The situation is different in the integral assessment of sustainability, which affords compromises between different scales, ratings and targets – in this area science tends to promote momentary trends that change with the seasons – this is the weak point for any consideration or comparison of sustainability. Against a backdrop of heightened awareness of limited natural resources, a shifting global economic situation and a changing legislative scene, the first decade of this millennium has seen increasing interest in the production of fibres and textiles in a more environmentally and socially acceptable manner. Any consumer would wish to cause minimal negative impact in their choice and use of sportswear, but as sustainability challenges are different for different materials, the picture can be complex and confusing.

19.9.1 Sustainability concerns with natural fibres

As renewable, recyclable and biodegradable fibres, the natural fibre family is well placed to take advantage of the current and ongoing concern with sustainability, but both cotton and wool have some ingrained negative issues to shed. As Ferringo (2012, p. 28) points out in his guide to the sustainability of cotton, the period from the 1950s to the 1990s was very much focused on volume growth through the intensification of farming practices, leading to cotton becoming at one stage the major user of pesticides and insecticides. Negativity over intensive farming practices has been hard to shake off, but industry has been forced to change, partly in response to heightened consumer awareness of global environmental issues, and also with regard to increased legislation, rising costs and efficient use of natural resources. Conventional production has had to improve, via precision use of fertilizers and pesticides and more drought-resistant plant varieties under cultivation. Genetic modification has also contentiously entered the frame, with products such as BT cotton now widely used, particularly in India. Alternative initiatives and pathways towards a more sustainable future already exist, linking environmental and social responsibility factors and certified by organisations such as the Global Organic Textile Standard (GOTS). For cotton, projects such as the Better Cotton Initiative, Fairtrade cotton and Cotton Made in Africa marry social responsibility with sustainable farming systems. Some of these are using 100% organic production methods while others seek to minimise the impacts of conventional production practices. Animal welfare issues can be of concern, such as sheep mulesing and pesticide use, as large-scale commercial grazing has relied on synthetic chemical inputs to increase profitability and to allow wool growing in regions and climates to which it is unsuited. Certified organically produced fibre comes from farms that practise good grassland conservation, create healthy ecosystems and pay attention to the best livestock practices without the use of any synthetic input. There is little competition with acreage for food crops as sheep can be reared on marginal areas of land that are difficult to cultivate. Supply chain transparency is also now possible, allowing the wearer to trace the garment right back to the farmer and the actual flock. For example, with Icebreaker brand's Baacode system, consumers can trace the origins of their garment right back to the sheep station in New Zealand where the Merino fibre was grown, to see the living conditions of the sheep, meet the growers who raised them and follow the production process through to the finished garment.

The main negative factor in silk production is filament degumming with detergents that are then discharged into groundwater. Wild silk is available in a limited manner, but the fibres will be short staple, having been damaged by uncontrolled emergence of the moth. Organic silk also exists, but in both cases high costs exclude use of the fibre in all but the most expensive of apparel.

End of life issues with natural fibres must also be considered with regards to overall environmental impact. Reuse by recycling can offer reduced energy, resource and chemical consumption. Fletcher (2008, p. 35) informs us that in a recycled pure wool fabric, energy consumption is thought to be half of that needed to produce virgin material. In most cases, recycling results in the downgrading of fibre quality and use in low-end products as the aggressive nature of the process breaks and weakens the fibres. Recycling potential is also dependent on the garment being made from 100% pure fibre, restricting blends, and ease and cost of collection and removal of trims and details. Biodegradability and composting are also options with natural fibres, but the chemicals used during processing cannot contain toxic elements, such as heavy metals used in some dyestuffs.

19.9.2 Sustainability concerns with synthetic fibres

In the twenty-first century, the one major negative aspect to this very versatile and useful family of synthetically derived fibres that outweighs all the attributes and could even threaten its future is that as a petroleum-based product it is not viewed as sustainable (WSA, 2006). For this non-renewable resource, estimates around the 'peak oil' situation vary. The 'ownership' and branding of synthetic textiles, as mentioned earlier in this chapter, also allow manufacturers to present reports and examinations of the sustainability factors of their fibres in the least damaging manner, thus making assessment and comparison more complex and unclear. Fletcher (2008) proposes two value systems that influence the approach we take to sustainability. Simplified, the first she describes as more of the same but more efficient involving making incremental changes to present-day institutions and practices. The second sees sustainability as necessitating something different, more fundamental, personal, social and institutional change. It is in this first category, of making small incremental changes, where we are seeing the main response from synthetic fibre producers. The second response is also taking place, to a lesser degree, in the exploration of alternative sources of feedstock. Both are equally valid, and both essential to ensure a diversity of materials and continuity of volume supply.

In general terms, water consumption is lower in synthetics than natural fibres. In the production of polyester and polyamide, aside from the key issues surrounding oil, toxic emissions and potentially carcinogenic elements such as antimony are of high concern. A more recent alarming discovery is the possibility of non-biodegradable microscopic plastic particles found in sand and shed during production and laundry entering into the food chain. Polypropylene, maintain the manufacturers, carries a very convincing positive ecological story. Because it has a low melting point and has to be melted only once to colour and convert it into a useable high-quality textile fibre, the energy usage is extremely low in comparison to other synthetic materials. The fabrics only need to be washed/relaxed with no harsh chemicals and are then low temperature dried, giving a low ecological footprint compared with cotton and wool. Within colour limitations, polypropylene is also truly recyclable, even back to textile-grade fibres.

One significant way forward lies with chemical polymer recycling, the remelting and reprocessing of plastic material, a route that has captured the imagination of both public and producer, keen to 'do the right thing' and react to popular demand with regard to the squandering of natural resources. In the United States, Wellman claim to have introduced its first polyester textile fibre, Fortrel EcoSpun, made from post-consumer packaging and bottles, back in 1993. Just as with virgin polyester, recycled polyester or rPET can be blended with other fibres such as wool and cotton. Other examples are Unifi's Repreve yarn, a recycled filament polyester. The range available now includes recycled staple polyester, recycled filament nylon 6.6 and recycled functional fibres with properties such as flame retardancy, moisture wicking, UV protection and stretch (WSA, 2010). The general perception is that fibres produced in this manner cause less harm to the environment through reducing oil consumption and preventing large numbers of PET bottles ending up in landfill sites. Reputable producers use the Global Recycle Standard issued by Control Union, a track and trace system that verifies recycled claims and examines critical issues of processing and workers rights, similarly as found with the GOTS. There can be performance limitations on reclaimed polymers, as with each recycling, the polymer chains break down, weakening the plastic and making it less suitable for high-end use. PET degrades after about five melt cycles, making recycled polyesters suitable for fleece-type products for mid- and outer layer sports.

Prior to the conception of the Cradle to Cradle – the C2C concept – in the 1990s, it was considered that the lifecycle of a product could be visualised in a 180° manner, from creation to disposal. McDonough and Braungart's revolutionary philosophy, that a full circle 360° loop was possible, forced some innovative thinking. C2C certification specifies that all synthetically derived material be part of a 'technical nutrient' cycle, with the overall and admirable aim of creating product while recycling non-degradable materials continually and completely, thus eliminating waste. One of the first companies to be certified, the Victor Group in North America produce Eco Intelligence® Polyesters containing no heavy metals (such as antimony), chlorine and PBTs and can be safely and perpetually recycled and remanufactured in a closed loop system. Being used at present for upholstery and interior use, issues can arise with how a product is assembled, using what additional materials and how simple and cost-effective it is to disassemble. With regard to garments, this has been a major stumbling block to an otherwise enlightened approach.

The use of mono-materials to create a garment – where all component parts are from the same polymer type, can facilitate recycling and the creation of new product. In 2005, US sportswear brand Patagonia in liaison with Teijin introduced a programme entitled The Common Threads Garment Recycling Program. With a long history of innovating to reduce environmental impact, Patagonia aims to collect worn-out, old Capilene base layer garments from customers and recycle them into new polyester filament yarns that will be used to make new polyester (PET), using the ECO CIRCLE™ recycling system from Teijin. The concept of taking responsibility for garments at the end of their useful life is a revolutionary idea in clothing, but as Patagonia state in their 2005 'White Paper Analysis' 2005 of the programme, it is not without problems, including the energy used to collect and return garments for processing. Patagonia continue to encourage their customers to 'Reduce, Reuse, Repair, Recycle' while also recognizing that recycling clothing is not a cure-all for environmental challenges.

As Yvon Chouinard, Patagonia's founder and owner states: 'The Common Threads Initiative addresses a significant part of today's environmental problem: the footprint of our stuff. This program first asks customers to not buy something if they don't need it. If they do need it, we ask that they buy what will last a long time – and to repair what breaks, reuse, or resell whatever they don't wear any more. And, finally, recycle whatever's truly worn out.' Since its inception, the programme has been expanded to take back synthetic garments from other brands, including those made from some polyamides, and most recently an innovative arrangement with online-auction Website eBay, to create a virtual swap meet for buying and selling used Patagonia gear. The underlying message from the brand remains clear – 'The greenest product is the one that already exists.'

19.9.3 Sustainability concerns with synthetic cellulosic fibres

Synthetic cellulosic fibres appear to have benefited from the increased acceptability of renewable materials derived from fast-growing plant sources that absorb carbon dioxide from the atmosphere during their natural growth (Fletcher, 2008). With soya production, though heralded as a 'sustainable' natural and biodegradable alternative, on closer examination, there appear to be issues relating to the environmental impact of large-scale agriculture. As Greenpeace reported in 2006, the demand for soya has led to destructive land clearances in the Amazon, increased pesticide use and genetically modified crops (GMO), as most soy crops are now GMO.

Traceability and transparency in the supply chain are currently difficult, both key issues for companies not wishing to be implicated in environmental destruction or the use of GMO crops. On the other hand, Lyocell's closed loop manufacturing system and its certification by Forest Stewardship Council (FSC) have increased interest from brands wishing to reduce environmental impact and improve corporate social responsibility performance. Lenzing AG maintains their production of Modal® has a very low impact on the environment if compared with cotton, for example. The yield by acre of Modal® is up to six times higher and the water required is 10–20 times lower than with cotton. Moreover, beech wood or eucalyptus forests grow in a completely natural way and do not compete with crops for arable land, and the stock of beech wood is obtained as a result of 'rejuvenation'. Because of the potential health risks and damage to the environment from the incorrect disposal of effluent, textile manufacturing processes for bamboo or other regenerated fibres using toxic process chemicals are not considered sustainable or environmentally supportable. Newer manufacturing facilities have begun using other technologies to chemically manufacture bamboo fibre that are more benign, similar to the closed loop process used for Lyocell.

19.9.4 Sustainability concerns with biofibres

'Bioplastics produced from a plant instead of from petroleum', states Jim Thompson (2008) writing in the *New Internationalist* 'is neither a new breakthrough nor a guarantee of ecological soundness.' He likens the process to developments with the earliest plastics such as celluloid, made from tree cellulose before petroleum proved itself a cheaper source. Today, with oil prices skyrocketing, Thompson suggests it is cheaper feedstocks, not green principles, that is driving chemical companies back to bio-based plastics. He claims that there is nothing sustainable or organic about most industrial agriculture feedstocks. At present GMO corn grown using pesticides is probably the leading source of starch for bioplastics, with a strong link between genetic contamination and bioplastics. There are certainly challenges pertaining to these new sources, and proof that they are actually better for the environment is debatable. It is, however, evidence of the industry questioning tradition and responding to customer demands, for whatever reasons, while the ability to replace some fossil fuels with plant-derived feedstocks has to be a good thing, provided the functionality and versatility are on a par. As Ferringo (2012, p. 120) rightly suggests, major challenges will continue, with water, energy, climate change, population growth, competition for land, prices, competition from other fibres, efficient and reduced consumption of raw materials, financing and policy all being issues in the coming years. But there will be no sustainable fibre production if farmers and producers cannot stay in business, and it will continue to be a delicate balancing act between commerciality and planet welfare to create sustainable practices, through from the fibre to the fabric production, dyeing, garment manufacture, distribution, laundry and final disposal by the user.

19.10 Conclusion

Many mature consumers have a close affinity with natural fibres, having grown up with them, in times that were largely pre-synthetic, and may well affect their sportswear clothing choices, providing the required function is satisfactory. Visual appearance and comfort will always be important aspects of consumer appeal, and natural fibres have much to offer in creating that appeal. There is a strong consumer interest in things natural, and high-performance products that meet all the requirements of this market and retain a natural focus have great potential. Importantly, natural fibres present the possibility of multi-purpose garments that cross easily from active to casual everyday wear, suitable for a variety of situations and environments. Synthetically derived fibres also demonstrate many different functions and a high level of durability for a long life – where longevity in use can be claimed as the prime sustainable feature. Their potential for extending and varying functionality via blends with other fibres is huge, and the ability, unlike natural fibres, to re-engineer the polymers could be endless, resulting possibly in the fourth generation of textile materials that are finer, stronger, multifunctional and innately adaptable to human requirements. However, plastics are the fastest growing segment of our waste stream, and unless practical, cost-effective reclaiming and reuse becomes more widespread, sustainability of the synthetic fibre area will continue to be a barrier. The experimental use of bio-fibres in sports textiles should be viewed as a positive response in introducing variety while reducing the monopoly held by polyester and polyamide. These fibres have much to offer the user in terms of function, comfort and potential for reduced environmental impact. The biggest hurdle may lie in communicating these factors to the user, and marketing new and unusual materials to a mature audience.

19.11 Future trends

As the average life span increases, so will our demands for a continued high quality of life. The recent linking of medical research and textile development will be of optimal benefit to those in the older generation wishing to stay active, independent and healthy for as long as possible. With higher aspirations, greater spending power and anticipation of a higher quality of life than their predecessors, this mature age group will expect more from textile technology, having greater prior knowledge and understanding of branded functions such as Gore-Tex®. As Bryson (2009, p. 335) states, they will also anticipate their needs to be specifically considered in the design process, which should be increasingly possible via individual customisation of products using techniques such as body scanning, 3D printing and the embedding of personalised electronics into clothing. As sportswear has traditionally been an early adopter of new technologies, we can certainly anticipate that sports attire of the future will include an array of properties that will integrate and interact with the wearer, adapt and respond to accommodate changes in the internal and external environment, while taking and transferring inspiration from nature and from the military, medical and aeronautical industries. Developments such as miniaturisation and nanoscaling, shape memory possibilities and the introduction of washable electronic circuitry into clothing will doubtless bring new 'well being' and interactive functions to this audience. Although smart and intelligent functions will be of great benefit, the choice of clothing will continue to be governed by comfort factors, visual appeal, fashion, price and personal preference. This generation are 'environmentally aware' and will wish to reduce their impact on the environment wherever possible. Thus they will choose brands demonstrating integrity and honesty, those they trust and expect to produce clothing that uses the best available technology to achieve this. Fibres and garments must function subliminally, as enduring crossover pieces suited to leisure and sport, to enhance and support both the wearer's mental and physical health and well-being.

References

Bartels, V. (2011). In G. Song (Ed.), *Improving comfort in clothing*. Woodhead Publishing.

Brunel, C. (2002). *The T shirt book*. Assouline Publishing.

Burke, M. (March 2008). *Green couture*. Chemistry World.

Dionysios, V. (2006). *Soybean fibre, a novel fibre in the textile industry*. University of Manchester.

Dyer, J. (2011). In N. Pan, & G. Sun (Eds.), *Functional textiles for improved performance, protection and health*. Woodhead Publishing.

Ferringo, S. (2012). *Cotton and sustainability*. MCL Global.

Fletcher, K. (2008). *Sustainable fashion and textiles*. Earthscan.
Gherzi. (2011). From *Lenzing Roadshow Presentation 2011*.
Hammerle, F. M. (2011). *The cellulose gap, the future of cellulose fibres*. Gherzi/Lenzing.
Hibbert, R. (2006). *Textile innovation, interactive, contemporary and traditional materials*. Line.
Holcombe, B. (2008). In N. Johnson, & I. Russell (Eds.), *Advances in wool technology*. Woodhead Publishing.
Holmes, D. (2000). In A. Horrocks, & S. Anand (Eds.), *Handbook of technical textiles*. Woodhead Publishing.
Honga, T., & Philips, G. (1997). *New fibres*. Woodhead Publishing.
Johnson, N., & Russell, I. (2008). *Advances in wool technology*. Woodhead Publishing.
Kapsali, V. (2009). In J. T. Williams (Ed.), *Textiles for cold weather apparel*. Woodhead Publishing.
Kosa. (1999). *Dictionary of fiber and textile technology*.
McCann, J. (2005). In R. Shishoo (Ed.), *Textiles in sport*. Woodhead Publishing.
Meyer, U. (2011). *Handbook for sustainable textiles*. Woodhead Publishing.
Nelson, G. (2002). Application of microencapsulation in textiles. *International Journal of Pharmaceutics, 242*.
O Ecotex report Plastics Part 1. (2010).
Purane, S.V., & Panigrahi, N. R. (September 2007). Microfibres, microfilaments and their applications. *AUTEX Research Journal, 7*(3).
Samanta, A., & Bhute, A. (2011). *Wool in sportswear*. Coulourage.
Shallard, S. (May/June 2008). How silk began to make an impression on fabric mixes for outdoor apparel. *WSA Article*.
Shishoo, R. (2005). *Textiles in sport*. Introduction Woodhead Publishing.
Solaz, J.-S., (2006). In H. Mattila (Ed.), *Intelligent textiles and clothing*. Woodhead Publishing.
Srivastara, S. (2012). *Natural & Man made Fibres*. Pushpa Publishing House.
Textile intelligence microencapsulation for enhanced textile performance. (12), 1st quarter 2005.
Textile world news. July 2009.
Thompson, J. (2008). New internationalist. *Bioplastics* (418).
Tobler-Rohr, M. (2011). *Handbook of sustainable textile production*. Woodhead Publishing.
World Sports Activewear. (November 14, 2006). *Polypropylene fabrics for sport*.
World Sports Activewear. (January/February 2007). *Five years on and PTT starts to make a mark*.
World Sports Activewear. (July/August 2007). *Fighting back against nature*.
World Sports Activewear. (July/August 2007). *New trends in polyester*.
World Sports Activewear. (November/December 2007). *Biomimetics – Alpacas go surfing*.
World Sports Activewear. (March/April 2008). *Cotton, pushing performance*.
World Sports Activewear. (November/December 2010). *rPet now and tomorrow*.

Websites

Materials www.sustainablebiomaterials.org.
www.merinoaustralia.com.au.
www.icebreaker.com.
www.eclipsetextiles.com.au.
www.nanowerk.com.

www.ramblers.org.uk.
www.swicofil.com.
www.oecotextiles.com.
www.thehenryford.org/research/soybeancar.
www.lycra.com.
GOTS – www.global-standard.org.
www.patagonia.com.
New Internationalist – www.newint.org.
www.cottoninc.com.

Further reading

Biologically Inspired Textiles published. (2008). *Woodhead Publishing*.
Common Threads Garment Recycling Report. (2005). Patagonia.
Functional Textiles for Improved Performance Function and Health. (2011). *Woodhead Publishing*.
McDonough, W., & Braungart, M. (2002). *Cradle to Cradle*. North Point Press.
McDonough, W., & Braungart, M. (2013). *The Upcycle – Beyond Sustainability, Designing for Abundance*. North Point Press.
Military Textiles. (2008). *Wodhead Publishing*.
Smart Clothes and Wearable Technology. (2009). *Woodhead Publishing*.

Designing base layers for apparel for the active ageing population: balancing technology and aesthetics

20

V. Haffenden, J. Smith
University of Brighton, Brighton, UK

20.1 Introduction

This chapter will discuss a range of different fibre, fabric and garment construction technologies which are currently employed in the manufacture of body and base layers, and their suitability for the older figure shape. It will next explore the body shape requirements and discuss the wearer needs from both body and base layer garment design, as was established through working with a user group of people aged 60–75. Thirdly it will look at the aesthetic requirements of the older wearer from the body and base layer garments in the context of garments produced for the Design for Ageing Well (DAW) project, in which the focus was on walking to maintain well-being.

The term 'base layer' has evolved to describe the innermost garment in a layering system which is, apart from underwear, closest to the skin. Indeed in male upper-body apparel the base layer often lies closest to the skin. During the DAW project, a male undervest was designed to carry the electronic components and was therefore designated the male 'body layer'. In female attire, however, because women in generally prefer to wear a brassiere when participating even in light exercise, bras were designed to carry the electronic components as the female 'body layer'. Consequently in this chapter, when discussing the challenges faced when designing technological base layers for older consumers, and those of structural bra design for the ageing female shape, the term 'body layer' refers to vests and bras, whilst 'base layer' refers to the garment worn immediately on top of these.

Design for older figure types requires particular attention to the physical effects of ageing, not only to the linear dimensions but also to the three-dimensional body shape. The advent of 3D body scanning technology has greatly enhanced the possibilities when working with these aspects of clothing design for this demographic. International sizing surveys, pioneered by SizeUK carried out between 2001 and 2002 (SizeUK, 2004) and apparel sizing research projects such as those of Ashdown, Choi, and Milke (2008), Daanen and Hong (2008) and Connell, Ulrich, and Ashdown (2003), are now mainly gathering body shape and size data through 3D whole body scanning.

Textile-led Design for the Active Ageing Population. http://dx.doi.org/10.1016/B978-0-85709-538-1.00020-1

As part of the DAW project, representative BMI groupings were extracted from the SizeUK database and body shape categories were established in order to select a user group. The user group was selected from volunteer walking enthusiasts by matching and morphing their 3D body scan shapes against these categories. The garments discussed here were developed for these body shape categories. This is discussed in more detail by Bougourd in her chapter, 'Ageing populations: 3D scanning for apparel size and shape'. Understanding of older body shape, not just size, was fundamental to the design process. It was not, however, the exclusive source, but was used in combination with manual measurements and user try-on and evaluation.

Co-design and user involvement were fundamental to the design and evaluation processes of garments discussed here, and the style, shape and construction of the garments are consequences of these processes.

20.2 Defining technologies

From the start of the DAW project it was clear that a number of technologies co-existing within the team, and the meshing of these technologies contributed to the outcomes. The main technologies represented were computing, communications, electronics, fibre, textile and clothing, but this chapter concentrates largely on the last three areas.

20.2.1 Fibre technology

As body and base layers generally have contact with the skin, they are most comfortable if they are fitted but softly extensible, allowing natural body movements without restricting or adding bulk. Because of this, the majority of these garments are made largely, if not exclusively, from some type of fine-gauge knitted fabric which has natural stretch properties.

Yarn technology is fundamental to satisfactory performance of a knit structure, and plays an important role in body and base layer design. Four major properties are required by the consumer from body and base layer fibres; wicking (moving moisture outwards and away from the skin), body heat retention, heat release during exercise and ease of care, the last of which is widely understood to be offered by synthetic fibres.

Within synthetic fibres there are two main sub-divisions that are widely used in apparel production: synthetic polymers such as polyester, polyamide (nylon) and acrylic, and natural polymers which include regenerated cellulose rayon, in particular viscose, cupro and modal (Tubbs and Daniels, 1993). Synthetic polymer fibres, because of their hydrophobic (polyester) and hydrophilic (nylon) properties, could become uncomfortable when the wearer perspired, and despite their washability, (which made them popular for other uses), were not truly successful for base layer use. However, since their early incarnations, synthetic fibres, and the yarns and fabrics made from them, have been progressively re-engineered to become hydrophilic and will now wick moisture to varying degrees (Troynikov & Wardiningsih, 2011).

Yarns with wicking properties therefore now include a number of natural, synthetic and blended fibres. A recent resurgence of interest in wool as a sportswear fibre has led to merino wool base layers becoming popular at the upper end of the market. In its natural state, 100% cotton does not offer ideal properties as a body or base layer fibre as it absorbs moisture, becoming damp and heavy. However, Cotton Incorporated have now developed a cotton with moisture management properties, creating a version that is more suitable for sports body and base layers.

Modern synthetic fibres that have a soft, natural handle and appearance, but offer the durability and performance of synthetics, include amongst others, Invista's Supplex®, and Nylstar's Meryl®, both of which are polyamide based. Regenerated cellulose fibres that may also be used for body and base layers include Lenzing's Tencel® (Lyocell) and MicroModal® (made from bamboo), both of which are environmentally friendly alternatives to cotton, with excellent wicking and anti-bacterial properties.

20.2.2 Knitted textile technology

Knitting has three fundamental stitch formations: knit – which makes a loop, slip – which leaves a float of yarn, and tuck – which forms a multiple loop. Combinations of these stitches create the majority of knit fabric structures, and further permutations providing different performance and design potentials are possible depending whether the knitting is worked over single or double sets of needles.

Two different, but related knitting machine technologies produce knitted fabric, those of warp knitting (also called Raschel) and weft knitting. Both technologies are employed across a range of applications from automobile interiors and medical appliances to fashion clothing.

Within weft knitting, sub-categories of flat and circular machine production exist, variously producing for the clothing market: yardage for cut-and-sew manufacturing, cut stitch-shaped garment pieces, and partly or fully shaped garment pieces. Over the last 20 years both warp and weft knitting technologies have developed variations of 'seamless' or 'complete' garments, offering improved wearer comfort as there are fewer seams to irritate the skin. Innovators in flat-bed weft knitting of complete garments include Shima Seiki Ltd. of Japan (Wholegarment®) and the German company Stoll GmbH (Knit & Wear®). Recent developments of integrally knitted, seamless athletic shoes by the two major sportswear market competitors Nike and Adidas have taken the application of flat-bed integral knitting technology to a new level of finesse (Wilson, 2012).

Circular 'seamless' machinery is largely dominated in Europe by the Italian Lonati-owned Santoni company (Knitting Industry News, 2010). Warp knitted 'seamless' garments (pioneered by Karl Meyer of Germany) are now also the province of Santoni technology (Hunter, 2012).

Circular produced 'seamless' garments are not truly seamless because although they are without side seams, (because the torso piece is knitted in a tube on a circular dial of needles), they are seamed elsewhere. This technology is particularly represented in base layer manufacture, where excellent body fit can be achieved by the strategic application of stretch yarn in combination with critical finishing methods.

20.2.3 Stretch yarns

Stretch yarn refers to extensible yarn manufactured from rubber or more usually synthetic polyurethane fibres, known generically as elastane or spandex. Since its invention by DuPont in 1959, elastane has become increasingly popular for sport and leisure wear due to its strength and durability, so that of the various types of stretch fibre it is currently the one most commonly used in sportswear. According to a Textile Intelligence report (2011), the demand for elastane doubled between 2000 and 2010, a fact reflected in the 2005 estimation that 35–40% of all clothing in developed markets contained a percentage of the filament form of the fibre (Holmes, 2005). Current brands include, amongst others, Invista's Lycra® and Asahi Kasei's Dorlastan®. Elastane fibres stretch to at least three times their original length, but in addition will rapidly revert to this length. These properties mean that garments made from fabrics containing elastane recover their shape well after movement and can provide additional support and compression to the body. Whilst most knit fabrics already have inherent stretch due to their structure, those that also include a percentage of elastane are ideal for base layers which are required to fit very closely to the body without restriction of movement and go through repeated washing cycles.

Alongside the obvious advantages of fewer seams, open meshes of varying sizes can be knitted alongside solid structures, thus enhancing heat dispersal at strategic points on the body. Similarly technical knitted garments offering improved performance and recovery through varying levels of support and compression can be achieved by altering fibre content, yarn type, knit structure and stitch size throughout the same garment. Developments are not limited exclusively to sportswear, but include lingerie, shapewear and medical textile products. Thus, dependent on the manner in which they are knitted and the fibres used, base layers can, to varying degrees, provide enhanced wearer comfort through reduced chafing, moisture management, heat management, support and compression.

The final level of stretch, support or compression of the knit fabric or knitted garment can be determined by the percentage of elastane in the fabric, stitch structures and the strategic positioning of combinations of these on the body surface, plus specific finishing methods. Therefore knit fabrics which contain elastane are ideal for base layers. In bras, however, areas that provide breast support should not stretch, and therefore non-stretch 100% polyester knits are particularly useful in these focused areas.

20.2.4 Sewn base layer construction

Knitted garments are manufactured in a number of ways: shaped on the machine during knitting, cut and sewn from knitted fabric and in various combinations of these (Figure 20.1). Traditional shaped, seamed knitwear is not, however, prevalent in the base layer market, as the fabric is generally of a heavier gauge than that achievable with circular knitting (although finer gauges are becoming available) and production costs are high. Base layers can be manufactured as garments cut and sewn from knitted fabric, but an increasing number are now made with seamless knitting technology

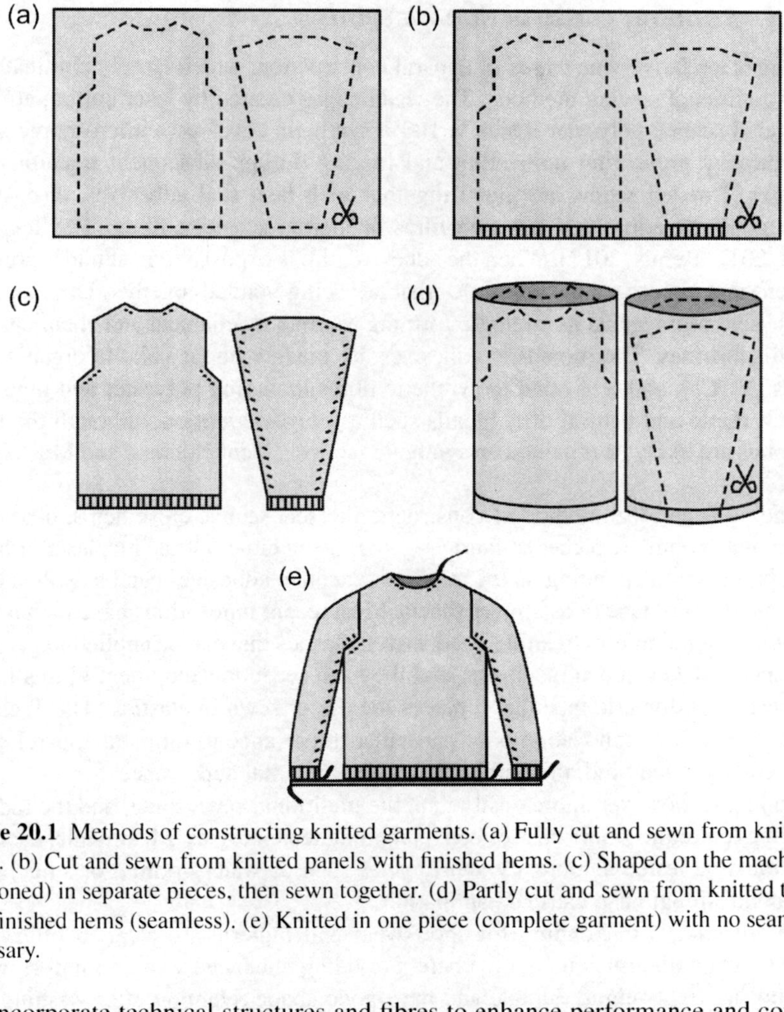

Figure 20.1 Methods of constructing knitted garments. (a) Fully cut and sewn from knitted fabric. (b) Cut and sewn from knitted panels with finished hems. (c) Shaped on the machine (fashioned) in separate pieces, then sewn together. (d) Partly cut and sewn from knitted tubes with finished hems (seamless). (e) Knitted in one piece (complete garment) with no seaming necessary.

and incorporate technical structures and fibres to enhance performance and comfort. This is particularly so in base layers for specific uses, such as skiwear, cycling and running and high-tech sports bras used during participation in all of these.

When manufacturing fitted base layers from fabrics containing elastane, the garment is generally made either to closely match the body measurements, with no ease added, or at a minus percentage of body measurement, depending on the closeness of the required fit (Branson & Nam, 2007, p. 269). Consequent to this, the base layers for the DAW project were all designed using minus 12% of the body measurement. It is important to note that manufacturing and sizing of knitted and stretch garments have unique needs in pattern drafting, cutting, construction and fit, and this relatively new clothing technology is an evolving process which is currently an international subject of enquiry (Abdessalem, Abdelkader, Mokhtar, & Elmarzougui, 2009; Branson & Nam, 2007; Brownbridge, 2011; Fellingham, 2007; Power & Otieno, 2007; Watkins, 2011).

20.2.5 Bonding construction methods

Bonding is a relatively new area of apparel construction, which largely eliminates the use of traditional sewing methods. The clean edges created by laser cutting are ideal for bonded seams; polyester which is 100% synthetic develops a microscopic sealed edge, thereby preventing unraveling and fraying during subsequent manufacturing processes. Bonded seams are glued together with heat seal adhesives, also known as thermoplastic adhesives, hot-melt films or thermo-sensitive films (The Technical Center, 2012. Bemis, 2012). When the adhesive film is exposed to heat under pressure, it softens and penetrates the substrates that are being bonded together. Upon cooling, the adhesive film regains its strength, forming a strong mechanical and chemical bond to both substrates. Thermoplastic adhesives are made without volatile organic compounds (VOCs), and will bond to synthetic fibres including polyester and nylon and some synthetic and natural fibre blends such as polyester/cotton, although the strongest bonds are likely be obtained on synthetic fabrics. (Jacubčionienė and Masteikatiė, 2010).

Bonding is an ideal method of construction to seal seams, close hems, bind edges and laminate complete pieces of fabric together using either ultrasonic, laser or hot air fusing processes. Depending on the required outcome, adhesives can be applied either as narrow strips of tape or as allover sheets. Most recent innovations have included the adhesion of bonding film to knitted and woven elastics that can be applied to garments using specially developed machines, and the most recent development in this field is an adhesive elastomeric tape that replaces the use of sewn in elastics (The Technical Centre, 2012). This innovation is of particular importance to intimate apparel where high recovery is required, in combination with a very flat appearance.

Bonding is, however, more costly than the traditional sewn route, and the industry is striving to produce more cost-effective bonding technologies. Meanwhile, each new fabric must be tested as porosity and finishes such as waterproofing or silicone can affect bond strength and cause delamination.

The advantages of bonding for sportswear and lingerie are clear. It offers low-profile seams with less potential to chafe, providing enhanced wearer comfort, whilst garments are lightweight, durable and have good shape retention after washing. The complete lack of needle holes creates completely waterproof seams that are ideal for outerwear garments. Bonding technology has the ability to create a modern, clean-looking garment that offers new aesthetic potential and outcomes to garment designers.

20.2.6 Electronic technology

The electronic components of the garment layering system developed during the DAW project are discussed in other chapters, but it is important to note that the aims were distilled to the final outcomes through user engagement, testing and feedback. Technology presentations and practical workshops with sample testing provided users with insight into the possibilities: for example, in 2010 a workshop was organised in which the users provided feedback on sample mobile phones as potential interfaces. Following similar workshops, ease of use, a simple interface and clearly readability because of ageing eyesight were found to be critical to potential user acceptance. In the final DAW body layer garments, the textile sensors in the base layer garment provided physiological feedback and sent a signal to a local device. The communication requirement of the electronics was to transmit wirelessly using Bluetooth between this device and a peripheral interface located elsewhere on the wearer's body, in this case a mobile phone.

Other products have already achieved this, for example, the Performance Sports Bra in the Women's Adidas by Stella McCartney which was unveiled just prior to the 2012 Olympics (Wilson, 2012). In the sports field, however, this type of garment and functionality had so far been promoted to the younger, active market until DAW addressed it in the context of the older body shape and the needs of older people.

20.3 The roles of body and base layers in a clothing system

The base layer, as already discussed, is generally a close-fitting garment designed in fibres and fabrics that effectively retain body warmth in cold weather and wick moisture away from the body. They also provide a comfortable layer which is unlikely to rub or irritate, and act as an interface between the body and mid- and outer layers which may of necessity have other overriding properties making them less pleasant next to the skin.

20.3.1 Female body layer

For the female physique, two additional functions are required of upper-body clothing that is worn next to the skin, those of supporting and reducing movement of the breasts during activity (Mason et al. 1999). These objectives are primarily achieved by wearing a brassiere designed to provide a combination of compression and encapsulation of the breast tissue (Mason et al, 1999).

In the case of designing technologically enabled garments for the Design of Ageing Well project, the secondary purpose of the female skin layer was to carry the necessary technological components in as comfortable and discreet a manner as possible.

20.3.2 Male body layer

The brief for this garment was to design a close-fitting skin layer garment of which the primary function was to enable sensors to maintain close contact with the skin and carry the signal from sensor to a wireless transmitter also located somewhere in the garment. At the same time, this garment should help maintain body temperature, wick moisture and not chafe the skin. As will be discussed in detail later on, e-textile sensors were incorporated into this garment in a similar manner to that of the female body layer, and the garment was designed to meet the aesthetic requirements of the older male users.

20.4 Designing for the older body shape

The ageing body shape differs from that of youth and even middle age, and therefore these changes must be accommodated when clothing is designed for older wearers. Wells has described some of the characteristic body shape changes that occur during ageing (Wells, Treleaven, & Cole, 2007, p. 422). His analysis of the results of the national sizing survey SizeUK (2002–2004) based on body shape data gathered from 3D body scans, rather than BMI (body mass index), found that BMI was not a good indicator of the distribution of weight, particularly in older people, but that weight tends to increase steadily with age (Wells et al., 2007, p. 423). Men at any age were found to have larger chest girths in relationship to their waist/hip girths, but as they age, this difference decreases. Well's analysis showed that men's body shape remains fairly stable through the life course, although there is a tendency for the waist girth to increase. Women's waist and upper arm girths increase relative to their thigh circumference, even after age 60, an effect that is emphasised because after age 50 their thigh circumference begins to decrease (Wells et al., 2007, p. 422). Although the overall body shapes of men and women become less dimorphic as physical, gender-specific characteristics become less marked, Wells established that there is a gender difference in the way ageing affects body shape.

Anthropometric studies relating to the postural changes experienced during ageing, particularly by women, have found the bust, shoulder slope and cross shoulder dimensions to be affected in conjunction with the development of a forward slope to the head and neck angle, and a forward roll to the shoulders with some upper back curvature (Goldsberry, Shim, & Reich, 1996b). Ashdown's research into postural changes during ageing confirmed this and also found a tendency toward increased right to left asymmetry in women beyond 55 years of age (Ashdown & Na, 2008, p. 301). It was established by Twomey that height in both sexes decreases with age, whilst Cutler and Katzman both associated ageing with a stooped posture, particularly in women (Cutler, Friedmann, & Genovese-Stone, 1993; Katzman, Wanek, Shepherd, & Sellmeyer, 2010; Twomey & Taylor, 1987). These manifestations of the corporeal aspects of the ageing process were similar to those found during visual analysis of the 3D body scans, and were supported by observations and anecdotal evidence from users, all of which informed the design process discussed here. When designing for the ageing body shape these physical manifestations of ageing must be taken into account during the technical design process of shaping the garment and should contribute to establishing a successful aesthetic for garments for older body.

20.4.1 The importance of new design thinking for the relevant body shape changes of ageing

In 1962, both Bartley and Shipley studied the clothing preferences of women over 65 years of age and drew attention to the problems they faced when purchasing clothing that matched their aesthetic aspirations (Bartley & Warden, 1962; Shipley & Rosencranz, 1962). In 2012, dissatisfaction with the clothing choices available to them was continuing to be expressed by the 60–75 year old user group in DAW, despite a number of other studies in the intervening years which have analysed older consumers' clothing preferences and made recommendations for improved provision (Chowdhary, 1997; Goldsberry, Shim, & Reich, 1996a; Horne, Campbell, & Scholz, 1999; Lee, Damhorst, Lee, Kozar, & Martin, 2012; Woodson & Horridge, 1990). Examples of these points of dissatisfaction and the reasons expressed were:

- tight pull-on garments – elderly stiff shoulders, openings were preferred
- lack of three-quarter sleeves in female tops – short and sleeveless show 'flabby' upper arms
- necklines – too high at front on women's round-neck tops
- necklines – too low and wide on women's scoop-neck tops
- wide use of synthetic fibres (this altered as a result of the project as can be seen in Section 20.5.2.), although wool was perceived as 'hot and prickly'
- small zip pullers – fiddly due to onset of arthritis in fingers
- zipped-in hoods were difficult to access quickly – stiff shoulders, arthritic fingers
- fashionable colours not available in sizes/styles older people preferred
- the importance of functionality as well as appearance

All of these points were addressed during the design and development of the body and base layer garments for the DAW project.

20.5 Technical and aesthetic design considerations and processes

20.5.1 Design considerations

User requirements are of particular concern when designing body and base layers for the older body shape. Considerations that were highlighted by potential users included problems when donning and doffing garments caused by stiffness in the shoulders, back and hips. This was in addition to dexterity issues concerned with small fastenings such as press studs that were difficult to open and close, zip pullers that were too thin and short to grip with arthritic fingers and a general dislike of hook and loop fastenings for both aesthetic and practical issues.

Older female users were especially concerned with wearability, seeking base layers that they could wear throughout the day, and both male and female users wanted longer body lengths.

Male users expressed preference for shoulder emphasis, and particularly liked the possibilities for ventilation offered by meshes in seamless garments at underarm and upper chest. They did not like high collars, preferring round-necked styles, although there was some interest in zip-fronted turtle-neck collars as they offer flexibility.

Both male and female expressed preferences for lightweight layers, preferably in subtle co-ordinating colours, but not with bold colour blocking.

20.5.2 User preferences for the base layer design

The design brief to cater for the base layer needs of this demographic was defined by a uniquely involved and informed user group. Initially, there was significant resistance amongst the user group to anything but natural fibres for the base layer garment.

However, following opportunities for wearer tests, and information sessions on fibre and textile technology, a wider variety of fabrics became acceptable. Overall, a preference was still held for a natural handle to the fabric, but this was balanced by the enhanced wearer comfort and performance of synthetic fibres.

Female users were particularly concerned that their base layer should be suitable in indoor and outdoor situations, their example being stopping at a cafe and being sufficiently satisfied with the aesthetics of their base layer to wear it confidently once they removed their outer and mid layers. The general requirements for this included a moderately lower scooped neckline than is common in a base layer, whilst a skimming fit through the side panels which would accommodate varying bust sizes was preferred for all layers. Sleeve length was of equal importance, and although sleeveless styles were generally disliked a variety of preferences for sleeve length were displayed (some of which was seasonally dependent), but overall three-quarter length sleeves were preferred. Colour was of particular significance to the women in the user group, and they disliked the pale pastels and beiges which were reported by Church Gibson as commonly associated with older women's clothing (Church Gibson, 2000). They equally found the harsh bright colours often used for sportswear to be unsympathetic to ageing skin tones, and strong resistance was expressed to the vivid, contrasting colour blocking found in sportswear. Finally a colour palette encompassing strong but sophisticated coral, red and turquoise hues, through to subtler shades of lilac and silver, was proposed. Black and white were also popular, although not necessarily used together. Contrast trimming and piping in combinations of these preferred colours was considered a more acceptable alternative to colour blocking.

Initially male users found it hard to analyse their preferences, but their awareness developed as they became more involved and informed through the range of workshops introducing them to choices in fibre and clothing technology, fit and colour. Their requirements were considerably less prescriptive than the female requirements but included that the garment should allow them to move without restriction, feature a round or high V neck (although some preference was expressed for a zipped turtleneck), and be available in both short and long sleeved styles. A clear preference for a body fit that 'skims not clings' was displayed by both sexes.

Men's colour choices were darker than the women's, with a mid green and shades of grey being the most popular. A longer body length was important to the men, and they preferred either high V or round necks. Sleeve lengths differed from the women's in that short sleeves were popular for warmer weather, and full length for winter wear.

Co-design sessions produced life-size, 3D templates of design details such as neck depth, ventilation panel positions and colour shading that flattered the body shape, an important consideration to both sexes (Figure 20.2).

20.5.3 A bra size matrix for older female body shapes

During data mining of the SizeUK databases, only four women in the correct age group had been found that could be classed as 'underweight'; therefore this category could not formally be included in the study. However, as there was one female that

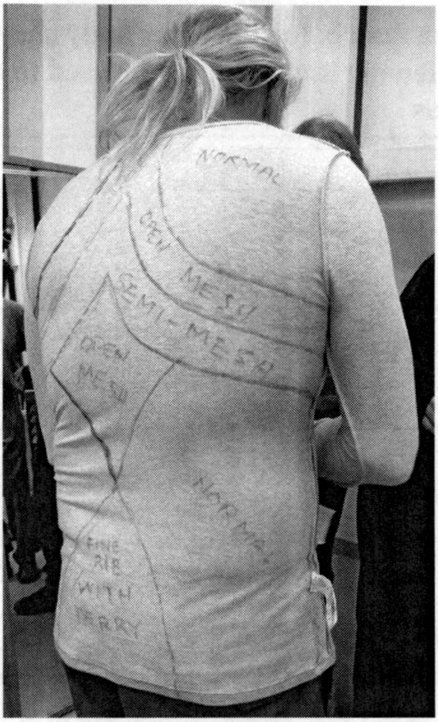

Figure 20.2 Co-design session for the male base layer. The required areas of compression (rib), stretch (normal) and ventilation (different meshes) have been mapped onto the older male body shape.

was in this category in the project's walking group, it was decided to include her in the scanning and design development process. Bougourd describes this in more detail her case study in Section 13.6.2. Having this category enabled population of the lower end of a bra sizing matrix and the design and development of sports bras for older users with thinner body shape, smaller busts and different control requirements from what had been found to be more the prevalent body shapes amongst the 60–75 year old female age group (Table 20.1).

Using extensive experience of bra design, in collaboration with the users, and taking body measurement data from the older aged female user group the project bra designer developed the bra size matrix for the older body shape. From this, four sizes ranging from 34AA to 40D, each needing a different level of either Low, Moderate, High or Maximum Impact support were selected for sampling and a family of bras that sit together visually and technically have been developed (Figures 20.3 and 20.4). As will be discussed further in Section 20.6.1, cut, bonded and sewn were the chosen methods of construction, but the choice of fabric, its positioning in the bra relative to the body area, and cut of the separate component pieces varied by body shape and support requirements.

Table 20.1 A bra size matrix for the older body shape was developed of the older aged female user group to be used to create four different sized bras

			28B	28C		
		30A	30B	30C	30D	30DD
	32AA	32A	32B	32C	32D	32DD
34AAA	**34AA**	34A	**34B**	34C	**34D**	34DD
36AAA	36AA	36A	36B	36C	36D	36DD
38AAA	38AA	38A	38B	38C	38D	
40AAA	40AA	40A	**40B**	40C		

	Low
	Moderate
	High
	Maximum

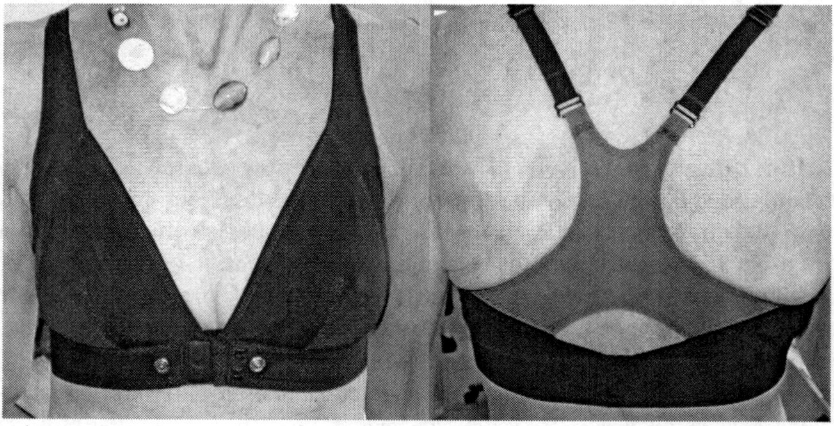

Figure 20.3 Low impact bra showing the low 'V' front cup shape, front fastening and racer back style preferred by the smaller body shaped user. This style allows freedom of shoulder movement as the fabric stretches and moves with the body whilst the straps remain in position.

20.5.4 Bra Design

Bra design has not been widely researched until recently, and none of these studies have been found to concentrate specifically on the older body shape. However, a common conclusion is that bra design is a process that above all requires expertise and

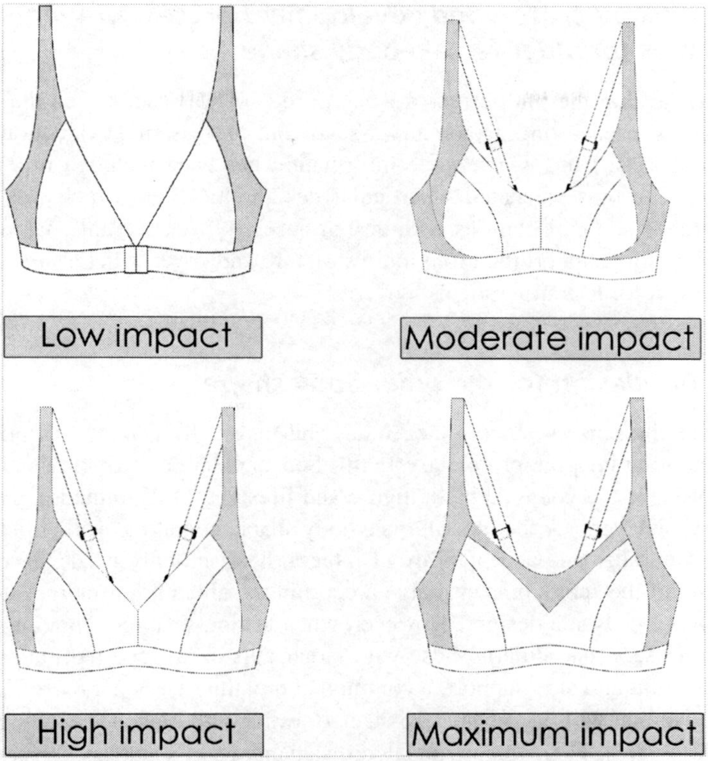

Figure 20.4 A family of four sizes of bra were created offering Low, Moderate, High and Maximum Impact support.

experience (Yu et al., 2006, p. 113; Hardaker & Fozzard, 1997, p. 311). There are numerous styles of bra on the market: push-up, balcony, minimiser, trainer, maternity, and, as Yu et al. (2006, p. 99) points out, all of these require distinctly differently cut patterns to fulfil their intended function. This complexity is described by Hardaker & Fozzard (1997, p. 313) who writes that 'the design process is unique for each bra', and as it is also critically dependent on fabric choice, 'it is the usual practice to design each new style of bra from first principles'. This process was indeed the case in the DAW project when designing styles of bras for the identified older figure types.

User workshops about bra choices and co-design sessions had resulted in a sports bra being the style preferred by the older users when walking. Within this category of bras there are two main types, pull-on ones that hold the breasts firmly against the body in order to reduce movement (compression), and ones that look more like a traditional bra (encapsulation), which provide extra support all around the breast through strategic use of fabrics in the panels. Yu et al, (2006, p. 172) observes that compression bras are designed for smaller breasted women in low impact use, whilst those with a C cup or larger have a better experience with the traditional encapsulated design. Both these styles were adapted for use for the older female users in the project.

20.5.5 Design/ prototyped development process of women's bras for older female body shape

In order to develop the bra designs a design protocol was established that included questionnaires, user testing, try-on and assessment, professional evaluation and user feedback sessions, from which some information has been included in the preceding section. The bras presented a particular design challenge, which was to match user preference and expectations with design necessity; for example, larger bra cup sizes require more support therefore more structure and cover, whereas the expressed preference was for less structure and cover.

20.5.6 Bra design for the older body shape

Bras with textile sensors, such as the Adidas NuMetrex® heart rate monitoring sports bra incorporating integral, proprietary Textro-Sensors®, are already on the market but are promoted for the young, athletic figure and lifestyle. DAW aimed to design bras as crucial base layers for the older female body shape, and at the same time to investigate the suitability and acceptability of integral heart monitoring devices for older users. Many of the sports bras available are seamless, and a brief internet search will return many models and designs. However, when testing seamless compression bras, a number of users (particularly those with larger cup sizes) reported the problem of insufficient front and side support, a common complaint that is paralleled in existing studies by women with larger bra cup sizes (Lawson and Mason in Chen, LaBat, & Bye, 2011, p. 697). This was obviously a disadvantage in seamless construction for older users, who had prioritised good support in preliminary design preference questionnaire responses. It was clear that in order to achieve the combination of stretch and support required in bras for older women with larger cup sizes, the bra would have to be designed along the lines of the encapsulated style, and contain several different fabrics with quite different properties.

In response to this a combination of cut, bonded and sewn methods were developed as this offered the most versatility for bra design for the older figure shape, enabling fabrics and fibres with different properties, such as firm non-stretch for support and strong stretch for close fit, to be engineered into neighbouring areas of the garment as required.

Responses to the questionnaire indicated that older users prioritised comfort, and wanted wide adjustable straps along with good support. However, they sought support without excessive structure or cover, and did not want their bra to define them as 'old', but equally it should not look too 'sporty'.

The female users expressed dissatisfaction with bra fastenings in general, and although alternatives were tested, the traditional two row hook and eye was preferred. Front fastenings were generally considered uncomfortable in combination with the lowered breast position common in older women, especially if the bra was not very supportive, although some preferences for zip front fastenings was expressed amongst the Moderate to Maximum Impact groups. Overall the back was the favoured position for fastenings, despite being reported by some as difficult to reach, particularly those with older, stiffer shoulders and the presence of a mild kyphotic curve in the upper back.

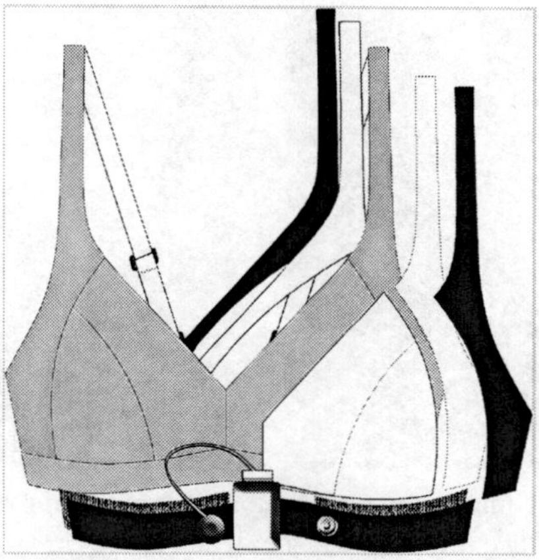

Figure 20.5 The bras are constructed of layers of bonded fabric to build up areas of support and stretch where required.

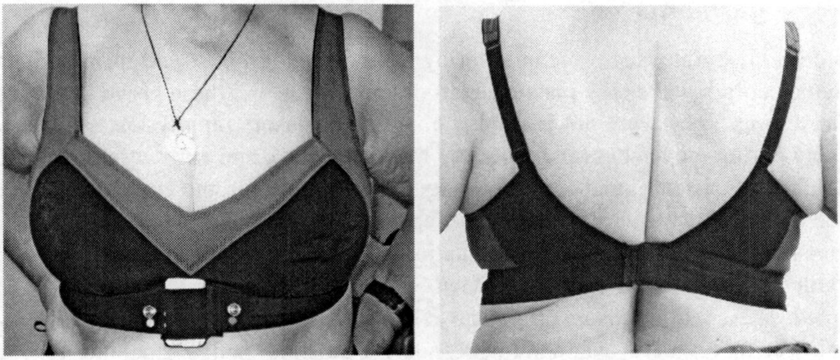

Figure 20.6 To provide additional support, the sides, neck edge and back wings of the High and Maximum impact bras were built up, and an inner rigid lining was included. The Maximum impact bra had additional neck-edge support incorporated to suppress vertical movement.

Colour preferences were quite surprising, as although the original range selected was quite broad, black and grey were chosen for the prototypes.

To provide additional support, the sides and back wings of the High and Maximum impact bras were built up, and an inner rigid lining was included, whilst the Maximum Impact bra had additional neck-edge support incorporated to suppress vertical movement (Figures 20.5 and 20.6). Strap design varied across the designs; the Low Impact bra was developed in line with younger styles with a 'racer back', which offers less support to the breasts but prevents the straps falling off the shoulder during activity, a common complaint reported by Chen (Chen et al., 2011, p. 699). The straps of the Maximum Impact bra were cut wider than those of the other designs to provide extra support, although all the designs incorporated wide straps as this was a clear preference amongst older users and, as can be seen in Figure 20.7, allowed for side support throughout the complete depth of the breast (Figure 20.7).

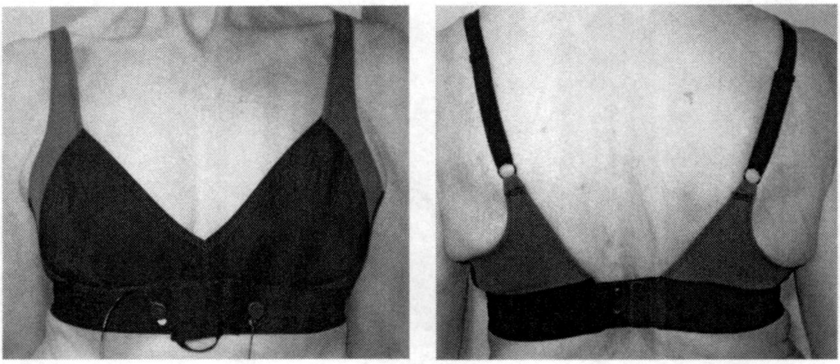

Figure 20.7 The Moderate impact bra allowed for side support throughout the complete depth of the breast while providing a low coverage garment.

20.5.7 Technical integration for body layer and base layer garments

In order to integrate electronic monitoring technology, early bra and male body later designs incorporated sticky plastic electrodes and wires, but these proved unsuccessful in testing. They were not portable enough, the sensors rapidly lost contact with the skin during movement (and therefore lost the signal), and wires strong enough to maintain a consistent signal were too bulky to be comfortable and could not be hidden inside the bra so were aesthetically unacceptable.

Eventually, after further experimentation, and in collaboration with Textronics®, e-textile sensors as used in their NuMetrex® range of bio-monitoring clothing were adopted. These fabric sensors offered an existing and proven sensor array which could be integrated into body layer and bra designs for the older body shape, thus to a large extent removing the barriers that had, until this point, compromised the aesthetic potential and wearer comfort of the earlier prototypes.

A major requirement of the technical design was to ensure that these textile sensors, which conduct the heart rate signal, could be held quite firmly in place around the body with optimum contact with the skin. However, as the ageing female body shape tends to have more and looser flesh than younger body shapes in the area where the bra band sits, it was important that the band should not be too tight, otherwise bulges of flesh form above and below the band. It was equally important that the band should sit at a comfortable level on the body and that the straps should not fall off the shoulders. Chen recommended that this and other aspects of improved bra design should be informed by postural information, which in this case was gathered through 3D body scans and verified by user feedback (Chen et al., 2011, p. 701).

Flat pattern blocks of a well-fitting bra, in a similar style to the proposed design and in the nearest size, were selected as a starting template for each bra. It is standard practise in industry to use this method of pattern progression for bras, rather than drafting fresh pattern blocks from measurements as is done with outer wear or base layers. Drafting a flat pattern directly from measurements would have required considerable adjustments before it was even a suitable starting point for such a sophisticated garment: therefore following this commercial practice circumvented lengthy and unnecessary stages.

Fitting is critical to bra design, as was recognised by Yu and Hardaker in their studies on bra design (Yu et al., 2006, p. 113; Hardaker & Fozzard, 1997, p. 325). Traditional measurements for determining bra size, the under bust and around bust measurement, do not provide the volume distribution differences found in all women of the same size. It is only during the fitting process that a bra pattern can be amended accurately to create the desired fit. It was clear during the fitting of the bras designed for the project that the necessary adjustments for the older body shape could only really be made at this stage, and would not have been possible with measurements alone.

Particular areas of contention were found at the underarm and around the under bust. All the female users had less flesh in the underarm area regardless of their size, which required excess fabric to be removed from the side wing panel. It was also evident that the bottom band, if made as tight as standard bras, was digging into fleshy areas, causing loose skin to be pulled on small body shapes, whilst rolls of flesh formed on larger body shapes. To alleviate these problems, the bottom bands of all bras were eventually, after testing, made at minus 12% of body measurement and the angle of the wing panel was increased more than on standard bras. Both of these adjustments had to be made with consideration for the position of the textile sensors on the body and their skin proximity. During bra development, it is often necessary to repeat the fit cycle several times; the more function and support is required, the more adjustments are required.

As the male chest shape is less complex than the female, there were fewer problems encountered in positioning the e-textile sensors. The male base layers, unlike the women's bras, did not require an opening; therefore it was necessary to ensure the garment was tight enough around the textile sensor area but could be comfortably pulled on and off over the head, and most importantly, without discomfort in the shoulders during this movement. After experimentation it was found that the same method used in the bras of a high modulus bonding film bonded to a band of fabric proved to have enough stretch and recovery to maintain close contact with the skin. Therefore these design elements were incorporated into a close fitting, sleeveless vest made of the same Eurojersey fabric as the woman's bra. In both garments the e-textile sensors were connected by conductive snaps to a slim 'Shimmer' wireless transmission device in order to relay the signal to the peripheral interface.

20.6 Manufacturing considerations: materials, methods and costs

Both seamless knitting and bonding, as discussed here, are valuable technologies for the older body shape and were found to be highly suitable for satisfying older user's requirements of both body and base layer. However, as with any new manufacturing technique, adopters will need to invest in new machinery and there will be additional costs of research and development of products to consider. Therefore it is important to recognise that the prices of garments manufactured by these new technologies will, at first, be higher than those made by traditional methods. However, despite this situation, the manufacturer of seamless knitting who collaborated to produce the commercial prototype garments expressed interest in developing this new market using the DAW project's co-design methodology.

20.6.1 Bra manufacturing considerations

Bras designed and produced for the DAW project utilised a combination of manufacturing techniques. These were carefully selected for both practical and aesthetic reasons based on informed user-centered, co-design methods. With the help of a highly experienced bra designer, the female users designed the styles based on their preferences. A two-dimensional method of bonding allover areas of fabric was used for particular sections of garments where firmness and figure support was required. This was also necessary in areas requiring strength with stretch such as the firm bottom bands of bras which housed the textile sensors. An additional three-dimensional method of adhering 12 mm strip-cut self-fabric to bonding film was utilized to close seams, which replaced traditionally sewn seam methods. This produced a low profile butt-join on the inner seams to make them appear almost invisible on the outside of the garment.

DAW project bras and male skin layers were manufactured from Nylon/Lycra Sensitive® fabric from Italian producer Eurojersey SpA. This light, soft, warp-knitted fabric has been used extensively for commercial bonded garments. Having a breathable construction, its high Lycra content with good shape retention and quick drying abilities makes it ideal for sportswear and underwear. As this fabric can be cut without fraying edges, it is ideally suited for the construction of garments using bonding technology.

20.6.2 Base layer manufacturing considerations

Fit was unanimously chosen as 'skimming' rather than clinging to the body. There was a cross-gender interest in the possibilities of strategic ventilation, and some female users favoured this on the upper chest area due to 'overheating' in this area.

These preferences were instrumental in the choice of Santoni seamless knitted garments for manufacturing the base layer garments, as this technology offers just such possibilities, in combination with a lightweight fabric. The two potential problems with this choice of production were firstly that seamless samples shown to the users were made in synthetic fibres, and secondly, although ostensibly the correct size these were found to be too tight fitting in what were described by users as the 'wrong places' (tummy and upper arms) for comfort. As seen in Section 20.5.5, resistance to synthetic fibres was reduced through a series of educational workshops, and 3D body scan data was applied when designing the final garments to reduce the effect of stretch yarns on the stomach. The arms were also cut wider to accommodate the larger bicep measurement often associated with the older body shape (Wells et al., 2007, p. 422).

Both male and female users had similar needs for cut and sew base layers. The users reported that when walking, any layering of garments above the base layer increased potential for base layer seams to irritate sensitive areas such as the underarm. Therefore seams around the underarm area were kept to a minimum. A raglan sleeve, in combination with a vertical panel up the side of the body which continued through the arm pit and along the underneath of the sleeve, was designed into the garment to address this issue. This detail successfully replaced and moved side and underarm seams from sensitive areas.

The final important requirement by both genders was that the back hem of the garment should fall lower than the front, dropping below the curve of the bottom.

20.7 Conclusion

Overall, base layers were preferred to be as seamless as possible for comfort, and once a fibre choice had been satisfactorily established, design considerations centred on fit and colour. For older female body shape, aesthetic concerns of cut focused on the neck shape and sleeve length, and the garment was required to fit smoothly over the bust area whilst adequate width was wanted lower down the torso to provide a looser fit across the stomach and hips. Colour was significantly important to female base layer design, and the garment was required to successfully act as activity and daywear.

For the male older body shape, the focus on a good fit was on the shoulder area, but also, as with the female version, there was a requirement to provide a loose fit across the abdomen. Men were more concerned with ventilation, but colour, although influential, did not carry the same importance for men as it did for women.

Designing the base layer prototypes, which were produced using seamless knitting technology, provided considerable opportunities for the older user group to satisfy their aesthetic requirements from a base layer, whilst the 3D body scans provided the necessary technical measurement data for the manufacturer.

The skin layer component of the layering system for men did not require as highly complex engineering as the women's bras, and construction methods were adapted from the female bra developments. This incorporated new bonding technology to join the seams and also to integrate the body monitoring e-textile sensors. It is clear that designing bras for older body shapes benefits significantly from knowledge of the 3D body shape, but it still requires a considerable amount of trying-on and adjustment to achieve the necessary levels of support and stretch in strategic positions. It was found in the prototype bras that apart from colour choice, the aesthetic of the design, although consciously accommodated where possible, was sometimes of necessity subordinate to technical design considerations. This was attributed to the considerable amount of 3D engineering required to fit, support and enhance the older body shape.

Co-design proved to be a valuable tool when working with older people, breaking down preconceptions on the side of both designer and wearer and providing insights into the requirements of ageing body shape.

20.8 Future trends

In terms of base layer design, developments are moving fast, and can be seen in the rapid deployment of cutting-edge innovations such as those developed for the 2012 Olympic athletes. Smart textiles such as the Team GB track cyclists Adipower muscle-warming 'hot pants' garments incorporating Fibretronic Heatwear FIR fabric heating panels (Anonymous, 2010, p. 19) will certainly become commercially available (Anonymous, 2012). These panels were considered by DAW users to be equally suitable for integration into base layers for older people to alleviate back strain when exercising. This smart textile technology, developed by Adidas, British Cycling and Loughborough University, also features a version of the seamless garment technologies found to be popular with older users.

Bra design is receiving attention, with lingerie companies such as the American-based Victoria's Secret investing in 3D body scanning technology to understand bust shape (Textile Clothing Technology Corporation, 2008). The two complementary technologies of 3D body scanning and computerised 3D pattern design are presenting new opportunities for experimental bra pattern design and construction techniques based on real breast shape. Yu (2006, p.104) reported that a number of companies, including Lectra, had developed mannequin-based 3D visualisation techniques for bra design. More recently, in combination with morphed avatars, CAD developments have enabled complex bra designs to be drawn onto the 3D avatar and then peeled off the body into 2D pattern form (Yu et al, 2006, p.106). TC^2 now include breast extraction software in their latest NX16 3D body scanner software suite, which enables exact measurement of the 3D breast shape (Figure 20.8). Thus true-to-life avatars can be created through a number of 3D body scanning technologies, 3D pattern cutting software is now available to develop bra patterns from real 3D body shape, and sophisticated

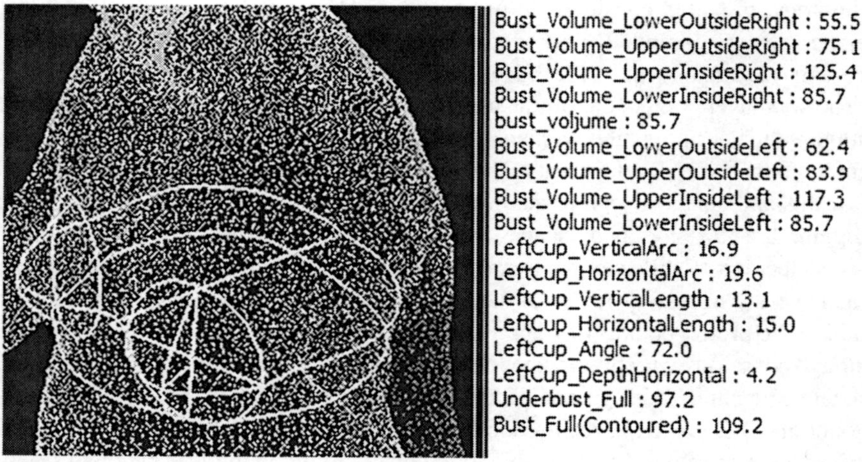

Figure 20.8 Image from TC2 KX16 3D body scanner software showing 3D breast extraction measurement lines.

Source: Image courtesy of TC2.

software offers the complete pre-production simulation of fitting, visualisation and colour and texture choices.

Older people do not consider themselves to be old. Sixty-year-old people of the early twenty-first century have purchased and worn fashionable clothing all their adult life and expect to continue to be able to continue to do so throughout their life. The 'grey market' spent over £12 billion on apparel in 2006 (Mintel, 2006); therefore, the opinions and requirements of older consumers are of considerable importance to the clothing industry. Design for older body shape requires careful negotiation between a pleasing aesthetic which embodies older people's perception of themselves, (rather than one imposed on them by a designer who may be 30 or 40 years their junior), and the possibilities presented by modern clothing manufacturing technologies.

The DAW project has found that there is opportunity for a change in older users' perceptions sports bras and sports clothing, and in the older consumer's understanding of the beneficial properties of modern fibres. It remains for the commercial sportswear manufacturing and retail businesses to embrace this growing demographic and cater to their wants and needs in this market. This project, the communication it has stimulated between leading sports clothing designers, manufacturers and older users, and the tangible knowledge base it has produced, will indeed lead to older consumers being able to purchase the functional and fashionable sportswear they so clearly wish to wear.

References

Abdessalem, S. B., Abdelkader, Y. B., Mokhtar, S., & Elmarzougui, S. (2009). Influence of elastane consumption on plated plain knitted fabric. *Journal of Engineering Fibers and Fabrics*, *4*, 30–35.

Anonymous. (2002). 100% cotton moisture management. *Journal of Textile and Apparel Technology and Management*, 2, 1–10.

Anonymous. (2004). *SizeUK – results from the UK national sizing survey*. London College of Fashion. http://www.fashion.arts.ac.uk/research/projects-collaborations/archive/size-uk/. Accessed 12.01.05.

Anonymous. (2010). *Knitting in weft or warp, Santoni Seamless wear, knits the way everywhere*. http://www.knittingindustry.com/articles/851. Accessed 30.04.12.

Anonymous. (2012). GB track cyclists wear muscle warming garments. *Knitting International*, *118*(1406), 19.

Ashdown, S., Choi, M. S., & Milke, E. (2008). Automated side-seam placement from 3D body scan data. *International Journal of Clothing Science and Technology*, *20*, 199–213.

Ashdown, S. P., & Na, H. (2008). Comparison of 3-D body scan data to quantify upper-body postural variation in older and younger women. *Clothing and Textiles Research Journal*, *26*, 292–307.

Bartley, L., & Warden, J. (1962). Clothing preference of women 65 and older. *Journal of Home Economics*, *54*, 716–717.

Branson, D. H., & Nam, J. (2007). Materials and sizing. In S. Ashdown (Ed.), *Sizing in clothing*. Cambridge: Woodhead.

Brownbridge, K. (2011). Complete fit. *Knitting International*, *117*(1389), 32–35.

Chen, C.-M., LaBat, K., & Bye, E. (2011). Bust prominence related to bra fit problems. *International Journal of Consumer Studies. 5*, 695–701.

Chowdhary, U. (1997). Clothing preferences of older consumers. *Perceptual and Motor Skills, 86*, 819–826.

Church Gibson, P. (2000). 'No one expects me anywhere': Invisible women, aging and the fashion industry. In S. Bruzzi, & P. Gibson (Eds.), *Fashion cultures: Theories, explorations and analysis*. London: Routledge.

Connell, L. J., Ulrich, P., & Ashdown, S. (2003). Body scan analysis for fit models based on body shape and posture analysis. In A. Knox, G. Hutton, & D. Woronka (Eds.), *NTC project: S01-AC27*. National Textile Centre.

Cotton Incorporated, http://www.cottoninc.com/product/Product-Technology/Moisture-Management/Wicking-Windows/Technology. Accessed 04.10.12.

Cutler, W. B., Friedmann, E., & Genovese-Stone, E. (1993). Prevalence of kyphosis in a healthy sample of pre- and postmenopausal women. *American Journal of Physical Medicine and Rehabilitation/Association of Academic Physiatrists, 72*, 219–225.

Daanen, H., & Hong, S.-A. (2008). Made-to-measure pattern development based on 3D whole body scans. *International Journal of Clothing Science and Technology, 20*, 15–25.

Fellingham, A. (2007). Seam-less knitting technology and the factors effecting final garment size. In *The 85th textile institute world conference. Colombo*.

Goldsberry, E., Shim, S., & Reich, N. (1996a). Women 55 years and older. II. Overall satisfaction and dissatisfaction with the fit of ready-to-wear. *Clothing and Textiles Research Journal, 14*, 121–132.

Goldsberry, E., Shim, S., & Reich, N. (1996b). Women 55 years and older. Part 1. Current body measurements as contrasted to the PS 42-70 data. *Clothing and Textiles Research Journal, 14*, 108–120.

Jacubčionienė, Ž., & Masteikatiė, V. (2010). Investigation of Textile Bonded Seams. *Material Science (Medžiagotyra), 16*, 1–4. 79.

Hardaker, C. H. M., & Fozzard, G. J. W. (1997). The bra design process - a study of professional practice. *International Journal of Clothing Science and Technology, 9*, 311–312.

Holmes, D. I. (2005). Giving effective stretch to polyester. *Knitting International, 112*, 43.

Horne, L., Campbell, L., & Scholz, C. (1999). Older females as a market segment for well-fitting clothing. *Journal of Fashion Marketing and Management, 3*, 236–244.

Hunter, B. (2012). *ISPO success for seamless specialist*. http://www.innovationintextiles.com/fibres-yarns-fabrics/ispo-success-for-seamless-specialist/ Accessed 17.02.2012.

Katzman, W. B., Wanek, L., Shepherd, J. A., & Sellmeyer, D. E. (2010). Age-related hyperkyphosis: Its causes, consequences, and management. *Journal of Orthopaedic and Sports Physical Therapy, 40*, 352–360.

Lee, Y.-A., Damhorst, M. L., Lee, M.-S., Kozar, J. M., & Martin, P. (2012). Older women's clothing fit and style concerns and their attitudes toward the use of 3D body scanning. *Clothing and Textiles Research Journal, 30*, 102–118.

Mason, B. R., Page, L., & Fallon, K. (1999). An analysis of movement and discomfort of the female breast during exercise and the effects of breast support in three cases. *Journal of Science and Medicine in Sport, 2*, 134–144.

Mintel. (2006). *Fashion for the Over-45s - UK-November*. London: Mintel.

Power, E. J., & Otieno, R. (2007). Study of anthropometric data in knitted garments. In *The 85th textile institute world conference. Sri Lanka*. Sewfree®, Bemis. http://www.bemisworldwide.com/products/sewfree/. Accessed 30.12.12.

Shipley, S., & Rosencranz, M. L. (1962). Older women's clothing preferences. *Journal of Home Economics, 54*, 854.
Stretch fibres and fabrics: reaching new levels of comfort.(2011). *Peformance Apparel Markets*. Textile Intelligence.
The Technical Center. (2012). *Bemis offers innovative film and tape, changing the way apparel is made and worn today*. http://www.thetechnicalcenter.com/Presentations/index.cfm?id=58. Accessed 30.11.12.
Tubbs, M. C., & Daniels, P. N. (Eds.), (1993). *Textile terms and definitions*. Manchester: The Textile Institute.
Textile Clothing Technology Corporation. (2008). *Size USA User Group Meeting. 2005, 2007, 2008*. http://www.tc2.com/newsletter/arc/071206.html#two. Accessed 15.11.2012.
Troynikov, O., & Wardiningsih, W. (2011). Moisture management properties of wool/polyester and wool/bamboo knitted fabrics for the sportswear base layer. *Textile Research Journal, 81*, 621–631.
Twomey, L. T., & Taylor, J. R. (1987). Age changes in lumbar vertebrae and intervertebral discs. *Clinical Orthopaedics and Related Research, 224*, 97–104.
Watkins, P. A. (2011). Designing with stretch fabrics. *Indian Journal of Fibre and Textile Research*, 366–379.
Wells, J. C. K., Treleaven, P., & Cole, T. J. (2007). BMI compared with 3-dimensional body shape: The UK National Sizing Survey. *American Journal of Clinical Nutrition, 85*, 419–425.
Wilson, A. (2012). Finest hour for flat knitting? *Knitting International*. Bradford: World Textile Publications Ltd, *118*, 22–23.
Woodson, E. M., & Horridge, P. E. (1990). Apparel sizing as it relates to women age sixty-five plus. *Clothing and Textiles Research Journal*.
Yu, W., Wang, J.-P., & Shin, K. (2006). Bra pattern technology. In W. Yu, J.-P. Wang, & K. Shin (Eds.), *Innovation and technology of women's intimate apparel* (pp. 92–113). Woodhead Publishing.

Further reading

Textiles and knitting

Fashion, http://www.lectra.com/en/industries/fashion/index.html. Accessed 10.11.12.
Innovation in Textiles: http://www.innovationintextiles.com.
Knitting Industry News: http://www.knittingindustry.com.
O'Mahony, M., & Braddock, S. E. (2002). *Sports tech*. New York: Thames & Hudson.

Accommodating figure differences when designing clothing

Ashdown, S. P., & O'Connell, E. K. (2006). Comparison of test protocols for judging the fit of mature women's apparel. *Clothing and Textiles Research Journal, 24*, 137–146.
Betzina, S. (2001). *Fast fit*. Newtown: Taunton Press.
Endgren, B., & Benkton, M. (2000). *Klädd: (Clothing for women with osteoporosis)*. Stockholm: Bilda (out of print).
Rasband, J., & Liechty, E. (2006). *Fabulous fit*. New York: Fairchild.

Social context and perceptions of clothing

Clayton, R., Lennon, S. J., & Larkin, J. (1987). Perceived fashionability of a garment as inferred from the age and body type of the wearer. *Home Economics Research Journal*, *15*, 237–246.

Jackson, H. O., & O'Neal, G. S. (1994). Dress and appearance responses to perceptions of aging. *Clothing and Textiles Research Journal*, *12*, 8–15.

LaBat, K. L., & DeLong, M. R. (1990). Body cathexis and satisfaction with fit of apparel. *Clothing and Textiles Research Journal*, *8*, 43–48.

Smathers, D. G., & Horridge, P. E. (1978). The effects of physical changes on clothing preferences of elderly women. *International Journal of Aging and Human Development*, *9*, 273–278.

Twigg, J. (2010). How does vogue negotiate age? Fashion, the body, and the older woman. *Fashion Theory: The Journal of Dress, Body and Culture*, *14*, 471–490.

Co-design development: design direction for the clothing layering system as a wearable technology platform

21

J. McCann
University of South Wales, Newport, UK

21.1 Introduction

This chapter offers direction for the design development of clothing as a functional layering system, with the integration of wearable technologies, for the active ageing. The performance sportswear market has been an early adopter of wearable electronics, in garments predominantly for athletic figure types and the youth market, with little consideration of styling for the changing physiology of the older body. Typically, wearable electronics are considered as medical devices for the old and ill while worn in relatively static postures in hospital environments. Active sportswear has become prevalent in the high street, with casual dressing increasingly adopted across all age groups. In both the fashion and outdoor sports sectors, current clothing range development is driven predominantly by last year's sales and trend forecasting. To date, there has been little information on active ageing buying habits and virtually no trend information to inform product development to address the style requirements of the rapidly growing ageing demographic. Clear design direction is needed to raise the profile and improve the image of active ageing, and to guide product development for this valuable and relatively predictable emerging market.

Recent cross-disciplinary collaborative research that begins to address the design needs and aspirations of the growing active ageing demographic is used as a case study. 'Design for Ageing Well: improving the quality of life for the ageing population using a technology enabled garment system' has been carried out within the RCUK-funded New Dynamics of Ageing programme (NDA). The aim of this practical research has been to bring smart clothing and wearable technologies to this neglected community (60–75 years) through the development of a sports-type clothing 'layering system' to encourage engagement in healthy exercise with a focus on walking. The layering system comprises a close-fitting base layer, a midlayer insulation layer and outer protective wear, with wearable electronics and their technology user interfaces integrated within the system. Co-design methodology (new to clothing development) has been adopted to bring together a team of older research participants with academic researchers, independent experts and industry stakeholders. The iterative design development of working prototypes has been informed through identification

Textile-led Design for the Active Ageing Population. http://dx.doi.org/10.1016/B978-0-85709-538-1.00021-3

and verification of the real life needs and aspirations of older users. The merging of cross-disciplinary language and methods for the elicitation of user needs is explained in Chapters 12 and 13. In this chapter, the author follows the co-design led practical prototype development phase through to the specification of a smart clothing layering system as a wearable technology platform.

A collaborative design-led approach has enabled the explanation of complex terminologies and practices across the range of disciplines, embracing technologists in electronics, clothing and textiles, and gerontology. An older user reference group contributed to technical and aesthetic design specifications with regard to body scanning to capture size and shape, smart textile selection, garment styling, placement of wearable technologies within the system and the usability of the wearable technology interface. This project has demanded a breadth of design competencies that have widened the scope of clothing design, in embracing visual communication, graphic and web design and specialist skills from product design and architecture. The evolution of a new shared language has enabled communication between academic and industry stakeholders and older research participants involved in the iterative co-design development process. Many of the collaborators within the case study have elaborated on the practical detail of the co-design development process as contributors to this publication (Greengrass, Bougourd, Stevens, Taylor, Timmins, Price and Lewis, Haffenden and Smith, and Burns). The resulting cross-disciplinary process has informed new practice in the technical design specification of the highly complex integration of wearable electronics in garment assembly.

21.2 Creating a hierarchy of emerging key design requirements

21.2.1 Resume: initial co-design phase: layering system workshops

The initial collaborative research phase involved all stakeholders in the identification of the needs, wants and aspirations of the active ageing in terms of functional outdoor clothing, for walking, with the potential for enhanced functionality through the application of wearable technologies. The sharing of plain language was necessary for communication across the disparate cultures of the academic disciplines, within the work packages, with industry experts (from clothing and electronics) and, most importantly, with older end-users. New developments in outdoor clothing, with innovations in smart textile attributes and wearable electronics, and the associated terminology, were introduced in a sequence of workshops, as described by the author in Chapter 12 (Timmins & McCann) and Chapter 13 (McCann). This co-design engagement was successful largely through the input of designers acting as facilitators, observers, translators and interpreters to enhance understanding between academic and industry contributors and to enable older participants to have the confidence to express their views. The semi-structured process involved a 'hands-on' review of current products, working systematically from base layer to mid- and outer -layers, including legwear, in separate sessions, with either direct industry stakeholder involvement or products and catalogues sent to the events.

21.2.2 Gathering user feedback

A wealth of user design requirements came from the co-design engagement as the process gradually uncovered user attitudes and preferences, as well as their perceived limitations, both with clothing and the usability of wearable technologies. Data were gathered by a variety of methods as follows:

- Video recordings – of stakeholder co-design workshops with audio transcripts
- Still photography – to record the key design details of user interest or concern
- 'Hands-on' user selection – to indicate user preferences for materials and garments
- Comparison of garments – with opportunity to try on/give critical comment on styles
- Observation by designers – while acting as translators, interpreters, visualizers
- Report sheets – to capture user written feedback on presented clothing products
- Collages of images and drawings with users – to communicate user style preferences
- Observation in workshops by technology team – to capture user needs in relation to functionality
- In-workshop visual questionnaire regarding preferences for positioning of technology
- Still photography against a measurement grid – to record posture regarding movement
- Dialogue between experts and users – to note technology usability/limitations
- Flip chart recording from resume sessions – to capture key design-related issues
- Written feedback on project from users – both requested and volunteered
- Comparison of smartphones – to identify user preferences and/or limitations
- Colour selection workshop – supported by trade explanation of timelines and minimums
- Development of collage 'design boards' to represent range plan overview

21.2.3 Triangulation: interpreting the design requirements

The breadth of findings gathered from the stakeholder engagement in the co-design workshops and related events was to be analysed with design requirements prioritised under key design issues. These issues were to be addressed during the practical design development process. The interpretation of both verbal and visual feedback required the overview of experienced designers. The way forward for analysis was based on triangulation (as described in Figure 21.1) between:

- Findings gathered directly from user views regarding clothing and technology needs
- Tacit knowledge and experience of designers and other academic team members
- Observation of user engagement with all stakeholders by specialist designers and technologists

21.2.4 Tacit understanding

Throughout the co-design process the tacit understanding and experience of expert stakeholders ensured a balanced discussion of design considerations with recognition of essential technical issues to be addressed in the subsequent practical prototype

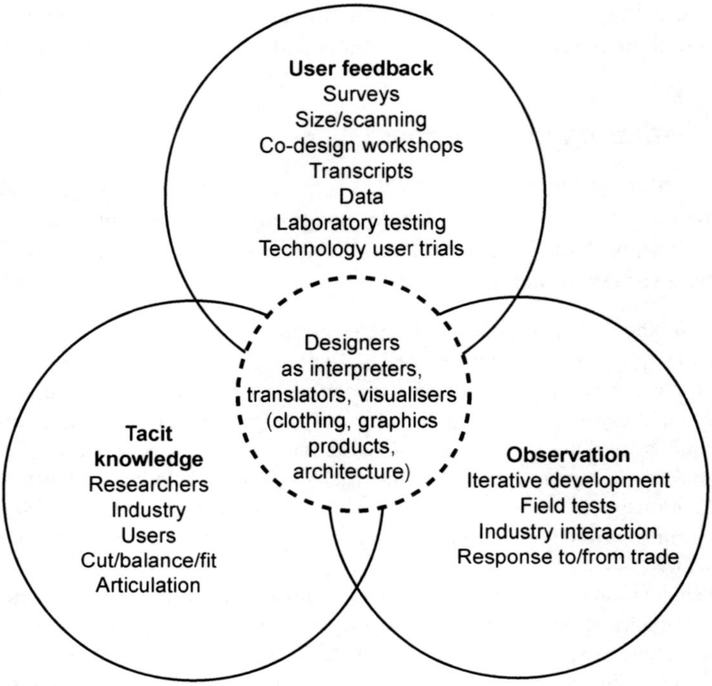

Figure 21.1 Triangulation. Interpreting the design requirements.
Source: Mike Timmins.

development phase. It became evident that unless designers anticipated the breadth of information needed to inform the three-dimensional development phase, there could be gaps in capturing the necessary data. Designers would either prompt discussion on a key topic or, in expert observation at a glance, make notes as a reminder of key technical adjustments to be made, often without making comments in order not to disturb the flow. It became clear that, following the design-related workshops, and in analysing the transcripts of the audio discussions, higher level executive design decisions would be necessary between the design experts, including expert industry representation, in order to progress prototype development. In recognition of the realistic constraints of the trade, the technical design team began to prioritise the end-user design and technology requirements. A range plan constituting an adaptable clothing system for varying levels of participation in walking was drawn up based on an overview of a breadth of user feedback to be realized into prototype garments during the co-design development phase.

21.2.5 Observation

A wealth of information may be captured intuitively at a glance by the expert eye to inform the three-dimensional product development process. The observational understanding required to inform the design of the functional clothing layering system that must address a range of movement and extreme postures is seldom considered when designing for the erect figure predominant in catwalk fashion. Within the Design for Ageing Well (DfAW) case study, the co-design process depended on the input of a multi-disciplinary team of designers from product, graphic, architecture, intimate apparel and user interface design, led by a fashion designer (with expert experience in performance clothing design). A product designer, with an understanding of the shaping of a variety of materials with associated technical language, and a graphic designer, expert in communication skills, contributed different perspectives in developing new design strategies to address the emerging, hybrid area of 'smart wearables'. The interdisciplinary design team observed the interaction with older research participants as they tried on, assessed and commented on the design and usability of the variety of garments and wearable technologies provided by industry stakeholders. Designers made notes and sketches, and took photographs of key aspects of interest, whether purely observed or with additional user comment. For example, users might not notice problems with garment balance or restricted arm-lift while discussing a closure or pocket detail. These problems could be noted and addressed, with or without comment, by the design team while the technology team captured feedback on specific aspects of functionality and usability.

21.2.6 Categorising the key requirements

The author in previous work proposes her methodology for the identification of end-user needs for the design of performance sportswear. Her original hierarchical tree of generic design requirements provides a starting point for the categorisation of requirements for the active ageing. The original tree is presented in the author's thesis: *Establishing the requirements for the design development of performance sportswear* (McCann, 1999) and further adapted in *End-user based design of innovative smart clothing* (2009). Some topics may be under more than one heading. For example, the colour of clothing is an important consideration from the perspective of *Aesthetics*, influenced by fashion or mood, but may also be listed as a sub-heading under *Culture*, associated with an activity steeped in tradition, or be specified under the *Demands of the Activity*, where colour is a safety requirement (Figure 21.2).

An amended Table 21.1 represents an overview of co-design requirements, for walking, gathered from older user feedback and observation during co-design engagement, and elaborated in response to the contribution of academic and industry stakeholders. Key generic issues remain categorised under *Form*, the aesthetic and cultural needs of the active ageing end-users balanced with *Function*, the needs of the ageing body and the demands of the activity (walking). In addition, findings to do with *Commercial Realities* were gathered resulting from dialogue between older research participants and industrial partners or through user feedback on their retail experiences.

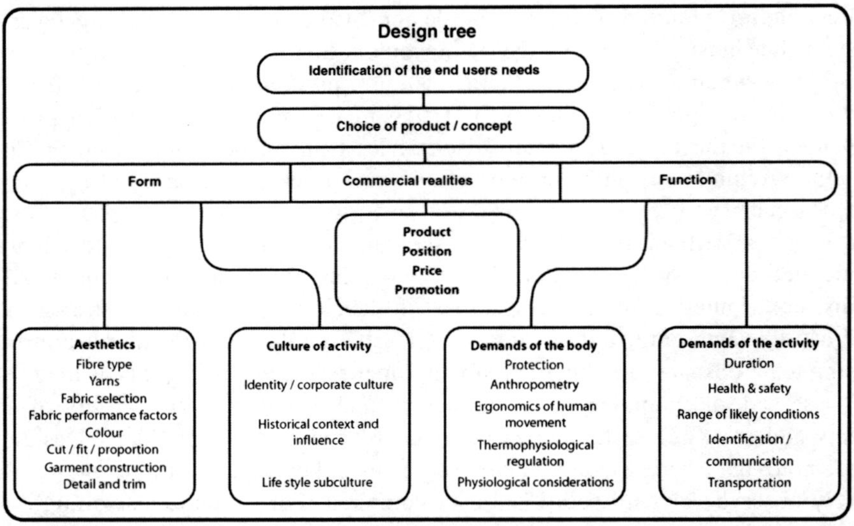

Figure 21.2 Identification of design requirements (McCann, 1999).

Table 21.1 Categorising older user design requirements: Design for Ageing Well, using the model from Figure 21.2

	Aesthetics	Culture of the activity	Commercial realities	Demands of the body	Demands of the activity
Choice of Product	Colour Fit and proportion Style lines Fabrication Design detail Non-transient style	Baby boomer generation Image of ageing Peers: e.g.: walking group or club Friends and family Multi functional or casual dressing	New consumer Majority Casual dressing Product Selection Branding Retail experience	Health and wellness Size / shape Movement Moisture manage-ment Thermal regulation Senses	Walking types or categories Participation levels Location Environment Terrain Weather with seasonal consider-ations
Technology	Positioning of electronic devices and user-inter-faces	Commu-nication and social engagement	Point of sale technology	Vital Signs monitor-ing and warming	Position-ing and distance covered

21.3 Sorting and elaborating the design requirements: form

21.3.1 Form: aesthetic requirements

21.3.1.1 Aesthetic requirements: colour

Colour, in combination with shape and fit, was found to be a key motivation in garment selection within the female members of the active ageing user group. Color is ubiquitous and represents what is first seen at a distance and is what many users are attracted to when they first enter the retail environment. 'Clothing choice can affect the way we feel; it gives a degree of personal identity; it can influence self image, and it also provides us with a sense of comfort and security' (Schlesinger, 2009). It was not long ago that clothes for older people were generally found to be conservative, plain and drab. However, the 1960s style revolution has had some lasting effect on baby boomer clothing choice. The ageing consumer does not want to be restricted to black, which continues to be popular in fashion because it is said to make people appear thinner as well as stylish and timeless. As Fay Schlesinger states, 'Designers will have us believe that it's the colour of the season and you must not be seen in anything else. But it can also bring out dark circles under the eyes and lines. The effect can be as serious as making women feel drained, self-conscious and introverted' (2010). The active ageing research participants confirmed that they enjoy wearing colours that are characterful and that flatter their ageing skin tones and changing hair colour. They do not want relentless, transient, swings in trend-driven colour. If they do happen to find an attractive, enduring style, women admit to often buying two or more garments and, if appropriate, in several colours.

21.3.1.2 Aesthetic requirements: fit

Garment fit was found to be another major consideration by older wearers in product selection, alongside colour choice. This is confirmed in recent survey carried out on attitudes to women's fashion with findings reported in *Clothes: A Manifesto*, where women state; 'Above all we would like our clothes to fit us. We want more petite and tall sizes; we want half sizes, in clothes and in shoes; we want legs and sleeves in several lengths' and, ideally in some clothing sectors, 'We want in-house alteration' (Willis, 2013). The chosen fit models within the case study represented BMI average measurements for the age group ranging from small to extra large men and women. Users complained that current clothing ranges often look very similar and with size and fit that is unpredictable. There was unanimous confusion around the topic of the commercial sizing of garments, with little standardisation or synchronisation of sizes between different brands and even between different styles within one brand. The lack of clarity is both in terms of measurements and in the nomenclature of sizes, exacerbated by the degrees of ease (tolerance) added to body measurements (that vary in the size charts from brand to brand) according the style and mood of the range. For example, the fit of casual wear may allow for more layers to be worn under a jacket in comparison with the more tailored and articulated fit of lightweight protective garments designed for extreme end use. In addition, there is the concept of 'vanity sizing' adopted by some brands; the practice of using relatively smaller size codes on garments to make customers feel better about themselves and become more inclined to buy. The lack of clear standards with respect to sizing, both within and between different countries, is an ongoing challenge.

21.3.1.3 Aesthetic requirements: style

Fashion styling is normally subject to seasonal changes in silhouette and proportion, as dictated by trend forecasting, often with a focus on the youth market. However, 'As the years go by having a "personal style" becomes more and more important. In fact this is one area where older women have an advantage over younger women… because older women have a full sense of who they are and they can dress accordingly' (Lindemann, 2014). In the recent survey findings reported in *Clothes: A Manifesto*, Willis states that 'style trumped fashion overwhelmingly. That's because style is personal, and – crucially – something that can't be bought. No one admits to striving to look "fashionable"; they would prefer to look "well-dressed", "considered", "confident", but above all "stylish"' (Willis, 2013). These comments were reflected in the case study where the active ageing were found to be more concerned with subtle, non-transient, age-appropriate style rather than fashion with design lines to flatter the changing body as opposed to following dramatic swings in fashion. Some users valued multifunctional styling, suitable for a range of everyday life activities, to encompass both countryside and urban environments, in garments cut with sufficient ease to accommodate varied inner layers. Those involved in more extreme hill walking required greater protection in more streamlined garments with co-ordinating functional inner layers.

Style lines were determined to be sympathetic to the full range of representative older user figure types (from small to extra large based on average sizeUK BMI measurements). Women were more precise in their requirements than men; however, both suggested subtle shoulder emphasis, such as softened saddle-shoulder lines, to take attention away from their less-defined ageing waistlines. In general, there was agreement for style lines that would profile or 'skim' the body, being not too tight or too 'baggy'. There was a preference for softer organic shapes, as opposed to angular seaming, with fluid contours to accommodate the ergonomics of movement such as arm-lift and elbow and knee articulation. They requested optional fit at waistlines through full or partial draw cords in garments of 'a decent length' (covering the seat) for the natural outdoors and with longer lines for urban walking. Side panels were proposed as flattering to both men and women to accommodate varied shape requirements (from bust/chest to waist to hip). Side panels would enable the elimination of side seams, and any associated friction in walking. Once style lines were determined, the patterns for the various layers would be refined to ensure that garment would move in tandem and without friction between the layers.

21.3.1.4 Aesthetic requirements: fabrication

This area of design is fibre and fabric led. The active ageing generation has seen the advent of modern fibres and fabrics during their lifetime. However, co-design engagement has revealed that the positive attributes of these materials were not initially well understood by this community. Some have had a suspicion of synthetics, since the early poor image and handle of Bri Nylon and Crimplene. They have retained the misconception that natural fibres, such as conventional cotton and wool, are considered 'good', while synthetic fibres are 'bad'. Co-design workshops enabled an introduction to the positive attributes of modern fibres and fabrics with the potential to enhance overall comfort within the layering system. There was little prior user awareness of the bad environmental impact of traditional cotton growing and processing, other than that related to organic cotton, or of the current wider debates with regard to ethical practices within the textile and clothing industries. In particular, the co-design stakeholders were made aware of initiatives within the outdoor trade for the introduction of more sustainable practices throughout the textile and garment chain, highlighted by the commitment of the European Outdoor Group's (EOG) Sustainability Working Group.

Since the end of the 1950s, the attributes of elastomeric fibres have revolutionised clothing design. Stretch fibres are key to the technical and aesthetic design and comfort of 'underwear' and intimate apparel incorporated in warp and weft knitted structures, laces and nets, woven constructions and narrow fabrics. Stretch is of particular relevance to garment design for older figure types. A varied percentage of elastomeric content as well as the direction of stretch may enhance comfort and fit in providing engineered areas of support or increased movement that enables the wearer to put on and take off the garment with greater ease. Stretch also contributes directly to garment fit and shape retention. The embedding of textile sensors and wearable electronics, currently found in close-fitting base layer garments and bras for athletic figure types, relies on stretch materials to position and hold the technologies in the designated locations within garments. Most recently protective spacer knit fabrics have been developed by Baltex (UK) with two-way stretch and YKK has recently launched two versions of a coil zip with vertical or horizontal stretch tape (2013).

Midlayer insulation fibres and materials provide thermal regulation in a variety of constructions to trap still air. Midlayers may be made up of more than one garment with thermal regulation involving the addition or removal of insulating layers and through adjusting ventilation by means of closures. Typically, the midlayer is made up of knit structure fleeces (brushed pile fabrics), and fibre pile (fake fur) constructions and of lightweight tightly woven fabrics incorporating synthetic down that may be varied in thickness and often quilted. Older people, who are susceptible to cold, may also benefit from adaptations of existing and emerging wearable technologies initially adopted in performance sportswear, such as heated fibres or flexible panels (such as products from exo2 and Fibretronic HEATwear) prevalent in clothing for extreme conditions such as motorcycling.

'Soft shell' materials incorporate the protective elements of outer shells combined with midlayer comfort and insulation. Soft shell fabric technology is normally a laminated assembly of two fabrics with a membrane in the middle. Typically the layers comprise a combination of polyester or nylon woven face fabrics backed by lightweight to high loft knit constructions with an inner membrane intended to allow air flow through the fabric but that prevents water from penetrating from the outside.

These fabrics are water repellent rather than waterproof and do not require seam taping. They are softer and less noisy than more protective 'hard shell' fabrics, often with a degree of stretch to enhance comfort and fit. The 'sandwich' constructions present a smooth, relatively stable, surface enabling bonded zip insertion and seam joining. As extreme protection is not a daily requirement, soft shell fabrics offer attractive, adaptable and comfortable protection for everyday use.

Outer layer fabrics provide protection for the clothing system microclimate by adopting variations on woven or knit structure protective textile assemblies with additional properties and finishes, such as abrasion resistance for more extreme requirements (McCann, 2009). Fabric innovation has enabled outer layer garments to become increasingly lightweight, fitted and stylish, with enhanced comfort, through moisture management, in hydrophilic or microporous waterproof 'breathable' membranes with stretch properties. The concept of 'Body Mapping' is adopted for the seam-free engineering of textile stitch structures, involving the selection of areas or 'zones' with appropriate fibre/yarn selection, in relation to comfort factors such as breathability, protection, ease of movement and articulation. When materials are selected with predominantly synthetic content, cut and sew manufacturing may adopt the heat bonding of seams, zip insertions and laminated panels to revolutionise appearance and reduce bulk.

As the co-design process progressed, with industrial partners presenting fabric innovations to the research participants, with trial demonstrator garments, the participants became increasingly enthusiastic about synthetic and man-made fibre types. The comfort-enhancing properties of modern textiles became readily accepted especially in materials, and fibre blends, with a 'natural' handle. Supportive materials for body layer garments were selected from Eurojersey's 'Sensitive' range. Base layer qualities of interest included fibre content from Tencel (Lenzing), DriRelease, (Optimer), Lycra and Coolmax® (Invista) and woollen developments for sport from different suppliers. Midlayer insulation included fleece constructions, sourced by industrial partners, engineered quilted wovens, with phase-change properties from PDA, and down-proof woven Pertex with synthetic filling from Primaloft (Japan). Soft shell assemblies include Energear (Schoeller) and thermo-regulating fleece from Aclimatise (APT) while outer layer lightweight fabric technologies included constructions with microporous laminates, such as Gore-Tex (W.L. Gore), two layer assemblies such as Dual-Protection (Keela).

In terms of textile-based wearable electronics, knitted sensors were supplied by Adidas (Textronics) for the body layer garments, as sports bras for women and vests for men, to connect a Shimmer wireless device for vital signs monitoring. Flexible heated panels for midlayer gilets, and potentially jackets, were supplied by Fibretronic and exo2. Both soft shells and outer protective garments incorporated textile-based cabling and soft switch controls to connect and operate the smartphone, which acted as the hub, for location and communications. Thus the breadth of interest, generated through co-design, involving stakeholders form industrial fibre, fabric and trims suppliers, and with the input of electronics providers, has enabled practical research and development to move nearer to market. The fibres or fabrics selected provide generic examples of properties that may constitute a comfortable layering system for the active ageing as a platform for the ongoing rapid development of wearable electronics.

21.3.1.5 Aesthetic requirements: design detail

Users indicated preferences for subtle discreet design detail to preserve clean lines that offer the opportunity to co-ordinate differing layering combinations in catering for individual taste. Women requested adaptable collar design that could be worn up, over the chin for warmth, or lie open to reveal inner garment detail or individual accessories. Both male and female walkers requested a varied selection of pockets including those big enough to carry maps and/or essential contents from a handbag (including a lipstick!) and with security pockets for valuables and car keys. Emphasis was put on ease of fastenings, especially in outdoor conditions, with single direction zips on jackets (easier access than double ended). The effective storing of bulky hoods in collars was considered a challenge in terms of reduced strength and dexterity as well as the choice of zip pulls, especially when wearing gloves in the outdoors. The use of Velcro was debated, by comparison with poppers on storm flaps and pockets, normally being considered noisy and damaging to textiles. In one context, however, on trouser or gilet security pockets, the noise of a Velcro fastening may occasionally be considered an asset as a security alert. Consideration of accessories included trouser hemlines cut wide enough to cover boots and adjustable hoods to accommodate hats, with general agreement on wired brims for enhanced visibility. The research participants were impressed with the slick appearance of new manufacturing methods as yet not commonplace in everyday clothing for the older market, with the use of heat bonding for zips and seaming (possible on fabrics with synthetic content) and seam-free knit construction for base layer garments.

21.3.1.6 Aesthetic requirements: wearable technology

Practical design issues involved in specifying the positioning of the electronic devices and user interfaces within the clothing layering system are a highly complex task still to be resolved on an industrial scale. The topic is introduced by the author in her paper delivered at the Textile Institute World Conference (McCann, 2012) and elaborated in greater detail by the designers, Price and Lewis, who have contributed to the 'hands-on' development of working prototypes within the DfAW case study. Their unique solutions to the specification of the integration of the electronic architecture within the garment layering system is described in this publication (see Chapter 22).

21.3.2 Form: cultural/social requirements

21.3.2.1 Cultural/social requirements: generation

The research participants, as the baby boomer generation, represent the active ageing community that has enjoyed being stylish since their teenage years in the 'swinging 60s'. Now in their 'Third Age', they are able to capture opportunities open to them after their main period of working life is concluded. Co-design engagement identified that the active ageing no longer want relentless, transient, swings in trend-driven fashion and colour or to have to resort to wearing black. In theory, they have disposable income but are dissatisfied with what is currently available in the retail environment.

In 'show and tell' sessions users sometimes referred to an individual garment as 'an old friend'. They would like to be able to replace favourite items of clothing and, if they do happen to find an attractive and enduring style, some admit to buying more than one garment and, possibly, in more than one colour.

21.3.2.2 Cultural/social requirements: casual dressing

'Not all women as they get older relish the prospect of dressing in baggy cardigans, comfort waistbands and fluffy slippers. It's true, though, that comfort does become more of a consideration' (Willis, 2013). Willis continues, 'What is surprising, despite all we hear about the West's ageing baby-boomer population bulge, is how poorly served older women still feel by the fashion industry'(Willis, 2013). Within the case study, male walkers had been more content to make do with relatively heavy traditional outerwear such as a tweed hunting jacket and an army poncho. During the lifetime of baby boomers, clothing for walking has developed from army surplus 'gear' to wax jackets and, more recently, lighter weight synthetics as innovations in fibres, materials, coatings and laminates have escalated. Many older wearers have now adopted sportswear in their everyday lives as sportswear has crossed over into street wear and the population, as a whole, has become increasingly casual in dress. Some use performance-oriented sports outerwear but without understanding the principle of the clothing layering system. Some are dressing younger, possibly because technical outdoor clothing is more available in styling for younger people but trans-generational style is limited in size and fit. The profusion of products and technologies related to the layering system is confusing to older consumers who often resort to compromise and bargain prices. Many older research participants wanted adaptable clothing acceptable for socialising 'in the pub'.

21.3.2.3 Cultural/social requirements: image of ageing

A German government study states that 'it is worth taking a new look at old age. The term "old age" should be replaced by "ageing" (Office of the German reports on the Elderly, 2010). In developing and designing products and their advertising and marketing, the consumer industry should give great attention to the development of individual needs over the entire life course. In designing technical products a "universal design" (i.e. product conceived in such a way as to meet the needs of all age groups) is a sound basis for lifelong brand loyalty'. However, clothing design, in respect of the needs of the changing older body, has to be more age specific. 'Images of ageing have a great influence on both the realization of development potentials in old age and on dealing with the restrictions of old age. The future of old age thus depends to a large degree on images of ageing. Yet the images of ageing that dominate our society often do not do justice to the diversity of old age' (Office of the German reports on the Elderly, 2010). Images of ageing differ widely according to their social context, with a whole repertoire of images competing with each other for dominance. 'The media obviously play an important role in images of the ageing…although older people are underrepresented in television films and advertising. In public debate, demographic change is often associated with negative attributes and depicted as a threat' (Office of the German reports on the Elderly, 2010). Some of the most commonly held negative stereotypes about the elderly include perceptions that the elderly are all alike, elderly people are isolated and lonely, that old means being physically disabled, they do not enjoy life, someone has to take care of them, they are a burden, and they have stopped learning new things (http://library.thinkquest.org/10120/cyber/extended/image.html. Accessed 07.12.12).

However, older people today are said to be relatively younger in attitude and more adventurous and more confident than previous generation, with the older research participants suggesting that more positive descriptors are adopted with regard to the ageing such as experienced, knowledgeable, informed and expert. Designers and marketers need to beware of age discrimination and be aware of mental age (perception) and physical age (chronological age). Chronological ageing (the number of years a person has lived) may be distinguished from 'social ageing' (cultural age-expectations of how people should act as they grow older) and 'biological ageing' (an organism's physical state as it ages). Ageing in humans refers to a multidimensional process of physical, psychological, and social change. Some dimensions of ageing grow and expand over time, while others decline. Reaction time, for example, may slow with age, while knowledge of world events and wisdom may expand. (http://en.wikipedia.org/wiki/Ageing. Accessed 07.12.12).

On the whole, ageing seems to be a positive experience for the active ageing, who do not think of themselves as old. Those who feel younger than their age are normally healthier than those who desire to be younger and who are relatively less healthy. However, changes to the body that often happen imperceptibly have to do with posture, fit and proportion. Some discussion on the image of ageing is taking place as designers and journalists join the club! 'The fashion industry has been prone to portraying "bone thin models with long white hair who looked as if they'd been starving on a desert island for two years" while Dame Vivienne Westwood has criticized the rise of "disposable fashion" and warned it makes us all look the same and that only 70-year-old women display any sense of style these days' (Whitehorn, 2012). Magazines discuss ageing but normally to do with skin care. Ageing has been the focus of the cover of French Vogue (November 2012). However, as people move into midlife and beyond, looking good to others, at least in the sense of making a favourable impression, begins to be less important. 'For the first time ever, most adults are in the years when the forces of self-actualisation needs exert decisive influences on lifestyle aspirations, buying decisions and overall consumer behaviour (Wolfe & Snyder, 2003).

From the perspective of the outdoor clothing sector, the active ageing community embraces early pioneers as icons, responsible for the introduction of authentic sportswear brands such as the sports practitioners Yvon Chouinard, Patagonia, (USA) and Ron Hill, Musto and Douglas Gill (UK). Bonnington's ascent of Annapurna (1970) coincided with the launch of colour television, raising awareness and improving the image of outdoor pursuits. This industry has been driven by fibre and textile innovation, often encouraged by practitioners, who have led the development of lighter weight, relatively breathable clothing that has enabled greater speed and agility. Many of the current sportswear technical developments began in the 1970s with the launch of microporous coatings (Gore-Tex) and polyester fleece pile fabrics (Calami and Malden), while nylon was relaunched as Tactel in the early 1980s.

21.3.2.4 Cultural/social requirements: communication and social engagement

The functionality desired by older users in terms of the selection and usability of the wearable electronic devices and their user interfaces within the garment layering system is discussed in this publication by Price and Lewis.

21.3.3 Function

21.3.3.1 Needs of the body: body size and shape

Physical capabilities are influenced by the size and shape of the body. From mid-adulthood, people begin to lose height, see Chapter 8 for more details. Other bodily dimensions also change with advancing age; feet become broader, waistlines thicken as the ratio of body fat to muscle changes. Physical strength is generally related to body size, with women physically weaker than men at all ages, and with weaker muscles and a poorer power-to-weight ratio. Their disadvantage is even greater on weight-bearing activities such as walking and stair climbing (Huppert, 2003). The findings of the SizeUK survey state that the average size of the female population, which was size 12 in 1952, had increased to size 15 in 2005. However, within the range of mannequins developed for brands, through mining data from this survey, the mature fit model is normally designated as 35-plus.

In the case study discussed, the user group was selected as walkers (in the 60–75 year age group) within a 30-mile radius of the lead university. The male and female participants were asked to provide their height and weight to calculate their BMI, from which 100 were chosen to be body scanned with the intention to select men and women as size models (S/M/L/XL) as described by Bougourd in chapter 10. To inform subsequent prototype design development, basic garment blocks were generated from the scans and made up as toiles (calico prototypes) to be fitted on the selected models. Blocks developed represented the starting point for the development of base, mid, soft shell and outer layer garments to be further amended, in terms of sleeve pitch and articulated cut to enhance movement, through the layering system. Parallel work was carried out by an expert intimate apparel designer in the development of user-specific body layers: as sports-type bras, designed and cut for individual levels of support, and as close-fitting vests for men, as discussed by Smith and Haffenden in chapter 20. As a generic comment, users requested a choice of garment lengths to better cater for the variation in height within this age group and in body length, sleeve and hemlines, and the need for more inclusive sizing to extend from extra small to extra large. In general, the demand was for fewer and less transitory styles to be available in a wider size range. Women, in particular, requested that waists be designed with adjustability, for changing older figure types, but not 'baggy', possibly through partial elastication. It was observed that garments must be cut with corresponding balance, sleeve pitch and degree of articulation throughout the layering system to enable movement and prevent friction.

21.3.2.2 Needs of the body: movement

Particular aspects of the body that decline with age, and which are important for functioning, include mobility, dexterity and the ability to reach and stretch. Flexibility and adequate range of movement are essential for many activities, with loss of range, especially in the shoulder joint, common in old age. Arthritis may cause swelling and pain in the joints and limited movement and weakness in the arms and hands, resulting in poor dexterity and restricted flexibility with dressing and undressing and difficulty in getting a firm grip or in making precise finger movements (Metz & Underwood, 2005). Double actions such as push and twist are particularly difficult for people with poor dexterity. Balance requires a complex integration of sensory inputs followed speedily by precise motor reactions. To restore equilibrium of the body mass, to avoid falling after tripping, requires rapid powerful movements, and these decline with age, resulting in a greater tendency to fall (Huppert, 2003).

Users were interested to learn of the more technical attributes of performance clothing in catering for extended 'arm-lift', elbow articulation and sophisticated hood design, in ranges often exclusive to extreme sports practitioners, due to limited size and shape, colours and inappropriate branding. Through co-design, the benefit of 'cutting in the round', to address the ergonomics of movement, was discussed in relation to the range of predominant postures associated with the activity of walking, as well as catering for the seated figure. Walkers require ease of movement to bend and stretch in adjusting footwear and putting on and taking off various layers of clothing. Arm lift would also reduce restriction in movement when holding onto the rail in public transport or putting luggage on a rack. Articulated cutting may be further enhanced through the placement, or 'body mapping', of varied fabric constructions and the use of the direction of stretch of a knit, the use of bias, or any quality with elastomeric content. The physiological needs of the end-user, and the likely range of environments, must be understood in order to interpret these needs into cutting lines related to the placement of appropriate materials around the body. Ease of movement is required in tandem with other textile attributes such as moisture management, thermal regulation and protection.

21.3.3.3 Needs of the body: moisture management

Older people will be at risk in cold environments if a high moisture level is absorbed within the garment base layer and the temperature drops. For enhanced moisture management, there are obvious advantages in the application of high moisture wicking, quick drying textiles. Base layers are normally of knitted construction with varied structures such as meshes and denier gradient areas, sometimes engineered to be seam-free, placed around the body to aid the transport of moisture. As perspiration increases within the clothing microenvironment, there is a reduction in the capacity of evaporation to remove heat energy, and this means that most of the sweat from the body becomes trapped within the fibres and textiles. Clothes saturated by sweat can affect the thermal characteristics and influence further heat transfer rates. Innovative 'wicking' fibre and fabric constructions may transport moisture away aided by the provision of ventilation within the garment system.

21.3.3.4 Needs of the body: thermal regulation

People need to regulate their body temperature in order to function well. With advancing age our ability to thermoregulate tends to decrease. The inability of older people to reduce heat loss during cold exposure is associated partly with alterations in the insulating properties of body fat. Better education is needed for the ageing 'in the possible advantages of high tech clothing materials' and their use, and 'manufacturers should consider adjusting their marketing policies to include the elderly in their target groups' (Havenith, 2001). Material innovations include insulating, moisture transporting, waterproof/breathable, phase change and smart properties, all with a direct relationship to adjusting to climate. Textile insulation to weight ratio has been improved using hollow and thinner fibres. Synthetic fibres (in down and other structures) are preferable to natural fibres where moisture absorption is usually high. Sweating can accumulate in natural materials during exercise, leading to 'afterchill' as the moisture evaporates. The build-up of moisture content may be reduced by wicking base layer materials, non-absorbent midlayers, and with the increased breathability of waterproof protective fabrics. Phase-change materials, when worn close to the body, may act as a buffer in absorbing or liberating heat but with limited overall effect.

21.3.3.5 Needs of the body: senses

Physiological changes occur in all our sense organs as we age, reducing our sensitivity to incoming information' (Huppert, 2003). The skin as the source of most of our tactile sensations declines in sensitivity with age, the same as other sensory receptors (Wolfe & Snyder, 2003). Touch is a combination of the stimulation of three types of receptors on the skin: pressure, pain and heat/cold. As ageing skin receptors die off, hands become less sensitive to pressure although older people are subject to less tolerance of extremes in temperature. Points touching the skin need to be further apart for an older person to recognize two sources of pressure rather than one, leading to a reduced ability to differentiate between shapes and textures (Huppert, 2003). In co-design development, the technology team liaised with users on the design of the technology interface, in terms of visibility, while the clothing team designed hoods with adjustments to ensure visibility. Garment detail was specified in response to the challenges of reduced dexterity for closures and for the appropriate positioning of fabric switching control buttons for the technology. The size and spacing of the soft keypad controls (from Fibretronic) was a compromise in this prototype phase that could be rectified in commercial development.

21.3.3.6 Needs of the body: vital signs monitoring/warming

Normal ranges of measurements of vital signs are liable to change with age and medical conditions. As arteries begin to lose a little of their elasticity, blood pressure may start to rise slightly. The aim of the research has been to introduce textile-based vital signs monitoring into the garment layering system that is acceptable, in terms of comfort and usability, to active older wearers who do not perceive themselves as ill. Initial co-design discussion revealed that users were unaware of textile-based wearable monitoring devices and unsure of the relevance. The technology providers that had previously introduced vital signs monitoring into both clothing for extreme sports and within products to measure fitness have normally catered to sports practitioners with athletic figure types. The challenge was to find the optimal positioning for textile-based sensors on a range of both male and female figure types within the 60–75 age range, in sizes representing small to extra large, with reference to BMI average data coming from the SizeUK survey. The development of the body layer garments and specification for the integration of the technology are discussed in detail by Haffenden and Smith in relation to garment design.

21.3.4 Function: demands of walking

21.3.4.1 Demands of walking: healthy exercise

Physical exercise can improve physical performance as well as aspects of intellectual performance in later life. Physical deterioration is often due to reduced levels of physical activity that may be halted and reversed through exercise. 'Older people have the greatest need to maintain their exercise levels, and those with some disease related impairment may have the most to gain'(Metz & Underwood, 2005). 'There is growing evidence that keeping fit and supple in old age is a good way to protect the body from a number of medical problems such as high blood pressure, heart disease and osteoporosis, as well as helping to prevent falls and broken bones' (Metz & Underwood, 2005). 'Surveys find that walking is by far the predominant activity reported in surveys of older people, half the men in their 60s and 40 per cent of women reporting this' (Metz & Underwood, 2005). This can range from more extreme hill walking to more moderate exercise. The motivation for active sport and healthy exercise will be increased if, at the same time, the activity provides joint or individual pleasurable experiences, an enjoyment of nature, a sense of adventure, education and entertainment as well as increased fitness and the prevention of ill health rather than enforced rehabilitation. There is the suggestion that the older people are taking more active holidays than the younger age groups, who seem more content with the beach. Older people want to experience the world for themselves, spending the money they have accumulated during a lifetime of work, before it is too late to enjoy it (Metz & Underwood, 2005, p. 107).

21.3.4.2 Demands of walking: categorising participation

Clothing is affected by varying levels of participation, with considerations that include seasonal weather conditions and environmental conditions, whether gentle or hostile and predictable or unknown, as well as the duration and distance to be covered. Designers need to appreciate the purpose of the walk, the level of social engagement and how serious the walkers are, possibly driven by peer group pressure or with friends and family wishing to stop for a picnic and enjoy the view. The DfAW case study, through co-design engagement, has identified four walking categories, which are discussed in more detail by the author in chapters 12 and 13. Different personas have been created to represent the different walking types although, in everyday life, some of the active ageing community may engage in several or all of the walking types. The walking types embrace differing degrees and levels of participation, as follows:

Type I. Day-to-day mobility, walking to shops, visit friends and relations
Type 2. Externally time controlled: work, doctor's appointment, school run, first stage of travel, pet walking, other
Type 3. Woods walking, beach walking, urban sightseeing, walking on holiday
Type 4. Hiking, fell walking, mountain walking, power walking

(This list is based on the original concept map of walking characteristics, (Timmins, 2013)).

21.3.4.3 Demands of walking: positioning, location and distance covered

Feedback from a pilot study on user technology requirements, and co-design engagement, informed both the development of the technology devices and the user interface, and influenced the design integration of the electronic components (Table 21.2).

Table 21.2 Design for Ageing Well pilot study: user feedback on their perceived technology requirements

Keep me warm	88%
Cool me	71%
Help me navigate	55%
Tell me the distance covered already	55%
Keep me in contact with the group	52%
Call for help if I should fall	52%
Protect me if I should fall	50%
Tell me where to find the nearest facilities	38%
Interact with my mobile phone	36%
Tell me where my group members are	26%
Tell me where to find the nearest bus stop	24%
Monitor my health for medical reasons	17%
Tell me if I over-exercise	5%
Interact with my portable music device	3%

Source: University of Westminster (WP1).

21.3.5 Commercial realities

21.3.5.1 Commercial realities: the New Consumer Majority

'Some great brands have come about by companies treating their customers as people, not as data sets' (Wolfe & Snyder, 2003, p. 31). The 'New Consumer Majority' emerged in 1989 as adults in their 40s and older became the adult majority for the first time in history. These baby boomers will add about 2 million extra older people to those aged 70 and over by the 2030s. This is the only adult market with realistic prospects for significant sales growth in dozens of product lines for thousands of companies. 'The marketer who understands them is more likely to connect most deeply with the more than 123 million people that make up the New Consumer Majority', in the USA, (Wolfe & Snyder, 2003, p. 16). This community is mostly 'experiencing ebbing of materialistic influences and rising influence of experiential aspirations on their behaviour' and 'consumer research and marketing have failed to realize the relationship between changes in the leading values, views, and behaviors of the marketplace and the New Customer Majority'. It seems that 'thus far marketing remains rooted primarily in the materialistic values that generally hold the most sway over people in the pre-middle-age years of adulthood. Because of this, many members of the new Customer Majority feel marginalised by companies and their marketers' (Wolfe & Snyder, 2003, p. 15).

The active ageing want less 'stuff' as they seek more balance in reprioritising their lives to become more self-reliant. They tend to be more concerned with value and the usability of the product (fitness for purpose) so consequently do not usually figure among the early adopters of products unless there is considerable advantage to them. It follows that new products are not designed initially for older people: they are expected to follow in the early or late majority, when the technology is becoming more mature and the selling points no longer stress the novelty of technology, rather other aspects such as ease of use (Metz & Underwood, 2005, p. 102). Clothing, however, cannot be easily customised in the way that, for example, a font can be altered on a mobile phone. Developments related to the mass customisation of clothing, in terms of the link between body scanning and pattern development, have begun but commercialisation within the global industry is still a long way off. However, 'as the centre of gravity of society shifts away from the young, and the new old become included, the prospects of the continued dominance of the youth culture are beginning to look rather uncertain. Marketing orientated towards the young may therefore be based on increasingly shaky foundations' (Metz & Underwood, 2005, p. 32). It may be that, with the concept of a 'new dawn' or the 'homecoming' of textiles and clothing production, industry leaders might consider this rapidly growing ageing market.

21.3.5.2 Commercial realities: product selection

The research participants made it clear that they would welcome fewer styles in a better fitting product, with a greater choice of lengths, in colours and textures sympathetic to the age group. As already discussed, research participants are confused with regard to garment sizing and the associated nomenclature and admitted that past assumptions with regard to the credibility of commercial sizing had now altered. Willis states that 'it's common for women to have at least three different sizes in their cupboards; one of our respondents pointed out that she 'owns and wear garments ranging from size 10 to XL' (Willis, 2013). As mentioned, if women do happen to find an attractive, enduring style that fits, they may buy two or more garments, often in several colours.

In terms of product selection, male feedback indicated that decisions are made on colour, and 'what am I going to wear it for', prior to checking sizing and fit for comfort and ease of movement. Some female participants go in to a retail outlet, look around and come out again if nothing 'catches the eye' while others go in and immediately consult the advice of a sales assistant, if they are sure of their intended end use. Most stated that 'if spending a lot, and you have found what you want, and are convinced, you will buy in the store despite the potential for a cheaper price elsewhere. If there is a problem you can go back to get it right!' The active ageing want to look good, and may be smart, and be comfortable, with accessories to suit the individual wearer. The importance of performance features will vary depending on the walking type, with some users participating in range of different walking types as described in more detail in the two previous chapters.

21.3.5.3 Commercial realities: branding

Strong feelings in feedback from the older research, within the case study participants, indicated that there should be *no* branding on garments! This community had grown up with the garment maker's label on the inside of the garment. They resent promoting stitched, embossed and printed brand logos, however discreet, on the exterior of garments. The message was clear: No external logos at all for women and, from choice, no logos for many men and, for others, if more tolerant, any branding must be very discreet. A brand stakeholder asked about tonal branding and this was dismissed. The active ageing would wish the brand label to be on the inside, and on the point of sale (POS) material. They consider that the style and 'look' of the product should provide credible and distinctive brand recognition. Their attitude is: 'Why pay for a garment to advertise the brand?' Women, in particular, resent branding as it prevents multifunctional garment end use, especially in the crossover of functional wear into everyday lifestyle, as the bold image of logos conflicts with the wearing of jewellery and other, sometimes, subtle accessories.

21.3.5.4 Commercial realities: retail experience: clothing

Research participant feedback indicated that the public does not always know what is available in the outdoor trade and its products. To enable older wearers to be more confident in product selection, and promote uptake, clear information, in plain language, is needed on the concept and attributes of the layering system in order to avoid suspicion of the marketing hype of brands. Even if older consumers have the confidence to go into the specialist retail environment, POS labels have print that is too small, with confusing technical information promoted with a plethora of complex diagrams and icons. Older consumers are suspicious of the claims of brands and the profusion of marketing hype. What type of shop should they go to: specialist, high street, warehouse store or department store? Are the garments displayed with hanger appeal, a particular problem for sports apparel, especially base layers such as Santoni knits? Is the sales staff trained to help older customers and do the retail outlets have suitable changing facilities? Customers must have the confidence to try on garments in relative comfort. Another frustration, highlighted by Willis in quoting survey respondents, is consumer resentment in being fed seasonal ranges that are 'way out of kilter with the weather' with the question, 'Why do shop assistants look so wounded when you ask for jerseys when it is snowing and they have their summer range installed already?' (Willis, 2013).

In co-design engagement it was found that the active ageing, who had become more aware of the attributes of the functional layering system through the co-design process, had gradually changed their attitudes as they had become more conversant with the design development process and the specialist terminology involved. Because of their continuing confusion with different claims of the brands, they had developed strong views about needing clear label indicators, possibly with a colour code, or 'traffic light' system, in terms of a measuring scale for degrees of functionality and related benefits: What are the technical features of a product and, for example, will it keep you dry? When making a purchase in the retail environment, they were clear that, if they had compared available products, with a helpful sales assistant, and made a decision, they would invest slightly more than going online to search for a lower price.

21.3.5.5 Commercial realities: retail experience: technology

When the purchase of wearable technology was discussed, older consumers said that they would expect to go to a specialist outdoor store. Taylor discusses the complexity of POS material for outdoor clothing even prior to the addition of technologies (Taylor, 2011). As older people tend to be more concerned with value and fitness for purpose they are not normally the early adopters of a product unless there is considerable advantage to them. 'New products are not designed initially for older people: they are expected to follow in the early or late majority, when the technology is becoming more mature and the selling points no longer stress the novelty of technology, rather other aspects such as ease of use' (Metz & Underwood, 2005, p. 2005). The question is, 'Whether demographic developments will change this in any way, as the relative numbers of young people fall, thereby changing the characteristics of the early adopter market. Does this mean that new products will have to move faster up the usability curve on introduction to attract the larger numbers of older purchases earlier in the product life cycle?' (Metz & Underwood, 2005, p. 102).

21.4 Co-design prototype design development process

21.4.1 Confirming the range plan in outline

The initial phase of the co-design process, in capturing the design requirements of the research participants, culminated in a co-design workshop to propose and verify an outline garment range plan (July 2011). Garment types were selected to accommodate the varied needs of the user group related to the identified range of levels of participation in the different types of walking, and contrasting outdoor environments. Design requirements had been identified and elaborated by means of iterative co-design workshops, through direct user feedback and in user observation. In a studio setting, users contributed to generating collages of images, sketches and notes as reminders of specific design details. Their needs and aspirations were interpreted through the knowledge and experience of designers, both from the academic community and from industry. Key design requirements were attributed to the range of garment types, of main interest to users to constitute the layering system. The plan was drawn up in outline by the design team based on the group interpretation of the user feedback. The graphic designer and an architect compiled the collages of sketches and photographs, with older participants' notes, captured within the co-design workshops, to illustrate quite specific user needs and wants.

The co-design process then moved from the requirements capture phase, which had embraced the wider team of participants, into the practical development phase to work predominantly with the selected fit models (originally four men and four women) in the translation of the outcomes of the co-design findings. The lead clothing designer created initial rough sketches that were converted into basic computer line drawings

Table 21.3 **Range plan style list**

Style No	Men	Style No	Women
M1	Body layer – vest	F1	Body layer – sports bra
M2	Santoni (knit) base layer	F2	Santoni (knit) base layer
M3	Short sleeve base layer	F3	Short sleeve base layer
M4	Long sleeve base layer	F4	Long sleeve base layer
M5	Long sleeve check shirt *	F5	Camisole base layer
M6	Gilet: (cargo pockets) *	F6	Gilet (optional heat panel)
M7	Fleece midlayer	F7	Fleece midlayer
M8	Soft shell jacket	F8	Soft shell jacket
M9	Outer shell hooded jacket	F9	Outer shell hooded jacket
M10	Long outer layer coat	F10	Long outer layer coat
M11	Cargo trouser	F11	Soft shell trouser
M12	Waterproof over trouser	F12	Waterproof trouser (stretch)
		F13	(Cropped trouser)*
		F14	(Skirt)*

*Designed but not developed into prototypes.

by the product designer. The garment categories, selected by both male and female users, are listed in Table 21.3. The co-design process progressed to initial toiles (calico trial garments) and the subsequent development of working prototypes. Consultation with the wider group was ongoing at key stages in the iterative prototype development.

21.4.2 Design competencies required within the clothing development process

The academic team embraced experts with a range of competencies in size and shape, pattern cutting for movement, knitwear for larger figure types and intimate apparel design. In addition, these clothing-oriented designers were supported by a product designer and an architect, skilled in the development of innovative working drawings and specifications. The team had external support in the computer aided design (CAD) plotting and storage of pattern development, essential to effective garment specification (Coleg-sir-gar, Camarthen College). The co-design development process involved the male and female size and fit models, in regular fittings. Ongoing consultation was maintained across the wider user group to verify the development stages. Co-design involvement with industry stakeholders provided input at key stages to support materials sourcing and selection and the provision of access to a range of specialist manufacturing processes. The practical iteration of garments, constituting the clothing layering system, benefited from the involvement of a leading UK outdoor brand (with production in China), a UK functional clothing manufacturer, with additional offshore production, and a regional small to medium sized enterprise (SME) with an interest in the integration of electronics into performance clothing. An overview of the prototype development process is outlined in Table 21.4.

Table 21.4 Co-design prototype design development process

Design development process
Range plan:
Outline range plan
Photographic collages to illustrate concepts
Initial sketches based on interpretation of user feedback
Computer record
Design lines:
Defining silhouette and style lines across the garment categories
Fit/proportion/pitch in relationship to needs of older body
Design detail: closures, collars (convertible), hoods, pockets, etc.
Materials selection:
Selection across clothing layers – directly related to garment categories
Collaboration with industry stakeholders re sourcing/availability
Technology specification:
Initial user choice re electronic functionality
Design of the technology user interface
Placement of the technology components
Refining detail of range plan:
With verification by user group
In liaison with industry 'partners' re feasibility
Technical 3D development: (designers key to technical translation and development)
Refining the pattern blocks: body/base/mid/outer/legwear, etc.
Developing design lines in the round
Regular co-design development with older size models
Positioning of garment features and design detail
Fittings to confirm/amend co-design decisions
Refining and recording final patterns
Specialist bra development (expert bra designer in liaison with size models)
Santoni seam-free knit development (co-design workshop)
Garment specification (collaboration between product designer and architect)
Recording of 3D pattern development in 2D (CAD)
Development of a generic specification sheet
Gender, sample size, garment reference number, etc.
Material selection and trims (Source, and references re quality, weight, etc.)
Flat garment drawings: F/B and side view etc. where necessary
Flat drawings re garment detail + photographic/visual references
Development of 'exploded' drawings to show garment detail and assembly
Electronic integration: Placement/positioning of technology
(Liaison between design and technology academics and industry)
Developing clear strategies to describe integration of electronic architecture
Methodology for placement of components, interfaces and powering devices
Incorporation of the powering devices
Provision of images of toile/prototype development as relevant
Testing and verifying clarity of new processes with industry collaborators

Continued

Table 21.4 **Continued**

Industrial collaboration in Prototype development: (re textile and garment sourcing)
Body Layer: male vest/female sports bra
Santana seam-free knit
Base Layer
Mid Insulation layers
Soft shell
Outer layers
Legwear
Colour selection:
Working with users to translating workshop findings into colour 'card'
Naming and referencing selected colours
Suggestion of colour levels across the layering system

CAD, computer aided design.

21.4.3 Design lines

The silhouette and style lines were defined across the garment categories, informed through user feedback on aspects of fit and proportion, with observed garment balance and sleeve pitch, consistent throughout the various layers. Organic, rounded design lines were determined through user feedback, with gently shaped side panels throughout the layers, to best accommodate a range of figure types in garments to 'skim' the body. Initial design detail was specified with regard to identified preferences for collar types (convertible), hoods (articulated), garment hems (shaped), pockets and closures (easy access). A seam-free knit workshop, based on Santoni technology, involved hands-on user engagement (Figure 21.3).

21.4.4 Materials selection

Fabric sourcing, throughout the garment layers, was based on co-design collaboration with all stakeholders guided by industry collaborators with regard to the suitability of fabric qualities for certain processes and their realistic availability. Final fabric selection was made in direct relationship to the identified garment categories chosen to constitute the co-ordinated layering system. The process involved technical and smart textiles selection for the placement, or 'body-mapping', of materials around the body, to address shape and fit, cutting for movement, and garment detail throughout the layers. Materials were selected to work in suitable combinations, co-ordinating in terms of cut, direction of stretch and surface friction to address the overall physiological demands of maintaining a comfortable system microclimate. The aesthetic handle and feel of the fabrics was of prime importance but preferred colour choices could not be satisfied due to the commercial constraints of materials sourcing. It was agreed that neutral tones, with greys and blacks for prototype development, provide the widest choice of material qualities. A separate co-design colour workshop was planned at a point when users could visualise the completion of the prototypes.

Figure 21.3 Original concept flat sketches.

21.4.5 Technology specification

The functionality of the electronic devices was determined through user questionnaires, co-design training sessions and practical workshops. The choice of the wearable electronics components was from products available or emerging with the potential to be adapted and suitable for older wearers. Participants, in the context of walking, were interested in communicating with their peers, friends and family, and in positioning and location and distance covered. In addition, female users were interested in garment warming. The potential for fitness monitoring and safety was less important, but, when users saw that the textile-based electrodes could be comfortable, they were willing to experiment with the technology. Burns discusses the placement of the electronic components and architecture in greater detail in chapter 25. Co-design development sessions involved the positioning of the devices, and their switches and powering components, in initial prototype fittings with the user fit models, to determine the positioning of the necessary electronic architecture within the garment assembly. The integration of the electronic 'architecture' was further amended, as necessary in hands-on development, with the academic designers and an industry partner, in the SME factory environment. Co-design engagement with the wider group of research participants involved verifying the design and usability of the technology user interface.

21.5 Technical 3D development

21.5.1 Refining the pattern blocks

The starting point for garment development was the adaptation of the pattern blocks that had been prepared from the body scanning of the selected size and fit models. As these blocks were based on erect postures, the author adopted her tried and tested design-led approach to cutting for movement for performance sports end use (McCann, Morsky, & Dong, 2009). In co-design development with older size models, she systematically adapted the basic pattern blocks into articulated shapes, with suitable sleeve pitch, and underarm-lift, and elbow and knee articulation, to accommodate ease of movement fit for walking. Blocks were adapted for Body, Base, Mid, Outer layers and Legwear, to constitute a finely tuned layering system. Toiles were made to test the co-ordination of fit and appropriate degrees of tolerance for ease of movement between the layers. The Body layer garments for men and women were developed by a specialist intimate apparel designer as discussed in this publication (Haffenden and Smith, see chapter 20).

21.5.2 Developing design lines in the round

The blocks developed already had shoulder seams and inset sleeves heads diverted into a rounded saddle sleeve cut, with the sleeve of a suitable pitch to enable a range of arm movements. The ergonomic placement of style lines through inner and outer elbow points created sleeve articulation while side panels provided flattering body shaping. The rounded lines of the saddle shoulder and side panels were echoed throughout the styles, in adaptations for all the layers, for both men and women. Base layer style lines were initiated on a mannequin in consultation with all users and then adapted in fittings with the models. Style lines were verified in fittings to ensure their suitability for use across all the sizes. The carefully honed style patterns, to fit the individual fit models, were plotted onto computer for safekeeping prior to further style specific detail being added. These patterns did not represent commercial sizing.

21.5.3 Positioning of garment features and design detail

Garment fittings took place with the user fit models and the interdisciplinary design team to confirm garment design detail. Calico toiles provided a basis on which to mark on and amend details such as inner and outer pocket positions, storm flaps, waist adjustments, collar heights, hood volume, cuff details, knee reinforcement and the like. Garment layers were cut to be compatible in terms of styling and ease to avoid impeding movement with lengths and proportions verified across the layering system prior to detail being transferred to flat patterns. In outerwear garments, the ergonomic design lines would also be used to accommodate the integration of the textile-based electronic architecture, and the interconnections between devices and their controls and powering devices. The final patterns were stored in CAD format.

21.5.4 Body layer development

The development of the close-fitting body layer styles to accommodate the positioning of vital signs monitoring devices demanded regular co-design engagement with the size and fit models, in close liaison between the intimate apparel designer, the size and fit specialist and representatives of the technology academic team. This is discussed by the specialist intimate apparel designer, Smith in chapter 20.

21.5.5 Santoni seam-free knit development

The co-design development of seam-free knit base layers took place in three phases. Seam-free knit was the focus of an early workshop where users were shown a variety of products. Different examples of body mapping were discussed with reference to a range of fibre types, and blends, and a breadth of stitch structures. The samples were of varying degrees of fit, dependent on the fibre type, and predominantly designed for athletic figure types. Lenzing's Tencel, in a wool blend, was of particular interest to the older participants, because of the looser knit construction and natural handle see Figure 21.4. The male and female participants, who represented size and fit models, were asked to put on slightly fitted, long sleeved, pale gray 'T' shirts while other users worked on mannequins. Users collaborated in small groups, with advice available from the design team, to draw their desired design lines directly onto the 'T' shirts. Notes confirmed selected stitch constructions for the areas as designated around the body. Some of the users also supplied flat drawings. The next phase involved three designers working, in an academic partner's knit studio (Haffended, Price, & McCann), to interpret and categorize information taken from the 'T' shirt 'toiles' into six selected stitch structures with drawings and measurements to accompany the toiles. The third

Figure 21.4 Co-design seam-free knit development using Tencel/Wool blend.

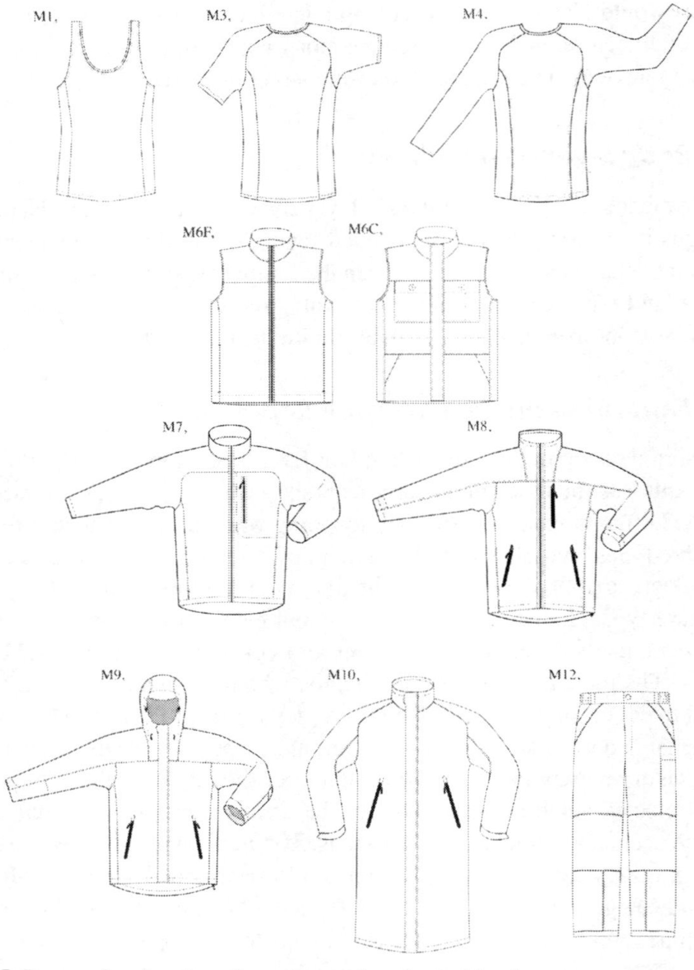

Figure 21.5 Range plan drawings for male and female clothing layering systems.

phase involved two designers (Haffenden & McCann) visiting the manufacturing base to hand over the more complex design specifications to expert technicians. Both fibre producer (Lenzing) and knit producer (Blum) value the importance of addressing the active ageing market (Figure 21.5).

21.6 Final prototype development

21.6.1 Importance of clear garment specification

There is growing concern about the environmental impact of the textile production chain and the traceability of materials, components and finishes, as textiles and garments are sourced globally (https://www.stringtogether.com) (EOG Seminar, Outdoor, 2012).

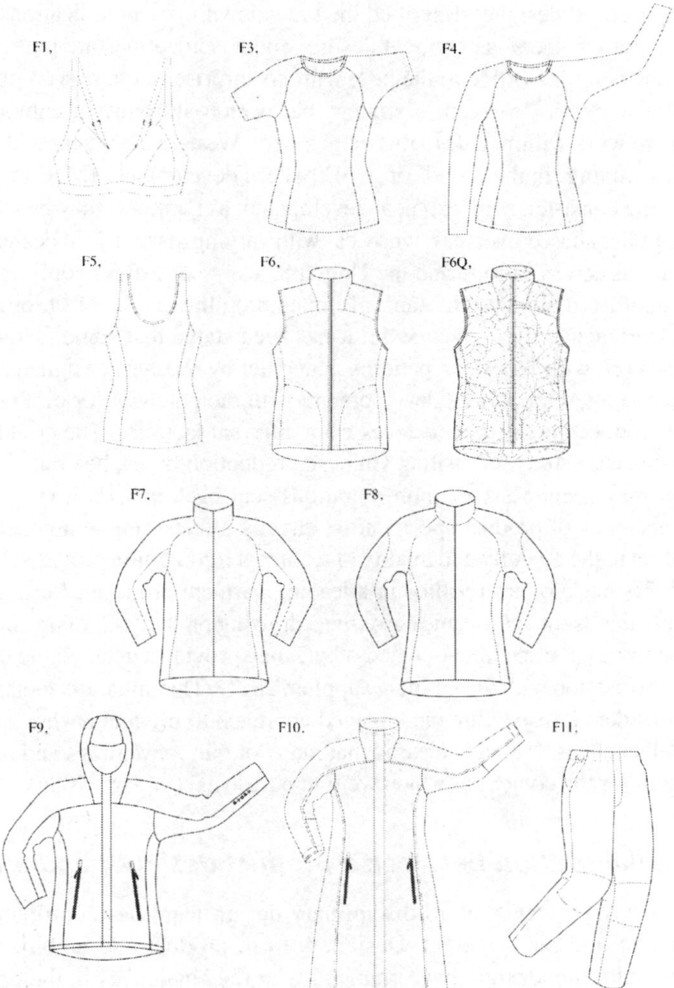

Figure 21.5 Continued.

However, there appears to be less attention to control over inaccuracies in clothing specification that often results in garment in landfills. A major challenge, from a technical design perspective, is that product developers have become increasingly detached from 'hands-on' involvement in the technical design process. In particular, designers are no longer involved in the generation of pattern blocks and style patterns. With the production base often overseas, designers are primarily involved in the generation of design concepts and design selection based on 2D computer-generated 'flat' line drawings. Flat drawings, with measurements, cannot communicate the subtlety of skilled pattern cutting, especially for sophisticated sportswear garments with tailored fit, intricate articulation to achieve arm-lift elbow and knee movement and hood adjustment. In a discussions with practitioners at the international sports trade event, ISPO, Munich in

2011, an experienced designer described the breakdown in communications as comparable to 'Chinese whispers' as garment designs and specifications are passed down the chain from one sourcing office to the next with no surprises as 'a curved line becomes straight in the process'. One brand's strategy has been to introduce common templates for designers to work within. Due to this detachment, Western designers seldom become involved in designing 'in the round' or in 3D pattern development (McCann, 2011).

There is little consistency in technical development as garments may be sourced from a range of suppliers based overseas, who deal with varying standards in design specification, from brands across Europe and the United States. Sizing across different countries adds to the confusion along with scant information on the degree of fit, or ease on top of body measurements, for a given style. It has been stated that some factories in Asia prefer not to work with blocks or patterns generated by Western designers, using their own initiative to interpret pattern development, with the potential for different versions of a product produced in different batches from different factories. The product designer involved in this case study, in visiting Chinese production bases, has found that design development may require six or more iterations sent back and forth (Price & Lewis, 2014). The accuracy of product specification directly affects sample numbers, and lead times. Variation in the quantity and quality of technical information provided by the West, seldom with 3D backup, has resulted in a leading garment producer, for multi-national brands, stating that 'some companies give me a drawing on the back of an envelope, others give me a three page specification and others may provide a detailed file of 25 pages' (Lewis, in conversation with the garment supplier, 2011). This situation demands a return to a more considered, respectful and engaged approach to product design and development. As Walker states, 'What is clear is that many of our conventions and practices are no longer valid for the context in which we find ourselves' (Walker, 2006).

21.6.2 Collaboration between a product designer and architect

Within the DfAW project, a multi-disciplinary design team has contributed to technical translation and development. Designers from product and architectural disciplines, new to clothing design, have engaged with the challenges in the development of shared technical language beyond that of traditional fashion design for the specification of smart clothing with wearable technologies. These designers had previously enjoyed the benefits of CAD with software programs such as Solidworks (product design) and AutoCad (architecture). In these programmes, decision-making is aided by a knowledge of industrial processes with constraints, programed into the software, that reflect limitations that occur in production, in order to verify realistic and accurate stages in the design specification.

These practitioners found a lack of equivalent, currently available CAD options related to garment specification. Clothing designers often use Illustrator software to produce line drawings, but this is unrelated to pattern digitising programmes. Flat drawings make reference to body scale measurements, often in limited detail, with patterns seldom created by designers in the West. Without patterns generated, and a digitised record made, there is no accurate reference from which garments may be made (discussed by Lewis and Price in chapter 22). This lack of clarity in specification would never be the case in product design development or in architectural practice where key dimensions, are drawn in plan, elevation and section and sent directly to be measured and made to scale. A growing interest in the capturing of body size and shape, through laser scanning e.g. Sizemic and the generation of customised 3D patterns and mannequins is gradually being linked to 3D pattern generation but, as yet, not available in the global supply chain (McCann, 2012).

In DfAW, effective communication has been required for effective design specification, including traditional drawing, computer aided illustration, 2D pattern cutting and 3D development on the mannequin, with a digitised record of patterns supported by sample prototypes. The product and architectural designers worked together as they engaged with the body scanning process and observed the development of style patterns and the pattern digitising process. Having gained an initial awareness of the clothing design process, the designers (new to the area) brought their methodologies and skills to merge with those of the experienced clothing designer/pattern cutter to help resolve the description, both visual and text, of novel material assemblies. They generated 2D computer illustrations, showing both full garments and enlarged technical details, with direct references linked to the digitised patterns, representing the 3D style development. Soft elements, such as textile-based conductive fibres and ribbons, knitted sensors and electronic components, had not previously been within the traditional realm of the fashion designer!

21.6.3 Electronic integration: placement/positioning of technology

The complexity of developing a clothing layering system as a 'platform' for textile-based electronics relied on transferring cross-disciplinary knowledge and skills between designers and both technology academics and industry technology providers. The mix of electronic components to be integrated into textiles and clothing demanded clothing specialists to work in liaison with electronics experts for the optimal positioning of both hard devices and soft technologies (textile-based sensors, cables and switches) within the garment layering system. The smartphone acted as the wearable computing hub within the system, requiring a choice of locations within the various layers. An ongoing concern was that some components, such as power for the heated panels, demanded relatively heavy batteries. Regular consultation between the user fit models and the academic team ensured that the electronic devices and their components were considered comfortable, in terms of their placement within the layers, and in the positioning and usability of the technology user interface. Electronic cables and soft switches were to be encapsulated, to be waterproof as necessary, while hard devices were to be removable for washing and/or upgrading.

The product designer and architect were proactive in resolving issues to do with integrating the interconnections and switches for the various devices. This involved visits to industry partners to verify the processes involved in, for example, the heat bonding of seams to join the various fabric assemblies, with their specialist coatings and finishes. The positioning to enable usability of the technology user interface was of particular concern in relation to garment assembly. Hybrid knowledge and skills from product design and architecture helped to describe the placement of components, interfaces and powering devices within the clothing layers. Once they confirmed the positioning of the electronic architecture and prototypes were assembled, they refined the drawings to describe the assembly of the garments with the intention that others might replicate the process. Imagery included photography, diagrams and 'exploded' drawings to clarify the detail of garment construction with references to toiles and working prototypes as relevant. The development of these new hybrid strategies for communicating the merging of elements within the garments is described by Price and Lewis in chapter 22.

The product designer liaised with industrial collaborators to verify the development of a generic garment specification sheet. The recommendation from the trade was to communicate with limited text, to avoid language translation problems, but with clear visual references. The provision of accurate digitized garment patterns that showed the positioning and integration of the technology components with electronic architecture and cross-referenced to flat specification drawings were supported by garment samples. In theory, this design 'package' would correlate with circuit diagrams to be produced by the technology partners. Predictably, the specification sheet was designed to provide information such as:

- Gender, sample size, garment reference number, etc.
- Material selection and trims: source, and references re: quality, weight, etc.
- 'Flat' garment drawings: front, back and side view as necessary
- Enlarged garment detail with photographic and other visual references
- 'Exploded' drawings to show electronic architecture and components
- Small-scale record of digitised pattern pieces with location of electronics
- Photograph of sample garment: toile and/or final working prototype

21.6.4 Industrial collaboration in prototype development

Industry collaboration has been effective in bringing the initial outcomes of the practical research to near market. The co-design stakeholder network embraced fibre and fabric producers and clothing companies, from a UK outdoor brand, with garment sourcing in Asia, to another UK brand with both national and offshore manufacturing plants, to a regional SME with an affinity to wearable electronics. All provided significant levels of support in textile sourcing, prototype development and project dissemination. This collaboration enabled the academic team to test and verify the clarity of the new prototype specification processes with feedback from industry collaborators. Industry feedback included face-to-face liaison with the regional SME, showroom meetings at the UK base of an outdoor brand, and email/electronic communication direct to offshore manufacturers. The SME was proactive in testing the wearable electronics in collaboration with the project's designers, while the brands progressed with greater access to materials but with longer lead times due to offshore production see Figure 21.6. The product designer (Lewis) gained experience of Asian factory practice.

Figure 21.6 Images of prototype development with regional SME.

Back in the university, co-design workshops, with designers and the technology team, enabled older participants to test and verify the placement of the electronic system such as the smartphone, with Bluetooth hub, PTT Module, the heat pad and its battery, and soft switch controls, within the garment system.

This project has succeeded in raising awareness of the importance of the active ageing market. Early co-design findings were adopted in commercial range development by one of the UK collaborating brands, shown to the trade in July 2012, with further design direction, influencing functional detail and colour selection, shown at ISPO, Munich, in February 2013. It was the brand's first experience of both co-design engagement and of designing specifically for the active ageing as discussed by Greengrass. The various industry stakeholders engaged with different product categories within the range plan and provided alternative interpretations in terms of fabric usage. Prototypes were tried and tested by the user group culminating in a final winter walk with both users and researchers. A list of the initial prototypes developed is given in Table 21.5 with a separate page of drawings of the range plan.

21.6.5 Colour

Colour has been identified as a key influence in garment selection, for the active ageing, in combination with size, fit and proportion (Schlesinger, 2010). Colour prediction was the subject of a co-design workshop with research participants gaining an understanding of the processes involved from national and international forecasting meetings to the minimum dye lots imposed and time scales that the global industry plans into the production critical path. This is discussed by Johnson, in chapter 4, from a forecasting view point, and by Greengrass, in chapter 18, from an industry perspective, in this publication. Following the workshop, both users and academics understood why colour choice often seems restricted and predictable. It would be difficult, therefore, to specify user colour selection within the fabric minimums of particular fabric qualities and the time scale for prototype development. It was decided to compromise in selecting classic neutral colours (such as greys and subtle marl effects) and to prioritise attention on prototype design lines and detail. In a separate co-design

Table 21.5 List of initial prototypes with materials and sampling suppliers

Style	Garment	Fabric	Company	Technology
F1	Body layer – sports bra	Eurojersey	Custom*	Sensors
F2	Santoni (knit) base layer	Lenzing	Blum	Santoni
F3	Short sleeve base layer	Tencel	Challenge	Drirelease
F4	Long sleeve base layer	Tencel/wool	Challenge	Sportwool
F5	Camisole base layer	Eurojersey	Custom*	Sensors
F6	Gilet (optional heat panel)	Aclimitise	Brenig	Fibretronic
F6	Gilet (quilted down)	Pertex	Sprayway	Primaloft
F6	Gilet (woven quilt)	PDA	PDA	Phasechange
F6	Gilet (fleece)	Aclimitise	Brenig	Fibretronic
F7	Fleece midlayer	Challenge?	Sprayway	
F8	Soft shell jacket	Schoeller	Brenig	Energear
F8	Soft shell	Aclimitise	Brenig	Fibretronic
F9	Outer shell hooded jacket	Aclimitise	Brenig	Fibretronic
F9	Outer shell hooded jacket	W.L. Gore	Sprayway	(Not made)
F10	Long outer layer coat	PDA	Sprayway	
F10	Long outer layer coat	Aclimiitise	Brenig	
F10	Long outer layer coat	PDA	PDA	Phasechange
F11	Soft shell trouser	W.L. Gore	Sprayway	
F11	Soft shell trouser	Schoeller	Brenig	Energear
F12	Stretch waterproof trouser		Sprayway	
F13	Cropped trouser	Not prototyped		
F14	Skirt	Not prototyped		
M1	Body layer – vest	Eurojersey	Custom*	Sensors
M2	Santoni (knit) base layer	Lenzing	Blum	Santoni
M3	Short sleeve base layer	Tencel	Challenge	Drirelease
M4	Long sleeve base layer	Tencel/wool	Challenge	Sportwool
M5	Long sleeve check shirt	Not prototyped		
M6	Gilet: (cargo pockets)	Not prototyped		
M7	Fleece midlayer	Challenge?	Sprayway	
M8	Soft shell jacket	W.L. Gore	Sprayway	
M8	Soft shell jacket	Keela	Keela	
M9	Outer shell hooded jacket	DualProtection	Keela	Fibretronic
M9	Outer shell hooded jacket		Youngone	
M9	Outer shell hooded jacket	W.L. Gore	Sprayway	(Not made)
M10	Long outer layer coat	W.L. Gore	Sprayway	(Not made)
M11	Cargo trouser	Keela	Keela	
M12	Waterproof over trouser	Keela	Keela	

*Prototyped by design consultant, Juliette Smith, intimate apparel designer.

workshop, the desired colours were chosen to constitute a colour chart, with yarn wrappings, colour names and Pantone references. This represented a relatively wide selection of enduring colours that could be coordinated to address different moods and levels throughout the layering system.

The final co-design colour workshop took place after the range plan was confirmed so that colours would be selected for use within known garment categories within the different layers. Users welcomed less transient and extreme colour swings than those driven by fashion trend. They were attracted to kinder and more enduring colours, from sober to more characterful tones considered sympathetic to ageing complexion and changing hair colour, and avoiding the current predominance of black within the fashion trade. For outdoor clothing, they showed a preference for natural colours, influenced by the countryside, from deep rich tones to powdered pales, with references to flowers, berries, fruits, lichens, autumn russets, forest greens, with neutral slates and stones, but not beige! The male representatives would normally select conservative outdoor colours such as forest greens, khakis, navies and greys but, when given the opportunity to consider more adventurous colours, were in agreement in introducing the richer tones from the countryside. Both male and female users wished for garment ranges, with less products, with more consideration of appropriate fit and proportion for ageing figure types, that you could go back and replace, but with a wider choice of colours and fabrics, in varied textures and blends. Designers do need to be aware that colours often have significant meanings in different cultures.

21.7 The way forward: design direction to help bring product to market

'Fashion is pleasant until the age of 60', writes a recent female survey respondent and 'After that we are forgotten' as stereotypes prevail (Willis, 2013). Within the clothing industry, product development continues with business as usual, driven by last season's sales and trend. However, there is scant information on last year's sales given that the active ageing community finds their clothing in a breadth of outlets, often in end of season sales, and sometimes in charity shops. The lack of current trend information for the active ageing (Slowfashion) sector is confirmed by Mode Information and by other fashion trend prediction companies, at ISPO and outdoor trade fairs, 2012/2013. 'Trend' is not entirely suitable as a term, suggesting the ephemeral and transient rather than timelessness and enduring style. There was one memorable season where positive trend information for baby boomers came from Promostyl's trend book 'Baby Boomer Generation' (BBG) produced in 2003, providing style guidance on proportion and detail for varied older figure types, and 'Rainbow Youth,' where *Viewpoint* (Issue 13, 2003) devoted a whole issue to raising the profile of older age with up-beat colourful photography.

David Shah, the publisher of *Viewpoint*, has since promoted 'Timeless Design', where products have attachment, being wanted because of their timeless function. He suggests that 'you do not throw away emotional design' (Beyond Green Conference, 2009) and also proposes 'Slow Fashion' that 'is about saving up and waiting to buy until you buy that one well-designed, well-made, beautiful piece that will be a staple in your wardrobe for years to come'. He suggests that 'we need to think harder and feel better about what we buy and how we care for it' (SLOWFASHIONHouse.com. Accessed 20.04.10). This more considered approach is described by Fuad-Luke: 'The process of slow design is comprehensive, holistic, inclusive, reflective and considered. It permits evolution and development of the design outcomes. It belongs to the professional and public arenas and emphasises the importance of democratising the design process by embracing a wide range of stakeholders' (www.slowlab.net/ideas.html).

'DfAW' research has revealed that the active ageing would welcome less transient styling, in designs fit-for-purpose, promoting longevity of use, and multi-use. It has confirmed that participants engaged in outdoors pursuits, such as walking and mountaineering, generally have respect for nature and do not wish, intentionally, to harm the environment. In this context wearers are often engaged in activities that promote health and wellness and predominantly in environments that are natural and often with protection enhanced through initiatives from companies such as Patagonia who, as early as 1974, began to donate money to a variety of grassroots environmental organisations and individuals (McCann, 2011). In his Enviro Essay Yvon Chouinard quotes conservationist David Brower 'No business can be done on a dead planet'. and that a company that is taking the long view must accept that it has an obligation to minimize its impact on the natural environment Chouinard, 2004. Fuad-Luke proposes that 'slow design requires stepping outside the existing mental construct of capitalism, where metabolism is driven entirely by economic imperatives, to consider other metabolisms and in doing so generate fresh awareness, possibilities and subsequently help create new societal values' (Fuad-Luke, 2007).

'Women want a smaller choice of better-made clothes' with 'more taste, less speed' and Willis suggests that 'at the high end, designers' stress levels would benefit from having to produce no more than two collections a year; on the high street, fewer unsold garments would end up being shredded. This frenetic, breathless experience is missing the point for most of us: we want better clothes, not faster ones' (Willis, 2013). The research case study confirms that older wearers seek enduring style, age appropriate shape, and in cheerful colours sympathetic to the complexion of the older wearer, and in garments that may come to be regarded as 'an old friend'. These findings are in tune with Fletcher who suggests that, in terms of Slow Fashion, 'in melding the slow movement's ideas with the global clothing industry, we build a new vision for fashion in the era of sustainability; where pleasure and fashion is linked with awareness and responsibility' (Fletcher, 2008).

In terms of trend promotion, DfAW research is timely. In January 2012, the front cover of the launch edition of *ISPO News*, the magazine that represents the leading international sports trade event, staged in Munich, Germany, depicted an upbeat image of active agers with the caption: 'Live Healthy, Live Longer'. The editor, Tobias Grober, quoted Professor Christoph Bamberger in suggesting that 'people who are physically active throughout their lives remain younger and healthier longer' and that 'our genes are not the only factor determining whether we live happily to an advanced age. Prevention and lifestyle are much more important'. In January 2013, ISPO introduced a prominent exhibition area, 'ISPO Healthstyle' that has incorporated 'ISPO Generation'. This initiative has emerged partly as a result of many seasons of 'Best Ager' market research, in raising the profile of the growing active ageing community, undertaken by Generation Sport (Germany) with input from the research findings from DfAW. For the first time, ISPO has recognised the importance of raising awareness in addressing the clothing requirements of the active ageing to encourage users to engage in healthy exercise. In this context, in the UK, the

Outdoor Industry Association is promoting their 'Britain on Foot' campaign, from Spring 2013, that aims to 'inspire and inform the general public to get active and enjoy the Great British outdoors in many different ways'. (www.outdoorindustriesassociation.co.uk).

Design intervention is required at the range planning stage. The collaborative design-led research discussed has developed new knowledge and methodologies, with practical outputs, and generic guidance that may be transferred to the wider industry and educators. Co-design practice, has embraced disparate academic and industry cultures (that were interdisciplinary in their own right) while engaging with active ageing research participants, as members of the public. This has resulted in the development of new hybrid processes and new 'shared language'. All stakeholders have realised the value of adopting a co-design approach to gain new insights into the real design needs and aspirations of this growing, and potentially valuable, market. Co-design approach has begun to break down barriers between the textiles and clothing cultures and that of electronics, in both academia and industry. This has wider implications for design guidance to inform, e.g., sector skills, standards, education, and how to address older age. Findings from this case study may inform the development of a design manual, in clear language, (both text and visual) to guide the design development process, supported by the examples of working prototypes as proof of concept.

This research has synergy with current industry discussions on bringing back textiles and clothing production nearer to 'home' that has been the focus of two industry conferences in recent months: 'Made in the UK' (ASBCI, Manchester, 15.09.12) and 'A New Dawn: Rebuilding UK Textile Manufacturing' (The Clothworkers' Company, The Dyers' Company & The Weavers' Company, city of London, 02.11.12). Addressing the changing demographic that embraces the rapidly growing active ageing community could be an ideal market to address by UK manufacturers. This consumer is already worried by transient, disposable fashion and anxious that, if they find something that suits them, they may not be able to replace it. Slower, value-added design would be appropriate for both the consumer and for more considered production. Fewer styles in better fitting products, with a choice of lengths, and in colours and textures that are sympathetic to baby boomers, could be personalised with meaningful accessories to suit the individual wearer. Operators could build up skills in shorter runs, of known styles, in manufacture with a focus on quality, fit and appropriate styling, in alluring but more enduring colours and in fabrics that will not date easily.

The clothing needs of the active ageing could represent a predictable market where value added, more considered and sustainable production could be developed to cater for price points ranging from elite to mid-market level. This market is not going away, is growing, and could be a less frenetic sector to be addressed by UK manufacturers. DfAW findings offer guidance for the realisation of design fit-for-purpose from functional, aesthetic and cultural perspectives. Cross-disciplinary methodology has enabled the realisation of the smart clothing layering system, as an integrated 'wearable technology platform' that enhances comfort, independence, and enjoyment for varying levels of participation in walking. Greater understanding between academic researchers and design practitioners has extended to beneficiaries within the international textile, clothing and related 'soft' electronics supply chain, with the potential to promote slower, more considered, responsible and, ultimately, more sustainable approaches to product development.

References

Chouinard, Y. (2004). *On corporate responsibility for Planet Earth. Enviro Essay.* http://www.patagonia.com/eu/enGB/patagonia.go?assetid=9082. Accessed 16.05.14.

Fletcher, K. (2008). *Sustainable fashion and textiles: Design journeys.* London: Earthscan.

Fuad-Luke, A. (2007). Re-defining the purpose of (sustainable) design: enter the design enablers, catalysts in co-design. In J. Chapman, & N. Gant (Eds.), *Designers, visionaries + other stories: A collection of sustainable design essays.* UK and USA: Earthscan.

George, H. (2001). Temperature regulation and technology. *Gerontechnology, 1*(1), 41–49.

Huppert, F. (2003). Issues in designing for older users. In R. Coleman, J. Clarkson, & S. Keates (Eds.), *Inclusive design: Design for the whole population.* London, UK: Springer-Verlag.

Lindemann, K. (2014). *Stylish Clothes for older women.* http://www.anti-aging-articles.com/Fashion-older-women.html. Accessed 27.06.14.

McCann, J. (1999). *Establishing the requirements for the design development of performance sportswear.* (MPhil thesis).

McCann, J. (2009). End-user based design of innovative smart clothing. In J. McCann, & D. Bryson (Eds.), *Smart clothes and wearable technology.* Cambridge: Woodhead Publishing Ltd.

McCann, J. (2011). Design fit for purpose: Sustainable functional clothing development. In *Design principles and practices, Rome 2011.*

McCann, J. (2012). *Translating the hybrid methodologies and practical outputs of smart textile-oriented research, in clothing for the growing ageing market, for the benefit of all stakeholders.* Textile Institute World Conference, Kuala Lumpur, Malaysia. May 15–17, 2012.

McCann, J., Morsky, S., & Dong, X. (2009). Garment construction: Cutting and placing of materials. In J. McCann, & D. Bryson (Eds.), *Smart clothes and wearable technology.* Cambridge: Woodhead Publishing Ltd.

Metz, D., & Underwood, M. (2005). *Older Richer Fitter: Identifying the customer needs of Britains ageing population.* UK: Age Concern Books.

Office of the German reports on the Elderly. (2010). *A new culture of ageing: Images of Ageing in Society. Berlin: Federal Ministry for Famiuly Affairs, Senior Citizens, Women and Youth.* http://www.bmfsfj.de/RedaktionBMFSFJ/Broschuerenstelle/Pdf-Anlagen/6.-Altenbericht-Eine-neue-Kultur-des-Alterns-englisch,property=pdf,bereich=bmfsfj,sprache=de,rwb=true.pdf.

Outdoor Industries Association. (2012). *Britain on Foot debated at the House of Lords.* Association News. http://www.outdoorindustriesassociation.co.uk/news_association_item.asp?item_id=188. Accessed 10.12.12.

Schlesinger, C. (2009). *Fashion and apparel choices as we age.* MSG, Workforce Development, VA Greater Los Angeles Healthcare System. http://www.cccfcs.com/uploads/Gerontology/Jan%2009/Gero-Fashion%20Article.pdf.

Schlesinger, F. (2010). *Black clothes can put years on you: The unwanted side effect of fashion's favourite colour. MailOnline January 5th.* http://www.dailymail.co.uk/femail/article-1240616/Black-clothes-years-effect-fashions-favourite-colour.html.

Taylor, D. (2011). Developing a visual language to enhance knowledge transfer in the design of smart clothes and wearable technology for the active ageing. In *5th International Conference on Design Principles and Practices, Rome, February 2–4, 2011.* (Paper published in Design Principles and Practices Journal.

Timmins, M., & McCann, J. (2013). *Overview of the Design Requirements of the Active Ageing, Walks and Walking Concept Map. New Dynamics of Ageing Programme.* http://www.newdynamics.group.shef.ac.uk/assets/files/OPRG%20End%20of%20Programme%20Report%20FINAL%202013.pdf.

Walker, S. (2006). *Sustainable by design: Explorations in theory and practice.* London: Earthscan.

Whitehorn, K. (2012). *Vivienne Westwood may think women are stylish at 70 – but where's the help?* http://www.guardian.co.uk/commentisfree/2012/feb/20/vivienne-westwood-stylish-70. Accessed 26.06.13.

Willis, R. (2013). Clothes: A manifesto. *In Intelligent Life, The Economist, March/April 2013,* 48–54. http://moreintelligentlife.com/content/lifestyle/rebecca-willis/clothes-a-manifesto?page=full. Accessed 26.03.13.

Wolfe, D. B., & Snyder, R. E. (2003). *Ageless marketing.* Chicago: Dearborn.

Websites

'A new dawn – Rebuilding UK textile manufacturing', conference, London's Clothworkers Hall, November 2, 2012. http://www.clothworkers.co.uk/getattachment/Textiles/Textiles-Conference/A-New-Dawn-FINAL_LR.pdf.aspx. Accessed 27.06.13.

Beyond Green III (2009). Good design beyond Green. http://modebewust.nu/?p=632.

Strauss, C. F., & Fuad-Luke, A. (2008). *The slow design principles: A new interrogative and reflexive tool for design research and practice.* www.slowlab.net/slow_design.html. Accessed 27.06.13.

http://fibretronic.com/products/heatwear.

http://www.exo2.co.uk/fabrocreg-ezp-10.html.

http://www.textronicsinc.com/.

http://www.shimmer-research.com/.

http://www.eurojersey.com/.

http://www.gore-tex.co.uk/remote/Satellite/content/technologies.

http://www.lenzing.com/en/fibers/tencel.html.

http://www.sizemic.eu/.

http://www.slowdesign.org.

http://www.ecoindexbeta.org.

http://www.outdoorindustriesassociation.co.uk.

https://www.stringtogether.com.

http://www.anti-aging-articles.com/Fashion-older-women.html.

http://www.newdynamics.group.shef.ac.uk/.

http://www.anti-aging-articles.com/Fashion-older-women.html. Accessed 25.04.13.

http://www.europeanoutdoorgroup.com/conserve-sustain-care/sustainability/.

Developing a strategy for the effective specification of functional clothing with integrated wearable technology

22

M.C. Price[1], *C.M. Lewis*[2]
[1]Wearable Futures Design Group, Bristol, UK; [2]University of South Wales, Newport, UK

22.1 Introduction

Clothing and technology sectors are both well-established but quite distinct industries. They have traditionally occupied separate supply chains and routes to market. The miniaturisation of electronics is now enabling innovation in consumer wearable technology products, but as these cultures merge, significant challenges exist for design and communication within the various stages of bringing products to market. This chapter addresses bridging *the gaps* between the cultures of clothing and electronics, and the particular challenges in bringing them together with garment design and assembly. The authors describe the development of a unique strategy for design specification that ensures successful manufacture of complex outdoor clothing products with integrated wearable electronics, highlighting the need for testing of prototype designs in order to verify the design process. The authors aim to provide a methodology that can be shared and transferred to other end-uses within the clothing industry.

The aim of the methodology described in this chapter has been the development of garments that meet the needs of active older walkers as part of the Design for Ageing Well (DfAW) cross-disciplinary collaborative research project, carried out within the RCUK-funded New Dynamics of Ageing Programme (NDA) with the aim for *improving the quality of life for the ageing population using a technology-enabled garment system*. Within this case study, the motivation for integrating wearable technology into everyday clothing for older walkers has been to maintain well-being through healthy exercise, independence and social engagement. One of the design challenges has been to investigate and utilise *existing* wearable technologies that could be adapted to meet the everyday needs of this neglected older user group. The challenge in this case has been to match specific end-user group requirements with current, available or off-the-shelf technologies that would be suitably robust, easily useable and ready to specify. Effective specification of complex functional garments with integrated technology has required development of new design strategies, bringing together a hybrid mix of skills and expertise.

Textile-led Design for the Active Ageing Population. http://dx.doi.org/10.1016/B978-0-85709-538-1.00022-5

22.2 Co-design team

For the specification of complex garments with integrated technology, a multidisciplinary co-design team is essential (McCann, 2012b). For the Design for Ageing Well case study, a *co-design team* was established, involving end-users in contributing to a team of experts from clothing design, wearable technology, garment sizing and textile technology, with representatives coming from both academia and industry. A *co-design process* involves specialists working together with end-users to collectively inform a design outcome.

The DfAW case study has brought together a range of specialists in a co-design team that was grouped into the following work packages:

1. *Behaviour*, involving gerontologists and physiologists in behavior research
2. *Clothing*, the garment designers and technologists
3. *Technology*, the hardware and software engineers

These groups worked closely together to share their knowledge and expertise throughout the co-design process, and crucially, an end-user group of active older walkers, who were involved from the selection and placement of technology through to final design specification, have been central to the co-design team to inform the whole process. Figure 22.1 outlines this interaction.

22.2.1 Additional competencies for design specification

The core competencies in the *Clothing* work package have been experts in performance clothing design, anthropometry with 3D body scanning, intimate apparel design, textile technology, and seam-free knit and garment engineering. However, there are still very few experts in the emerging area of integrating electronics into textiles and clothing, since it is still not well understood by fashion and textile designers. Similarly, electronics and software developers have not traditionally needed to have knowledge of clothing and textile design engineering. It became apparent that additional competencies would be needed outside of traditional clothing design and technology to bridge these professional knowledge gaps. This role has been gradually defined through the process of a *product designer* and *architect* working together in collaboration with the garment design engineer and technology developers. They have merged their technical insights and prior experiences to communicate the complex design integration of wearable technology (see Figure 22.1).

The Behavior and Technology work packages have embraced many competencies, but these are not the focus of this chapter; competencies related to the Clothing work package are discussed in other chapters of this publication. The remainder of this chapter focuses on the core competencies that have been brought together for the hybrid design specification of functional clothing with integrated wearable technology.

The three specialisms that have worked together to develop new strategies for the specification of functional clothing with integrated wearable electronics were

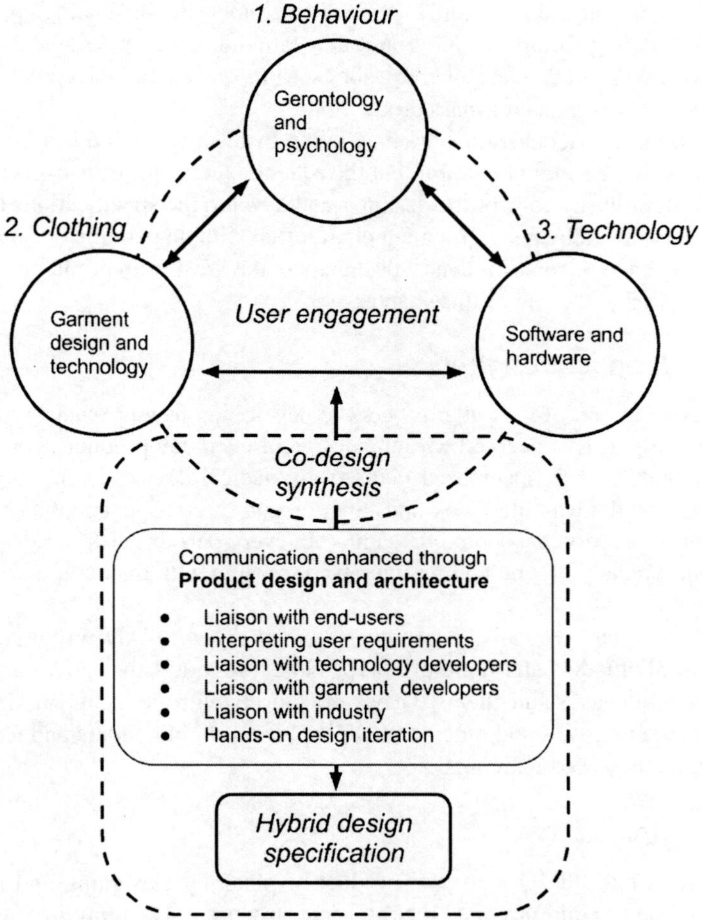

Figure 22.1 Co-design synthesis. Role of product design and architecture in bringing new communication strategies to guide the hybrid design specification and realisation of functional clothing with integrated electronics.

a garment designer, a product designer and an architect. These competencies are described in the following sections.

22.2.1.1 Garment designers

Garment designers work in the context of a rapidly changing global fashion industry to the brief of the sales and marketing team. This is often based on last season's sales figures in combination with design trend forecasting. They liaise with buying and production teams to ensure the item suits the customer, the market and price points. Traditionally, garment designers would consider the design requirements of garments in terms of the size, shape and style for different consumers. They would then develop patterns that were cut and sewn into toiles (calico); supervise the making up of sample garments (including fitting, detailing and adaptations); develop technical specifications for designs; source, select and buy fabrics, trims, fastenings and embellishments; and adapt existing designs for mass production.

However, these practical skills are now being lost as production has moved overseas. It is now increasingly the norm that the role of a technologist is overseas to take the 'flat' design illustrations of the designer and develop the specifications for manufacture, accompanied by a measurement chart, often with no patterns or samples. The problems related to the detachment of designers in the West from production overseas will be discussed at the end of this chapter.

22.2.1.2 Product designers

Product design is part of the wider process of developing new products, typically for mass production. It is concerned with the design of everyday products from chairs to computers, and of more specialised items from medical devices to military ones. It aims to improve the way they look and function and often to reduce the cost of production. It is a process that should dovetail with every part of the wider development process, but typically it is much more involved to begin with than at the end (Design Council, 2012).

Product designers typically use *computer-aided design* (CAD) programs to generate virtual 3D models and rapid prototypes to express, test their ideas and discuss designs with colleagues and clients. They work closely with engineers, model makers, sales and marketing staff and other skilled people and should understand technology, production methods and materials.

22.2.1.3 Architects

Architecture is both the process and product of planning, designing and managing the construction of buildings and other physical structures. The term *architecture* has been adopted to describe the activity of designing any kind of complex system, and is commonly used to describe systems in information technology.

An architect is comfortable in consulting across an expert team, which would traditionally include structural and systems engineers, for the integration of building technology, lighting and services. The experienced architect is used to dealing with associated challenges of working across a multidisciplinary expert team in order to successfully integrate systems and structures into a design. They work at a variety of scales with reference to accurate dimensions in plan, section and elevation to consider their 3D design in context (in this case study, of the body), whilst in parallel developing construction details that utilise a material knowledge.

An important aspect of architectural practice, with design relevance to wearable technologies, is the creation of a plan or convention for the construction of an object or a system, such as an engineering drawing or system diagram (or in clothing, flat working drawings with patterns). This approach leads to the production of accurately dimensioned drawings for the design and construction at macro- and micro-scales.

22.2.2 Co-design development process

The authors of this chapter, as practitioners in product design and architecture, brought new technical skills to clothing design in the case study. The aim of this involvement was twofold.Firstly, to interpret user requirements in terms of comfort and usability in order to resolve the technical challenges for integrating electronic components and systems into textiles and clothing. This was achieved through hands-on prototyping in the factory environment. Secondly, the aim was to generate clear design specifications that would be appropriate for larger-scale overseas production and that would be transferrable to other wearable technology products. These design roles were key to successfully interpreting, explaining, facilitating and subsequently visualising outcomes from the co-design development process. This contribution depended on close liaison with end-users, technology developers and clothing industry partners, as summarised in Figure 22.2.

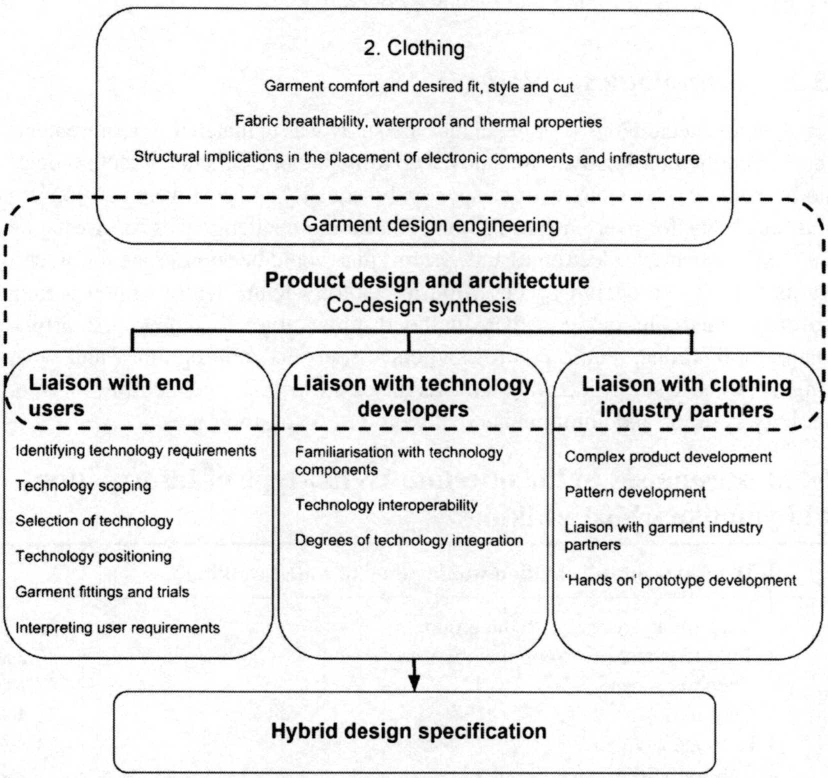

Figure 22.2 Co-design synthesis. Practitioners from architecture and product design worked closely with the garment design engineer and in liaison end-users, technology developers and clothing industry partners to verify their design and construction details, and to refine working drawings for hybrid design specifications.

22.3 Co-design development process: liaison with end-users

22.3.1 *Identifying technology requirements*

The selection of technology for integration within the garments was made through workshops with an older user reference group (URG) with a keen interest in walking. The group provided information to the team about technology they currently use when walking and indicated additional information they might like to be made available to them. Data was collected by quantitative and qualitative means, involving questionnaires and workshops. Findings from this revealed that compasses, cameras, global positioning systems (GPS), and mobile phones were already widely used by older walkers; Table 22.1 summarises types of additional information that were found to be desirable (University of Westminster, 2009, p. 2). The subsequent scoping of available wearable technologies would need to include communication, warming, emergency alarm, navigation, positioning and distance covered.

22.3.2 *Technology scoping*

The challenge of technology scoping in the case study was to match user requirements that had been identified during the pilot study with available or *off-the-shelf* technologies that would be suitably robust and ready to specify. Practicality and cost were key concerns, as well as suitability for overseas production. The design challenge was to investigate and utilise existing wearable electronics and systems that would be appropriate for older users.

As part of the co-design process, the technology team demonstrated a range of appropriate wearable devices that included physiological sensors, smartphones, software applications, tactile control switches, speakers, microphones and wearable heating systems. User group trials and feedback informed the selection of wearable technology systems and components that would best suit their needs.

Table 22.1 Responses to the question 'What type of information would you like whilst walking?'

	What type of information would you like whilst walking?	
1	Keep me in contact with the group	52.4%
1a	Tell me where my group members are	26.2%
2	Keep me warm	88.1%
3	Cool me	71.4%
4	Help me navigate	54.8%
5	Call for help if I should fall	52.4%
6	Protect me if I should fall	50.0%
7	Tell me the distance covered already	54.8%

Source: Data extracted from University of Westminster Pilot Study Data Report, which shows requirements that scored 50% or more from the user group of older walkers.

22.3.3 Selection of technology

Table 22.2 summarises wearable technology systems and components that were co-selected by the active older walkers for integration into the DfAW case study garments. Based on this, garment construction details for the integration of each individual component would need to be developed and detail drawings produced for design specifications.

22.3.4 The clothing layering system proposed as a platform for wearable technology

A *clothing layering system* consists of supportive *body layer* garments, such as vests for men and bras for women; *baselayer* garments, which are tight fitting and worn close to the body for moisture and thermal regulation; *midlayer* fleeces and gilets for insulation and protective *outer layer* waterproof and windproof jackets and trousers (McCann, 2012a, Smart Textiles for Protection).

22.3.4.1 Wearable technology throughout the layering system

Close-fitting body layer garments provide an ideal platform for the integration of the physiological sensors (in the case study, the *Shimmer* ECG device) with textile knitted electrodes (supplied by Textronics/Adidas), which can be positioned close to the body for optimum functionality. Midlayer garments were well suited to the incorporation of HEATwear systems (produced by Fibretronic) to enhance thermal insulation. Outer layer garments were proposed as the main technology hub to house the smartphone (supplied by HTC), the *Bluetooth module* (BTM) and soft control switches (both produced by Fibretronic) (Table 22.3).

22.3.5 Wearable technology positioning

22.3.5.1 Liaison with the technology work package team

Communication between academic collaborators has been essential throughout the development of the garment design specifications. This was an iterative process with the exchange of relevant knowledge of terminology, processes, technology and materials. It enabled development of the clothing, technology hardware and software to progress in parallel. Merging these cultures demanded compromises in timescales and overlapping phases of development, but has resulted in shared recognition of the value of a co-design process for effective product development.

22.3.5.2 Initial 3D toiles of the garment layering system

Prior to prototype development, the Clothing team conducted *3D body scans* of selected individuals within the user group who were representative of the average *body mass index* (BMI) for the age range, in order to collect a full dataset for their body shape. The data was used to generate pattern basic blocks for development of the articulated pattern blocks for 3D garment prototypes (toiles). The result constituted a

Table 22.2 **Wearable technology systems and components co-selected by active older walkers for integration into DfAW garment range**

Component	Description	Image
Physiological sensing		
ECG physiological sensor Supplier: Shimmer Research	Shimmer is a small wireless sensor platform that can record and transmit physiological and kinematic data in real time. Designed as a wearable sensor, Shimmer incorporates wireless ECG, EMG, GSR, Accelerometer, Gyro, Mag, GPS, Tilt and Vibration sensors (Shimmer, 2012).	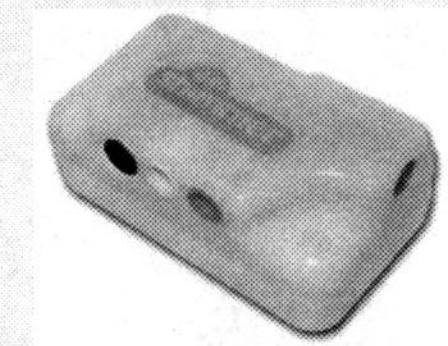 Sensor dimensions: 53 x 32 x 23 mm
5-function linear keypad Supplier: Fibretronic	The Fibretronic 5-function linear control keypad connects to an electronics module to form a complete solution for use in softgoods. The keypad is designed for easy integration into garments, bags or other wearable accessories using a variety of construction and attachment methods (Fibretronic, 2013).	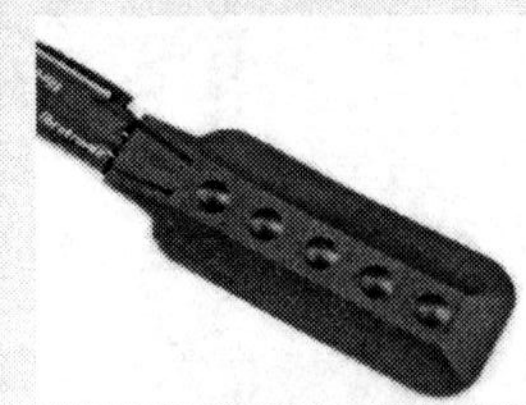 Keypad dimensions: 120 x 45 mm Cable length: 800 mm
Bluetooth phone controller Supplier: Fibretronic	The Bluetooth music phone controller links wirelessly to a mobile phone, allowing the user to make and receive calls as well as controlling all the audio functions of any mobile phone with stereo Bluetooth capability. The module allows the user to control music and it alerts when a phone call comes in so that the user can answer the call directly from a CONNECTED*wear* keypad or joystick (Fibretronic, 2013).	 BTM dimensions: 53 x 47 x 11 mm

HEATwear system		
HEATwear panel Supplier: Fibretronic	The core element of the HEATwear system is the fabric heat panel, which is designed to be stitched or bonded to garments and softgoods. The heat panels are bonded, breathable, washable, and have a 100% textile structure (Fibretronic, 2013).	 Heat pad dimensions: 280 x 130 mm Cable length 400 mm
Battery pack for HEATwear panel Supplier: Fibretronic	Mains rechargeable Li-ion battery ideal for powering standard HEATwear heat panels. Designed specifically for wearable electronic applications with a soft EVA moulded casing, this battery weighs 160 g and has a capacity for up to 8 h of heating (Fibretronic, 2013).	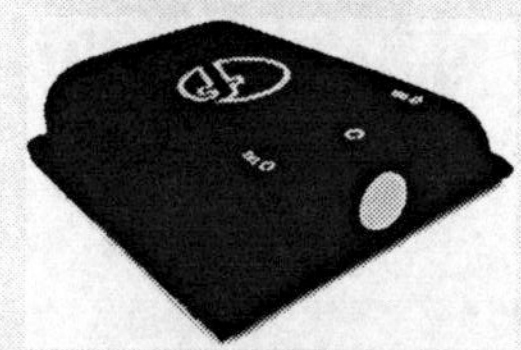 Battery dimensions: 90 x 70 x 23 mm
HEATwear LED extension control switch Supplier: Fibretronic	The LED control switch is supplied on a textile/PU extension cable allowing the control switch to be sewn or bonded to the garment in a position easily accessible to the user, for example on the chest or sleeve. The textile cable terminates in a DC jack plug that connects to the battery pack. The textile cable can be supplied in different lengths to suit a range of finished products (Fibretronic, 2013).	 Control switch dimensions: Diameter 40 mm Cable length: 800 mm

Continued

Table 22.2 Continued

Component	Description	Image
Push-to-talk radio communication		
PTT module Supplier: Fibretronic	The Fibretronic's push-to-talk (PTT) system is a communications interface for operation of consumer and professional 2-way radio handsets. The system is designed as a hands-free accessory and incorporates a microphone, speaker, push-to talk switch, and an interface circuit enclosed within a weatherproof soft fabric module (Fibretronic, 2013).	 PTT dimensions: 110 x 60 x 12 mm
Smartphone hub		
Smartphone HTC desire Supplier: HTC	HTC desire smartphone was utilised as a mobile computer, wearable database, and control hub. The smartphone can be wirelessly connected to a tactile control keypad positioned within any garment when connected to a Bluetooth receiver such as the Fibretronic BTM	 Smart phone dimensions: 119 x 60 x 11.9 mm

ECG, electrocardiogram; GPS, global positioning systems; BTM, Bluetooth module.

Table 22.3 Distribution of wearable technology components throughout the DfAW clothing layering system

Layer	Function	Technology
Body layer e.g. vests, bras	Next to the skin layer, supportive garments	Physiological sensing • Textile electrodes (Adidas) • ECG sensor (Shimmer)
Baselayer e.g. T-shirt (long/short sleeve)	Close to skin for moisture, management	Not applicable
Midlayer e.g. gilet/fleece	To provide insulation	HEATwear (Fibretronic) • Flexible HEATwear panel • Battery pack for HEATwear • LED control switch • Cabling
Outer layer e.g. shell waterproof	To provide protection from rain and wind	Smartphone (HTC) control sleeve • Linear keypad (Fibretronic) • BTM (Fibretronic) PTT communication (Fibretronic) • Microphone • Speaker

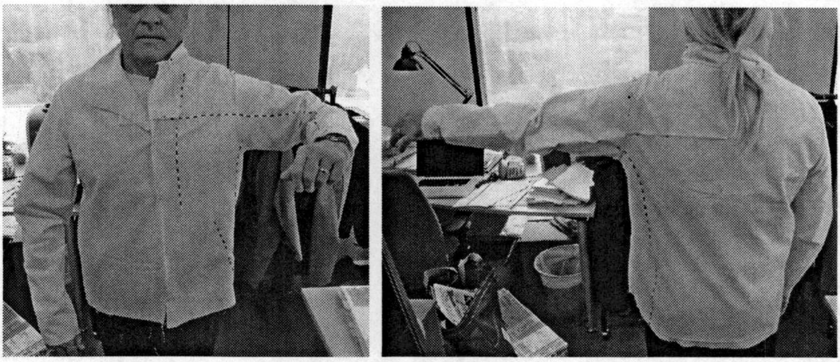

Figure 22.3 3D toiles developed by the clothing design engineer to verify fit, proportion, articulation and ease of use.

garment layering system that was coordinated in terms of fit, proportion, articulation and ease of use (see Figure 22.3). Representative end-users were selected to model the 3D toiles at regular fittings throughout the development process.

22.3.5.3 Technology positioning within the garment layers

Preferred positioning of the technology components was discussed with the user group through a programme of co-design workshops. The location and placement of the smartphone, BTM, heat pad, battery pack, soft controls, and speakers were defined by the user group through sketching their preferred placement onto diagrams. During user fittings, this information was transferred onto 3D calico toiles and aspects of garment comfort and the usability of the technology interface were addressed for the entire garment layering system.

22.3.5.4 Examples of technology positioning

22.3.5.4.1 Outer layer example

The user group specified the location of the keypad to align with the radius bone in the forearm for better support in engaging the control buttons. This position proved effective from a structural point of view because the control pad remained on the wrist as the arm is raised and lowered in use. These are illustrated in Figures 22.4 and 22.5, in which the following are labelled: (1) BTM; (2) soft controls: 5-function linear keypad and (3) *push-to-talk* (PTT) module.

Figure 22.4 Male outer layer example – user group technology positioning preferences.

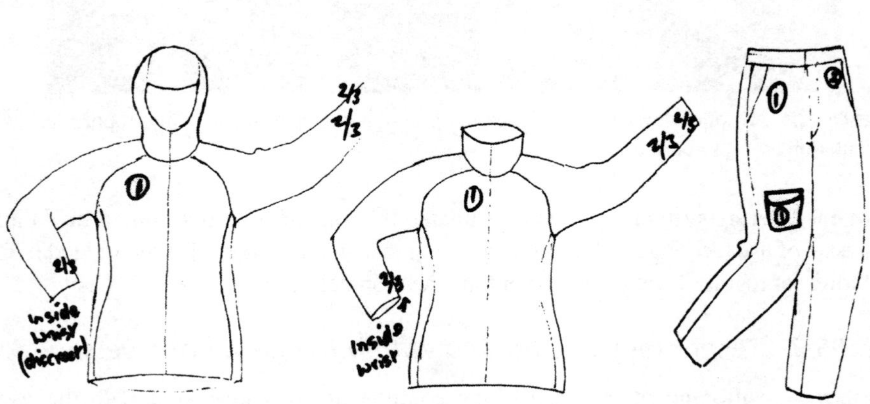

Figure 22.5 Female outer layer example – user group technology positioning preferences.

22.3.5.4.2 Midlayer example

The user group specified the location of the battery pack on the waistline between the back and side for larger, more curvaceous women and a second position just above the hip line for smaller physiques. This is illustrated in Figure 22.6, in which the battery is labelled (1) and the HEATwear LED extension control is labelled (2). It was important to locate the pack where its bulk went unnoticed during activity such as sitting and walking, to prevent it from interfering with other equipment such as a backpack. Two locations were considered necessary because neither one proved effective for comfort across all body shapes.

22.3.5.5 Design challenges for technology positioning

Members of the older user group were reluctant to carry heavy, bulky power supplies for heated garments and to accept the physiological sensors due to their awkward, bulky hard form factor worn close to the body. They wanted to minimise fully integrated, non-removable components due to concerns about reliability of technology hardware and high associated costs for replacing a garment in the likely case that its technology would need to be upgraded.

22.4 Liaison with technology developers

22.4.1 Familiarisation with technology components

The designers needed to establish close collaboration with technology developers and industry suppliers in an ongoing relationship throughout the co-design specification process. For the successful integration of technology within the garment layering system,

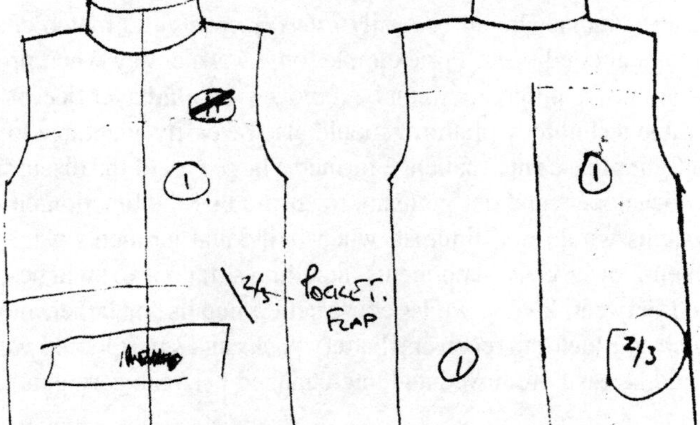

Figure 22.6 Midlayer example – user group technology positioning preferences for heat panel and battery pack within the female gilet garment.

the designers needed to familiarise themselves with the final suite of selected technology and the relationship between various components that make up their systems.

Technology specification documents were exchanged in order to clarify the function of each component, powering requirements and connectivity via physical jacks and cabling or by means of wireless (Bluetooth) receivers. Discussions concerning electronic circuitry and system specifications informed the final positioning of each technology component. Familiarisation with these particular requirements as well as feedback from user group trials defined the appropriate degree of integration of technology components during final garment assembly. This ranged from fixed componentry to removable devices. Finally, the designers developed an understanding of the *interoperability* between components to successfully distribute the devices throughout the clothing layering system.

A combination of *hard* and *soft* components is required for the functionality of wearable technology systems, and electronic circuitry or wireless networking to connect the devices. For example, the Fibretronic HEATwear system comprises a soft, flexible heat pad, rechargeable battery pack and remote user control switch that is connected by physical cabling within the garment. A smartphone application used to provide information about the user's heart rate during a walk depends on a *physiological sensor* worn close to the body, which provides data wirelessly via Bluetooth. Control switches that can operate the smartphone application wirelessly are connected to a BTM via physical cabling within the jacket. Familiarisation with these systems through liaison with technology developers and industry suppliers has been essential for the effective specification of technology integration (see Figure 22.7).

22.4.2 Interoperability

Understanding possible *interoperability* between technology components and devices that were distributed between separate garments within the layering system was an important design challenge. The clothing layering system is a range of separate garments that, when worn together as layers, build up to protect the body from a variety of outdoor conditions. Garments should be easily *interchangeable* to match changing external and physiological conditions. For example, on a warmer day when protective outer layer garments are not required, they can be removed and midlayer fleeces worn. In this respect, a wearable technology platform should also be easily adaptable to suit a variety of garment configurations. This challenge formed a large part of the discussions between the technology developers and design team, to ensure that the functionality of wearable technology systems would not diminish when individual garments were removed. To solve this problem, lower-cost components including soft control switches and speakers were duplicated between layers, whilst construction details for larger, more expensive components such as Bluetooth receivers, battery packs and smartphones were developed so that they could be easily removed and interchanged between garments.

22.4.3 The smartphone hub

'Smartphones will be a major part of the transition to our brave new connected world and so connected accessories will be the first wave of mass-market wearable tech' (The Guardian, 2013a).

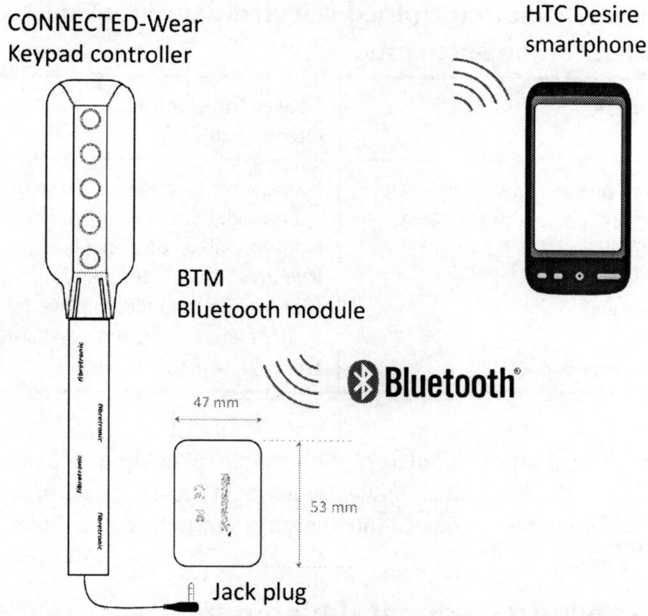

Figure 22.7 System diagram drawn by the designer following conversation with a technology developer and supplier (Fibretronic), showing the components, including physical cabling and wireless connections, required for the control of a smartphone application via soft control switches.

Smartphones provide familiar, robust and powerful mobile computers for storing and processing *data*. This data can be collected by components that are integrated into products, or garments, including sensors that collect *inputs* such as a user's heart rate. When these components are connected to a smartphone computer, wirelessly or by physical cabling, inputs or readings can be processed and translated via *software applications*, which *output* useful information for the user, communicated via audio or displayed on the screen.

The DfAW layering system utilises the HTC smartphone as a wearable technology *hub*, which collects information from a Shimmer Electrocardiogram (ECG) sensor (positioned within close-fitting baselayer garments) and GPS (embedded within the smartphone device) to provide the wearer with information about their current heart rate and distance travelled. The smartphone can be easily interchanged between garment layers as it is connected wirelessly via the BTM (Fibretronic BTM).

22.4.4 Degrees of technology integration

Different components that make up a wearable technology system are suited to different degrees of integration along a spectrum from loosely to closely integrated design details. An example of a loosely integrated detail would be a discreet additional pocket, designed to house removable components such as the smartphone or power supply, whilst closer integration would include channels for electrical wiring sewn

Table 22.4 Factors that determined selected degrees of device integration within the garments

Closer Integration (Fixed)	Looser Integration (Removeable)
Lower-cost components can be duplicated between garments in layering system.	Higher-cost components to be moved between garments in layering system.
Washable components	Non-washable components
Improved comfort	Requires removal for charging
Improved appearance	Likely to be upgraded/replaced due to rapidly changing technology hardware
Likelihood of loss or damage	Easier end-of-life disposal

into the sleeve seam. Fully integrated or embedded technologies include knitted or woven textile electrodes that collect inputs for heart rate sensing. Factors that affected the specification for degrees of device integration are introduced in Table 22.4.

22.5 Liaison with garment developers

22.5.1 Complex product development

New and complex product development, such as the design of clothing with integrated wearable electronics, demands an intensive, collaborative working relationship with the prospective industry partners, as complex designs add ambiguity and uncertainty to the process (Liao, Liao, Hutchinson, 2010).

Integrating wearable technology into functional outdoor clothing during garment assembly adds significant complexity to the product development process. Despite this, the performance sportswear and outdoor clothing industries have recognised the potential of wearable technology products and become early adopters of available technologies. There are examples of products that have been successfully brought to market, including the Lyric and Opus iPod and radio jackets from Sprayway, targeted primarily at athletes and the youth market. However, increased cost, time and complexity experienced in integrating electronics into garments during assembly overseas is leading to brands becoming wary of incorporating technology into new garment ranges (Greengrass, personal communication, September 20, 2011).

But the trend for wearable technology is set to continue, with 'next year set to be a watershed for wearables, creating a market worth more than $1.4 billion by the end of 2014' (Financial Times, 2013). This confirms the need for a new breed of designer to bridge the gaps for getting products to market, and for the development of new strategies for effective hybrid design specification. Few garment design teams and production facilities have the experience needed to bring a wearable technology concept from initial prototype through to mass production, and minimising ambiguity and uncertainty is key to ensuring successful overseas sampling and manufacture.

22.5.2 Pattern development

Providing the manufacturer with accurate garment *patterns* is key to maintaining control of the garment design, size and fit, eliminating the potential for misinterpretation. Within the DfAW case study, the clothing range patterns were digitised using Gerber AccuMark specialist equipment and software through a regional textile technology centre called Coleg Sir Gâr. The pattern-making software has enabled the designers to add indication of the position of a pocket, popper or cable position, to scale, to eliminate any potential misinterpretation. Accurate 2D digital pattern information provided a critical starting point for the development of hybrid design specifications and subsequent hands-on 3D prototyping (Figure 22.8).

22.5.3 Liaison with garment industry partners

Key industry partners in the development of the DfAW garment range specifications included Sprayway for guidance on the production of sample garments overseas (China); SME Brenig for hands-on production of prototype garments in the UK (Wales); and Keela for the production of garment samples in the UK (Scotland) and overseas (Sri Lanka).

Close communication between the clothing team, fabric suppliers and industry partners throughout the DfAW case study helped the authors (with their product design and architecture competencies) to familiarise themselves with the possibilities for technical fabrics and finishes; joining techniques and assembly methods; as well as in developing a knowledge of factory processes, capabilities, and

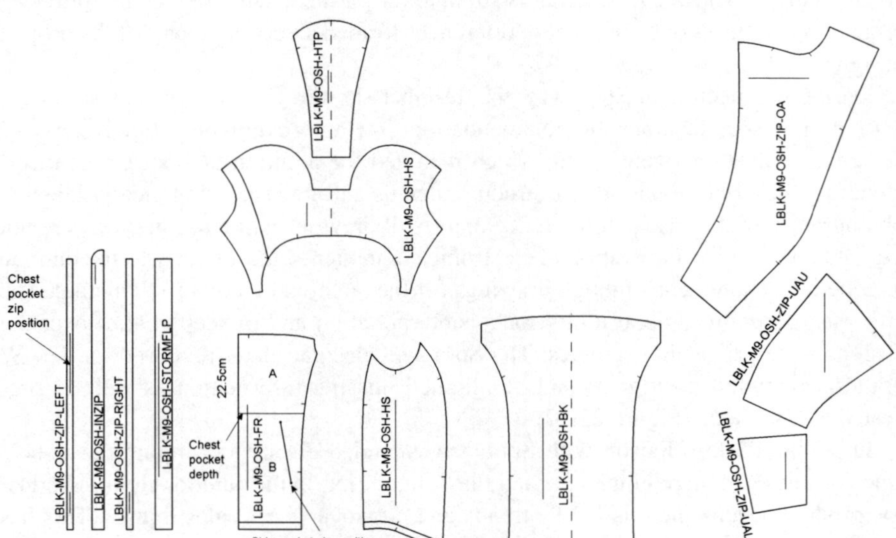

Figure 22.8 Accurate 2D digital pattern data for male outer layer garment. CAD pattern drawings can be printed or scaled at 1:1 for the addition of accurate technology placement information. CAD, computer-aided design.

methodologies. The design team have been able to maintain control over the production of garment samples by describing their design intent verbally, supported by accurate digital pattern information and initial 3D prototype toiles, whilst they worked up final technical drawings for the integration of electronic components. Design details were adapted where necessary based on the capabilities of a particular production facility.

22.5.3.1 Sprayway: UK brand, production in China

The UK brand Sprayway produces a wide range of outdoor clothing products that utilise innovative fabrics and new technologies. The brand has been an early adopter of wearable electronics, bringing iPod and radio jackets to market in 2007. Industry liaison with Sprayway was established early in the DfAW co-design process and the academic design team worked closely with an experienced designer from Sprayway who attended early co-design workshops.

The Sprayway designer developed new understanding of the needs of older consumers and helped to provide design guidance from an industry perspective. This involved demonstrating past Sprayway styles for reference, along with the selection and specification of innovative fabrics, details and finishes suitable for production. In addition, the academic design team attended meetings at the Sprayway offices in Manchester, UK, where researchers worked through the toiles of each of the individual 3D prototype styles in relation to user group feedback. Sample fabrics and colour ways were selected with reference to Sprayway products. Design details were referenced from commercial samples and included finishes in terms of seams, hems, hoods, collars, cuffs, pockets and zip guards suitable for garment sampling using Sprayway production facilities in China. These references formed an essential part of the project design specifications (Figure 22.9).

During the meeting at Sprayway, the designer showed examples of industry standard design specifications. Recommendations for improving the effectiveness of design specifications were given, which included the addition of final garment references to describe important construction details and finishes using clearly labelled photographs. In the case that reference images of previous garment samples were not available, i.e. for the integration of electronic components, the advice was to continue to generate detailed construction drawings for the positioning and infrastructure. The aim was to minimise opportunity for misinterpretation and to keep control over the design of the production samples. The Sprayway designer later reviewed the DfAW draft specification documents and highlighted missing information as well as suggested amendments (Figure 22.10).

In addition, close liaison with Sprayway included discussion around the importance of sales and marketing in driving the design brief. In the outdoor clothing industry, product development is led by trends and previous year's sales figures. This has led to continued neglect for older consumers (there is no trend for this demographic) and few garments for reference, as well as a hesitance to develop complex products that utilise wearable technology due to poor past sales figures (Greengrass, personal communication, September 20, 2011).

Figure 22.9 Initial 3D prototype toiles helped to demonstrate garment styles for the selection of fabrics, colour ways and construction details at Sprayway.

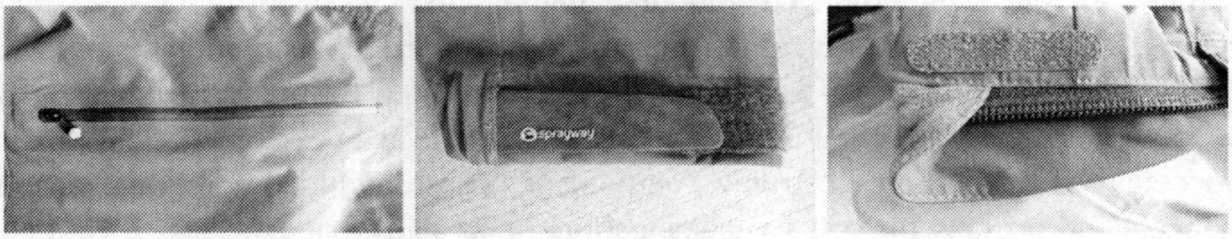

Figure 22.10 Photographs of sample garment construction details recorded to provide clear reference for detail design layout sheets (as shown in **Figure 22.16**).

22.5.3.2 Brenig: SME, production in UK

Brenig (SME), based in North Wales, UK, is specialised in producing extreme functional apparel developed for polar expedition, outdoor and countryside pursuits (Figure 22.11). The factory is equipped with sewing, taping and bonding capabilities as well as water-resistance testing equipment (called a *Suter Tester*) that obtains precise water pressure resistance levels of waterproof fabrics and taped seams. Owner and technologist Graham Ogle recognises the potential of wearable technology and has embraced it in developing an early prototype compass sleeve for use during extreme polar expeditions. In this, a microcontroller linked to a GPS triggers an LED to indicate when the jacket sleeve is pointing north.

Figure 22.11 Graham Ogle, owner of the Brenig Weather Proof Clothing Company (SME), at his factory in Holywell, North Wales, UK. Ogle has developed an early prototype compass sleeve for use during extreme polar expeditions and worked with the DfAW design team to develop construction details and efficient factory processes for the effective integration of wearable electronics. DfAW, Design for Ageing Well.

Brenig provided an ideal industry partner for the hands-on development of garment prototypes with integrated electronic componentry. The DfAW design team visited the factory for an introduction to the machinery, processes and capabilities. The team talked through the construction and patterns during this visit, which contributed to final specification for the production of the garment samples.

Initial samples were produced without technology in order to verify the size, shape and fit of the garments, to test the 2D patterns, and to clarify construction details. Following initial sampling from a distance, the team revisited Brenig to work alongside the factory operators to refine initial construction details and develop efficient factory processes for integrating electronic components. These included additional panels, channels and pockets for soft switches, cabling, heat pads, battery pack, BTMs and speakers. During the visit, the team produced fully functioning prototype garments for a heated gilet and outer layer jacket. The hands-on construction of garment prototypes is described in Section 22.6.3.

22.5.3.3 *Keela: UK brand, UK sampling and offshore manufacturing*

Keela is a design and manufacturer of outdoor clothing, based in Glenrothes, Scotland. The company specialises in waterproof clothing that utilises innovative fabrics developed in-house. Product prototyping is based in Scotland with mass-manufacture in Sri Lanka. The company is equipped with fabric laminating and testing capabilities, along with sewing, taping and bonding for garment assembly. Currently the brand successfully sells garments with integrated electronics, specifically heated midlayer garments.

The academic design team visited the factory in Scotland to discuss the production of selected garments. In particular, the female gilet style was selected with a view to adopting the heating technology *EXO*[2] with which the company has expertise. Further discussion took place on fabric selection and appropriate construction methods for the prototypes. This exercise established that the garment layering system, developed as a wearable technology platform, proved to be adaptable to technology components from a different provider.

22.6 Design communication

22.6.1 Garment specification

Communication through clear design specification is key to the reduction of errors made during sampling and manufacture (Greengrass, personal communication, September 20, 2011). In the garment industry, design teams normally provide their manufacturer with a design specification document that describes how a style within a new clothing range should be constructed. This document typically includes a design reference, size and grade, diagrammatic back and front design views, fabric selection, construction sequence and details (Keech, 2012, pp. 118–119). However, the specification will vary between design companies in the level of detail and communication techniques.

There are currently no industry standards for specification and so they can vary from a '20 page detailed spec to sketches on the back of a napkin' (Fowler, personal communication, November 11, 2010). Increasingly, garment patterns are no longer included as part of design specifications and 'eventually only sales, marketing and the designer's first sketch will remain in our home importing countries' (Birnbaum, 2012). The design team relies on the factory to work from a series of flat, illustrated sketches with measurements. As a result, the information gaps between the design team and factory inevitably leads to poor interpretation and inaccurate prototype samples (Fowler, personal communication, November 11, 2010). A single prototype can be expensive, time consuming and wasteful; this issue is discussed in Section 22.8.

22.6.2 Introducing new practices

In the clothing industry, it is standard practice to use 2D graphic programs such as Adobe Illustrator, Corel-DRAW, or industry specific equivalents such as Gerber Kaledo Style or Lectra Vision Fashion Studio for representing 2D details; 2D exploded detail drawings are not produced as standard. In general, vector-based software programs, such as Adobe Illustrator, are limited in their capability for producing detailed construction drawings for specifications. In particular, they do not provide appropriate scale or dimensional drawing facilities, which are necessary as part of a standard design information package.

There are a number of industry-specific 3D software programs being developed by Gerber, Lectra, Browzwear and OptiTex, amongst others. These combine pattern development software with 3D visualisation. However, research suggests that such software is still in its infancy and not yet suitable for detailed design, although is useful in the marketing stages (Easters, 2011).

The challenge for the design team was to produce 2D drawings and diagrams to clearly describe 3D constructions for the complex integration of electronic componentry. The team employed a number of techniques using Adobe Illustrator, which are familiar in product and architectural design, to communicate detail assemblies using exploded 3D representations, alongside detailed dimensioned drawings (Figure 22.12).

The design team illustrated key construction details including additional pockets for removable components and cable integrations, through 3D exploded drawings (see Figures 22.13 and 22.14). This technique is common practice to describe a construction, or *build up,* in product design and architecture. 3D diagrams were combined with conventional clothing seam profiles (see Figure 22.15) and garment illustration *flats*.

Photographs of sample construction details from reference garments provided clarity for chosen stitching, edge finishes, collar details, pockets and hoods. Final design information included a combination of technology system diagrams, key dimensions, 3D exploded views as well as seam profiles and reference images. Such detailed design information aims to reduce the potential for misinterpretation when working with overseas production facilities. For successful communication with overseas manufacturers, it is essential to provide a detailed set of design information that goes beyond conventional 2D line sketches and vague annotation (Greengrass, personal communication, September 20, 2011) (Figure 22.16).

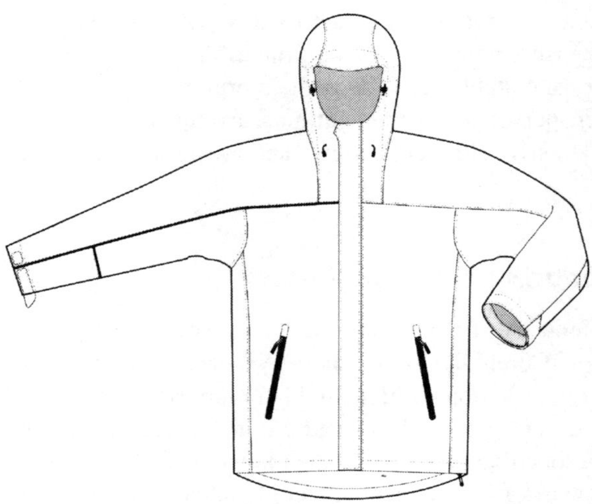

Figure 22.12 Male outer layer jacket drawn in Adobe Illustrator as a basis for developing technology positioning and detailed construction diagrams.

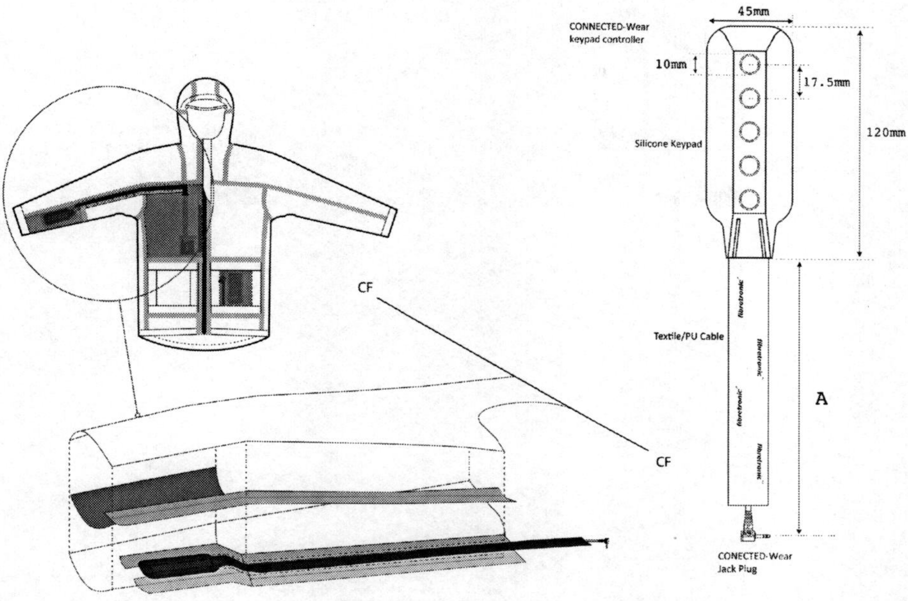

Figure 22.13 3D exploded drawing of a suggested soft switch and cabling integration detail. Switch pad dimensions are provided as well as a garment key to indicate technology positioning. Additional pattern pieces and accurate switch pad positioning were marked onto associated digital pattern pieces and provided as part of final design specification.

22.6.3 Hands-on prototype development

22.6.3.1 Outer layer jacket

Together with the Brenig machinists, the design team fabricated an outer layer jacket with integrated technology including the Fibretronic BTM, soft control switches and PTT system (see Figures 22.17–22.23). The aim was to test the construction and assembly details that had been developed through the co-design specification process. This relied on information from accurate digital patterns, initial 3D toiles and findings from user-led workshops that defined technology placement, usability, garment comfort and fit. Construction details and processes were adapted to align with Brenig factory capabilities.

22.6.3.2 Heated gilet

In parallel with the development of a prototype outer layer jacket, the team developed final construction details and a technology architecture for the female midlayer gilet (see Figures 22.24–22.26) with integrated HEATwear system, supplied by Fibretronic (see Table 22.2 and Section 22.3) (Figure 22.27).

Accurate digital pattern information was used for the placement of a soft heat panel, battery pack and remote LED controls, with reference to the initial 3D toiles.

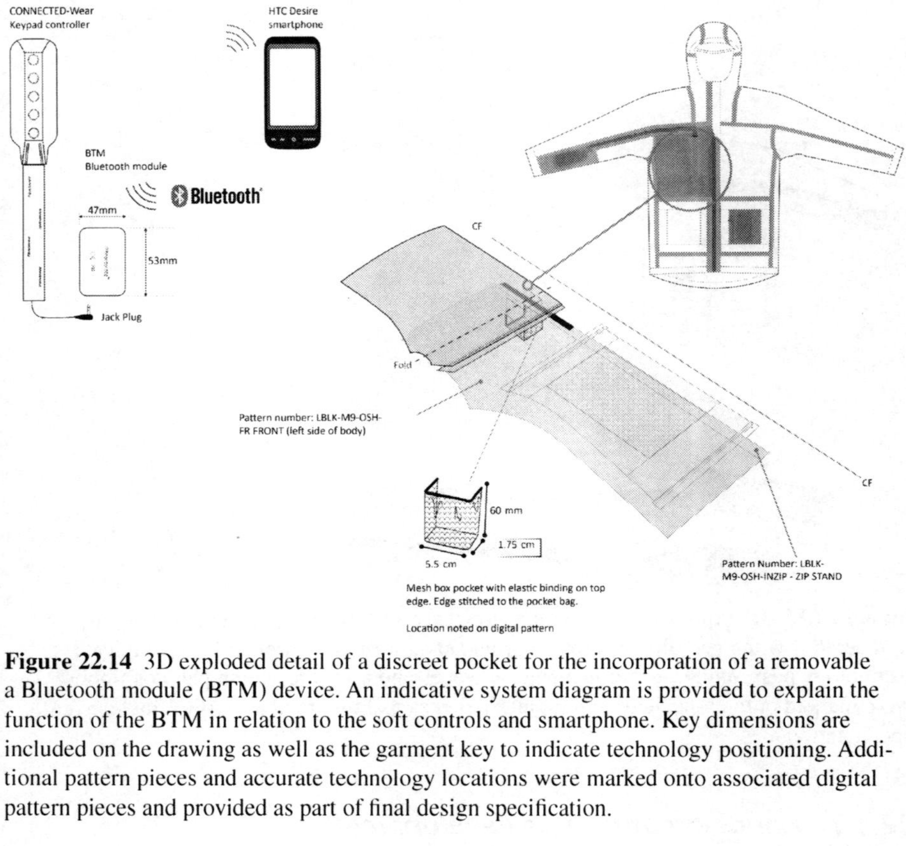

Figure 22.14 3D exploded detail of a discreet pocket for the incorporation of a removable a Bluetooth module (BTM) device. An indicative system diagram is provided to explain the function of the BTM in relation to the soft controls and smartphone. Key dimensions are included on the drawing as well as the garment key to indicate technology positioning. Additional pattern pieces and accurate technology locations were marked onto associated digital pattern pieces and provided as part of final design specification.

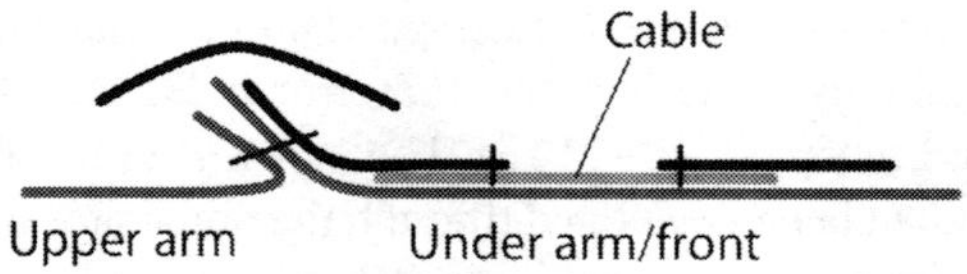

Figure 22.15 Conventional seam profiles were drawn and incorporated into design specification layouts.

Effective specification for the integration of wearable technology components for large-scale production at Brenig had been developed through hands-on prototyping in close liaison with the factory. Solutions developed were discreet and maintained their optimum weatherproofing, desired aesthetics, comfort and fit. Through hands-on prototyping, the team were able to finalise their garment construction details that combined end-user needs with simple manufacturing details. This highly collaborative approach enabled creative design decisions to be made immediately and in-situ, leading to the creation of successful, proof of concept wearable technology platforms.

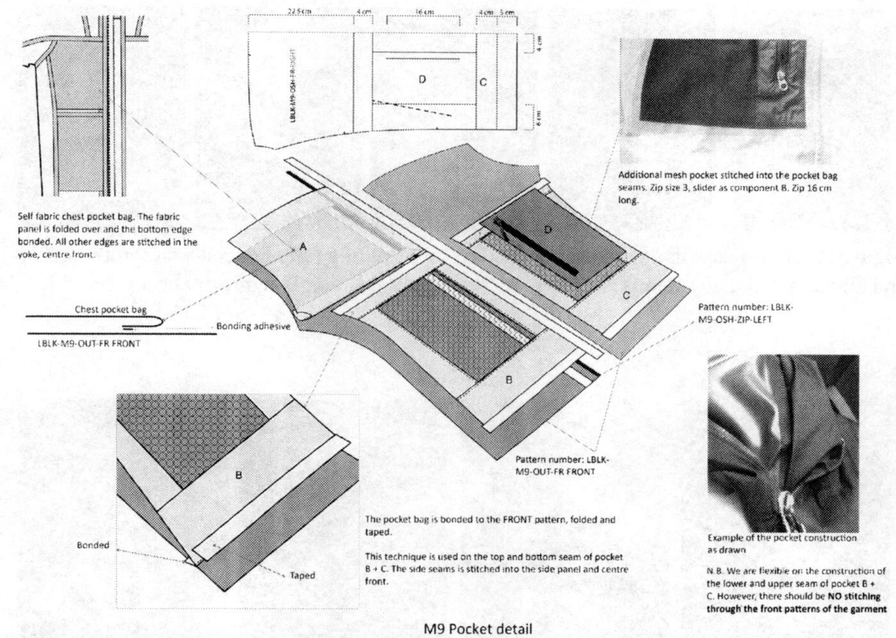

Figure 22.16 Detailed design information. 3D exploded construction drawings, key dimensions, seam profiles and garment 'flats' were developed, and presented as detail layout sheets alongside photographs of garment samples.

Figure 22.17 The DfAW design team worked with Brenig factory machinists to produce 3D garment prototypes with their new architecture for integrating technology components. Digital garment specifications, garment pattern pieces and detail layout sheets were amended as new construction details and factory processes developed. DfAW, Design for Ageing Well.

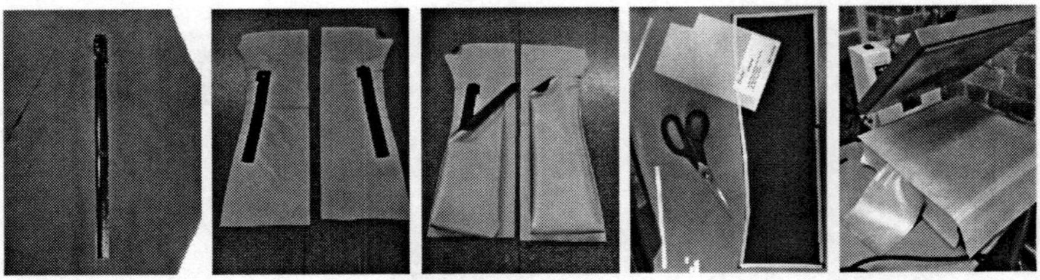

Figure 22.18 Process images showing the development of a fully bonded waterproof pocket and zipper detail using Bemis 'Sewfree' technology and Brenig's industrial heat press.

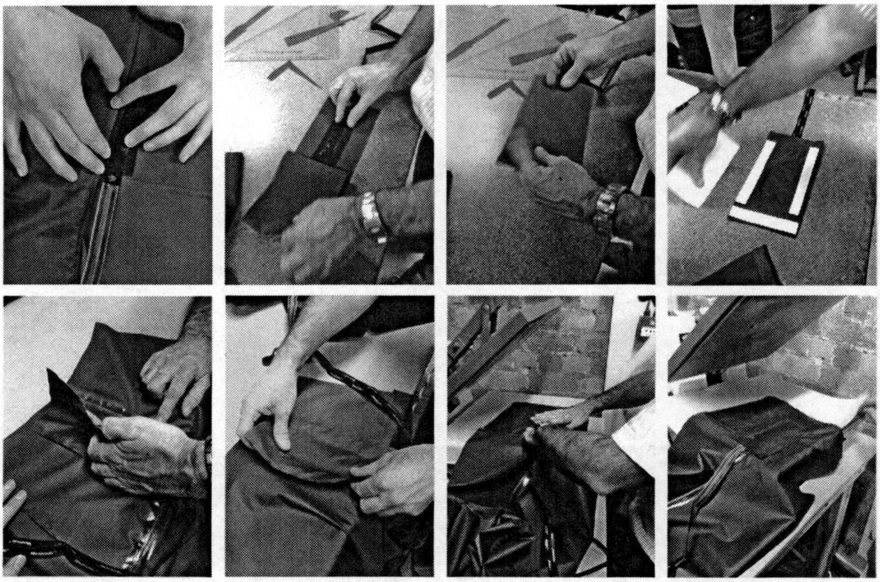

Figure 22.19 Prototyping the integration of soft control switches based on draft construction details. The controls are first positioned in relation to the sleeve seam and then stitched between two additional fabric panels and bonded to the assembly. The assembly maintained sleeve waterproofing when tested on the Suter Tester, and the control pad was held firmly in place.

22.7 Example: hybrid design specification

The following example specification, shown in Figures 22.28–22.34, describes the male outer layer garment consisting of illustration flats, fabric selection, diagrammatic construction details, electronic component integration, technical pattern register and bill of components. A complete set of digital patterns were supplied to the manufacturers. The clarity of this hybrid specification was verified in liaison with an industry designer experienced in regular liaison with a breadth of overseas and technical garment production facilities.

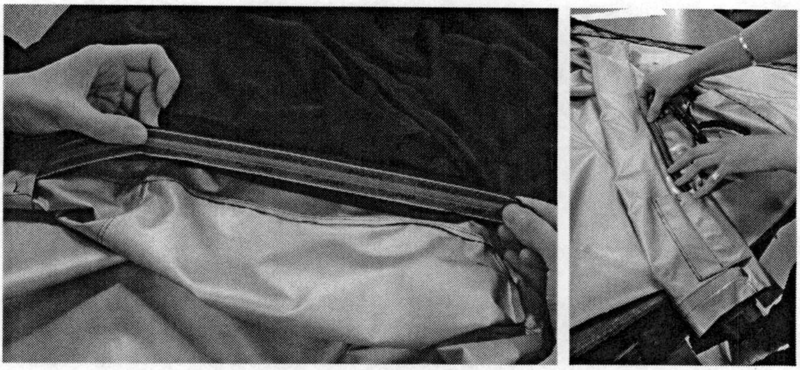

Figure 22.20 Cable channels were stitched along the inside sleeve seam to integrate cables required to connect the control pad to a Bluetooth module positioned in the jacket front panel (see **Figure 22.23**).

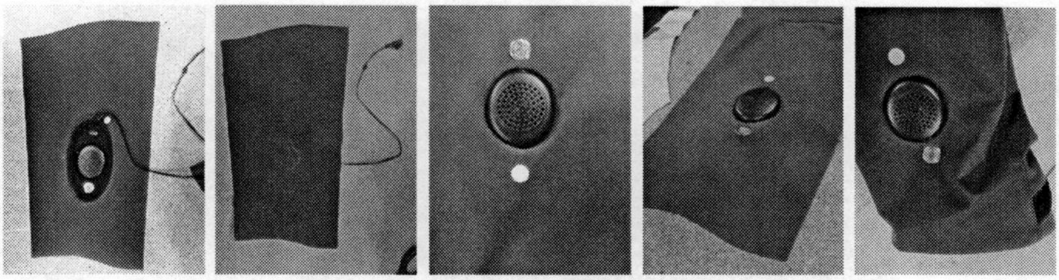

Figure 22.21 A push-to-talk (PTT) system was carefully positioned and discreetly integrated into the forearm panel. A simple circular graphic cut from 3M reflective tape would locate the 'push' function. For higher quantity production, the forearm pattern piece could be laser cut using digital pattern information to accurately locate the PTT module.

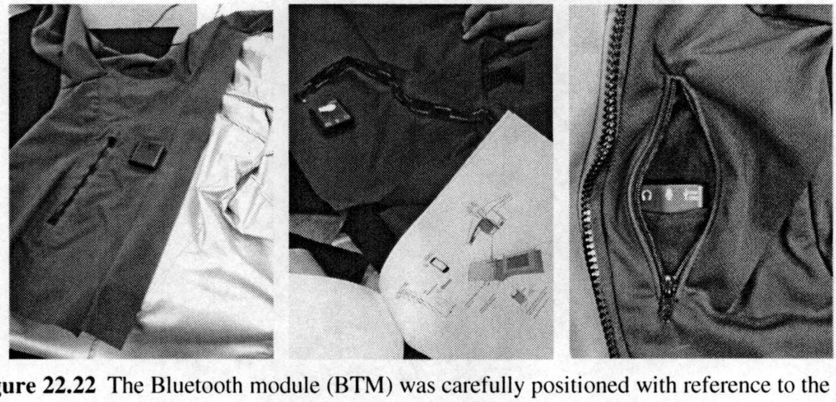

Figure 22.22 The Bluetooth module (BTM) was carefully positioned with reference to the garment key indicating technology positioning. A custom elasticated pocket was designed for the BTM, integrated into a secure internal pocket on the jacket front panel. The control pad cable and jack were pulled through the interior jacket lining to connect with the BTM socket. Since the BTM cannot be laundered, it was designed to be easily disconnected and removed.

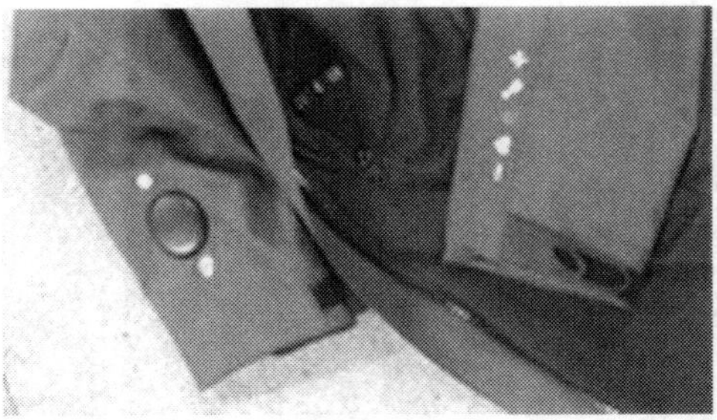

Figure 22.23 Final working prototype of an outer layer jacket transformed into a wearable technology platform with integrated soft control panels, push-to-talk (PTT) system, and Bluetooth module (BTM). The jacket and controls are linked wirelessly to a smartphone application via the BTM wireless receiver. Users can make a phone call via the PTT or to call emergency aid via the soft controls. They can also hear their distance travelled and current heart rate via the Shimmer sensor located in a body layer garment. Discreet graphic icons were designed, cut from 3M reflective tape and bonded to the sleeve to locate the volume, step count, heart rate and emergency functions.

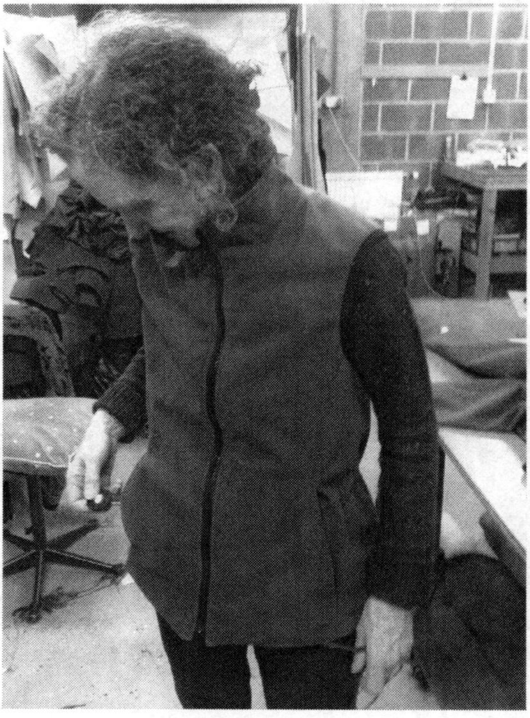

Figure 22.24 Working prototype for the female heated gilet.

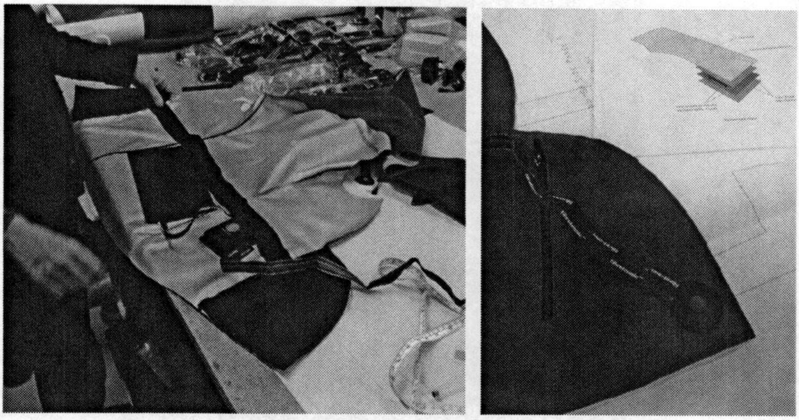

Figure 22.25 Positioning the soft heat pad panel, battery, cabling and LED controls with close reference to the design team's draft garment specifications.

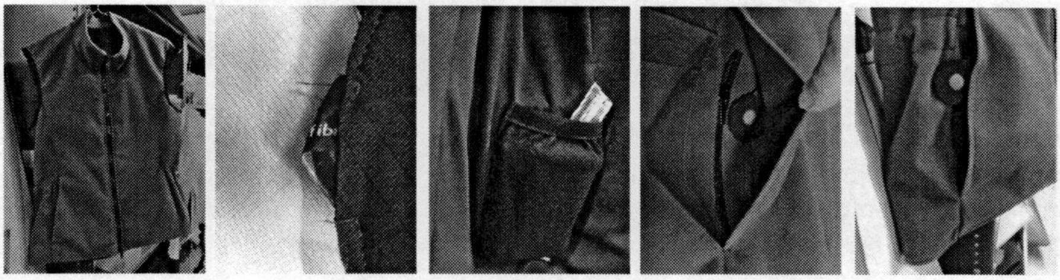

Figure 22.26 Discreet technology architecture for the heated gilet. Cables run through internal seams; an elasticated pocket within the garment side panel holds the battery pack and allows easy removal; and LED controls are hidden discreetly in a fleece-lined pocket for easy control of heat settings.

22.8 Challenges in the global clothing supply chain

The clothing supply chain is one of the most globalised industries. Typically, labour-intensive manufacturing is sourced overseas, leaving the design, marketing, sourcing, supply chain and logistic-based management within the developed West (Tyler, 2008, pp. 288–298). Separation of design cultures in the West and production in the East adds significant complexity to the effective manufacture of functional outdoor clothing and high-performance sportswear. Those involved in the process need to overcome language and cultural differences, large distances and time zones, varying capabilities from one factory to another and global issues related to evolving economies or natural disasters. With the multitude of people and barriers through the supply chain, relationships and communications between designers and the factory are becoming more important than ever before.

The outsourcing of pattern-making and product development skills to overseas manufacturers is causing the creative process to become gradually even further

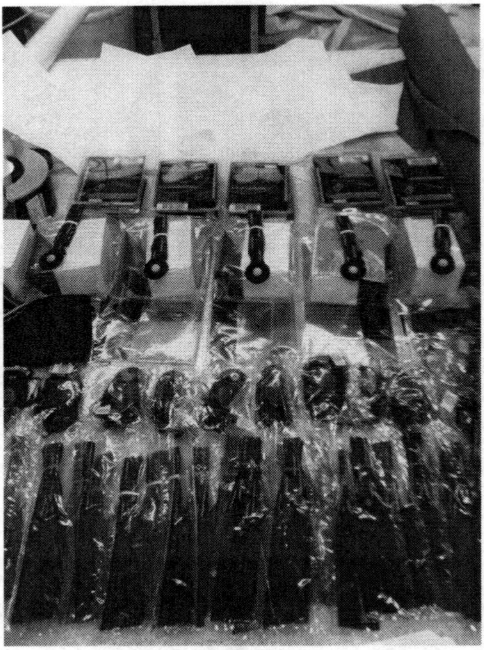

Figure 22.27 This shows the full supply of Fibretronic soft controls, PTT systems, BTM and Heatwear systems to Brenig for the full production of DfAW male and female outer layer garments, outer soft shell and heated gilets. PTT, push-to-talk; BTM, Bluetooth module; DfAW, Design for Ageing Well.

detached from production. Loss of skills for mass manufacturing in the West has led to a lack of first-hand experience in production among designers (Greengrass, 2012). A lack of consideration for the practical aspects of production can lead to significant problems later in the process; this is often during the development of prototype samples.

The disconnection between design and production is further compounded by factory workers' lack of design training and understanding of a product's intended end use (Ayub, personal communication, October 21, 2011). Designers talk the language of aesthetics, creativity and fashion trends, whilst the manufacturing culture in the East is concerned with processes, resource allocation, material management and unit costs (Tyler, 2008, pp. 288–298). Lack of technical skill from designers in the West leads to ineffective design specifications, which are then lost in translation. For instance, between two and ten prototype iterations will be necessary in the current design development process (Sayem, Kennon, & Clarke, 2010). This number is further increased when samples are sourced from more than one production facility.

Figure 22.35 describes the garment-sampling process of a sizable Western sportswear company manufacturing in China, with a buying agency based in Hong Kong. It provides an overview to demonstrate the complexity and phases involved in

M9 male mid-layer outer shell specification notes

Date	19/02/2012	Season	N/A	Product Name	M9 Outer
Garment reference	M9	Style/block	N/A		N/A
Garment	Male OUTER	Gender	Mens	Sample size	LARGE
Manufacturer	Sprayway	Contact	Chris Lewis	Liaison facilitator	Julie Greengrass

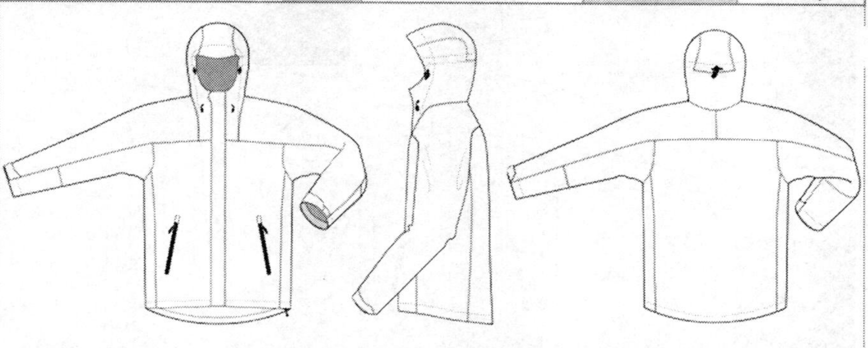

The garment is designed, as part of a range, to fit the shape and size of the 'active aging' - the active market aged 60-75 years old. The shape of the garment is cut to provide arm lift and arm articulation for urban and outdoor walking.

General notes	Patterns supplied. Note : no seamtallowance included on the patterns.

Main fabric

Supplier	Gore Tex	Article/quality ref	SDO80TC/TR 3L	Weight	176 g/m²	Width	142cm
Composition	Shell: 100% PA. Functional layer: Gore-Tex Membrane. Backer: 100% PES. Laminate content: 53% PA, 20% PES, 27% ePTFE			Minimum order quantity			
Releant performance test information							

Second fabric facings

Supplier	Youngone	Article/Quality ref		Weight		Width	
Composition	Tricot			Minimum order quantity			
Relevant Performance Test Information							

Third fabric

Supplier	Youngone	Article/Quality ref		Weight		Width	
Composition	Mesh 100%polyester			Minimum order quantity			
Relevant Performance Test Information							

Colour Pantone Reference

Main Fabric	n/a	Fabric 2	n/a	Fabric 3	n/a	Fabric 4	n/a	Fabric 5	n/a

Pattern pieces	Quantity	Refer to attached pattern drawings	Grain Direction	Fabric
1	2	LBLK-M9-OSH-FR (FRONT)	Grain	SDO80TC/TR 3L
2	1	LBLK-M9-OSH-BK (BACK)	Grain	SDO80TC/TR 3L
3	1	LBLK-M9-OSH-OA (OVER ARM)	Shown on patterns	SDO80TC/TR 3L
4	2	LBLK-M9-OSH- UAU (UNDER ARM UPPER)	Bias	SDO80TC/TR 3L
5	2	LBLK-M9-OSH- UAL UNDER (ARM LOWER)	Grain	SDO80TC/TR 3L
6	1	LBLK-M9-OSH- HS (HOOD)	Grain	SDO80TC/TR 3L
7	1	LBLK-M9-OSH- HTP (HOOD TOP PANEL)	Grain	SDO80TC/TR 3L
8	2	LBLK-M9-OSH- SP (SIDE PANEL)	Grain	SDO80TC/TR 3L
9	1	LBLK-M9-OSH- STORMFLP (STORM FLAP)	Grain	SDO80TC/TR 3L
10	1	LBLK-M9-OSH-INZIP (INNER ZIP GUARD)	Grain	SDO80TC/TR 3L
11	1	LBLK-M9-OSH-ZIP-LEFT (ZIP STAND)	Grain	SDO80TC/TR 3L
12	1	LBLK-M9-OSH-ZIP-RIGHT (ZIP STAND)	Grain	SDO80TC/TR 3L

Figure 22.28 Specification notes for male outer layer jacket (shell) for manufacture, which includes fabric selection, pantone colour reference and a technical pattern register.

Manufacturer	Sprayway	**Designer**		Jane McCann	
Season	n/a	**Garment Tech**		Chris Lewis	
Date	20/02/12	**Garment Ref**	M9	**Iteration**	1

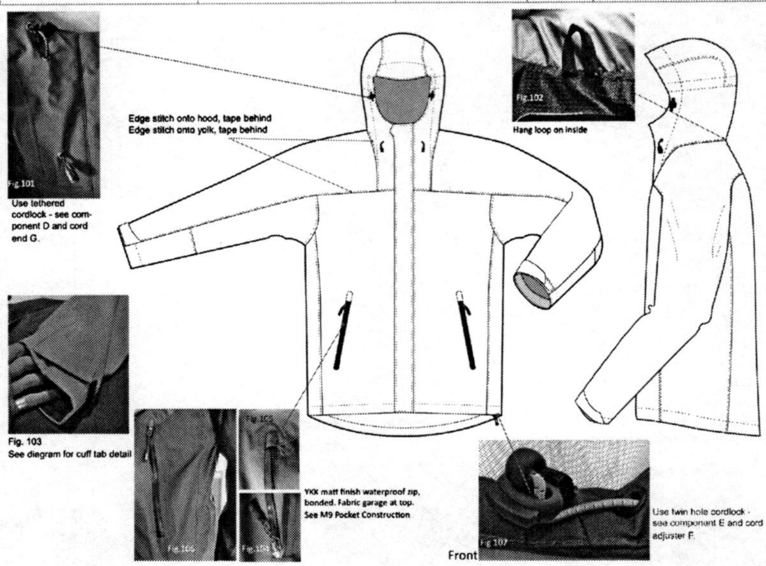

Manufacturer	Sprayway	**Designer**		Jane McCann	
Season	n/a	**Garment Tech**		Chris Lewis	
Date	20/02/12	**Garment Ref**	M9	**Iteration**	1

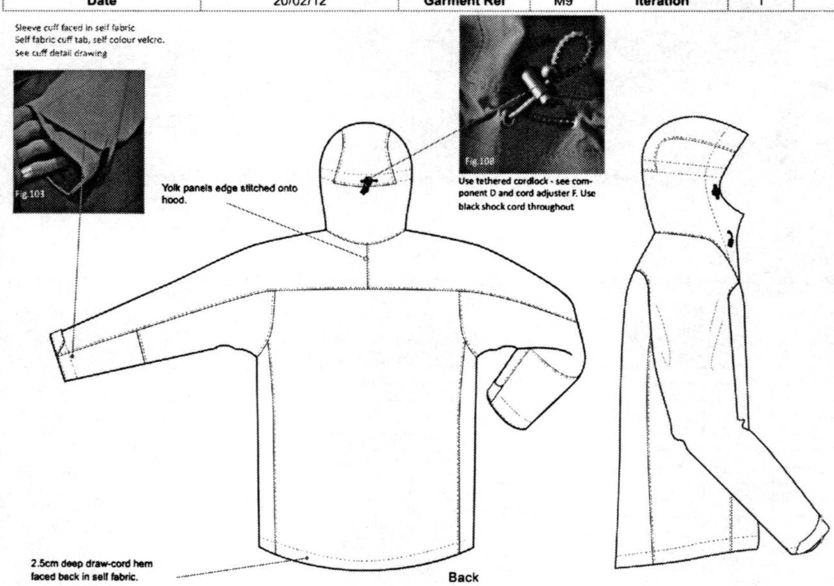

Figure 22.29 Front, back and side views together with clear photographic references provide clear guidance for garment detailing to include stitching, hood detail, zipper, drawstrings and cuff.

Manufacturer	Sprayway	Designer		Jane McCann	
Season	n/a	Garment Tech		Chris Lewis	
Date	19/02/12	Garment Ref	M9	Iteration	1

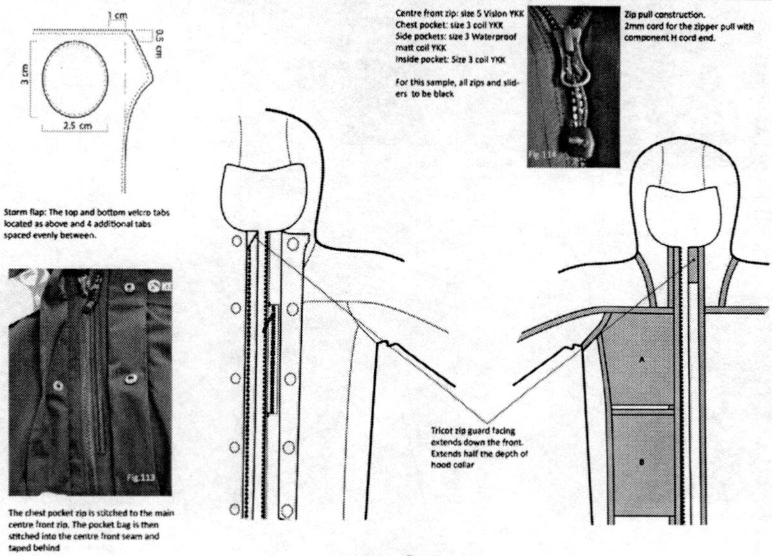

Storm flap

Manufacturer	Sprayway	Designer		Jane McCann	
Season	n/a	Garment Tech		Chris Lewis	
Date	19/02/12	Garment Ref	M9	Iteration	1

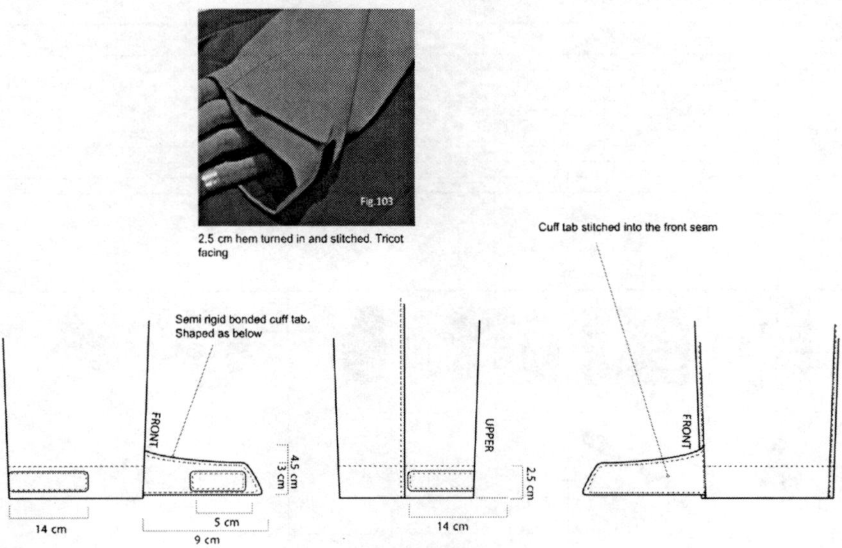

Cuff detail

Figure 22.30 Details 1 and 2.

Manufacturer	Sprayway	**Designer**		Jane McCann	
Season	n/a	**Garment Tech**		Chris Lewis	
Date	19/02/12	**Garment Ref**	M9	**Iteration**	1

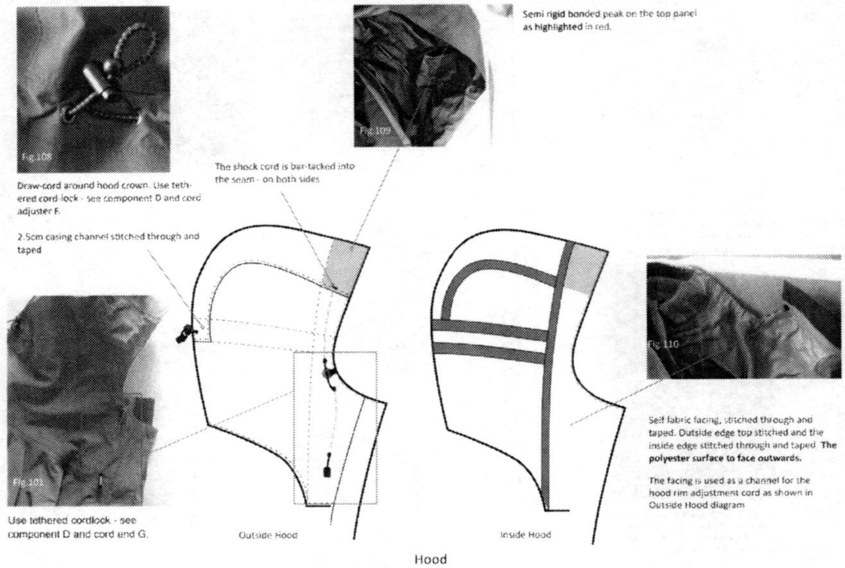

Manufacturer	Sprayway	**Designer**		Jane McCann	
Season	n/a	**Garment Tech**		Chris Lewis	
Date	20/02/12	**Garment Ref**	M9	**Iteration**	1

Trims				
	Component	**Position**	**Supplier**	**Article Reference**
A	Centre front zip slider	Centre zip	YKK	As pictured - black
B	Side pocket zip slider	Side pocket	YKK	As pictured - black
C	Chest+ inside pocket zip slider	Chest pocket	YKK	As pictured - black
D	Hood single cord lock	Hood compression around hood rim and back	Duraflex HK LTD	P185
E	Hem drawstring twin hole cord lock	Hem - left side of garment	Duraflex HK LTD	P871
F	Cord Adjuster	Hem and hood back cord pulls	Duraflex HK LTD	P386
G	Cord ends	Hood rim compression cord ends	Duraflex HK LTD	P-YL475
H	Zip pulls	All zips	Duraflex HK LTD	P494
I	Elasticated cord	Hood and hem adjustment	Youngone	
Notes	Centre front zip: size 5 Vislon YKK Chest pocket: size 3 coil YKK Side pockets: size 3 Waterproof matt coil YKK Inside pocket: Size 3 coil YKK For this sample, all zips and sliders to be black			

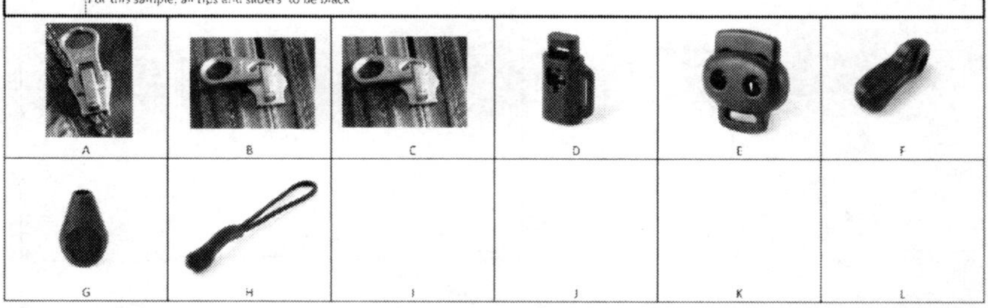

Figure 22.31 Detail 3 and 4.

prototyping samples overseas, as a design specification passes between many individuals in a number of languages.

The typical sequence begins when a specification is sent to the manufacturer via an agency and received by factory representatives. The specification is translated into

Manufactuer	Sprayway	**Designer**		Jane McCann			
Season	n/a	**Garment Tech**		Chris Lewis			
Date	19/02/12	**Garment Ref**	M9	**Iteration**	1		

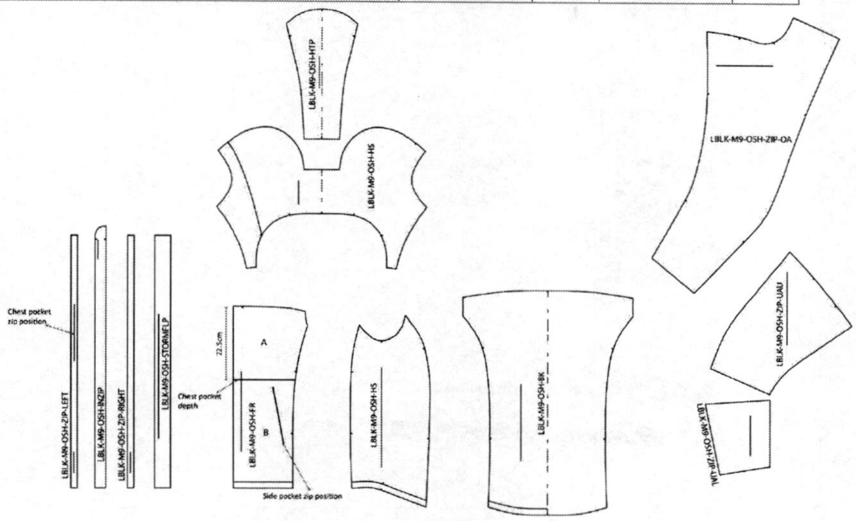

Current technical patterns

Manufactuer	Youngone	**Designer**		Jane McCann	
Season	n/a	**Garment Tech**		Chris Lewis	
Date	19/02/12	**Garment Ref**	M9	**Iteration**	1

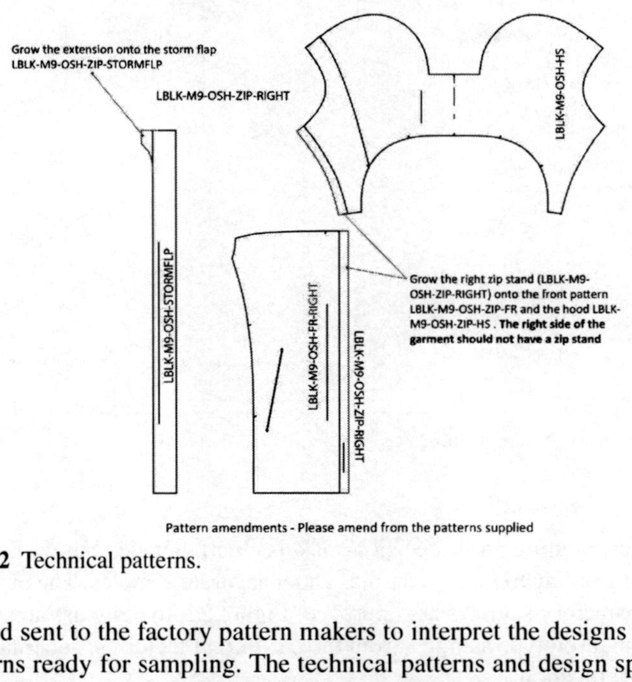

Pattern amendments - Please amend from the patterns supplied

Figure 22.32 Technical patterns.

Chinese and sent to the factory pattern makers to interpret the designs and draft technical patterns ready for sampling. The technical patterns and design specification are finally passed on to the sampling team, who plan and produce the garment prototype. Any discrepancy or queries related to the designs have to be fed back up through this

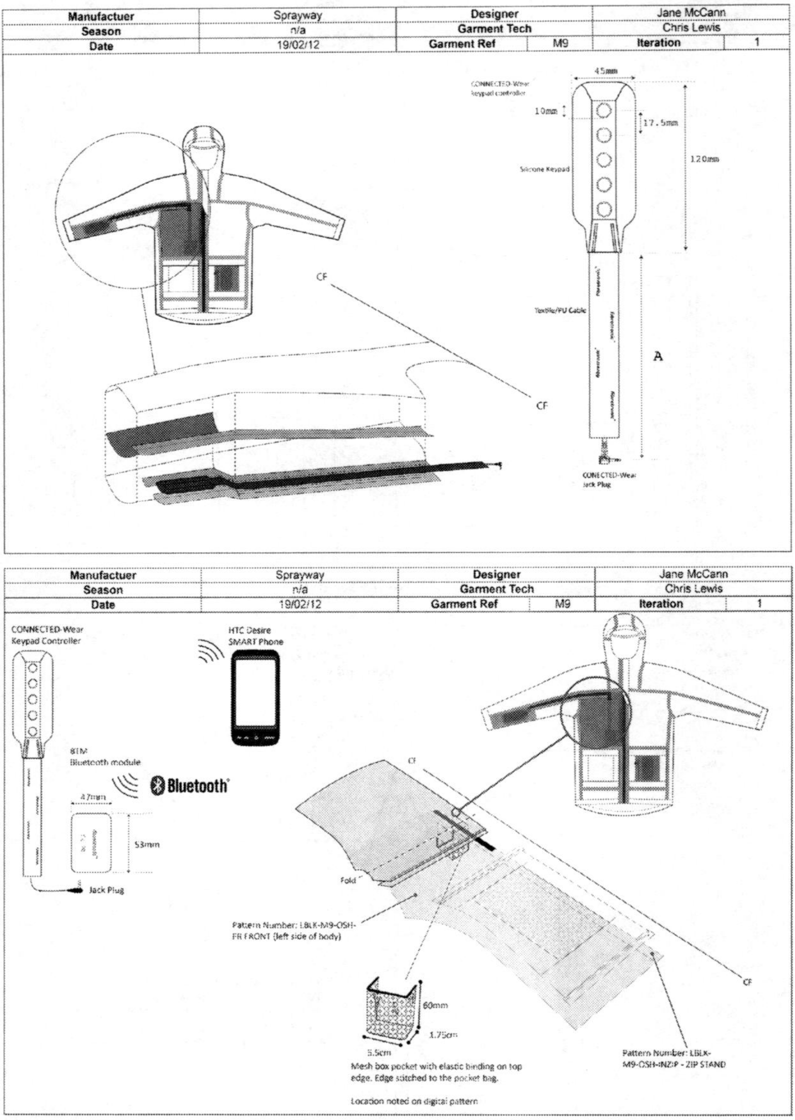

Figure 22.33 Technology integration.

hierarchy to the design team or decision-makers. Inefficiencies build up throughout the process, contributing to long lead times and inaccurate samples. The process has to repeat until satisfactory samples are approved. Figure 22.36 demonstrates this inconsistency, showing prototype samples from three different factories, generated from the same design specification.

Manufactuer	Sprayway	Designer		Jane McCann	
Season	n/a	Garment Tech		Chris Lewis	
Date	19/02/12	Garment Ref	M9	Iteration	1

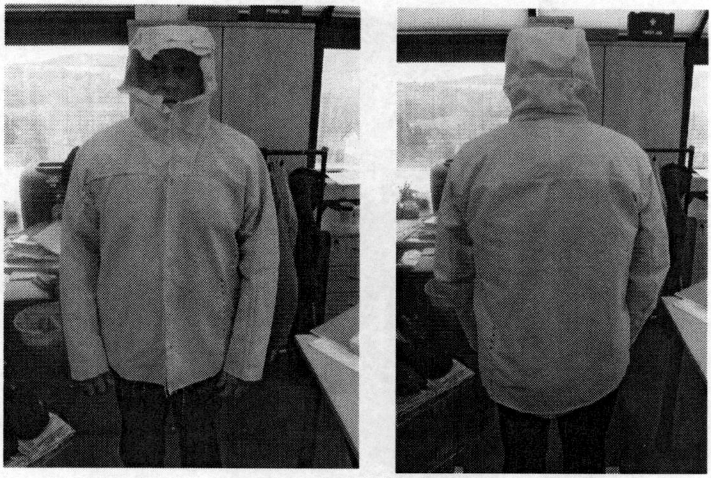

M9 toile images

Product background
The garment is designed, as part of a range, to fit the shape and size of the 'active aging' - the active market aged 60-75 years old. The shape of the garment is cut to provide arm lift and arm articulation for urban and outdoor walking.

Figure 22.34 3D toile.

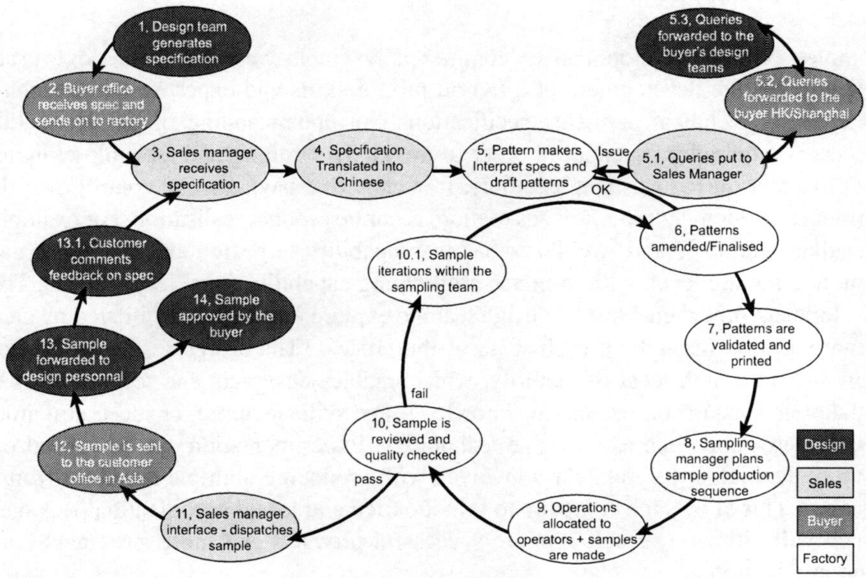

Figure 22.35 Factory sampling process across a complex global supply chain for a large Western sportswear company manufacturing in China.

Figure 22.36 Initial samples for baselayer garments received from three different factories using the same specification document.

22.9 Conclusion: more sustainable garment development

Complex product development of commercially viable wearable technology products requires the development of a hybrid mix of skills and expertise for producing increasingly technical garment specifications. An understanding of the production processes required is essential in order to merge the clothing and technology industries together during garment assembly. It is clear that having more control over the garment development process leads to more accurate product realisation. For example, Canadian outdoor brand Arc'Teryx has the capability to design and develop a garment to a mature level, with in-house prototyping capability (Arc'Teryx, 2012). This development model enables the design team to explore innovation, facilitated by clear communication throughout each stage of the process. The design team can develop a product to a high level of maturity, which enables designers and technologists to collaborate from the outset, sharing knowledge and skills required for successful products in new areas such as wearable technology. Experimentation can be carried out without the high costs and delays involved with producing multiple iterations from a distance. This allows the company to provide tried and tested specification packages, which will ultimately result in more successful products and more sustainable offshore production.

However, an in-house prototyping capability is costly and many design-led companies now rely on the manufacturer (typically overseas) to interpret design specifications from 2D flat drawings and size charts. The waste resulting from inaccurate interpretations signals a clear case for change, but 'people will only change when they perceive benefits and where there are cost drivers' (Tyler, 2008, pp. 288–298). In addition, the development of a product to a mature stage through *co-design* will add to this expense, in recruiting and working with users. Tyler (2008, pp. 288–298) encourages the principle of *concurrent product development* whereby the multiple disciplines in the design process work simultaneously, with the technologists involved from an early stage to increase collaboration and understanding throughout the process. Concurrent design can be considered similar to co-design with the use of a multidisciplinary design team to inform more effective specification of end-products and manufacturing process.

For the development of novel production processes in the absence of an in-house prototyping facility, it is important for the design team to build a close relationship with their production base. This enables decisions to be made quickly. Provision of digital patterns to accurately describe technology placement and clear technical drawings to describe complex constructions should be a minimum requirement. Today there are various software tools that are facilitating a more effective design specification process, such as pattern development software and Product Lifecycle Management (PLM) software, which claims 'right first time' samples. However, these programs still require design teams to describe a product assembly in 2D software programs since there is no industry standard as guidance. In particular, there are very few precedents for methods relating to the integration of wearable electronics.

The 3D design software is not yet established as a standard tool for the fashion industry (Easters, 2011), but 'innovative integrated processes from virtual product development to production now have the potential to close the gap to enhance process control, saving time and cost' (McCann, 2013). As the trend for wearable technology continues to increase, design-tool software companies will inevitably take on the challenge of incorporating new facilities into their programs. This may include the development of technical CAD libraries and product design software that support new types of rapid prototyping and digital fabrication techniques such as 3D printing or digitally printed/embroidered electronic circuitry with the potential for integrating wearable electronic componentry more efficiently, accurately and at a lower cost.

22.10 Future trends

> *If technology is to start becoming a common part of everyday clothing, then it should be integrated in a way that it provides a useful additional function while being visually understated.*
>
> *(Financial Times, 2013)*

In a recent article for the *Guardian*, Moritz Waldemeyer, who worked for 10 years in Phillips Future Labs, acknowledges that it is hard to make wearable technology work in a way that's elegant and reliable (The Guardian, 2013a). There is endless further potential for meaningful, wearable products that prompt and empower us to become healthier, safer, more socially responsible, more comfortable and better connected to our surroundings.

But in order for wearable technology to gain widespread popularity as it enters this new phase, it seems critical to design products that really add value to users; after all, people have to *want* to wear technology. This requires an equal balance between design and functionality, aesthetics and innovation, as well as understanding of specific end-user contexts.

At the time of writing, there has been continuous ongoing debate, hype and trepidation surrounding the future of wearable technology. Predicted to be the next major technology battleground, statistics show that the market will be worth more than £ 887 million by the end of 2014 (Financial Times, 2013) and that wearable technology products will increasingly become part of our daily lives. The active ageing market or baby boomer generation will undoubtedly benefit from the 'booming intersection between wearable technology and the self-improvement industry, as a generation that devoted unprecedented energy toward understanding their psyche, their emotional complexion and their physical wellbeing' (Future Human, 2013).

A recent article on the Fast Co-Design Website (Fast Co-design, 2013) observed that wearables are creating a paradigm shift in the way we interact with technology, using the body as a platform for interaction. However, this must be supported by a consideration for the aesthetic value of what it means to wear, rather than carry, something. For this to happen, the technology industry needs to move away from the wearable technology in its current 'brick phone' phase and place design and clothing at the centre of new innovations to make them desirable, wearable and therefore commercially viable. However, encouraging the technology industry to place *design* at the centre of the development of new products is a difficult challenge, one which will demand a new breed of designers and technologists who are prepared to bridge the gaps between their knowledge and professional cultures. Figure 22.37 shows sketches from a conversation between a designer and technology developer during the DfAW garment development, debating the form factor of an ECG sensor to be integrated within a close-fitting baselayer garment. The sketch design and prototype configurations demonstrate that rather than stacking components into a hard, box-like casing, the battery and electronic circuit board could be reconfigured side by side and packaged into a longer, slimmer, possibly

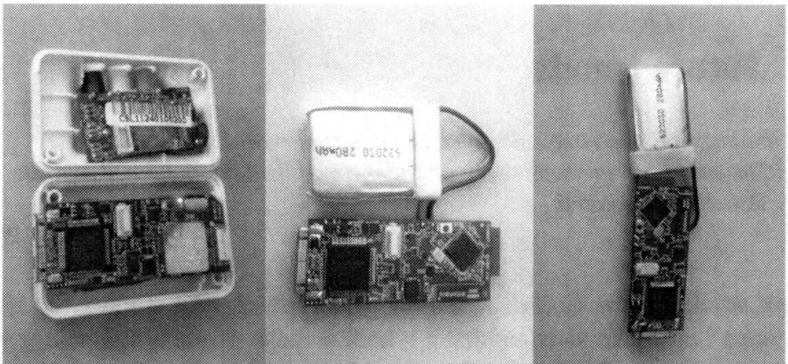

Figure 22.37 Prototype options for reconfiguring the Shimmer electrocardiogram sensor to achieve a slimmer form factor for discreet integration and improved comfort.

3D-printed, soft rubber casing. This would reduce bulk for the wearer, improve comfort by following the contours of the body and facilitate more *discreet* technology integration.

Wearable technology will develop out of close collaboration between designers and technologists who are willing to familiarise themselves with each other's processes and priorities. Importantly, they should share a motivation for finding out what people really would find useful and feel good wearing.

It appears that major technology companies are realising the importance of embracing design for the success of future wearable technology products, most recently during CES 2014, the consumer technology trade show where 'design or lack thereof, was a big debate throughout the week' (Telegraph, 2014). The microprocessor giant Intel argues that as technology becomes increasingly embedded into our lives, we need to focus on how we use it and 'approach these things from the point of view of what technology needs to be invented and made production-worthy' (The Guardian, 2013b). The vice president of Intel's New Devices Group now recognises that 'if we want the premise of wearable technology to come forward we really have to think about going back to the drawing board with the technology hardware, moving beyond the idea of a square block on your wrist' (Telegraph, 2014).

The Design for Ageing Well functional garments with integrated *wearable* technology provide a proof-of-concept for outdoor clothing that promotes *discreet* integration of technology. The product development and design specifications were closely informed by a group of end-users who made significant contributions to the final design. The positive outcomes from the case study demonstrate clearly that embracing a collaborative, co-design approach is a promising strategy for the development of more meaningful, attractive and commercially viable wearable technology products that will ultimately add value to our lives and be more successful, helping to drive future development and ways forward for the industry.

References

Arc'Teryx. (2012). *Manufacturing statement*. [Online] Available at http://arcteryx.com/Manufacturing.aspx. Accessed 23.09.12.

Birnbaum, D. (2012). *Where has globalisation left the garment industry?* [Online] Available at http://www.just-style.com/comment/where-has-globalisation-left-the-garment-industry_id116479.aspx. Accessed 07.07.13.

Design Council. (2012). *Introducing product design*. [Online] Available at http://www.designcouncil.org.uk. Accessed 09.10.12.

Easters, D. J. (2011). Global communication part 1: the use of apparel CAD technology. *International Journal of Fashion Design, Technology and Education*, *5*(1), 45–54.

Fast Co-design. (2013). *3 ways to make wearable tech actually wearable*. [Online] Available at http://www.fastcodesign.com/1672107/3-ways-to-make-wearable-tech-actually-wearable. Accessed 29.06.13.

Fibretronic. (2013). [Online] Available at http://fibretronic.com/ Accessed 28/04/13.

Financial Times. (2013). *On-trend tech, 'wearable' technology needs to be subtle.* [Online] Available at http://www.ft.com/cms/s/2/67a40bfa-9d3a-11e2-a8db-00144feabdc0.html#axzz2RfmlniQ1www.ft.com. Accessed 29.06.13.

Future Human. (2013). *Future human smart self.* [Online] Available at http://www.futurehuman.co.uk/2013/02/future-human-smart-self-wednesday-march-13/. Accessed 24.03.2013.

Greengrass, J. (2012). *Talk on design innovation at wearable futures: Inclusive design conference.* London: Wellcome Collection, [27.03.12].

Keech, P. (2012). *The freelance fashion designer's handbook.* Singapore: Trade Gothic by MPS Limited. pp. 118–119.

Liao, Y., Liao, K., & Hutchinson, R. (2010). A conceptual framework for prototyping outsourcing in new product development: a knowledge based view. *Journal of Manufacturing Technology Management, 21*(1), 122–138.

McCann, J. (2012a). *Smart textiles for protection.* Woodhead Publishing.

McCann, J. (2012b). *Translating the hybrid methodologies and practical outputs of smart-textile-orientated research in clothing for the growing ageing market, for the benefit of all stakeholders.* The 88th Textile Institute World Conference, 2012.

McCann, J. (2013). Techtextil texprocess 2013. *Textiles, the Quarterly Magazine of the Textile Institute*, 17–20.

Sayem, A. S. M., Kennon, R., & Clarke, N. (2010). 3D CAD systems for the clothing industry. *International Journal of Fashion Design, Technology and Education, 3*(2), 45–53.

Shimmer. (2012). *Wireless sensor platform for wearable applications.* [Online] Available at www.shimmer-research.com. Accessed 21.08.12.

Telegraph. (2014) *Why is wearable technology so damn ugly?* [Online]. Available at http://www.telegraph.co.uk/women/womens-life/10569007/Why-is-wearable-technology-so-damn-ugly.html. Accessed 16.01.14.

The Guardian. (2013a). *Wearable technology – Separating fact from science fiction.* [Online] Available at http://www.theguardian.com/media-network/media-network-blog/2013/apr/03/wearable-technology-machine-to-machine. Accessed 03.04.13.

The Guardian. (2013b). *Wearable tech: Why Intel thinks we should own our data.* [Online] Available at http://www.theguardian.com/technology/2013/jun/26/wearable-tech-intel-data. Accessed 29.06.13.

Tyler, D. J. (2008). *Carr and Latham's technology of clothing manufacture* (4th ed.). Oxford: Blackwell Publishing Limited. pp. 288–298.

University of Westminster. (2009). *Pilot study data report – Technical.*

Websites

http://www.newdynamics.group.shef.ac.uk/design-for-ageing.html.

http://www.adobe.com/uk/products/illustrator.html.

www.bemisworldwide.com/products/sewfree/.

www.shimmersensing.com.

www.fibretronic.com/.

www.brenig.co.uk/shop/.

www.sprayway.com/.

www.keela.co.uk/.

www.ttpwales.co.uk/index.php/textiles-technology-project/.

www.exo2.co.uk/.

www.corel.com/corel/.
www.lectra.com/en/solutions/design-room/product-information.html.
www.gerbertechnology.com/.
www.browzwear.com/.
www.optitex.com/.

Developing footwear for the active ageing population

23

M.J. Head[1], *C.S. Porter*[2]
[1]Inov-8 Ltd, Staveley, UK; [2]Loughborough University, Loughborough, UK

23.1 Introduction

With fertility rates dropping and people living longer, the worldwide population is ageing. In Japan, by 2025, for every 100 people aged 15–60 years old, there will be 66 people aged 60 and above (World Health Organisation, 2002). For those living longer, it is important that they can remain healthy, active and independent and, as a consequence, this demographic is a growing market in the active wear industry (Bieker, 2010). Many older people have money to spend and time to participate in the activities of their choice (Wolfe & Synder, 2003). A popular pastime for this age group is trail walking (Bieker, 2010), allowing the participant to raise their heart rate without subjecting their body to the stress that activities such as running exert. Their choice of footwear when walking is essential, as it enables them to participate effectively.

As people age, the muscles responsible for the movement of the foot and ankle become weaker (Spink, Fotoohabadi, & Menz, 2010), bone density will more than likely decrease (Nyman, Reyes, & Wang, 2005; Wang, Sheng, Li, Mauli, & Agarwal, 2002) and an individual's sense of their foot position lessens (Robbins, Waked, & McClaran, 1995). These are just some of the factors that make an older person more susceptible to foot and related lower limb injuries and likely to face a longer recovery period if injured (Rubenstein, 2006). Of those over 65 years of age, it is estimated that up to a third suffer a fall (Campbell, Borrie, & Spears, 1989), with a large percentage of these injuring themselves when walking (Berg, Alessio, Mills, & Tong, 1997).

It therefore becomes critical for an older person who wants to remain active that they wear the appropriate footwear in which to do so. For anybody, an injury can be frustrating, but for an older person staying active, it is often of critical importance to his or her overall physical and emotional well-being (Dunne, Bergman, Rogers, Inglin, & Rivara, 1993), enabling them to exercise and socialise. After suffering an accident, older people often lose confidence and restrict their activities in an attempt to avoid a similar occurrence (Dunne et al., 1993).

In addition to being of great importance to the wearers, this is also a great opportunity for sportswear manufacturers, who, as with clothing, seemingly largely ignore the active ageing when developing new footwear and related technologies. It is often the first to target a market effectively that subsequently dominates the market (Tellis & Golder, 1996) and, with this age group rapidly expanding, attention towards their needs and desires may be prudent. This chapter explores how walking shoes may be best developed to keep older people active, minimising their injury risk.

Textile-led Design for the Active Ageing Population. http://dx.doi.org/10.1016/B978-0-85709-538-1.00023-7

23.2 Footwear requirements for older people

In this section the author explores the key footwear requirements for older people found within the current literature. Figure 23.1 displays two of the most common silhouettes used for walking shoes with the necessary components identified; Table 23.1 provides further details about these different components to facilitate understanding of the literature. A large percentage of the literature the author identified focuses on the prevention of falls.

23.2.1 Secure footwear

Sherrington and Menz (2003) showed that wearing unsecure footwear is often a contributing factor as to why an older person falls. When surveying 95 people, average age of 78.3 years, that had recently sustained hip fractures in Sydney, Australia, many, at the time of the fall, were wearing shoes that were either backless or had a flexible heel counter. The need for secure footwear is supported by the results of a survey conducted by Hourihan, Cumming, Taverner-Smith, and Davidson (2000), where 68% of 107 people that had sustained hip fractures were wearing shoes with flexible heel counters. Menant (2008), who explored required characteristics for footwear to maximise balance and gait with 70 older people split over three different studies, also agreed that secure footwear was important to minimise falls. Securing footwear has, however, been identified as problematic in two studies (Dunne et al., 1993; Robbins & Waked, 1997) with some older people finding it difficult to fasten laces and straps and, in addition, reach down to secure the shoes.

23.2.2 Influence of sole properties on stability

There are contrasting findings regarding the required properties for the footwear's sole. Another result of Menant's studies (2008) was that to maximise stability it is important to have a shoe that has a neither too soft or too hard sole. This finding contradicts previous studies, as summarised in a literature review conducted by Menant, Steele, Menz, Munro, & Lord (2008), where thin hard soles were considered the most appropriate for older people. Waked, Robbins, and McClaran (1997) found that a

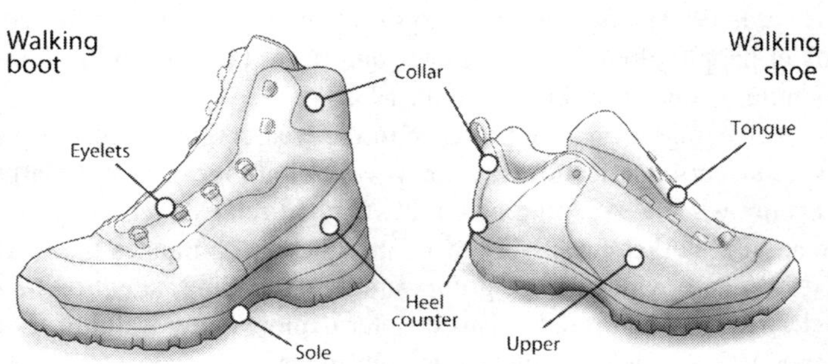

Figure 23.1 Basic anatomy of typical trail walking footwear types.

Table 23.1 Basic components found on walking shoes

Component		Role	Typical materials
Upper		Protect and support the foot while maintaining a comfortable environment (breathable and warm)	Leather, Suede, Nubuck, synthetic material (e.g. Gore-Tex)
Sole	Midsole	Provide cushioning, absorb and redirect shock. Will usually include an embedded metal or plastic plate (shank) to increase stiffness of the sole	Ethylene vinyl acetate (EVA), polyurethane (PU), rubber
	Outsole (tread)	Provide friction and avoid slipping when walking. Help to absorb and redirect shock	Carbon/blown rubber
Collar		Protects the Achilles tendon and ankle from chafing	Memory foam
Tongue		Ensures that water debris and dirt do not enter the shoe and protects foot from rubbing by, where employed, laces	As upper, often with foam for padding
Eyelets		Used to secure the employed fastening method. May also utilise d-rings, hooks etc.	Range of metals/ plastics
Heel counter		Wraps around the heel, holding it in place during movement. Embedded within the upper	Hard plastic

thin and hard sole enhanced the proprioceptive input the wearer received, improving their positional awareness and stability. The results of another study (Lord, Bashford, Howland, & Munroe, 1999), where 42 women between 60 and 92 years old conducted a series of static and dynamic balance tests in five different shoes, indicated that sole hardness did not affect balance. Munro, Mickle, and Steele (2011) supported the use of flexible soles to improve foot function and strengthen muscles in the older foot after conducting a walking program with 49 older men and women. Maximising the plantar sensation between the foot and the ground was identified as important in maintaining balance in a study conducted by Menz, Morns, and Lord (2005) where 176 participants, average age of 80.1 years, had a series of lower limb characteristics measured. The authors stopped short of making recommendations in shoe selections but did state that foot-strengthening exercises may be beneficial in reducing the likelihood of falls and other clinical conditions such as fallen arches. A good grip on the sole and non-slip properties have been reported as important requirements for older people's footwear (Menant et al., 2008). In the study conducted by Lord et al. (1999), a high collar that covers the ankle was also found to improve balance, a finding supported by Menant (2008).

23.2.3 Importance of fit

The notion of having a shoe that is fitted correctly to the wearer is reported as essential for older people in several studies to minimise the risk of falling (Helfand, 2003; Menant, 2008; Menant et al., 2008). Poorly fitting footwear can also contribute to clinical conditions such as hallux vulgus (Menz, Roddy, Thomas, & Croft, 2011). As people age their feet change shape; in a study by Chantelau and Gede (2002), where 668 patients with a mean age of 64 had anthropometric measurements of their feet taken it was found that two-thirds had feet too broad for standard sized footwear available, making it difficult for them to find suitably fitting footwear. This is supported in a study by Dunne et al. (1993); nearly 500 older people were interviewed over the telephone about why they did not wear 'sturdy' shoes, defined as athletic or walking shoes, with 12.5% of participants stating that fit issues prevented them from wearing such shoes.

23.2.4 Other factors

In the study by Dunne et al. (1993), a number of participants felt that appropriate sturdy shoes were too expensive, not stylish enough or did not think to wear them, indicating perhaps a lack of awareness.

From the limited literature that could be found on footwear requirements for older people there is little consensus as to what is required other than that it should be secure, fit and improve their balance. Most of this is centred around preventing falls, with none concerning solely walking shoes. No recent studies could be found where older people were asked directly about their influences when purchasing and experiences with their footwear, this information would provide a good starting point in understanding the requirements for walking shoes.

23.3 Meeting individual footwear requirements

Menz and Sherrington (2000) produced a form in an effort to assess footwear's suitability for older people. Criteria were listed for specific characteristics: heel height, heel counter stiffness, tread pattern, sole flexion point, sole rigidity and sole hardness, with values established from a previously conducted literature review, many of which were based on limited data. This form is a useful starting point for classifying suitable footwear; however, as the authors note, it does not take into account the individual foot's anatomy, of which there is so much variance across a population. In addition, as people age, their feet change shape (Jung, Lee, Jinhoo, & Park, 2001), not necessarily in a uniform fashion, often becoming too wide to fit standard footwear (Chantelau & Gede, 2002). If the footwear being assessed does not fit, then the form is of little use. Chantelau and Gede (2002) suggest the collection of anthropometric data for older people to define suitable footwear sizing for their age group. Another important factor not taken into account by this form is the wearer's personal perception; if it does not pique their interest, it does not matter if it is technically suitable. With all the potential variances, the question is whether footwear requirements can be standardised, or whether, because of older people's anthropometric differences, a solution is required that caters to the individual.

23.3.1 Personalisation of footwear

By personalising footwear, in theory, older people can receive a product which, functionally, fits their specific foot's geometry and performance requirements and aesthetically, matches their personal taste. Many companies currently offer primarily aesthetic personalisation, allowing consumers to select their desired colours, patterns and materials of leisure footwear, for a varying premium, usually between 10% and 100%, when compared to standard equivalents (Head, 2012). Additive manufacturing, a collection of production methods based on layer by layer manufacturing directly from a three-dimensional computer-aided design model without tooling (Hague, Campbell, & Dickens, 2003), however, is seen as a potential enabler of functional footwear personalisation. The concept of personalising functional footwear was explored in the Elite to High Street project (Loughborough University, 2012).

23.3.1.1 The Elite to High Street project

The Elite to High Street project was a multi-disciplinary project based at Loughborough University, conducted in association with New Balance and UK Sport, amongst others, that explored the possibility of personalised sports footwear for everyone from elite athletes to 'High Street' consumers. To produce a feasible yet desirable concept for the general consumers, a series of research studies were undertaken, both qualitative and quantitative, from focus groups to questionnaires to gain a better understanding of consumers and their relationships and attitudes towards their running shoes. One of the outputs of this research was the YourStep personalised running shoe product (see Figure 23.2) and service concept (see Figure 23.3) and counterpart digital tool (Head, 2011). The concept was that those who used the YourStep service would receive running shoes personalised to both their functional and aesthetic requirements (see Table 23.2).

An additive manufactured midsole would provide the key difference between currently available footwear options, manufactured after using data acquired by scanning the consumer's foot. The insole would match the plantar surface of the wearer's foot which has been shown in a longitudinal study to potentially reduce discomfort (Salles & Gyi, 2010) and would also provide the increased plantar sensation suggested as potentially desirable by Menz, Morris, and Lord (2005). Combining this with chosen shoe widths and upper heights allows for a fit far in advance of those available with current running shoe options.

Additive manufacturing processes also offer benefits to the implementing organisation when compared to traditional manufacture methods (Hague et al., 2003; Hopkinson, Hague, & Dickens, 2006; Ruffo, Tuck, & Hague, 2006):

- Parts can be produced on demand, minimising the need to carry stock
- No need for additional tooling or moulds

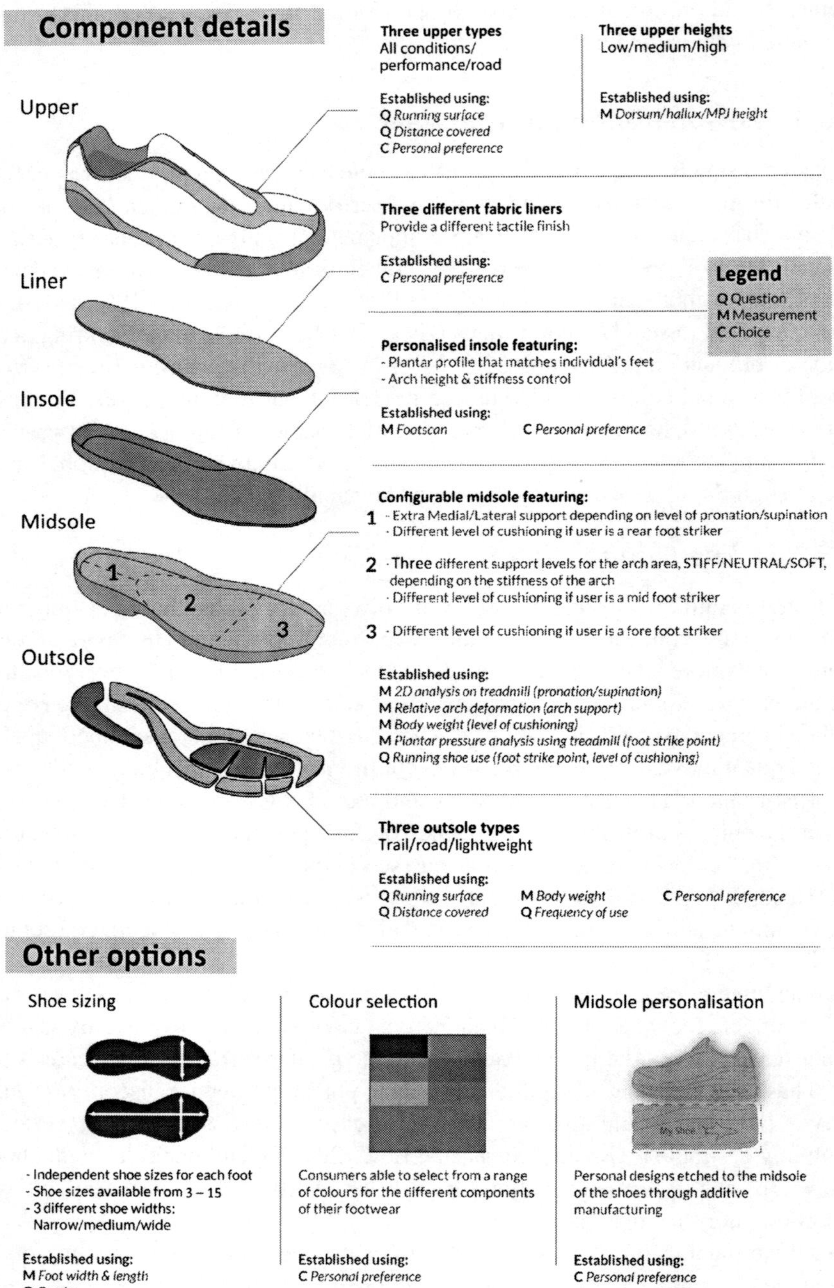

Figure 23.2 The YourStep product concept. A running shoe concept that can be personalised to suit the consumer's needs and desires.

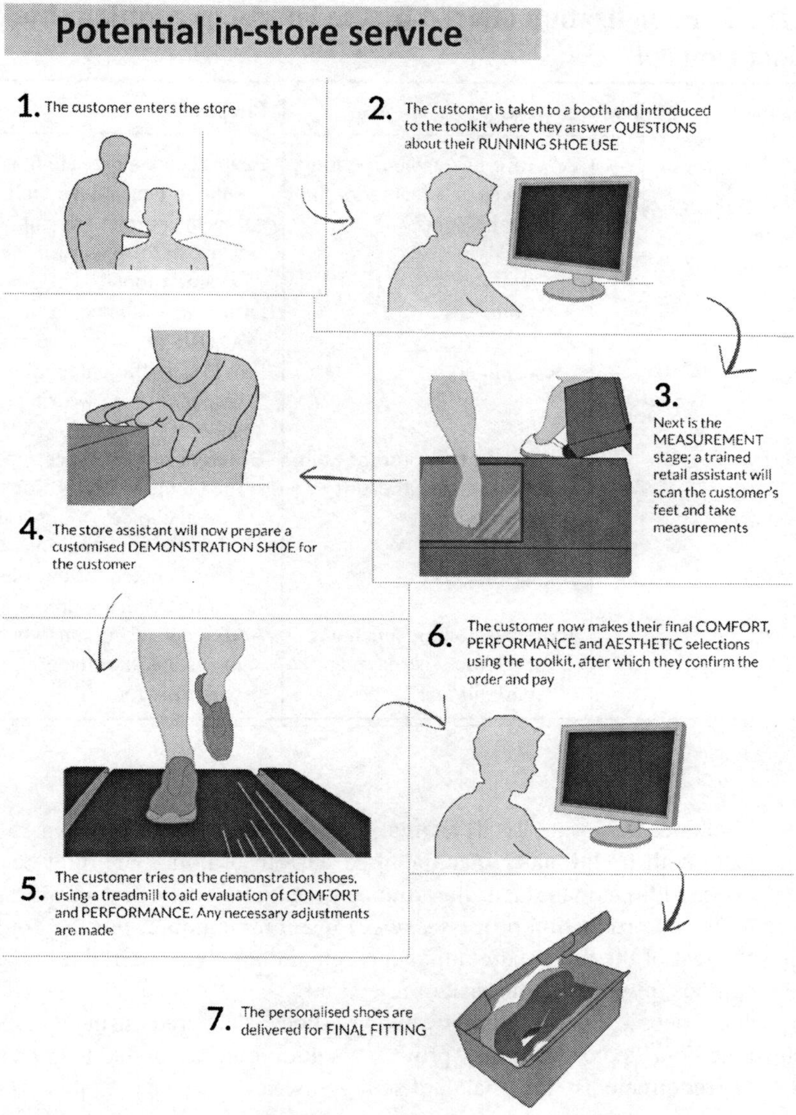

Figure 23.3 The YourStep service concept. An in-store service concept to deliver personalised running shoes.

- The ability to produce parts of high complexity
- Very accurate processes minimising material wastage

Without the need for tooling, each part produced can be different, lending itself to the personalisation of running shoes and also reducing the waiting time for the shoes to be manufactured. The ability to produce complex and accurate parts enables the better fitting footwear. A cost analysis of additive manufactured custom foot

Table 23.2 Personalisation offered by the YourStep running shoe product concept

Type of personalisation	Aspect	Purpose
Fit	A 'glove-fit' insole matching the wearer's plantar profile	Enables an appropriate fit and support beneath the arch area
	3 × upper heights	Allow the correct amount of space in the upper for the wearer's foot
	3 × widths	Different width choices to allow a better fit
Comfort	3 × sockliners	Provides a different tactile response when wearing the shoe
Support	3 × support types for the arch area – stiff/neutral/soft	Different support types depending on the wearer's arch type
Performance	3 × upper types	Different choices of upper to suit different conditions
	3 × outsole types	Different choices of outsoles to suit different conditions
	Extra cushioning depending on the foot's strike point: front/mid/rear	Additional cushioning where the user strikes the ground with their foot

orthotics (Saleh & Dalgarno, 2009) estimated production costs of between £50 and £70 per pair. With traditionally manufactured custom orthotics retailing between £50 and £200, additive manufacturing offers a potentially cost-effective alternative to popular current production processes, e.g. injection moulding. For the YourStep concept, the rest of the footwear components remain largely unaltered in their production methods, minimising the risk of rejection.

For older people, personalised walking shoes delivered using a service similar to the YourStep concept may provide a solution more suitable than trying to define the requirements for standardised footwear targeted at this specific demographic.

23.4 Researching walking footwear for older people

Personalisation provides an interesting concept; however, to produce suitable walking footwear that appeals to older people, it is important to work with them directly to gain a better understanding of their attitude towards currently available options and identify potential areas of improvement. An in-depth questionnaire and semi-structured interviews were planned to enable initial exploration of the topic.

23.4.1 Research objectives and methodology

To explore older peoples' relationship with their walking shoes, the following objectives were identified:

- To define and understand older people's walking shoe purchases
- To determine aspects of importance to older people in walking shoes
- To identify older people's walking shoe usage
- To discover issues with current walking shoes.

23.4.1.1 In-depth questionnaire

A questionnaire was chosen because the standardised format ensures the capture of the required information (Robson, 2002), enabling identification of areas of interest for deeper exploration with the interviews.

Ten participants, five females and five males all over 60 years old, completed the questionnaire, hosted online. They were selectively sampled from walking clubs across the United Kingdom to ensure a broad range of data; a larger sample was not collected as generalisable results were not required for this exploratory study. The questionnaire was piloted and revised; a mix of question types were employed, including open, ranking and rating to retain interest and minimise pattern answering (Robson, 2002).

23.4.1.2 Semi-structured interviews

Following the completion and initial analysis of the questionnaire, a series of eight semi-structured interviews were conducted to provide a better understanding of older people's relationship with their walking shoes. Semi-structured interviews were employed as they enabled the researcher to explore avenues of interest as they occur whilst retaining a structure that ensures the required topics are covered (Greenbaum, 1998). Four males and four females were interviewed, all members of walking groups in the East Midlands of the United Kingdom, enabling, in all but one instance, a face-to-face interview. The interviews were conducted either at the participant's home or at the end of a scheduled walk, allowing the researcher to gain further insight to the context of use. A script that covered the required topics was piloted and refined. The interviews were recorded where possible and with permission, otherwise notes were taken. Participants were asked to have the shoes they used for walking with them to assist the topic exploration.

23.4.2 Analysis and results

The data was transcribed where necessary and studied several times with areas of interest highlighted. A process of axial coding was then employed, identifying possible connections across the data allowing key themes to be subsequently drawn. Affinity mapping was used as a visual method to assist this process (see Figure 23.4).

The results are categorised with reference to the objectives identified in Section 23.4.1.

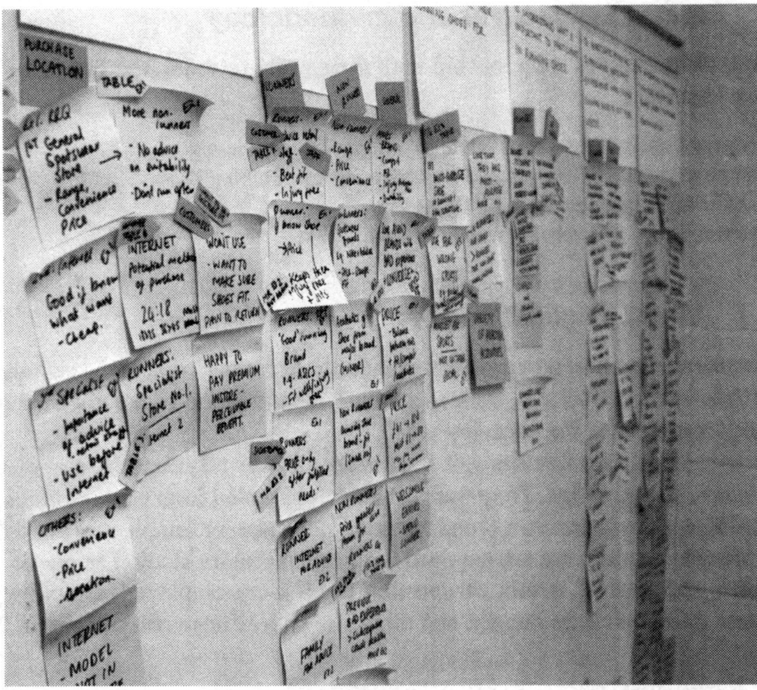

Figure 23.4 An Affinity Map, produced to assist analysis of gathered data concerning consumers' relationships with their walking shoes.

23.4.2.1 Current walking shoe purchases

Factors affecting purchasing decisions include:

- Brand: most of the participants purchased brands well known for producing walking shoes, for example, Berghaus, Brasher, Hi-Tec and Karrimor, although two of the participants, who walked less frequently, did purchase 'bargain' unbranded footwear. Participants did not tend to purchase a specific model.
- Upper material: leather and Gore-Tex were the most common upper material choices as they were trusted to perform well. Many participants would purchase only leather shoes because it was considered the highest quality material, one that would last, is easy to clean and kept their feet waterproof. Others by contrast would sometimes avoid leather shoes because they were too heavy and stiff, causing discomfort. Some people were dubious as to the quality of shoes with synthetic uppers, whilst some preferred them because they were lightweight and more flexible, though most participants would purchase shoes only with a trusted synthetic upper material such as Gore-Tex. Shoes with a mix of leather and synthetic materials were popular with a number of participants as they presented a mix of the desired qualities from both.
- Sizing: many participants reported purchasing their walking shoes in bigger sizes than their other shoes, stating that they believe this is what everyone did, wearing thick and often multiple pairs of socks to ensure their feet were secure.

- Colours: questionnaire participants all purchased dark colour walking shoes, the majority preferring shoes that were primarily brown. The interviews identified potential reasons why:
  - The predominance of this colour's availability
  - Dirt is not as visible as on other colours, reducing the amount of cleaning
  - Brown is a natural colour in keeping with the common context of use.
- Location: shoes were purchased from a range of different locations from specialist outdoors shops to department stores to a stall at an exhibition. The reasons people went to these locations varied from specialist knowledge and trust (specialist shop, exhibition stall) to price (supermarket) and convenience (department store). One participant avoided specialist stores because they felt that the service provided was biased due to factors such as brand loyalty and commission. The internet was not considered a suitable purchase location primarily because fit was of critical importance to participants.
- Advice: participants did, however, use the internet to read reviews before purchasing shoes, as well as talking to family, friends and fellow members of clubs.
- Cost: spending varied amongst participants; many would spend below £50 on their shoes, with others spending close to £100. Once they found a shoe they liked they may try to purchase it again the following time. Some of those who found if difficult to purchase suitable footwear stated that they would have no problem spending more money if they could find an appropriate shoe.

23.4.2.2 Aspects of importance in walking shoes

The priority for most participants when purchasing walking shoes is comfort and fit. Injury prevention, performance and durability are also important. The weight of the shoe is an important consideration to many, with lighter generally considered better, although some perceived a heavy shoe as a well-constructed shoe. Technology was considered of importance to a number of participants; for example, some stated that they would purchase shoes only if they had an outsole made by Vibram, a brand famous for producing rubber outsoles. In addition to the required characteristics, questionnaire participants also stated that getting value for money was essential, which for a number translated to a low cost.

The specific brand was not of high importance to most; however, many participants would not consider purchasing a non-established brand. Aesthetics was similarly thought to be a low priority, but several participants stated that they did have a preference for a simple, non-technical appearance to their footwear, perhaps signifying that they did not care for their shoes being 'styled'.

Participants liked their shoes to be low maintenance; not needing to clean or polish often was considered a benefit. One participant liked to purchase buckskin brown shoes as when they became dirty they still looked acceptable. A thick-cushioned sole to absorb impact was selected as important by another participant. Waterproof shoes were considered essential for winter, with many feeling that only leather could serve this purpose, others treated their synthetic uppers with a waterproofing wax to ensure their feet remained dry.

23.4.2.3 Walking shoe usage

Participants generally used their walking shoes at least once a week, especially during the summer months, with a large number using them more often. For the majority it

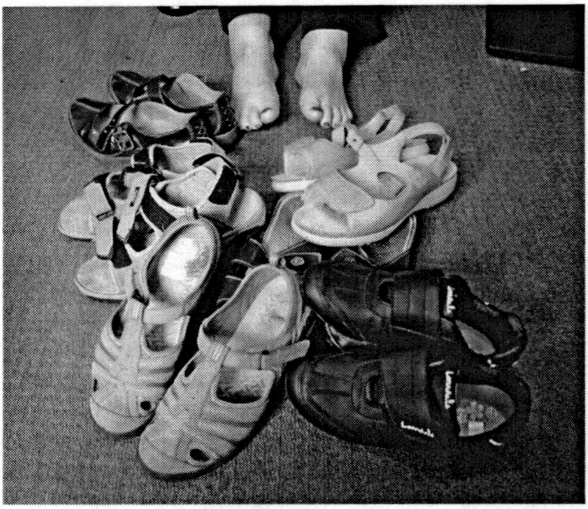

Figure 23.5 Exploration of active agers' footwear purchases revealed, for example, that one participant had bought multiple pairs of shoes in a short period of time, some completely inappropriate for walking, in an effort to stay active.

was their only form of exercise. If participants found a suitable shoe they would, generally, try and keep it as long as possible; most questionnaire participants stated that they would keep theirs over two years. Interviewees often kept their shoes until they fell apart, sometimes over five years, getting them repaired if necessary. Participants appeared hesitant to change shoes if they found one that works. Conversely those who struggle to find suitable shoes purchased new pairs often, every six months in certain cases; staying walking was the priority to them. Some even knowingly purchased the wrong kind of shoes, not intended for trail walking, in an attempt to walk without pain (see Figure 23.5).

Participants generally preferred wearing a walking shoe with a high collar in the winter or more adverse conditions, providing more lateral support. Due to the extra weight of these shoes, which often had leather uppers, many participants had different shoes made out of softer, lighter, non-waterproof materials for walking in fine conditions.

23.4.2.4 Issues with current walking shoes

Many participants experienced issues with the fit of their walking shoes. As detailed earlier, it is common practice to purchase larger sized shoes to accommodate what participants perceived as their broad feet; however, this then left them with shoes that were oversized in certain areas. To compensate for this they would often wear additional pairs of socks which would, in turn, lead their feet to overheat and rub. One participant, because of their small shoe size, had to purchase from a limited range of shoes and therefore had difficulties finding an appropriate fit. Others bought new pairs often, every six months, when their shoes quickly became uncomfortable. Common complaints from participants were rubbing, blisters, aches and pinching across the tops of their toes after long walks; many believe this is just a consequence of walking. Some of the participants believed the fit of the shoes was inconsistent, needing to buy different sizes of the same shoes. This may be a result of their foot changing shape as they age (Chantelau & Gede, 2002), but it is also possible to find variances amongst certain properties of supposedly identically sized standard footwear (Toon, Majewski, Zarringhalam, Hopkinson, & Caine, 2008). A number of participants felt that the support of their arch was not enough as their arches would ache after long walks; buying off-the-shelf insoles to compensate was not uncommon.

Many participants experienced issues with the weight of their shoes, particularly those with high-collared leather shoes, but seemingly accepted that they needed heavy shoes to withstand adverse conditions in winter. A number of the participants also avoided high-collared shoes because of the discomfort they caused around the ankle area.

For those using shoes with a synthetic upper, it was common to treat them with a waterproofing wax to ensure their feet remain dry. For some of the participants, who already avoided leather because of its stiffness, this treatment made their synthetic shoes too stiff and caused discomfort.

Fastening was problematic for some of the participants, aligning with the results in the literature (Menant et al., 2008); laces are either too long or too short, difficult to tie and often would break on the eyelets. Velcro was not considered a suitable alternative as it felt 'horrible' to use and was considered ineffective.

Concentrating on the aesthetics, some participants stated that they did not like that walking shoes looked 'clumsy'; often having a larger silhouette than other forms of footwear to provide increased stability on uneven surfaces. Simplicity was another common theme with some distrustful of shoes that looked complex and flashy; despite this a number of participants did not purchase shoes unless they contained the required technology.

When asked what improvements they would like to see in their footwear, participants gave varied answers, primarily related to functional aspects of their footwear:

- Better fitting: as detailed earlier, few participants achieved optimum fit
- Longer lasting: a number of participants stated that they felt that shoes did not last as long as they used to, with 5 years of usage often expected
- Footwear tailored to their requirements, custom levels of support: many had to buy replacement insoles to provide additional comfort, frustrating after making the initial purchase
- More choice: participants found the general choice restrictive
- Lighter shoes: the weight of shoes was problematic for many, especially over long distances
- Softer upper materials: the rubbing from stiff materials made wearing shoes uncomfortable
- A continuous tongue: stopping moisture from entering the shoe in wet conditions.

23.5 Discussion: key requirements for walking shoes for older people

It appears that some older people will actually go to great lengths to remain active: spending often and even potentially increasing their injury risk, for example, keeping their shoes far beyond their functional lifespan or knowingly buying the wrong kind of footwear. This underlines how important walking can be to an older person.

Sacrifices seem to be commonplace when participants purchase their walking shoes. People wanted comfortable shoes that fitted well, prevented injury and protected, yet rarely appeared able to satisfy all of these requirements adequately. A walking shoe that provided stability with a high collar may be uncomfortable, a low-profile walking shoe that was lightweight and felt comfortable offered little ankle support. The fit of shoes appears to be the primary area where participants are not getting what they desire; it was common to encounter someone that perceived their foot as too broad and even once they have purchased shoes that they are 'happy' with, they often have to wear additional pairs of socks for a satisfactory fit which, in turn, causes additional problems. Further research is required to understand whether this is an isolated issue; research by Chantelau and Gede (2002) suggests that current footwear sizing methods may be at least partially responsible.

Perhaps wearing shoes that are too large is also partially why some people think their shoes look too 'clumsy', although, historically, walking shoes do have a larger silhouette than other shoes to provide increased stability. With certain studies (Munro, Mickle, & Steel, 2011) showing that slimmer, more contoured profiles may be better for balance it may be time to reassess what a walking shoe should look like.

Trust is a key word in these results, guiding the purchase decisions of many of the participants; whether it is the brand, the technology, the material, the shop or the person, people went with what they trusted. It may be difficult to get people to switch purchase habits, so providing an effective product with demonstrable benefits in the areas older people require them, through a well-delivered service, is essential. In this instance participants that do make the switch may be more likely to stay loyal (Polyani, 1958).

Participants also search for value and do not feel pressured to purchase the latest models; if a shoe provides the materials and, in some cases, the technology that they need, many will happily purchase the older model. Participants will certainly shop around to find this value but will not purchase on the internet: the fit and comfort is too critical. One issue with this behaviour is when they make the purchases based on misguided information; for example, those looking for a more flexible lightweight shoe yet suffer with heavy leather shoes on the premise that nothing else is suitable. Better communication of the required information to older people can help them make informed purchase decisions and may enable them to continue to search for value, safely.

As with active footwear for younger age groups, technology is a major selling point; however, it appears for older people that its visual impact should be more subtle. Instead of assuming a complex design that attempts to embody the technology (see Figure 23.6), potential buyers should perhaps be aware that the shoe incorporates the desired technical materials, i.e. Gore-Tex and Vibram, but the styling remains restrained, echoing traditional walking shoe styles.

There are also areas where product development may be desirable to produce alternative solutions; fastening is one such area, providing a reliable easy to fasten alternative to the sometimes problematic laces may be of benefit to many. Beach shoes often feature adjustable toggles (see Figure 23.7); better fitting footwear may require

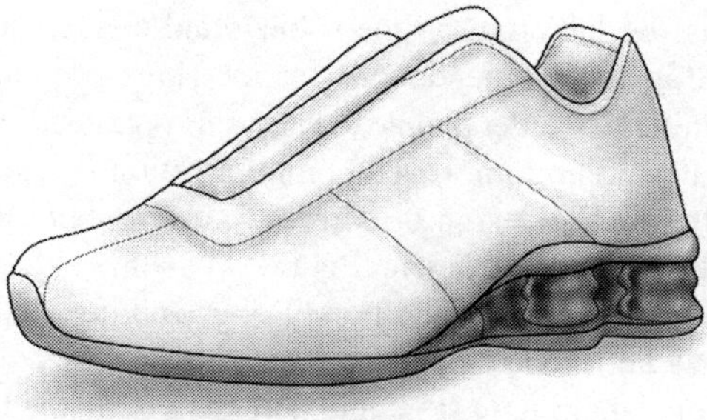

Figure 23.6 Sportswear manufacturers often focus on producing footwear with a highly technical appearance that appears to embody the technology, for example, the spring-like sole. Older people appear less interested in these types of footwear, favouring subtle styling.

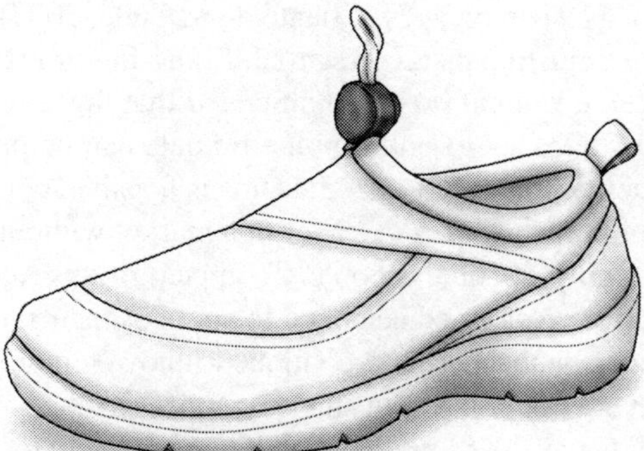

Figure 23.7 Beach shoes with a toggle offer simple fit adjustment and an alternative to laces, often problematic for older wearers.

less fastening and a simple method like this could provide a suitable alternative. Developing more walking shoes with integrated tongues is another area, to minimise moisture entering the foot, although again, their availability may just require better communication.

23.6 Conclusion

As people above 60 become one of the largest demographics in the world population (World Health Organisation, 2002), it is essential that we cater to their needs. Staying active is important to many, with trail walking popular (Bieker, 2010), but they have to be careful because as an older person's body ages they become more susceptible to injury and slower to recover (Campbell et al., 1989; Rubenstein, 2006). To minimise their susceptibility to injury and enable their comfort, it is critical that older people wear appropriate footwear; however, some studies suggest that many do not (Dunne et al.,

1993). After exploring the literature on the subject and conducting some preliminary research, we make design recommendations for the short, medium and long term to maximise the opportunities for older people to wear appropriate footwear when walking.

In the short term, developing improved communication is a key recommendation. This is not strictly altering the design of current footwear, but with better communication, as suggested by Menz and Sherrington (2000), some older people may be able to make more informed decisions when purchasing walking shoes, looking past the branding and what they believe to be appropriate. This may be employed through better product information, labelling in store and perhaps a dedicated Website or publication, targeted at older people that helps to guide their selection process. The footwear assessment form provides a good starting point; however, once footwear has been assessed, it is the communication of its capabilities that is essential. Improving the communication is unlikely to help address one of the main apparent problems with walking shoes: they do not fit appropriately. They are designed based on the measurements of a younger person (Jung et al., 2001) and therefore do not allow for the broadening of the foot as a person ages (Chantelau & Gede, 2002). As a medium-term goal, therefore, sizing adjustments are essential. Some initial measurements of older people's feet have been taken but it is recommended that these should be extended to understand if a standardised set of anthropometric data can be produced upon which walking shoes for older people can be sized. If this is feasible, companies may then be faced with the issue of how they may cater for this market without inducing stigma by producing shoes especially for older people. To appeal to this market, manufacturers may need to apply a more restrained aesthetic to some of their footwear.

In the long term, personalisation and continued innovation are the most important aims. As people's feet are not uniform in shape or appear to change shape uniformly as they age (Chantelau & Gede, 2002) the concept of personalising footwear to the individuals is the most desirable. The problem up to now is that it has not been cost-effective or technically feasible to produce effective, affordable personalised footwear. Additive manufacturing paired with foot scanning is a potential enabler of such footwear production (Hopkinson et al., 2006). A service concept such as the YourStep service (see Section 23.3.1) adjusted for walking shoes (appropriate footwear styles, colour and material choices) would allow an older person to receive footwear fitted to their feet and tailored to the style they desire. There have been promising prototypes of footwear produced using additive manufacturing (Head, 2012); however; further work is required with the materials to produce wearable, testable footwear. If footwear design is implemented effectively, those that wear personalised footwear may well be loyal to the service, firstly because of the fitting process they have been through and secondly because many older people appear to stick to shoes they've found most suitable.

In addition, as this age group expands and becomes more educated about their footwear purchases, it is essential that companies continue with innovative research and development, producing better fitting and performing shoes, otherwise people will find other purchase avenues.

An important point to note is that this research represents a starting point when exploring requirements for walking shoes for the active ageing in the United Kingdom. Further research is necessary before making generalisations about such requirements.

References

Bieker, C. (2010). *Sports Tracking Europe 2010*. Sonthofen: Generation Sport.

Berg, W. P., Alessio, H. M., Mills, E. M., & Tong, C. (1997). Circumstances and consequences of falls in independent community-dwelling older adults. *Age and Ageing*, *26*(4), 261–268.

Campbell, A. J., Borrie, M. J., & Spears, G. F. (1989). Risk factors for falls in a community-based prospective study of people 70 years and older. *Journal of Gerontology*, *44*(4), M112–M117.

Chantelau, E., & Gede, A. (2002). Foot dimensions of elderly people with and without diabetes mellitus – a data basis for shoe design. *Gerontology*, *48*(4), 241–244.

Dunne, R. G., Bergman, A. B., Rogers, L. W., Inglin, B., & Rivara, F. P. (1993). Elderly person's attitudes towards footwear – a factor in preventing falls. *Public Health Reports in Brief*, *108*(2), 245–248.

Greenbaum, T. L. (1998). *The handbook for focus group research*. California: Sage.

Hague, R., Campbell, I., & Dickens, P. (2003). Implications on design of rapid manufacturing. *Proceedings of the Institution of Mechanical Engineers, Part C: Journal of Mechanical Engineering Science*, *217*(1), 25–30.

Head, M. J. (2011). *The YourStep personalisation tool*. Available from: http://www.yourstep.co.uk. Accessed 20.06.12.

Head, M. J. (2012). *Developing a service for the personalisation of running shoes (Ph.D. thesis)*. Loughborough Design School, Loughborough University.

Helfand, A. E. (2003). Basic consideration for geriatric footwear. *Clinics in Podiatric Medicine and Surgery*, *20*(3), 593–605.

Hopkinson, N., Hague, R. J. M., & Dickens, P. M. (2006). Introduction to rapid manufacturing. In N. Hopkinson, R. J. M. Hague, & P. M. Dickens (Eds.), *Rapid manufacturing: An industrial revolution for the digital age* (pp. 1–4). Chichester: Wiley.

Hourihan, F., Cumming, R. G., Taverner-Smith, K. M., & Davidson, I. (2000). Footwear and hip fracture-related falls in older people. *Australasian Journal on Ageing*, *19*(2), 91–93.

Jung, S., Lee, S., Jinhoo, B., & Park, J. (2001). A classification of foot types for designing footwear of the Korean elderly. In E. Hennig, & A. Stacoff (Eds.), *Proceedings of the 5th symposium on footwear biomechanics, Zurich, Switzerland* (pp. 48–49).

Lord, S. R., Bashford, G. M., Howland, A., & Munroe, B. J. (1999). Effects of shoe collar height and sole hardness on balance in older women'. *Journal of American Geriatrics Society*, *47*(6), 681–684.

Loughborough University. (2012). *Personalised footwear: E2HS – From elite to high street*. Available from: http://www.lboro.ac.uk/microsites/enterprise/e2hs/index.html. Accessed 20.07.12.

Menant, J. C. (2008). *Effects of footwear on balance and gait in older people (Ph.D. thesis)*. School of Public Health and Community Medicine, University of New South Wales.

Menant, J. C., Steele, J. R., Menz, H. B., Munro, B. J., & Lord, S. R. (2008). Optimizing footwear for older people at risk of falls. *Journal of Rehabilitation Research and Development*, *45*(8), 1167–1182.

Menz, H. B., & Sherrington, C. (2000). The footwear assessment form: A reliable clinical tool to assess footwear characteristics of relevance to postural stability in older adults. *Clinical Rehabilitation*, *14*(6), 657–664.

Menz, H. B., Morris, M. E., & Lord, S. R. (2005). Foot and ankle characteristics associated with impaired balance and functional ability in older people. *Journal of Gerontology*, *60*(12), 1546–1552.

Menz, H. B., Roddy, E., Thomas, E., & Croft, P. R. (2011). Impact of hallux valgus severity on general and foot-specific health-related quality of life. *Arthritis Care and Research, 60*(3), 396–404.

Munro, B., Mickle, K. J., & Steel, J. (2011). Can flexible shoes improve function in the older foot? *Footwear Science, 3*(1), S116–S117.

Nyman, J. S., Reyes, J., & Wang, X. (2005). Effects of ultrastructural changes on the toughness of bone. *Micron, 36*(7–8), 566–582.

Polanyi, M. (1958). *Personal knowledge: Towards a post-critical philosophy*. Florence, KY: Routledge.

Robbins, S., & Waked, E. (1997). Hazard of deceptive advertising of athletic footwear. *British Journal of Sports Medicine, 31*(4), 299–303.

Robbins, S., Waked, E., & McClaran, J. (1995). Foot position and awareness as a function of age and footwear. *Age and Ageing, 24*(1), 67–72.

Robson, C. (2002). *Real world research*. Blackwell: Oxford.

Rubenstein, L. Z. (2006). Falls in older people: Epidemiology, risk factors and strategies for prevention. *Age and Ageing, 35*(S2), ii37–ii41.

Ruffo, M., Tuck, C., & Hague, R. (2006). Cost estimation for rapid manufacturing – laser sintering production for low to medium volumes. *Proceedings of the Institution of Mechanical Engineers, Part B: Journal of Engineering Manufacture, 220*(9), 1417–1427.

Saleh, J. M., & Dalgarno, K. W. (2009). Cost and benefit analysis of fused deposition modelling (FDM) technique and selective laser sintering (SLS) for fabrication of customized foot orthoses. In P. Bartolo (Ed.), *Proceedings of innovative developments in design and manufacturing: Advanced research in virtual and rapid prototyping conference, Leira, Portugal, October 2009* (pp. 705–710). Cleveland: CRC Press.

Salles, A. S., & Gyi, D. E. (2010). The specification and evaluation of personalised footwear for additive manufacturing. In W. Karwowski, & G. Salvendy (Eds.), *Advances in human factors, ergonomics, and safety in manufacturing and service industries* (pp. 355–366). Cleveland: CRC Press.

Sherrington, C., & Menz, H. B. (2003). An evaluation of footwear worn at the time of fall-related hip fracture. *Age and Ageing, 32*(3), 310–314.

Spink, M. J., Fotoohabadi, M. R., & Menz, H. B. (2010). Foot and ankle strength assessment using hand-held dynamometry: Reliability and age-related differences. *Gerontology, 56*(6), 525–532.

Tellis, G., & Golder, P. (1996). First to market, first to fail? Real causes of enduring market leadership. *MIT Sloan Management Review, 37*(2), 65–75.

Toon, D., Majewski, C., Zarringhalam, H., Hopkinson, N., & Caine, M. (2008). A novel approach to personalising the mechanical properties of sprint footwear (P179). *The Engineering of Sport, 7*(2), 20–213.

Waked, E., Robbins, S., & McClaran, J. (1997). Effect of footwear midsole hardness and thickness on proprioception and stability in older man. *Journal of Testing and Evaluation, 25*(1), 143–148.

Wang, X., Sheng, X., Li, X., & Mauli Agrawal, C. (2002). Age related changes in the collagen network and toughness of bone. *Bone, 31*(1), 1–7.

Wolfe, D. B., & Synder, R. (2003). *Ageless marketing: Strategies for reaching the hearts and minds of the new customer majority*. Wokingham: Kaplan.

World Health Organisation.2002). *Active ageing: A policy framework*. Spain: World Health Organisation.

Design for ageing: a focus on China

24

W. Lu
China Women's University, Beijing, China

24.1 Introduction

China now faces the same ageing population problem as do Japan and Western countries. This phenomenon is being researched in China, in terms of medical and social implications, but not in clothing design. In this chapter I will look at the relatively recent changes that have taken place in clothing design education in China and the new demands for clothing design for emerging markets as a result of demographic changes. I will explain how, prior to China's reform and opening-up policy, little clothing design existed in China either in education or in the retail market. However, since the 1980s, there has been a global shift of Western fashion production to the Far East and, in particular, China. The Chinese fashion and textile industry has evolved into two distinct sectors: the domestic market and foreign production. I will briefly describe both of these sectors in order to provide the context for discussing the new clothing needs of China's ageing society.

In the 1980s, the design of Western fashion ranges, produced in China, was dictated predominantly by Western brands and buyers with local training for industry production predominantly skills based. Fashion design education was introduced at that time, and, in the retail market, a limited choice of fashion clothing also became available. In the mid-1990s, elite fashion emerged in China without consideration of the demands of normal society. This was followed by consumer fashion bringing ready-to-wear design to the high street. Since the mid-2000s, consumerism has increased significantly with the introduction of youth-oriented fast fashion. This has led to a current increase in creative design in China as the local market develops. Demographic change now demands the consideration of a rapidly growing older market that is neglected by both design education and fast fashion.

I will introduce the new ageing consumer as China's equivalent to the Western "baby boomer" generation. I will look at the importance of design process to guide designers in catering for this growing market from the perspective of functional everyday clothing that has the potential to enhance well-being. In addition, I will explain how this process has been informed both from my postgraduate study in performance sportswear design and, subsequently, in my role as a visiting scholar involved in design for ageing research. In my current role as Dean of Art and Design at the China Women's University, I have attracted Chinese government funding to introduce design for ageing education and research at this university. I will discuss initial live design project work involving senior students in consulting with

Textile-led Design for the Active Ageing Population. http://dx.doi.org/10.1016/B978-0-85709-538-1.00024-9

older end users and how design prototypes have been produced in collaboration with industry. I will conclude with how this initial practical research has provided a focus for discussion with academic partner institutions and with industry trade associations. Finally, I will discuss the future for alternative approaches to fashion design education to address the clothing design needs of this valuable older market sector in China.

24.2 Background to clothing design in China

24.2.1 *Clothing design in China prior to the 1980s*

At the beginning of the reform and opening up in China in the 1980s, daily life products such as household appliances, furniture, textiles, and clothing were very basic and aesthetic design was rare. However, with the improvement of people's living standards, the quality and design of the products for daily life required more design development. In the 1980s, a small number of clothing enterprises began to design and develop new products, and some clothing brands began to appear within China's clothing market.

24.2.2 *The Chinese fashion industry from the 1980s*

Since the 1980s, the Chinese fashion and textile industry has been made up of two distinct sectors: the domestic market and foreign production. China has gradually begun to accept aspects of Western lifestyle, including an emphasis on aesthetic values and fashion trends. Partly as a result of the influence of Western films, music, and other cultural exchanges, the domestic clothing market began selling suits, jeans, and other Western style clothing.

The development of the Chinese clothing and textile industry changed with government policy encouraging expertise to be developed in different regions where industrial zones would cater to different types of production. Foreign trade clothing factories were opened up in the eastern coastal areas, such as Qingdao, Shanghai, Shenzhen, and Guangzhou, to make products for Western brands. For example, Guangdong and Fujian provinces became the main production areas for sportswear and sneakers, while knitted clothing production has been mainly concentrated in the Jiangsu and Zhejiang areas, and the production of synthetic textiles is located in Jiangsu and Shandong provinces.

24.2.3 *The origins of fashion design education in China from the 1980s*

Textile design had already begun in China in the1950s because of the textile tradition and the domestic market demands but with no emphasis on clothing. By the1980s, Chinese clothing design education was still a new subject in a few art colleges. It initially grew out of textile design, with the teaching of clothing drawing, pattern cutting, garment making, and other basic technical skills. China's fashion education is now at a comparable stage with that of the industry and the retail market.

24.2.4 The emerging Chinese fashion market

As China's reform and opening-up policy progresses into its fourth decade, China has become the biggest clothing industry in the world. However, China does not have its own globally famous clothing brands. With well-developed clothing manufacturing technology and well-equipped facilities, China produces high quality products, with relatively low cost production, for many of the famous brand names from across the world. On the one hand, China owns a huge domestic apparel market, with high level manufacturing and impressive young designers, but, on the other hand, it lacks famous Chinese brand names that are recognized in both local and international markets. This weakens the Chinese local clothing producers' power of competing with the challenge from European and North American companies. "Made in China" for foreign brands, has been the main focus of the nondomestic Chinese fashion industry and other related product sectors.

24.2.4.1 The emergence of elite fashion in China since the 1990s

Emerging elite fashion in China has had little relevance or consideration for the average person. Since the 1990s, some elite fashion brands such as Christian Dior began to sell luxury items in Beijing and Shanghai. Because of the higher price level and Western sizing and proportions, those brands have had relatively little influence on the Chinese fashion market. However, Chinese customers have accepted Western sportswear brands, for example, Adidas and Nike, for more casual everyday fashion demands and for sport.

24.2.4.2 Growth of Chinese consumerism, since the Mid-2000s

The growth of Chinese consumerism since the mid-2000s has brought youth-oriented fast fashion to ready-to-wear clothing, and, in addition, an increase in creative design as the Chinese domestic fashion market develops. Some fast fashion brands, like Zara and Mango, were launched into the Chinese market in the mid-2000s, with their youth-oriented designs. In the beginning, this concept was like a breath of fresh air for Chinese consumers, with styling following the top designers, but at more affordable prices. However, the use of cheap fabrics and poor manufacturing quality has become a problem as more and more customers, with higher consumer ability, would rather pay more to have longer-lasting and better-designed products. Better quality local brands have now begun to emerge such as Exception and JNBY. Another challenge to the fast fashion brands, within the retail environment, has been from online selling, where local brands are supplying cheaper-priced products to young customers.

24.2.5 Current clothing and textile design education in China

During the 1990s and 2000s, as the domestic market sales have improved, the demand for better product design has become a pressing issue. With the rapid development of China's economy, the level of income of the consumer base has increased, along with their purchasing power. Because of an initial lack of designers to meet the new design requirements, design education has become one of the fastest-growing professions since the 1990s.

Some 900 universities in China offer courses in art and design, with 70% of them offering degree-level clothing design. Based on the data from the China Education Ministry, there are currently 239,582 students in the first year of art and design study, with 167,707 of these students studying design. Each year now, China has about 100,000 design major graduates who will join professional design teams. However, despite 200 universities offering fashion design and clothing engineering training, the clothing industry is still struggling to find professional designers with the necessary skills. There is, therefore, conflict between the numbers of design graduates available and the lack of suitably talented designers that meet the real needs of industry.

The typical university fashion design curriculum and teaching content focuses on women's wear and is predominantly youth oriented. A few colleges and universities have set up courses directed to men's wear and accessories design, but sportswear design education is rare. The result is that China's garment enterprises still lack designers trained in the design of men's wear, sportswear, and everyday clothing. In the student graduation shows, one rarely sees design specialities other than in women's wear. In their final shows, fashion students are keen to show their creative catwalk styles, with some not even conforming to the basic needs of clothing. For many of those students, the aim of becoming a fashion designer is to be another Alexander McQueen. The skills that current fashion graduates possess cannot fully meet the needs of clothing industry. This fact highlights the disparity and misunderstanding between academia and industry.

24.2.6 The lack of design education in functional/performance clothing

The establishment of academic design courses in China has become separated from real market needs. Clothing design training is divided in two directions: engineering and fashion. In fashion education, the training has focused on the aesthetic value, appearance, and effect, following the trend of catwalk shows, with little attention to the user's daily needs. Little consideration has been given to the aesthetic design of functional clothing despite the gradual crossover of sportswear with street fashion. This situation is very common both in China and the UK. On the other hand, education in clothing design engineering focuses on pattern making, the manufacturing process, but ignores the aesthetic factors. The students' knowledge is not broad enough because of the separation between these two aspects of training. This makes it difficult for students to have a comprehensive overview in their design practice. As a result, most students only focus on the appearance of clothing and do not have the knowledge or skills to combine other design elements to meet consumer demands in blending aesthetic and functional needs. In particular, there has been minimal consideration of design and research into the functional clothing design characteristics for users with special needs.

24.2.7 Establishing performance sportswear design in China

Because of the gaps in Chinese education, I took the opportunity to study for a Master's degree in Performance Sportswear Design in the UK in 2001–2003. This unique programme introduced a multidisciplinary mix of generic topics, including human physiology, textile technology, and materials sourcing, marketing, and identification of user needs; each student investigated a specialized area. For example, some studied outdoor sportswear while others focused on racquet sports, water sports, travel, and extreme work wear. The overall philosophy established by the Programme leader, McCann, was to address the real needs of end users from aesthetic, technical, and cultural perspectives (McCann). For this singular Master's degree (the only one in the world) addressing sports clothing design, students produced their final practical work in industry, or with industry collaboration, that led to employment opportunities with international brands. Some Chinese graduates returned to take up senior design positions in China, both with branches of Western brands (Adidas, Shanghai) or with local brand leaders such as LiNing, in Beijing.

My experience as a mature postgraduate student provided me with the background methodology for a user-centered approach to the application of technical textiles in functional clothing. Now, in China, at the undergraduate level, I am introducing the methodologies that Professor McCann has contributed to postgraduate design education and research. I have adopted the essence of this design process, supported by my former tutor (McCann) for subsequent course development at the China Women's University. My colleague, Professor Xiang Dong, has also contributed to this process. While a visiting scholar in the UK, she worked with McCann on the development of rules for pattern cutting for movement. Collaboration between the UK research team and Chinese colleagues has led to a unique provision at the China Women's University involving important relationships with the Chinese industry and, in certain cases, liaison with their Western buyers.

The School of Art and Design, at the China Women's University, launched the first sportswear design course in China, in 2004. Students learn the "user-centered" design process discussed, as well as gain an understanding of the performance of textiles, 3D pattern cutting for movement, human physiological responses, and the psychological impact in relation to the design of both fashion and functional clothing ranges. This course has attracted industrial collaboration with domestic and overseas companies, providing students with practical project work leading to employment in this rapidly growing clothing sector. "Live" project work, to support design practice and teaching, has already achieved great results that are of benefit to both students and industry. The performance sportswear design speciality gained recognition in the Chinese Fashion Awards by the China Fashion Association during Graduate Fashion Week in 2013.

Currently, these user-centered design methods have not been recognized and applied in the field of clothing design elsewhere in China. As the first design program to adopt this system for functional clothing design, it is now being extended to clothing design for special groups. I believe that this type of design philosophy should be introduced widely to Chinese clothing programs in order that our future designers may put these ideas into practice. For example, a group of our current young designers at the university are now applying "co-design" process to functional clothing to address the clothing design requirements of the ageing demographic.

24.2.8 Introducing the new ageing consumer in China

Huge demographic changes across the globe are creating a new market sector that is being ignored by the fashion industry. China is currently affected by this demographic change, as well, moving from a relatively young aged population structure to an ageing structure. The year 2030 will be a turning point in this transformation. In the first phase, from 2001 to 2020, the ageing population will rise rapidly, reaching 248 million, representing 17.17% of the entire population (Commission on Ageing in China). Population ageing will have a significant impact on China in economic, social, cultural, and other fields in the near future. In the next decade, the Chinese baby boomers, born in 1960s and 1970s will be in their fifties, representing a new ageing consumer majority within the population. These people will be selective and wise customers with the highest purchasing power. They will raise the level of need, requiring us to speed up the preparation of design education; "Design for tomorrow" needs to be started today!

24.2.9 Involvement as a visiting scholar involved in Design for Ageing Well research

In recognizing the importance of this challenge in China, I received funding for a one-year visiting scholarship to join my former colleague, and MA programme leader, in the UK to work alongside the Design for Ageing Well research project (2010–2011). The focus of this collaborative research looked at how the attributes of performance sportswear design, with integrated wearable technologies, might enhance the everyday lives of the active ageing (60–75 years). With my knowledge of functional clothing design, I was well prepared to work alongside the research team. This cross-disciplinary research brought together technology and care of the ageing with functional clothing design, with the introduction of a co-design process being new to fashion design.

In clothing design, the aesthetics of fit, comfort, ease of wear, ease of personal care, as well as price are common concerns for the end user. Meeting the needs of users is the basic characteristic of *inclusive design*, and the starting point of design for ageing was researching and confirming older users' needs. By comparison with the traditional trend-led fashion design approach, the purpose of design for ageing is to design clothing to enhance the well-being of older people. The focus of the Design for Ageing Well research was on design to promote healthy exercise, and in particular, encouraging the ageing population to participate in walking. Work packages in Behaviour, Clothing, and Technology, combined sociology, physiology, psychology, clothing, and wearable technology to explore functional clothing design requirements for older people. The Clothing research team had experience in areas such as size and fit, for a range of figure types, the language and terminology of technical textiles, and the sophistication of outdoor clothing design ("Design for Ageing Well" Newsletters 1–6, New Dynamics of Ageing project, 2010–2012).

As a visiting scholar I became involved in the practical co-design process as academic and industry stakeholders worked with older participants to identify and address their design needs. The designers and the end users worked together through the stages of the co-design and development process to produce a functional clothing layering system for the active ageing. This practical research experience, that built on the philosophy of Performance Sportswear Design, provided me with a clear way forward for ongoing research activity in China.

24.3 Introducing Design for Ageing Well in China

24.3.1 Design for the active ageing market in China

China now faces the same ageing population problem as in Japan and Western countries. This phenomenon is being researched in China, in terms of medical and social implications, but not in clothing design, especially in the area of smart textile applications for functional clothing (Wang, 2011). Faced with a huge ageing population, how design may affect all aspects of life for old people becomes an important issue. Design for this growing active ageing market may be looked at from the perspective of functional everyday clothing that has the potential to enhance well-being. Designers and manufacturers who meet the challenge of ageing will find a receptive and growing mature market (Huppert, 2003). The challenge for the clothing industry is the lack of qualified designers with the appropriate knowledge and skills for ageing design. The design educators must rethink education strategies and make necessary adjustments to the existing system. My current aim is to train new designers in more considered design process for new emerging markets. I have begun to introduce co-design process to education and research at China Women's University that builds on the user-driven approach already central to performance sportswear design. In this chapter I will report on co-design-led research that is beginning to bring innovative textiles in functional clothing to enhance the lives of the growing ageing demographic in China.

24.3.2 Understanding Chinese culture: symbolic clothing

The first step in the design process should be the careful observation and understanding of how older people act in their daily lives, in terms of their habits and social activities, and how traditions and culture impact them. There is an annual festival, within the Chinese agricultural calendar, dedicated to older people, the Chong Yang Festival.

During such traditional holidays such as Spring Festival, the younger generation needs to buy clothes for their parents and/or grandparents, which represents an important purchasing season in China. When buying for the older people, the younger generation need to consider clothing symbolism and the style of clothing; for example, when they select clothing for Spring Festival, the main features that normally attract customers are traditional decoration styles, with a blessing and a symbol of the meaning of long life.

The choice of color in older people's product consumption has been influenced by traditional Chinese culture. Natal year tradition in China means that this year, your life will be rough, unlucky, therefore on the first day of this Chinese New Year, you and all your relatives, parents, spouses, and children are expecting you to wear red clothes, since Chinese people believe that the color red in the new year can ward off evil spirits and bring good luck. In China, when a person turns 60 years old, in the Chinese fifth natal year, called "Huajia," the grand birthday celebration is very important in Chinese tradition as a symbol of a long life and good health. In this year, a "Huajia" old person will receive clothes and other accessories from his or her family members and friends, and these high quality gifts will be in the symbolic color red, sometimes mixed with yellow. These gifts present blessings from the people's heart, with love and best wishes. Red was selected for the first mobile phone specifically designed for the elderly in Beijing, May 2010, as Chinese people believe red brings happiness and joy. This red mobile phone, designed for the older people, soon became the top choice on the list of gifts to parents.

24.3.3 Contributing to the Design Age Research Exchange summer school

In facing the huge market demand from older people, China must now attach importance to ageing design research and design practice. On September 12 to 19, 2010, in Beijing, the School of Art and Design at Tsinghua University and The Royal College of Art, UK, co-hosted a summer school, Design Age Research Exchange (DARE), with a series of presentations and workshops on ageing design. The professors and design educators from both China and the UK had a breadth of discussion and communication on many design topics related to older people's real needs, especially related to their living circumstances and daily activities. Inclusive design for ageing was one of the most discussed issues during the program.

This event took place during my year as a Visiting Scholar attached to the 'Design for Ageing Well' research project.

Returning to China, to attend the DARE Summer School, I assisted the principal investigator (McCann) and her resident assistant (Katy Stevens) in the delivery of a workshop on co-design process. This involved my recruitment of older Chinese members of the public to participate in assessing the appropriateness of the design of performance sportswear garments produced by Chinese industry. I arranged for garments to be presented to the user group by one of the China Women's University industry collaborators, who is working with Western brands (Youngone). Participation in this design research for ageing has now had direct impact on the development of research theory and practice to guide design education in China to address the needs of the growing community of older people.

24.3.4 Establishing research: "Facing the demand of the Chinese ageing society: Research into the theory and practice of inclusive design"

At the conclusion of my role as visiting scholar, I proposed that "the purpose of clothing design education and practices should meet the requirement of the older user in China, and improve new designers' skills in product development" (Wang, 2011). My involvement, in both observing and becoming involved in the co-design process of the Design for Ageing Well research confirmed the importance of the emerging global active ageing demographic change and the significance for the Chinese market. This has led to my subsequent successful application for three years of research funding associated with Design for Ageing project research in China, funded by the Chinese Government's China Cultural Administration, Office for China Art and Science Policy Management (September 2011–2014): "Facing the demand of the Chinese Ageing Society: Research into the theory and practice of inclusive design."

In terms of the rationale for government funding, the current focus is on Chinese social and cultural, educational, and economic development, namely, Chinese heritage with traditional art and design education. The funded project is investigating the demand of the Chinese ageing society, looking at how the theory and practice of inclusive design may inform Chinese designers and researchers in finding a new design process to address the aesthetic and functional clothing demands of the emerging older market. In bringing together other academic partners, this project offers a unique opportunity for integrating different disciplines such as clothing engineering and social science to collaborate on design for ageing issues.

24.3.5 Involving all stakeholders in research and live project work

Academic involvement came from staff at the China Women's University, in collaboration with a number of other participants, including Donghua University and the Chinese Academy of Social Science. An industry perspective was provided by a number of collaborators including garment manufacturers and fabric producers, *Outdoor* magazine (http://www.lvye.cn) and a number of trade associations, such as China Fashion Association (CFA) and the China Outdoor Commerce Alliance (COCA).

CFA is a national organization registered at Ministry of Civil Affairs of PRC. CFA was founded in 1993, and its headquarters are located in Beijing D Park. CFA has both individual and group members, including fashion designers, professionals, fashion brands, fashion media, and model agencies. It was honored as a "4A Social Organization" by the Ministry of Civil Affairs in 2010.

COCA was approved by the Ministry of Civil Affairs of PRC, and the community, to engage in outdoor goods circulation, production, scientific research teaching, exhibitions, promotional activities, enterprises and units. Voluntary associations and individuals related to the composition, the nature of the national nonprofit organizations.

COCA members include, for example, The North Face, Toread, Vaude, Ozark, Gore companies, Sanfo Outdoor, and other famous enterprises. In support of the entire membership, COCA will become a strong driving force in the future of the outdoor market.

24.4 Case study: student project

24.4.1 The brief: "Beauty at Sunset"—Sportswear for active ageing

In order to encourage fashion design students to look at the older users' demands related to the Chinese sportswear market, we set a brief for a student research project named "Beauty at Sunset"—sportswear for active ageing, to run as a one-year project from Spring/Summer 2013 to Spring/Summer 2014. This project title provides a positive view of ageing, with older age often referred to as "sunset" in Chinese culture. The project is being carried out in association with the China Cultural Administration funding research project. Its focus is on the application of a co-design approach, involving a sportswear design process for the design of functional clothing for older end users. The co-design process involves student designers, with academic guidance, working with older members of the public, with support from the Chinese outdoor industry. Senior students, as young researchers, have been working directly with older members of the public on the identification of design requirements for outdoor sports and exercise activities, with final prototypes being made in collaboration with industry.

24.4.2 Overview of ageing demographic: statistics, health issues, changing body, etc.

At the conclusion of my research in the role of visiting scholar, I stated that "Demographic data should be more diversified and plentiful to provide more detailed information for designers regarding end users and their needs. China still lacks such research, with data collection and analysis superficial and not reflecting the actual problems" (Wang, 2011).

In Europe, people older than 55, known as "baby boomers," born in the peak years 1946–1960, went through the third scientific and technological revolution in human history. This generation has experienced fashion and product design throughout their lives and will have higher expectations for products designed to suit their lifestyles (McCann, 2009). In comparison, by 2020, the Chinese baby

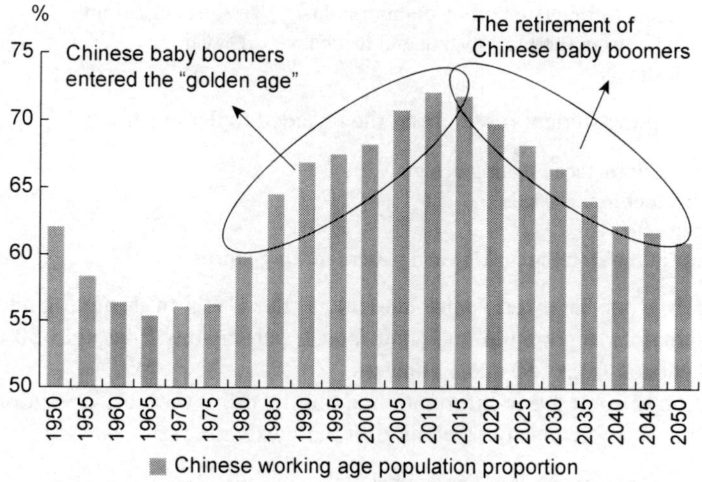

Figure 24.1 The proportion of baby boomers in the Chinese working-age population.

boomers, who were born in 1960–1970s, will represent a huge population that will enter old age (Figure 24.1). This generation has experienced the reform and opening of China, access to higher education, and integration into the global economy. Under the influence of the older-aged people in China, the expectations of this new group in terms of clothing are different from the previous generation's philosophy of life and consumption patterns, and they have a higher spending power for their higher consumption requirements. The new generation of large numbers of elders entering the market will not only affect China's overall spending power but will also include a new range of older user requirements to be addressed by Chinese designers.

24.4.3 Design process: introducing a co-design methodology

On concluding my period as a visiting scholar, I proposed that "Co-design theory and methods will need to be developed with more practice and experiments to be done in the field. This will involve more techniques such as how to research the end users' real needs, how to analyze the conversations, and descriptions recorded, and how and when end users can participate in the design process" (Wang, 2011).

Co-design is a relatively open design process that demands a new skill set and underlying philosophical approach from designers. In particular, end-user participation in co-design process has been an effective method to help the designers, and other stakeholders, understand the needs of the targeted customers.

Our process involved:

- Questionnaires
- Market research
- User group recruit
- End users' requirement workshop

- Current product testing (support by outdoor industry)/design suggestions
- The students' design samples' testing and feedback
- Design adjustment

When selecting the student researchers, they needed to demonstrate:

- Interest in performance clothing design
- Interest in older users' needs
- Design capacity
- Awareness of the clothing making and pattern making processes

The research group involved senior students at the CWU in designing and sending out questionnaires to potential male and female participants from ages 50 to 65. We received feedback from 150 questionnaires.

In addition to some basic information about the older user, the questionnaire also includes some specific questions about user daily exercise, such as:

- Which sports do you like at ordinary times?
- How long have you been involved in the sport?
- Your time is about every day?
- Your average monthly applied to sports equipment?
- What are considerations of you buying sports clothing?

In a face-to-face interviewing process with 60 older participants, we recruited two groups in the age range 55–65. One participant group was involved in exercise for 2–3 h per day and the second group exercises twice a week. These targeted customers were invited to participate in the co-design process.

Researchers went to older people's exercise areas such as the Beijing Olympic park for roller-skating, a residential park for dancing, and to the outskirts of Beijing for walking. It was part of the recruitment process. They talked to and became involved in exercises with older research participants, to gather the appropriate information by questionnaire (Figure 24.2).

24.4.4 Co-design workshops

Co-design workshops involved our research group, with the academic team, student designers, and older participants, reviewing the products supplied by our Chinese outdoor industry collaborator. The industry supplied examples of garments in different ranges as well as fabric samples for user selection and testing.

24.4.5 Outputs: final prototypes produced in industry

As an important stage in part of the on going collaborative design process, academic staff and student designers visited an outdoor industry garment factory. Once the students had finished the basic design prototypes, in collaboration with users in the university environment, they then spent two weeks in the industrial collaborator's factory to make the final prototypes guided by technicians in the factory.

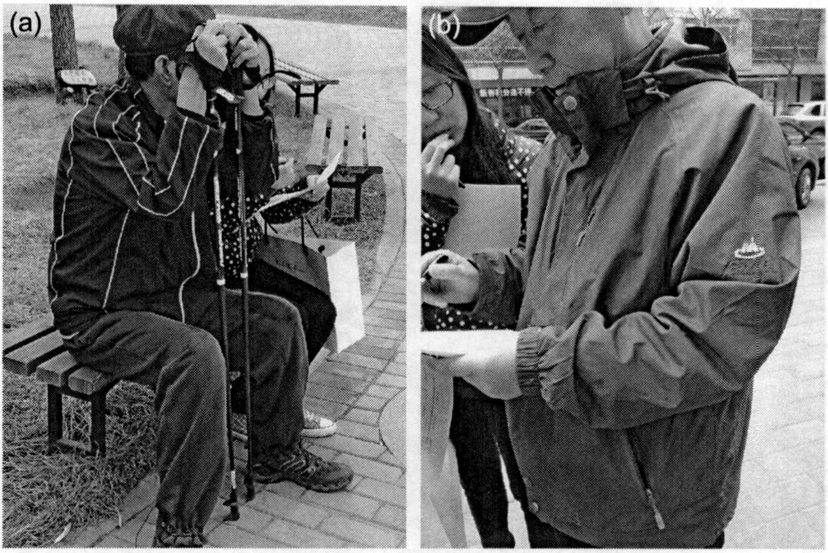

Figure 24.2 (a) and (b) Student researchers help users complete the questionnaires.

24.4.6 User feedback

After sample testing, our older users provided much feedback on the design details, fabric, and color. Figure 24.3(a–c) shows some of the products tested, along with feedback on some of the feature details.

24.4.7 Dissemination to academic and industry trade associations

I am closely involved with the Chinese Fashion Association's Academic Commission and Fashion Industry Commission who are aware of my commitment to bring functional clothing design to education and research and, more recently, design for the ageing demographic. The current practical research outcomes have been presented to a group of trade representatives to explain the size of the potential market and to encourage them to take up the opportunity of planning strategies for product development for this valuable sector. A seminar took place with the trade representatives on November 20, 2013, to disseminate research findings from both the UK and CWU as follows (Figure 24.4):

Seminar title: "Beauty at Sunset: Clothing design direction for older Chinese market".

Agenda:

Overview of UK "Design for Ageing Well" project to introduce the context

Presentation of the senior student practical research project and findings

Guests included:

China Fashion Association Chairman, Li Dang Qi, professor at Tsinghua University.

The China Textile Trade Organization,

Outdoor Industry Sector, secretary, Li Chang Fa.

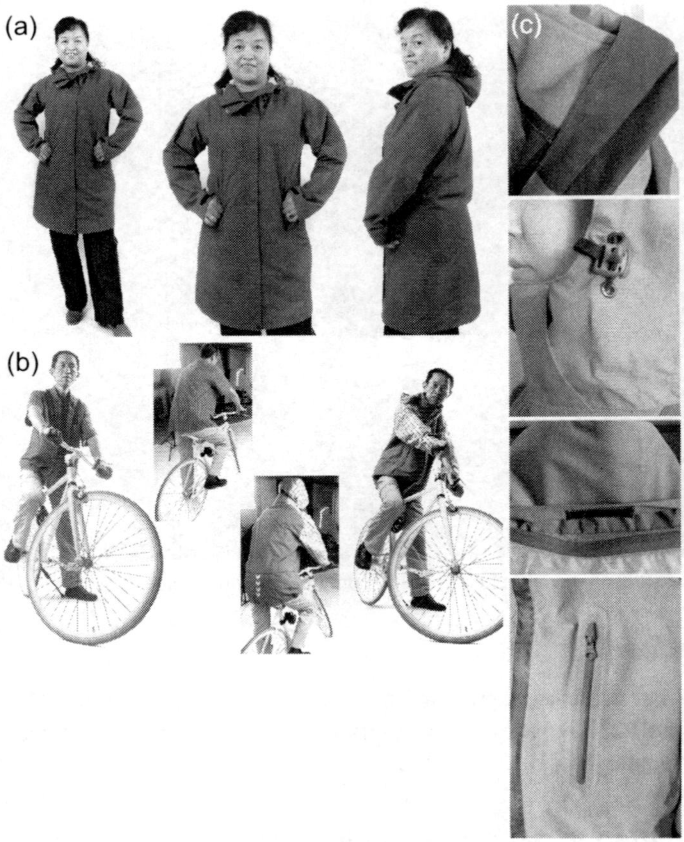

Figure 24.3 (a–c) Products tested by end users and feedback on some of the product feature details.

Figure 24.4 Seminar: "Beauty at Sunset: Clothing design direction for older Chinese market" (November 20, 2013).

24.5 Design direction: merging key findings

This section will provide a brief comparison of Chinese project findings with NDA Design for Ageing Well findings. I will summarize important issues in terms of design guidance with regard to user needs based on conversations between McCann and myself, based on research findings with some comments from Fowler, an industry collaborator aware of the need for a new generation of designers to be aware of the issues raised by practical research project collaboration (Figures 24.5 and 24.6).

Figure 24.5 Co-design process with end users for product testing.

Figure 24.6 End-user testing walk-wear as part of the co-design process.

24.5.1 Size/Shape/Fit

The importance of fitting the changing size and shape of the older body was a clear research outcome both in China and the UK. Older users in the UK had general confusion due to inconsistency in sizing in the retail market. They requested fewer styles with emphasis on age-appropriate fit and a greater choice of lengths for body, sleeves, and legs. The introduction of body scanning in the UK research produced customized fit (McCann, Chapter 13 in this publication). The body scanning process adopted is discussed by Bougourd within this book. The acquisition of this type of technology is being considered by CWU to inform future research.

During the research for the student project, student teams recorded many older people's body measurements, for both female and male bodies. Those measurements were considered when designing sample sizes. Typical sizing for older Chinese people is outlined in Figure 24.7 and Tables 24.1 and 24.2.

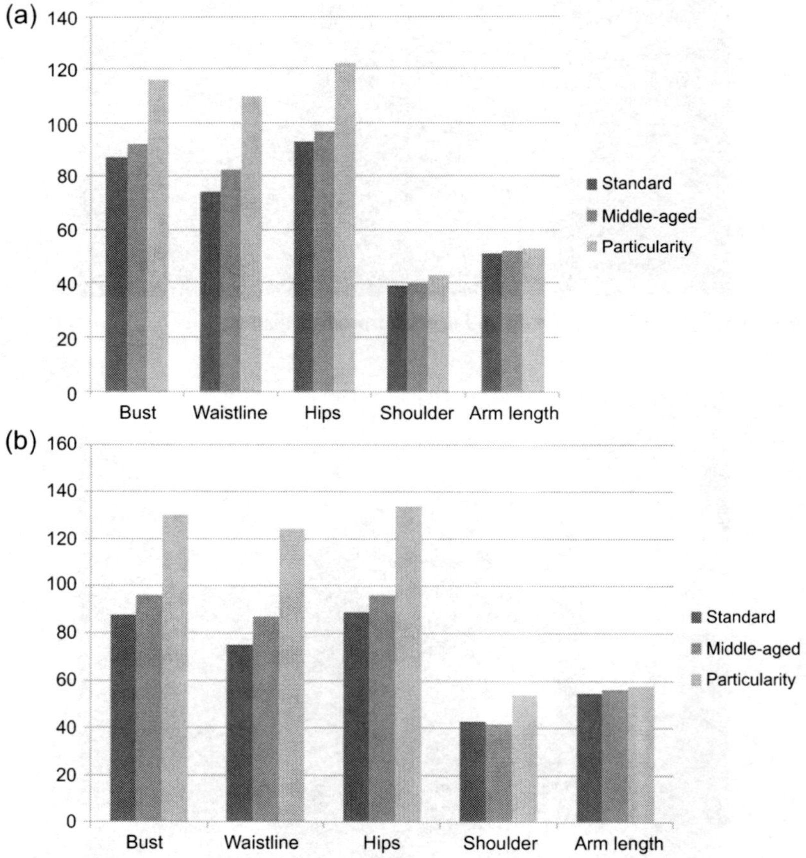

Figure 24.7 (a) Typical measurements for older Chinese women (cm), defined by age group. (b) Typical measurements for older Chinese men (cm), defined by age group.

Table 24.1 **Typical measurements for older Chinese women**

Ages/measurement	Bust	Waistline	Hips	Shoulder	Arm length
Standard	87	74	93	39	51
Middle-aged	92	82	97	40	51
Particularity	116	110	122	43	53

Table 24.2 **Typical measurements for older Chinese men**

Ages/measurement	Bust	Waistline	Hips	Shoulder	Arm length
Standard	90	75	85	43	55
Middle-aged	92	87	92	42	55
Particularity	130	124	134	54	58

24.5.2 Materials and trims

I have studied how previous researchers have identified five product design requirements concerned with aesthetic, symbolic, functional, quality, and ease-of-use aspects. However, research also shows that older people, in buying durable products, pay more attention to three types of functional aspects, namely functionalities, ease-of-use, and quality.

Both in China and in the UK, there is a problem in that, due to the rapid changes in technology, older users, and many users of other ages, still lack understanding of new textile technology and terminology. At the beginning of the research they were not aware of modern fibres and materials or the concept of the sports layering system (McCann, Chapter 13 in this publication).

For example, in the initial UK "Clothing" workshop looking at "base layer" garments the definition of the garment "layering system" (understood in the professional sportswear community) was barely known and was confusing to older research participants. These participants understood very little about base layer function, typically wearing cotton as they generally considered natural fibres to be superior to man-made fibres. However, it was found by Wang Lu that if they understand the properties of a layering system, they become interested in new fabrics and new technologies in design, as in base layer garment styling intended for cyclists. Such a design, using a body mapping concept, places different fabric qualities in different parts of the body's clothing for different performance properties (Wang Lu commenting on her UK experience). UK users were found to like Lyocell (branded, Tencel) for base layers due to their natural handling. For outerwear, UK users originally believed that waterproofness was the most important factor, but, when introduced to the concept of "soft shell" protection they appreciated the comfort factors and smart appearance of less waterproof but more breathable garments, with slight stretch, that may be used for a range of activities (McCann, Chapter 13 in this publication).

24.5.3 Styling/proportion

Much of the product development, in the outdoor technical clothing market, rarely considers a design process that takes into account the characteristics and needs of older users. Issues concerning clothing size, styling, and detail are designed mainly to fit professional participants and cannot completely meet the requirements of older users. In terms of styling, the older users explained their everyday experiences related to design needs concerning the fit of clothing including sleeve length, collar height, collar, zipper length (Wang Lu & China Textile, 2011).

UK users requested fewer styles in age-appropriate fit to be available in a range of colors and textures rather than fast-moving fashion design. They chose softer design lines, with rounded raglan sleeves, with side panels to allow for fit adjustment to different figure types (McCann, Chapter 21 in this publication).

24.5.4 Design detail

In terms of design detail, and possibly because UK baby boomers are relatively older, there were issues concerning reduced dexterity regarding fastenings. Users had difficulty with zips, cuff adjustment, certain pocket openings, and putting hoods into collars etc. Hood design was an important factor in terms of visibility. They wanted lots of pockets (especially the men) but with consideration of easy access (McCann, Chapter 21 in this publication).

During the UK co-design process, related to wearable technology, the older user group proposed that technology devices should have interface designs with options that are simple, related to applications, of interest, in health monitoring and information exchange (Wang, 2012).

24.5.5 Color

Both in the UK and China users requested color to be appropriate for older people's changing skin and hair color. Interesting prints on fabrics and natural tones are considered more acceptable than black and grey (Wang, 2012) UK users asked for more choice of colors and textures for nontransient styles (McCann, Chapter 21 in this publication).

24.5.6 Industry perspective

As the Chinese grey market size is increasing rapidly, the outdoor clothing industry has been willing to support CWU's research and to become involved in the sportswear design project. Their perspective on ageing is altering as they see the results of the research and become aware of the benefits to industry of CWU training young designers in an effective design process to address older user needs (Wang, 2011).

There has been an initial impact in the UK industry as companies have adopted some of the color direction and have altered garment lengths and improved accessibility with respect to pocket openings etc. However, the introduction of commercial sizing would require further near market research (McCann, Chapter 21 in this publication).

24.5.7 Effective commercial product specification

The Western outdoor industry is now obliged to begin to measure the effects of textile processing and garment manufacture leading to environmental damage and carbon footprint. Accuracy of product design specification is a major concern within the clothing supply chain. My experience with the UK research group, in collaboration with UK industry product developers, highlighted difficulties in communication between the UK base and production in China. For example, Sprayway's designer confirmed that garment patterns are no longer sent with specification packages from the UK to China. Youngone (dealing with US brands) said that specifications are inconsistent and inaccurate with, sometimes, a sketch on a napkin! The result is many stages of prototyping that waste time and resources (Fowler). At CWU, I am often approached by these Western companies requesting Chinese designers to interpret the design ideas of brands more effectively.

24.6 Way forward

As Dean of Art and Design at the China Women's University, I am in a strong position to implement new approaches to fashion design education. The future consumer market will be more diversified than at any time in the past in terms of age range and, therefore, the physical ability of certain users (Inclusive Design Toolkit). Where possible, both Chinese and UK designers should consider inclusive design/design for all age groups: young, middle age, and elderly customers. However, clothing design is problematic in terms of the demands of the diversity of body size and shape research. This is a particular concern when looking at the application of wearable technologies (such as textile-based electronic sensors) as the clothing must fit the body where body sizes, shape, and the fit of clothing cannot be standardized. It seems that one way forward must be concerned with mass customization through the use of new technologies such as body scanning.

This chapter has discussed how a co-design approach may be applied to functional clothing design and that older customers should be brought into the mainstream with better design solutions. Initial research has resulted in positive support from the Chinese industry for co-design-led process, a positive attitude shown by young designers, and better informed and satisfied older research participants. I propose that Chinese design education should continue to develop co-design theory in further collaboration with industry. Initial research, in both the UK and China, has established strategic links to industry and trade associations at a time when China is beginning to develop new domestic markets as well as offer more sophisticated levels of production to the West. This is a good time for Chinese academia to benefit from collaboration with Design for Ageing Well research and to adapt and build on important findings as the focus of future academic collaboration between the UK and China. Potential exists for developing further collaboration as follows.

1. **Co-design process leading to more responsible "slow" fashion**
   One of the most significant outcomes of Design for Ageing Well has been the request of older users for slower, more considered design with fewer styles but more choice of sizes, colors, and greater fabric selection. This is in contrast to young fast fashion. Professor McCann's recent presentation of research findings included the concept of "Slow Fashion," which is an idea that I plan to study in future academic projects. It is also a proposal to be discussed with trade associations in order to share better practices as we aim for more responsible and sustainable design methods.
2. **Strategies for more effective product specification**
   Improved product specification is needed to enable more effective garment sourcing within collaborative networks between China and the UK. The accuracy of product design specification has been identified as a major concern within the global clothing supply chain. In my academic role, I am often approached by Western companies requesting designers for product development teams in China to interpret the design ideas of Western brands more effectively. The technical garment sector is more demanding than the fashion sector because of the many innovative technologies using new materials and manufacturing techniques. UK researchers worked with industry to resolve methods for more accurate specifications of the functional garment layering system. CWU has already benefited from the project's practical research findings. Further collaboration, between academic research and industry, in the UK and China, could improve design process from concept to final prototype.
3. **Future collaboration between UK and CWU's Chinese academic partners**
   My final conclusion at the end of my period as a visiting scholar was that "The practice of cross-disciplinary research and teaching will need time to develop because it challenges the boundaries of existing disciplines and teaching management restrictions. Researchers with different background and experiences will need to work together and overcome these boundaries to find a common shared language in design for ageing. In the field of the management, systems must be adjusted to support this cross-disciplinary model of teaching and research" (Wang Lu & Tsinghua University, 2012).
   Potential exists for collaboration in clothing and textiles research with expertise at the Fashion and Art Design Institute, Dong Hua University, Shanghai, and with ongoing collaboration with members of the UK Design for Ageing Well research network. The Inclusive Design area of clothing research may have potential as a focus for future collaborative work with research partners. For example, the new Inclusive Design program in Shanghai has no current expertise in clothing design, and the University of Dong Hua program has little current experience of wearable electronics. There is much more to be done in moving forward with design for ageing.
4. **Involvement with trade bodies as a potential route to market uptake**
   Involvement with trade bodies such as the Chinese Fashion Association's Academic Commission and the Chinese Outdoor Industry Commerce Alliance offer potential routes to market uptake. (Chinese Fashion Association)
   Such bodies are at the center of fashion industry promotion, where there is no current recognition of the importance of functional clothing design. The CFA Technical Committee can, through technical cooperation with high-tech textile enterprises such as Tory, encourage design innovation by promoting new textile materials and enhancing consumer understanding and awareness of new technology and materials. China Fashion Week is held by CFA twice a year at the end March and end of October and could be an appropriate platform for potential market development.
   According to COCA figures, China's outdoor leisure clothing consumers are mainly aged 40 to 60 years. CWU can cooperate with COCA on performance sportswear design courses;

students' clothing design works for the older user can be undertaken as production in the enterprise, while CWU can work with Chinese outdoor companies to design leisure and performance apparel for Chinese older people.

5. Recognition of importance of this research by CWU
   My university confirmed me as a professor on November 23, 2013. The professorial board was particularly interested in my developing area of functional clothing education and research and of my recent success in attracting design for ageing funding. Such interest is key to the future success of design for ageing in China.

References

China Fashion Association (CFA): www.english.fashion.org.cn.

Huppert, F. (2003). *Design for older user inclusive design. Inclusive design: Design for the whole population*. London, UK: Springer, 31–49.

McCann, J. (2009). The garment design process for smart clothing: From fibre selection through to product launch. In J. McCann & D. Bryson (Eds.), *Smart clothes and wearable technology* (pp. 70–94). Oxford, UK: Woodhead Publishing.

McCann, J. (2009). Smart Clothing for the ageing population. In J. McCann & D. Bryson (Eds.), *Smart clothes and wearable technology*. Cambridge: Woodhead.

Sheffield University. *New Dynamics of Ageing (NDA) Newsletters*. Accessible at http://www.newdynamics.group.shef.ac.uk/the-nda-newsletter.html.

Wang, L. (April 2011). The importance of researching the requirement of older users for clothing design education in China. In *Presentation at Included 2011*.

Wang, L. (2012). With cross discipline and co-design for the characteristics of the British aging clothing design research model. *Tsinghua University Journal*, (May 2012).

Experiences in the design, iterative development and evaluation of a technology-enabled garment for active ageing walkers

25

W. Burns
University of Ulster, Londonderry, UK

25.1 Introduction

As a result of increased life expectancy and falling birth rates worldwide, an ageing society is emerging (Hogan, 2005). Within the Western world, older persons typically suffer from at least one chronic condition (Vergados, Alevizos, Mariolis, & Caragiozidis, 2008). Coupled with chronic medical conditions, older persons typically display a number of more general characteristics that come with the ageing process. These include diminished eyesight and hearing, memory loss and reduced dexterity (Hogan, 2005).

The World Health Organization defines health as 'a state of complete physical, mental and social wellbeing and not merely the absence of disease or infirmity' (WHO, 1948). As a result it is important for older persons to maintain their health levels by engaging in physical activity, and maintaining social contacts, as walking in later life is considered to significantly reduce the risk of pre-clinical disability and increase aerobic capacity and physical function (Moore-Harrison, Speer, Johnson, & Cress, 2008).

This work presents the methodologies used and the results obtained from a research study that aims to promote and maintain the physical and mental fitness of older walkers in the age range of 60–75 years by using a smart clothing system.

25.2 Background

Physical activity (PA) may play an important role in maintaining health and physical function while reducing the health care burden (Moore-Harrison et al., 2008). Physical activity has a myriad of effects that stem from physiological adaptations that may transfer to improvements in clinical outcomes such as reducing the risk of falls (Nike+). Individuals who begin a regular PA programme early in life and

Textile-led Design for the Active Ageing Population. http://dx.doi.org/10.1016/B978-0-85709-538-1.00025-0

maintain this over the years will likely have high physical performance throughout the lifespan, although a decline in physical function is inevitable.

PA benefits older people's health: preventing disease, and improving well-being, reducing disability. Adults, including older adults, are advised to perform at least moderate intensity PA (leading to increases in breathing, heart rate and temperature) for >30 min (in >10 min bouts) on at least 5 days weekly (Hogan, 2005; WHO, 1948). Other guidelines suggest that at least 10,000 steps should be taken in order for the PA to be effective. Nevertheless, the media-promoted 10,000 daily steps target ignores the guidelines' intensity component (MIThril). Within the older population, only 18% of men and 14% of women aged 65–74% and 8% of men and 4% of women aged 75 or over achieve this (Moore-Harrison et al., 2008).

Ageing is clearly associated with a decline in most physiological systems that culminates in limited physical capacity. Regarding the cardiovascular system, ageing is associated with a dramatic decline in maximal aerobic performance that is due to a decrease in cardiac output (i.e. the delivery of oxygenated blood to muscles) and oxygen uptake at the muscle (Design for Ageing Well Project).

There is an increasing awareness of the need to maintain a certain level of physical activity and fitness. This is reflected in the number of personal fitness devices available on the consumer market, specifically for the sports enthusiasts.

Devices such as *Nike+* use an accelerometer located in the wearer's shoe, to monitor the wearer's activity levels, specifically walking and running, and provides feedback via an iOS device (iPod, iPhone etc.). Nike+ have also released a wrist worn 'FuelBand' that, without the need for additional devices, monitors the wearer's activity levels and provides real-time feedback via a small display on the device (Nike FuelBand).

Garmin have a suite of wearable devices for tracking the amount of activity of the wearer, such as the 'Forerunner' devices (Garmin Forerunner). Unlike the Nike+ products, the Garmin products use GPS to track the location and movement of the wearer and can provide distance information. The Garmin products can also monitor the wearer's heart rate either using a wrist-worn device, peripheral chest straps or tight-fitting shirts with textile electrodes.

These devices typically target walkers, runners and athletes who typically belong to a younger age group. The devices themselves are not suitable for older persons who display the ageing characteristics mentioned above. For example, the devices are often too small for older users to operate and view data.

The lack of suitable clothing has been identified as a potential barrier for maintaining physical activity in older persons. The Design for Ageing Well Project aims to enhance the autonomy of the 60–75 year old age group through the application of smart clothing and wearable technologies. The project designed, implemented and evaluated a functional clothing layering system, with integrated technology, designed specifically for the needs of the older population, with particular focus on walking in different environments from towns to mountain trails.

25.3 Examples of research projects in health care monitoring

The research projects, as seen in Table 25.1, measure vital signs information for a number of reasons, such as health monitoring, protection and safety, and context extraction. Analysis and feedback of the information gathered from these systems are analysed and feedback provided to the wearer via a health care professional. As such, the wearer of the system does not know what is the current status of their vitals.

The Design for Ageing Well project aims to develop a wearable smart garment for use of older persons who suffer from no adverse health conditions, but are not active enough to use athletic/sport systems. Using similar technologies as in the aforementioned research projects, i.e. textile electrodes, ECG, accelerometers, etc., a smart clothing system will be developed.

As a result of the ageing process, a loss of mobility, dexterity and physical function is expected. Simple tasks, such as dressing, are now a major problem (McCann, 2009). Within the older age range, conventional tight-fitting garments would not be feasible, as such a garment would be deemed uncomfortable and impractical.

25.4 Research methodology

In this section we present the methodologies used throughout the duration of the two-year research project. These methodologies include the elicitation and mapping of user requirements to available technologies, the development of the technological solutions and clothing design. Finally we detail the evaluation procedures used in three evaluation phases.

25.4.1 *Elicitation of user requirements*

To obtain user requirements for older persons engaged in walking activities, questionnaires were sent to the University of the Third Age (U3A). These questionnaires asked the participants their thoughts on a number of issues, including clothing and technology in general. They were also asked what features they liked in clothing and technology systems. Due to the age of the participants, it was unlikely that they would have much knowledge on the available technologies.

In addition to the responses from the questionnaires, the researchers had a number of interactions with the target cohort in the form of workshops. Each workshop had a particular theme. This ranged from clothing design, to fabrics, to colour selection and also included technology sessions. The workshops comprised presentations of potential benefits and question and answer sessions. The average number of participants was 10, with usually more males than females.

The attendees were asked for their feedback on what they thought of the presented technology from an aesthetic, usability and functionality perspective. A predominant

Table 25.1 Research projects in wearable health systems

Project	Authors	Description
Wealthy	Paradiso, Loriga, and Taccini (2005)	Coupled textiles with wearable technologies to provide real-time health and well-being monitoring
eCAALYX	Boulos, Wheeler, Tavares, and Jones (2011)	Monitor the health of older persons with multiple chronic conditions, at home and on the move, using smartphone technologies, wearable sensors and textile electrodes
MyHeart	Villalba and Salvi (2009)	A personal health assistant that aims to detect early atrial fibrillation (Villalba & Salvi, 2009). The sensors used are either embedded inside the garment or a piece of clothing. It uses conductive yarn woven into the garment. The system can detect and classify, using accelerometers and ECG sensors, activities such as lying down, walking and running
Biotex		By developing textile sensors to measure the chemical composition of bodily fluids, such as sweat, as well as physiological signs. Sensors included in the system are ECG, respiration and blood oxygen sensors
MagIC	Meriggi and Castiglioni (2010)	Uses textile-based sensors to measure vital signs. By integrating textile sensors into a vest, the system can measure ECG and respiration rate as well as activity levels. Signal pre-processing is carried out and information is transmitted via Bluetooth to a PC or Personal digital assistant
ProeTEX	Magenes, Curone, Caldani, and Secco (2010)	Developed a smart garment for use with firefighters. The system consists of a wearable garment with embedded CO_2, accelerometer, temperature, heart rate and SpO2. Data collected by the garment are transmitted wirelessly to a controlled centre for analysis. If dangerous events are detected, an alert can be raised
Mermoth	Online (Mermoth)	Medical remote monitoring of clothes is a project that aims to develop a garment with embedded ECG, respiration, skin temperature and accelerometers to monitor the wearer's health status. The garment transmits the data via a radio frequency link to a PC for data visualisation and analysis
MIThril	Online (MIThril)	Developed a context-aware wearable solution. The system integrates a range of sensors to record contextual information such as grocery lists and movie recommendations. Unlike the previous projects, MIThril redirects wires around the garment as opposed to using a wireless solution

view emerged indicating that while there are many devices on the market that offer complementary functions, none offered all of what they needed in one easy-to-use, device.

There views were matched with the differing levels of activity in this age group. These levels were classified as *Easy*, *Leisurely*, *Moderate*, *Moderate+* and *Strenuous*.

Because of these differing levels, it is important that any developed solution have the ability to be personalised. For example, a *Leisurely* walker did not want GPS recording, maps or heart rate; however, they would like to know when they reached their recommended daily activity levels.

These workshops provided the researchers with the chance to educate the target cohort on the capabilities of the technologies available. At these workshops, it was useful to get feedback from the attendees on the technologies presented.

Based on the feedback from the questionnaires distributed to the U3A, the following requirements, shown in Table 25.2, were elicited from a total of 50 responses. These requirements were then used to map available technologies to the user need. This provided a foundation to begin development of the technological solution.

25.4.2 Technology mapping

The elicited requirements were ranked by their popularity and mapped onto possible technological solutions. Due to the target cohort for the project, the requirements elicited in relation to hikers were disregarded. During this mapping process, the limitations of the wearable systems, cost and usability, especially for older users, were all taken into consideration. Some of the requirements were not capable of being addressed using 'off-the-shelf' technologies; however, solutions are available.

Table 25.2 User requirements as elicited from questionnaires distributed to the University of the Third Age

Requirement	Walker %	Hiker %
Keep me in contact with a group	54.2	46.7
Tell me where my group members are	26.2	13.3
Keep me warm	88.1	93.3
Cool me	71.4	33.3
Help me navigate	54.8	40.0
Call for help if I should fall	52.4	53.3
Protect me if I should fall	50.0	20.0
Tell me if I over-exercise	4.8	0
Tell me the distance covered already	54.8	13.3
Tell me where to find the nearest bus stop	23.8	13.3
Tell me where to find the nearest facilities	38.1	13.3
Monitor my health for medical reasons	16.7	13.3
Interact with my mobile phone	35.7	13.3
Interact with my portable music device	2.4	13.3

Results were separated depending on the type of activities being performed, $n = 50$.

Requirements such as 'Cool me' and 'Protect me if I should fall' can be addressed using advanced textile-based approaches such as breathable fabrics for cooling and dilatant materials for impact protection (e.g. D3O). The results of the technology mapping are shown in Table 25.3. As part of the technology mapping process, the requirements were broken down into four categories, Health, Social, Safety and Functional.

25.4.2.1 Health

Of the aforementioned requirements, two have been identified as belonging to the health category. The first is the 'Monitor my health for medical reasons', and the second is 'Tell me if I over-exercise'. In order to address these requirements we need to acquire the user's vital sign information and gross movement levels. The Shimmer platform (Shimmer Research) was identified as a solution, as it provides three-lead

Table 25.3 Results of the technology mapping exercise detailing the possible technologies that could be used to address the requirements as well as the specific product to be used in the developed system

Requirement	Technologies available	Technology used
Keep me in contact with a group	Telephone, Walkie-Talkies, Social Networks	Smartphone
Tell me where my group members are	GPS, Google Latitude	Smartphone
Keep me warm	HeatWear (HEATwear)	HeatWear
Cool me	Fabric	–
Help me navigate	GPS, Electronic Maps	Smartphone
Call for help if I should fall	Accelerometer, Smartphone, GPS	Smartphone
Protect me if I should fall	D3O	–
Tell me if I over-exercise	Accelerometer, Heart Rate Monitor	Accelerometer, ECG sensor
Tell me the distance covered already	GPS	Smartphone
Tell me where to find the nearest bus stop	Maps, GPS	–
Tell me where to find the nearest facilities	Maps, GPS	–
Monitor my health for medical reasons	Heart Rate Monitor, Temperature	ECG sensor
Interact with my mobile phone	Touchscreen, Textile Switches	Textile Switches
Interact with my portable music device	Touchscreen, Textile Switches	Textile Switches

At the time of the technology mapping process, Google Latitude was an Application Programming Interface (API) used to track the locations of a user's friends/contacts. It is no longer available as a standalone API.

ECG recording coupled with a triaxial accelerometer and Bluetooth module. The form factor of the Shimmer device was also deemed wearable.

25.4.2.2 *Social*

This category consists of the following three requirements: 'Keep me in contact with a group', 'Tell me where my group members are' and 'Tell me where to find the nearest facilities/bus stop'.

Contact with group members can be achieved in a number of ways using a technological solution. The options identified were the use of two-way radios, or voice/text communication using a smartphone. The smartphone option was chosen as it reduced the amount of hardware the user would have to carry on their walk. By harnessing the in-built sensors of the smartphones, specifically GPS, it is possible to send and receive the locations of group members.

The location of local facilities is considered a part of this category. This would enable users to use facilities as a start, end or rendezvous point. Nevertheless, the display of facilities could be limited by the underlying infrastructure of the area. If the facility information is publicly available, specifically their GPS coordinates and descriptions, and use an Application Programming Interface (API), then this information could be integrated into the solution. If not, another possibility is the use of user-generated content, by way of tagging landmarks and facilities for storage and display on the system.

25.4.2.3 *Safety*

Safety is paramount for any vulnerable target cohort. Within this category, 'Call for help should I fall' was the highest ranking. This is not surprising, taking into consideration the age of the target cohort and the activities they will be undertaking.

This requirement can be addressed using the accelerometer located in the Shimmer sensor and coupling that with the processing power and communication capabilities of the smartphone. By detecting a fall via the sensor, the smartphone could automatically initiate a call to a pre-defined emergency contact. To remove the need for the user to do anything with the hardware, the smartphone would be placed into speaker mode and have the volume set to its maximum level.

25.4.2.4 *Functional*

The requirements within this category address the use of the technology within the system, including 'Interact with my mobile phone/media player' and other functions within the system itself. To address these issues, careful consideration must be undertaken in the design of the user interface (UI) of the application. Other technologies can be used to improve the overall User eXperience (UX) of the proposed solution. These include the use of Fibretronic textile switches (Fibretronic ConnectedWear) and Sony Ericsson's LiveView micro-display (Sony Ericsson LiveView). Both of these devices enable the user to interact with the functionality of the smartphone. Using textile switches allows the controls of the system to be integrated within the developed garment.

25.5 Prototype iterative developments and evaluations

Before development began, the user requirements were again presented to users to ascertain a prioritised list of requirements, shown in Table 25.4. This list was then used as a guide to begin the development of a prototype, aiming to address the highest priority requirements first.

The smartphone application was developed for the Android OS, as it provided a mature development platform as well as allowing access to a number of APIs, such as location, Bluetooth and telephony, all of which would be required to meet the user requirements.

The smartphone would primarily act as a processing computer and a telephone secondly. The Shimmer sensor was integrated, allowing the collection and processing of ECG and accelerometer data. This information would then be recorded and presented to the user in real time as a heart rate and step count value, respectively.

To address the 'Tell me where my friends are' requirement, an Apache web server was set up with a MySQL database. This would allow the smartphone application to send the user's current location and receive the last known locations of their friends/group members.

All data recorded by the system were stored on the smartphone's SD card in an XML file. To provide a user-friendly way of reviewing walking data, a desktop application was developed. This application allowed the user to plug the smartphone into their home computer and review their previous walks. The presentation of the data came in the form of textual information, in the case of heart rate, step count, etc. An interactive map was used to overlay the route the user had taken on that particular walk.

As seen in Figure 25.1, the system can have any number of users, each accessing data from the server and each having access to individual Desktop Viewer software running on their own computers.

A three-stage iterative evaluation process was undertaken to obtain user feedback and re-develop the proposed solution. The first two evaluation phases were used to

Table 25.4 Prioritised list of user requirements from an additional 18 users, these requirements were used to guide the development of the prototype

Requirement
Call for help if I should fall
Monitor my health
Keep me in contact with the group
Help me navigate
Distance covered
Tell me where my group members are
Tell me if I over-exercise
Interact with my mobile phone
Interact with my portable music device

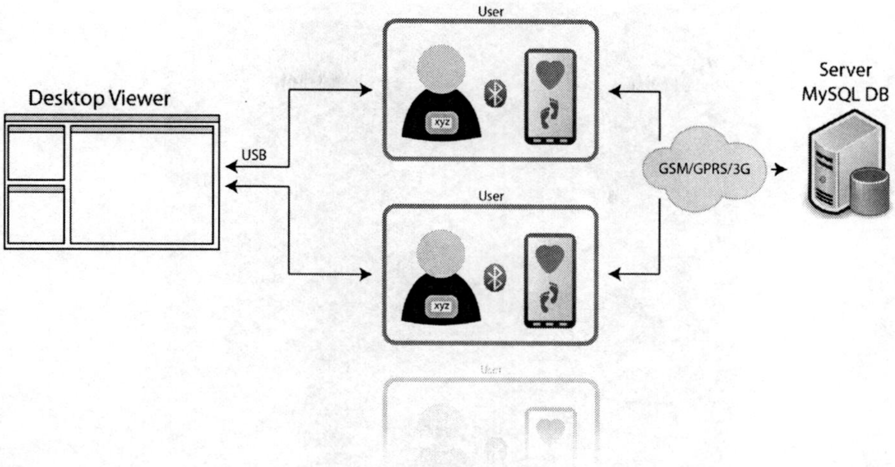

Figure 25.1 System architecture outlining the three main components of the system, including the Desktop Viewer, the User and their wearable components and Server with MySQL database containing location data.

evaluate the system to produce a robust solution for evaluation in a longitudinal third evaluation.

Each evaluation followed the same protocol, with a few modifications where significant changes to the prototype were required. Users were asked to evaluate the system during a walking exercise, and this was followed by the completion of questionnaires by each participant. The results, user feedback and researcher observations were used to modify and re-develop the technology solution for the next evaluation phase.

In each of the following subsections we outline a summary of the evaluation methodology, outline the current state of the technological prototype and present the results from the evaluation.

25.5.1 Evaluation 1

The first of the three evaluations consisted of six participants (three male and three female) between the ages of 62 and 73. During this evaluation phase, the participants were asked to undertake a short outdoor walk lasting approximately 15–20 min whilst using the prototype, in order to ascertain aesthetic, functional and UI/UX feedback. Figure 25.2 shows the main components of the prototype to be evaluated in this evaluation.

The main functionality of this version of the prototype were:

- Recording of walk data (GPS)
- Heart rate
- Step count
- Walk details displayed on wearable micro-display

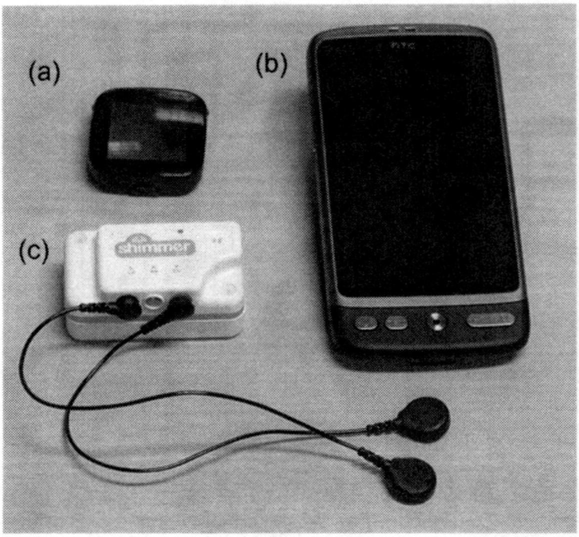

Figure 25.2 Prototype components for testing during Evaluation 1, consisting of (a) Sony Ericsson LiveView micro-display, (b) smartphone and (c) Shimmer sensor.

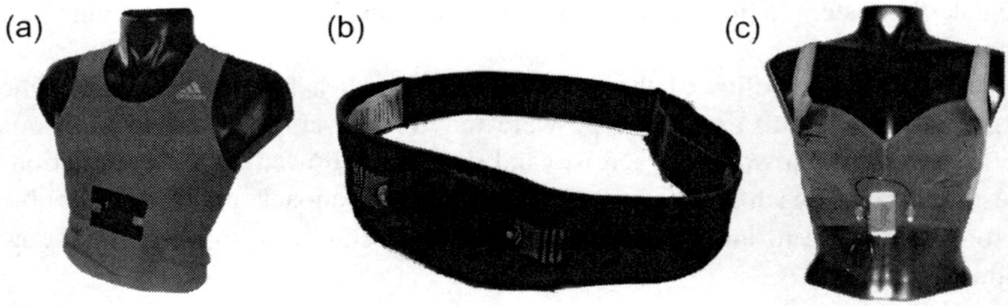

Figure 25.3 Base layer garments used in the evaluation. Males wore (a) modified Adidas base layer, two of the three females wore (b) modified Textronic chest straps and the third female wore (c) a custom designed bra.

The male participants were asked to wear an Adidas base layer, modified to include a pocket for the sensor and conversion of the female snaps to male snaps. Two of the three females wore Textronics chest straps and one female wore a custom designed bra. All base layers for all participants included textile electrodes, shown in Figure 25.3.

Figure 25.4(a) shows the interface of the Desktop Viewer application, used in this evaluation, that allows users to configure the system. In the current version of the application the customisation is limited to User Information, Shimmer Information (device ID), Health and Wellbeing information (maximum steps, heart rate, distance) and Emergency contact details. In order to present the data collected by the system, the smartphone is connected via USB and displays a graphical representation of the user's walk, health status and activity information shown in Figure 25.4(b).

Following a brief overview of the prototype functionality, all participants were asked to undertake the walking activity to evaluate the system. Following the walking

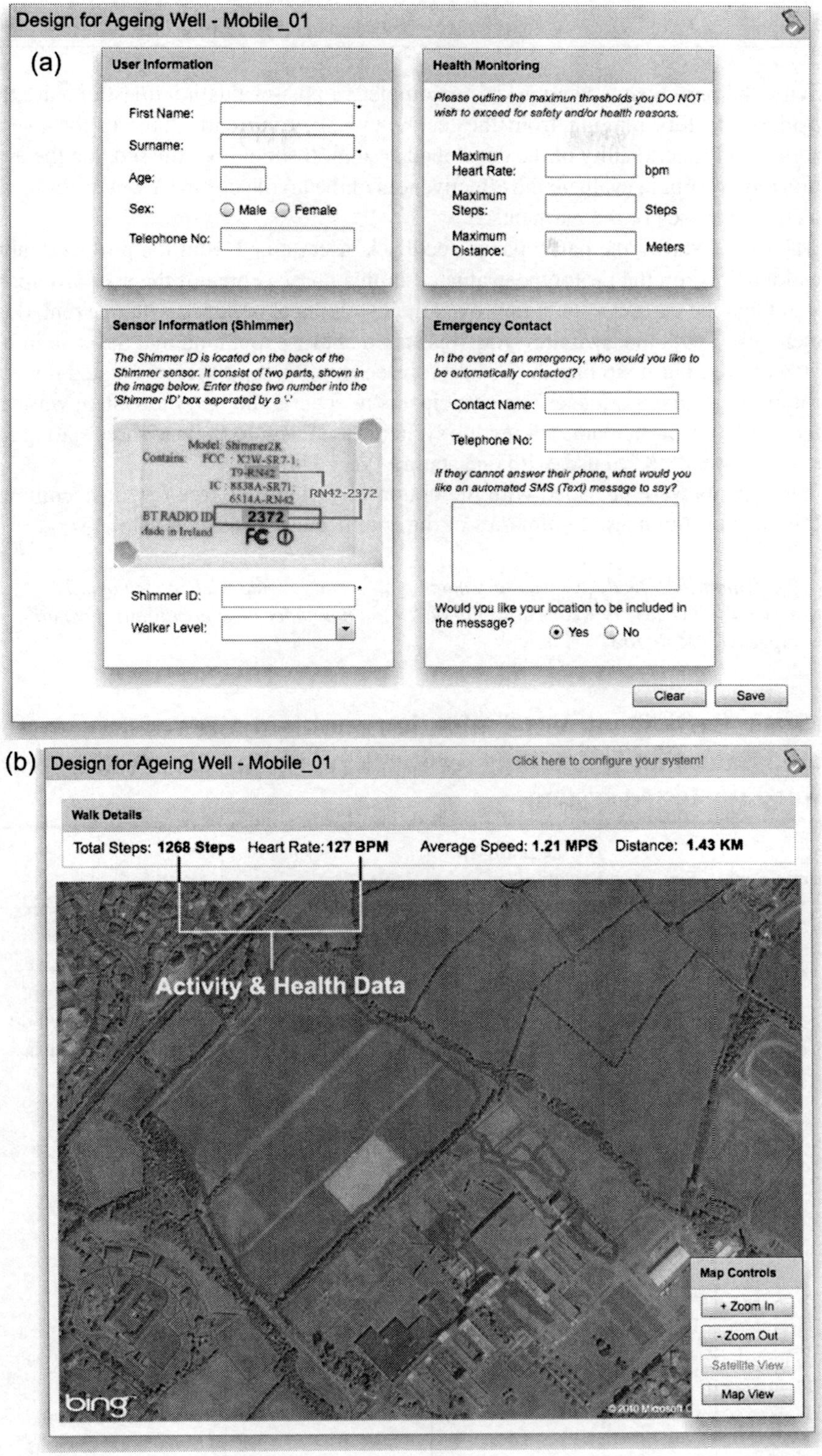

Figure 25.4 User Interface of the Desktop Viewer showing (a) configuration screen and (b) an interactive map with the user walk route overlaid and health and activity data displayed.

activity, the participants were asked to complete a post-evaluation questionnaire that considered factors ranging from the technology they currently have to the overall usability and functionality of the developed system. It should be stressed that the evaluation did not aim to evaluate the effectiveness of the textile electrodes, step count, fall detection accuracy or the garments.

Table 25.5 shows the participants' feedback, ascertained from the post-evaluation questionnaires, on the prototype evaluated in this phase. Three of the six participants thought that the current system improved their walking experience with one remarking 'Potentially', with the user clarifying this statement by explaining that the system was difficult to use but if simpler they could foresee using it. A further two stated it would not improve their experience. The participants in general thought the system was easy to use, with two participants stating the system was 'Easy' to use and three stating that the system was 'OK' to use, with one stating 'Very Difficult'.

Participants were provided with the option of including a free response comment within the questionnaire. Summaries of the comments are outlined below:

Participant 1: 'The device was very much of use, and I would wish to own one. The device itself is easy to wear but would be far too hot under most conditions. I would suggest a "Male Bra" version'.

Table 25.5 Participant post-evaluation questionnaire results from Evaluation 1, detailing feedback regarding user feelings toward the tested system

Participant	Sex	Like using the system?	Likes	Dislikes
1	M	Yes	'The vest was easy to use as was the basic technology'	'The vest was not to (sic) restrictive when actually on, but it was very hot; fine in summer in Greenland'!
2	M	Yes	'Clear screen use'	–
3	F	Yes & no	'Interesting if working properly'	'Unreliable'
4	M	Yes	'I liked the communication links but the map was slow to react'	–
5	F	No	'Sewn in underwear, not uncomfortable'	'Not accurate in measuring distance or heart rate'
6	F	No	–	'It didn't work and made me anxious'

Participant 3: 'It would be good to know that the functions are reliable. I would not wish to have a phone worn externally on a sleeve but would want it balanced (weight wise) of in a pocket. I would want to move technology between layers. I like the Fibretronic soft switch technology if interface clear. This is due to being able to update and/or move around technology'.

Participant 5: 'Technology did not prove to be accurate on the test, needs more work. Whilst it is useful to be able to contact people and services, independence is also important, and it is not a good idea to make a walker feel they are being tracked'.

Whilst the feedback from the questionnaires was helpful to ascertain the usability of the system and identify any issues, further feedback was elicited from a Q & A session carried out by the researcher.

Once the participants had completed their questionnaires, researchers asked the three male participants a number of questions regarding the overall system and the functionality it provides. A brief summary of the conversation follows.

The researcher asked the participants what they thought of the warning and alerts when they are approaching and have exceeded a number of thresholds such as Heart Rate, Distance and Steps. Participants did not like the idea of having to stop just because the system said so. It was suggested that perhaps these 'Thresholds' should become more like 'Milestones', where the user is alerted when they reach that particular milestone, i.e. every 5000 steps, 1000 m, etc.

The participants were asked what they though of the Heart Rate display and did they understand the meaning of it. Participants acknowledged that the Heart Rate was displayed clearly and they could recognise what it was; however, they did not know what a healthy heart rate was for them. Metaphor icons were suggested, outlining that when their heart rate was within normal parameters for their age, it would display a Green Heart, when it was within 20% of their maximum heart rate a Yellow Heart would be displayed and finally when they were within 5% of their maximum heart rate, a Red Heart, supplemented by some other form of feedback, would be displayed. The participants liked this idea.

During the walking exercise one of the participants stated that the map was wrong as it displayed a rugby pitch on his left, when in actual fact it was on his right. This error is due to the non-rotation of either the map or the 'Me' icon.

When asked what other features they would like in the system, participants remarked that they would like the option to set a destination that they would like to walk to as well as the ability to add a starting location to their walk. This is said to be useful because the users do not necessarily start a walk as soon as they leave the house, but may walk, cycle or drive to an unknown location and begin from there. Other functions included the ability to see where they had just been walking in real time and on the smartphone map, as well as some form of notification in half-hourly/hourly intervals to keep the user informed of the amount of time they have been walking.

Participants responded favourably to the idea of a dedicated ICE (In case Of Emergency) button that would initiate a phone call and send an SMS message to a pre-defined contact. While this functionality is added to the 'Fall Detector' function, participants wanted the ability to activate this function voluntarily.

It should be noted that this question and answer session was with the male participants only as the female participants were unavailable at that time. Also only one of the three male participants classed himself as a leisurely walker, whilst the other two said they were strenuous.

One major problem with this prototype was the unreliability of the system. When cycling through walks data, using the LiveView, on a number of occasions the LiveView would freeze. This in turn led to the smartphone freezing. This resulted in the smartphone needing to be turned off then on again, as the OS itself seemed to freeze. Further testing by the developers identified the LiveView as the cause of this unreliability, as the prototype worked well when the LiveView was not connected or used.

Based on the feedback from the questionnaires, walking activity and Q & A session, a number of changes to the prototype in terms of aesthetics, functionality and feedback need to be made:

- GPS should start recording once the application is loaded
- LiveView interface should be replaced
- Thresholds should now become milestones
- Heart rate information should be contextualised
- Plotting of a destination should be displayed on the smartphone; this can be achieved using the Desktop Viewer

25.5.2 Evaluation 2

This second evaluation consisted of six (three male and three female) participants aged between 63 and 73. The participants were asked to evaluate the prototype during a walking exercise to ascertain aesthetic, functional and UI/UX feedback. The components of the prototype to be evaluated in the second evaluation are shown in Figure 25.5.

As a result of the previous evaluation, a number of changes to the prototype were made (Figure 25.6). One major change was the removal of the LiveView display component and replacing with Android's Text-To-Speech (TTS) engine. This would mean the participant could 'hear' the system provide 'spoken' feedback on their walk, on demand. But using the Fibretronic Bluetooth Module (BTM), this iteration would allow textile switches to control the smartphone application. Earphones could also be attached to the BTM and this would allow the participants to hear on-demand feedback, despite not interacting with the smartphone itself.

The participants, in the previous evaluation, raised the issues of not knowing what the heart rate reading meant to them; i.e. they did not know if a heart rate of 178 was good or bad. This iteration of the prototype remedied this by changing the colour of the heart icon on the screen depending on the participant's heart rate. If their heart rate is less than 60% of their maximum, the icon will be green. Between 61% and 89% it will be yellow and ≥90% red.

The male participants were asked to wear a custom-made base layer, i.e. vests, and female participants were asked to wear custom-made bras. Both of these custom garments consisted of textile electrodes and a small pocket that would hold the Shimmer devices, shown in Figure 25.7. All participants were asked to wear a modified outer

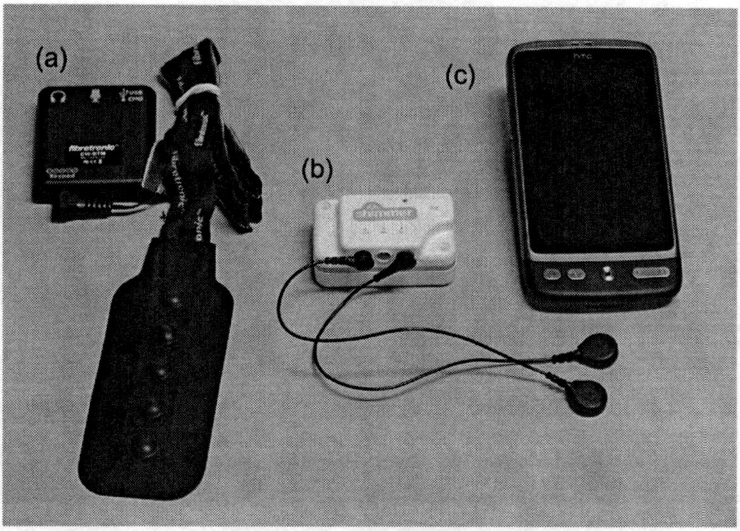

Figure 25.5 Prototype components for evaluation during the Evaluation 2, consisting of (a) Fibretronic BTM Module, (b) Shimmer sensor and (c) smartphone.

(a)

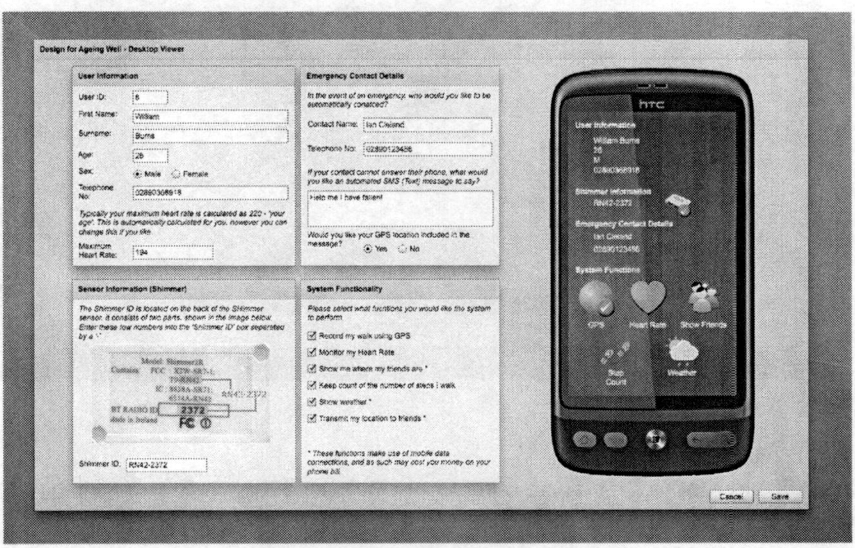

Figure 25.6 User interface of the updated Desktop Viewer showing (a) configuration screen, (b) a walk configuration screen allowing the users to enter milestones and select a walking destination and (c) an interactive map with the user walk route overlaid and health and activity data displayed.

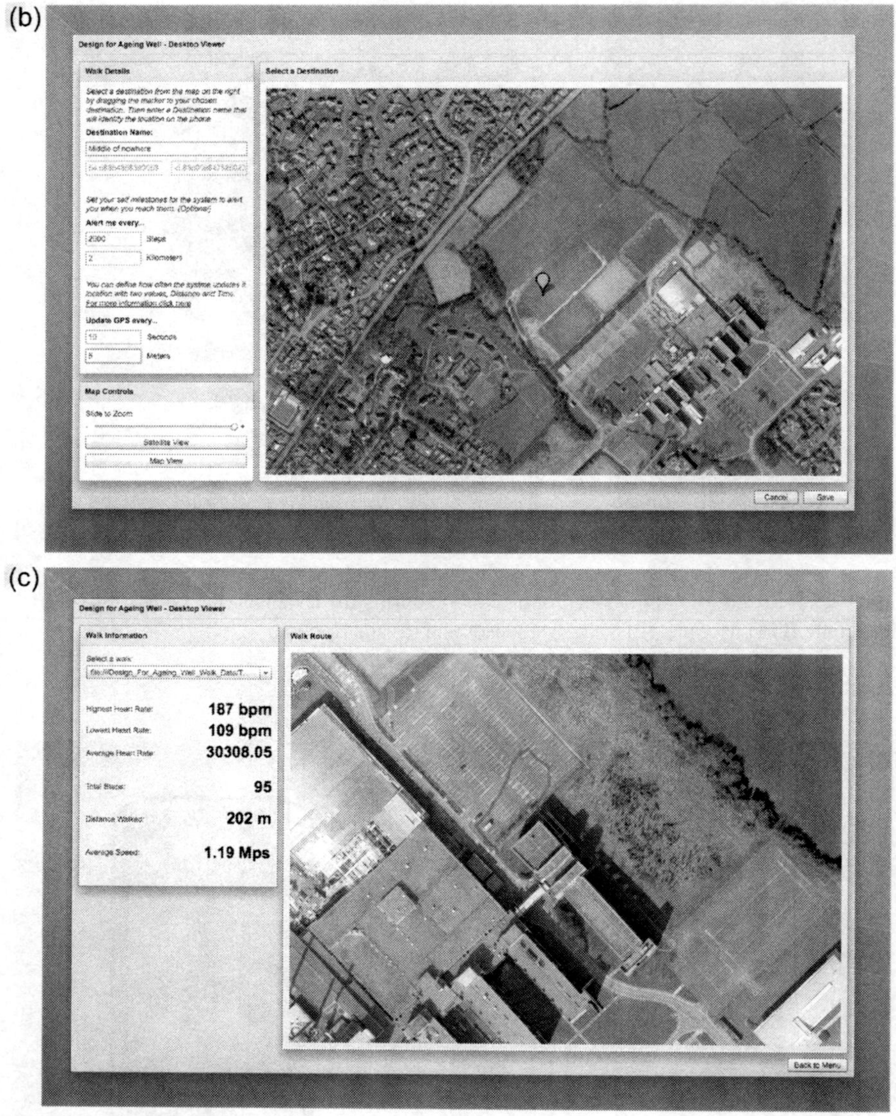

Figure 25.6 Continued.

layer (waterproof jacket). These outer layers included Fibretronic soft switches, headphones and a microphone.

The evaluation consisted of three stages, configuration, activity and review stage.

25.5.2.1 Configuration

During this stage of the evaluation, participants were asked to connect the smartphone to a laptop and enter some information via the Desktop Viewer software. This

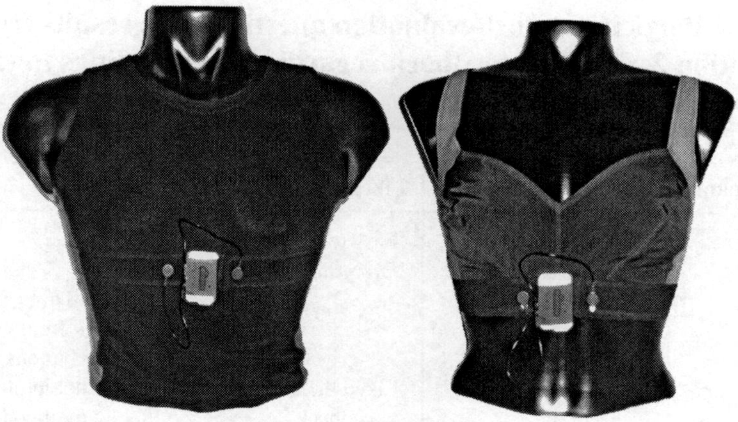

Figure 25.7 Custom-made base layer and bra. Each garment was custom designed for the participants of the evaluation and consists of textile electrodes, male snaps and a pocket to hold the Shimmer sensor.

information was used to personalise the system and consisted of parameters such as name, age, sex and emergency contact details (name and number). The age was used to calculate the maximum heart rate for the user and would allow the system to provide personalised feedback based on this.

Other information entered included the Shimmer sensor ID, used to ensure the smartphone connected to the correct device, destination coordinated, selected from an interactive map, and milestones such as alerts when the user has walked a certain distance or number of steps.

25.5.2.2 Activity

Upon completion of the configuration stage, participants were asked to wear their custom-made base layer (vest/bra), a mid layer garment of their choosing (fleece, shirt, etc.) and finally the modified outer layer. When suitably clothed, participants were asked to turn on the smartphone and start the Design for Ageing Well application. The application connected automatically to their Shimmer sensor. Once set up, participants were asked to perform a walking activity that lasted approximately 30 min.

25.5.2.3 Review

When the participants returned from their walking activity, they were asked to use the Desktop Viewer software to review their recently completed walk. This allowed the participants the opportunity to review the number of steps they had taken and their highest, lowest and average recorded heart rate and the distance travelled. The route they had taken on the walk was also overlaid on an interactive map.

The final part of this stage involved the participants completing a questionnaire asking questions relating to the current technology they have and use, how they use them and the overall usability and functionality of the current system.

Table 25.6 Participant post-evaluation questionnaire results from Evaluation 2, detailing feedback regarding their feelings toward the tested system

Participant	Sex	Like using the system?	Likes	Dislikes
1	M	Yes	'Easy to use, simple'	–
2	M	Yes	'Everything'	–
3	M	Yes	–	'I could not get feedback from using the two buttons'
4	F	Don't know	'Liked the audio feedback'	'Still not intuitive – possibly due to bugs & inconsistent icon – could rely on images rather than text'
5	F	Yes	'It records the walk and could make normal walks more ??/ challenging'	–
6	F	Don't know	'Would like to become more familiar with the system before deciding'	–

Table 25.6 shows the participants' feedback, ascertained from the post-evaluation questionnaires, on the prototype evaluated in this phase. As with Evaluation 1, participants were given free space on their questionnaires to provide any additional comments for feedback:

> *Participant 1: 'Ear communication sometimes comes out'. 'The "Tell my friends where I am" is the least used. Is it worth the "effort" in electronic (??) terms. Other functions much more useful'.*

> *Participant 5: 'I am not happy using headphones – earpieces fall out but I'm assured it would be built into the jacket hood'. 'The technology interests me and I feel it is or will be useful for my age group in particular. Could also give confidence to younger people with health problems and help towards greater independence and levels of fitness'.*

> *Participant 6: 'The system should be as simple as possible to use, and less obvious shimmers'.*

After the initial evaluation was complete, the researcher initiated a conversation with the participants. From these conversations it was noted that all the participants liked using the system in varying ways. Some were considering using it for other activities, i.e. while kayaking or riding their bike.

From observing the participants using the system, a number of bugs were identified:

- The 'Show me where my friends are' function significantly reduced the performance of the Maps screen and overall system
- The on-demand TTS feedback did not work when the user was in the Maps screen
- When pressing the 'Start Walk' button in the Maps screen, if the user exited this screen and re-entered it, the menu option would still display 'Start Walk' despite the walk already being started and recording.

One issue that must be addressed is the connection of the Fibretronic Bluetooth module. This process can be quite difficult for the target cohort. Participants would also have preferred headphones embedded into the hood as opposed to wearing an earpiece.

On one of the walking activities, the researcher accompanied the participant simply to observe how they used the system when outside walking. Over the course of the two days of evaluation, all participants performed the walking activity in bad weather conditions, principally rain, some hail. When walking with the participant, the researcher noticed that the participant was worrying about interacting with the phone due to the weather conditions and damaging the smartphone.

Other issues noted that may have affected the participants' perceived usability of the system was the position in which the participants were holding the smartphone. The participant was trying to activate a function on the smartphone screen; however, their action was not performed. What transpired to be the problem was that the participant was trying to activate the function with their right index finger; however, whilst holding the smartphone in their left hand, their left thumb was touching the notification area of the screen, hence negating the touch event being performed by the right index finger. Other issues similar to this were the use of a fingernail to activate the functions. This was a problem because the screen was capacitive. It is hoped that extended use of the technology would alleviate this issue.

Based on the feedback from the questionnaires, walking activity and discussions, a number of changes to the prototype in terms of aesthetics, functionality and feedback must be changed:

- Ensure TTS feedback is global and can be accessed from any screen
- Fix issues with the slow performance of 'Show me where my friends are' function (increase the time between HTTP requests)
- Have the 'Lock Screen' option available in the Maps screen
- All prompts should be audio/TTS enabled—Display walk start, end times and duration in Desktop Viewer.

25.5.3 Evaluation 3

This evaluation was sandwiched between a pre- and post-evaluation workshop with six participants, three male and three female, between the ages of 60 and 75 years. Participants were recruited during the early stages of the research project and have been involved in the two previous evaluation stages. A 3D body scanner has previously scanned the participants and custom clothing has been designed and manufactured for them. These garments have taken into consideration their feedback on fabrics, design, patterns and fittings (zips, pockets, etc.).

The pre-evaluation workshop lasted one day and consisted of system training, hands-on interaction with the technology, a troubleshooting exercise and walking activity. The system training consisted of a presentation by the researcher outlining the functionality, interface and limitations of the system (primarily focused on the technology component). This also involves training the participants on how to use the technology, such as switching on/off devices (smartphone, sensors, BTM).

During the hands-on interaction with technology session, each participant had each of the technology components in order to interact and experiment. At this point, the participant could follow the presentation on the device if they so wished.

During the troubleshooting session, participants were shown some troubleshooting techniques that would alleviate common problems, i.e. adjusting screen brightness, charging technology and lost connections with Bluetooth devices.

After the previous sessions, the participants were asked to wear the complete clothing system consisting of a base, middle and outer layer and leave the laboratory and perform a walking activity. This activity took approximately 20–30 min. This allowed the participants to interact with the system in the environment and context for which it was designed. If, after all of the previous sessions, there still arose issues that were small enough to fix with additional training, there was time for these to be addressed.

After the pre-evaluation workshop, participants were given a complete clothing system custom made for their size and shape, one of each technology component, software on USB drive, User Manual and pre-credited PAYG SIM card.

The evaluation ran over eight weeks (7th March–31st April, 2012). The smartphone kept an anonymous log of all interaction the participant had with the technology. This information included when the application was started, which buttons were pressed, either on the phone screen, hardware or textile.

Post-evaluation consisted of one-on-one interviews with the participants and the researcher who performed a semi-structured interview to ascertain the participant's opinions, views and thoughts on the system. This process was recorded for transcription.

All participants were asked to contribute to a general discussion on the overall system. This process was moderated and driven by the researchers. At the end of the post-evaluation workshop the participants were asked to complete a final questionnaire to grade the system using a series of Likert scale questions.

Previous evaluations consisted of six participants, three male and three female. One participant dropped out of the final evaluation, and another participant did not receive the clothing in time to evaluate the system for eight weeks. Although she attended this evaluation, because she had not used the system since the last evaluation, her comments have not been included, see Table 25.7.

During a group discussion, it emerged that all participants had some difficulty with the system, specifically turning on the Shimmer sensor and reviewing the recording on the Desktop Viewer. Participants stated that it was difficult to turn the Shimmer on and

Table 25.7 Participant post-evaluation questionnaire results from Evaluation 3, detailing feedback regarding their feelings toward the clothing/technology system after eight weeks of use

Participant	Sex	Like using the system?	Likes	Dislikes
1	M	Yes	'The idea of recording a walk and building up a library of walks on the computer'	'Too complex and not reliable'
2	M	Yes		'I had user and connection problems'
3	M	Yes	'The screen on the phone'	–
4	F	No	'Possibility to record route etc.'	'HTC problematic. Switch on/off worked sometimes, not others. Visibility in bright light. Not reliable'

when it was on it was difficult to tell if the device was on, as the LED was internal and had to be viewed through a very small hole in the device. Another issue was whether or not the Shimmer had a low battery or needed to be charged.

There was also difficulty in using the HTC Desire smartphone. Some users reported difficulty with turning the device on and off. Some found that they had turned the smartphone off only to find the battery was flat when they tried to turn it on a few days later. It transpired that the users were not turning the phone off, but merely putting it to sleep. This meant that the hardware such as GPS, GSM and the Design for Ageing Well application were still on and using additional power.

The aforementioned problems could not be resolved from a development perspective, but would be improved with continued use of the devices. Nevertheless, the developers could resolve some problems, such as the connection problems between the smartphone and the Desktop Viewer, reported by two participants. This was a result of the smartphone not being placed in 'USB Mode' when connected to the user's PC. This issue is resolved by changing the smartphone preferences. However after this problem was raised with the developers, the online help manual was updated with steps to remedy this problem.

Another issue was the cost of some functionality. The 'Show where my friends are' and 'Maps' functions used the data allowance on the phone. This meant that the £5 credit that was on the devices when the participants received the system was very quickly used. The participants were advised, midway through the evaluation period, to remove this function from the system by modifying the configuration setting using the Desktop Viewer.

From an application perspective, the Design for Ageing Well mobile application performed well and reliably in recording walk data. Some inaccuracies were reported with the Heart Rate detection, but this was to be expected based on the use of textile electrodes, and the physical activity the participants were performing. No 'Force Close' errors were reported.

25.6 Discussion

The functionality of the developed smartphone application was evaluated and shown to work as required. From the list of user requirements detailed in Table 25.4, all requirements were delivered upon and worked well. However, the acceptance and use of the functionalities varied by participant. One example of this was the 'Show where my friends are' function. The participants used in the evaluations either walked alone or walked in a group and tended to stay in that group for the duration of the walk. As such, during the longitudinal evaluation phase, this function was universally disabled.

All participants were able and happy to use the smartphone application, with positive feedback on the design of the UI. The application was designed with the target cohort in mind and as such performed a number of tasks automatically, such as connecting to the Shimmer sensor and BTM. Participants were impressed with the ability to request on-demand TTS feedback via the soft switches located on the outer layer of the clothing system.

From a hardware perspective, there were a number of usability issues. The one that was universal throughout the three evaluations was turning the Shimmer device on and off. The sensor was turned on with a paper clip, inserted into a small hole on the device. While there were LED indicators on the device, they were only visible through the small hole. It was difficult for participants to see if the device was on/off.

Another problem that arose throughout the longitudinal evaluation was the connecting of the smartphone to the participant's home computer to transfer their walk data to the Desktop Viewer. The problem seemed to lie with the smartphone. When the participants connected the smartphone to their computer the phone presented them with a dialogue asking how they would like the device connected, i.e. Charge Only, Mount SD or Internet Tethering. Because the participants either missed this dialogue or selected the charge option, the smartphone was not recognised and hence their data were not transferred. Nevertheless, the walk data were still saved on the smartphone's SD card. The participants raised this issue during the final evaluation, and the researcher addressed this issue by sending all participants step-by-step instructions on how to change the smartphone settings.

As a result of the project as a whole, the authors have identified a number of key points that should be taken into consideration in similar future projects.

25.6.1 Male vs female acceptance of technology

It was noted that, despite the problems identified during the various phases of evaluation, the male participants were more accepting of the prototype, with each participant stating they liked the system in all three evaluations. During the longitudinal evaluations, it was the male participants that maintained contact with researchers and asked questions and raised issues such as the connecting of the smartphone to the Desktop Viewer. This may be because the researcher was also male and perhaps the male participants felt more comfortable raising issues.

25.6.2 Hardware usability

Despite the best efforts of the developers and the consideration that went into the design of the developed application UI and UX, there is little that can be done to address hardware usability issues. Issues such as turning on/off the Shimmer sensor would fall into this category. More time should be taken to address these issues. The transmission of data from one device to another (smartphone to Desktop Viewer (PC)) should be simplified. If possible the data transmission should take place in the background via wireless technologies if possible.

Hardware components should be easy to turn on/off and charge if possible. However, on some devices, such as the smartphone, the on/off button requires one press to turn on, but one long press and an on-screen selection to turn the device off. This is confusing to older persons with no experience with smartphones.

25.6.3 Cost

Cost is a factor to the target cohort, and they did not understand how they were using money on a smartphone without making a call, i.e. 3G data to download maps and transmit location data. Once the participants used the allocated amount of money put on the smartphones before the final evaluations, all but one decided against adding more money. As a result those participants used the Maps screen without a map being displayed. Nevertheless, the walk data (GPS, heart rate and step count) were still recorded.

25.6.4 Technology maintenance

It was noticed during the evaluation phases that when the participants were finished using all of the technology components they felt the need to turn them off completely. Despite the researchers stating that the device can be charged and switched on at the same time, participants continued to turn devices off completely when not in use. Participants turned off the smartphone and Shimmer, which can be put into a sleeping mode, before charging.

25.6.5 Feature fatigue

Despite following a user-centred design approach and eliciting user requirements from a number of potential users, researchers need to take careful consideration before adding all requirements to the proposed solution. After evaluating the system with the participants, some features, despite being included throughout the duration of the prototype development, were not required by the participants. The developed solution addressed all of the requirements elicited, but on the evaluated cohort, the system offered too much. Despite the ability to personalise the solution, feature fatigue occurred.

25.6.6 User involvement

The researchers were more than happy to interact with the participants on a number of occasions. It allowed the rapid collection of feedback for prototype development. Nevertheless, a rapport between the researchers and participants grew throughout the project, during the large number of workshops. This may have resulted in skewed feedback from participants, as they did not wish to offend the developers who were also present at all of the evaluations.

25.7 Conclusions

In this chapter we have presented our methodology for the design, development and evaluation of a smart garment for active ageing walkers. We have shown the three-stage evaluation methodology and the re-development of the prototype and subsequent user feedback on all three evaluations.

Overall, the technology solution developed was accepted and seen as beneficial to the participants who used the system. However, personalisation of the solution should be encouraged. Not all of the participants wanted or used the same functionality.

In addition to the methodology and results of the evaluations of the prototype, a number of methodological issues were identified and should be considered in future projects developing technology solutions for the active ageing.

Acknowledgements

This research was funded by the ESRC under the UK's joint research council's New Dynamics of Ageing Program (http://www.newdynamics.group.shef.ac.uk/).

References

Boulos, M. N. K., Wheeler, S., Tavares, C., & Jones, R. (January 2011). How smartphones are changing the face of mobile and participatory healthcare: An overview, with example from eCAALYX. *Biomedical Engineering Online*, *10*(1), 24.

D3O Impact Protection. [Online]. Available: http://www.d3o.com/ Accessed 27.07.13.

Design for Ageing Well Project. [Online]. Available: http://newdynamics.group.shef.ac.uk/design-for-ageing.html Accessed 20.07.13.

Fibretronic ConnectedWear. [Online]. Available: http://fibretronic.com/connectedwear Accessed 28.07.13.

Fibretronic, HEATwear. [Online]. Available: http://fibretronic.com/products/heatwear Accessed 27.08.13.

Garmin Forerunner. [Online]. Available: http://www.garmin.com/uk/forerunner10 Accessed 28.07.13.

Hogan, M. (January 2005). Physical and cognitive activity and exercise for older adults: A review. *International Journal of Aging and Human Development*, *60*(2), 95–126.

Magenes, G., Curone, D., Caldani, L., & Secco, E. L. (January 2010). Fire fighters and rescuers monitoring through wearable sensors: The ProeTEX project. *Conference Proceedings: Annual International Conference of the IEEE Engineering in Medicine and Biology Society. IEEE Engineering in Medicine and Biology Society. Conference*, *2010*, 3594–3597.

McCann, J. (2009). Design for ageing well: Improving the quality of life for the ageing population using a technology enabled garment system. *Advances in Science and Technology*, *60*, 154–163.

Meriggi, P., & Castiglioni, P. (2010). Polysomnography in extreme environments: The MagIC wearable system for monitoring climbers at very-high altitude on Mt. Everest slopes. *Cardiology*, *2010*, 1087–1090.

Mermoth (Medical Remote Monitoring of Clothes). [Online]. Available: http://cordis.europa.eu/search/index.cfm?fuseaction=proj.document&PJ_RCN=7089127 Accessed 29.07.13.

MIThril. [Online]. Available: http://www.media.mit.edu/wearables/mithril/ Accessed 29.07.13.

Moore-Harrison, T. L., Speer, E. M., Johnson, F. T., & Cress, E. M. (2008). The effects of aerobic training and nutrition education on functional performance in low socioeconomic older adults. *Journal of Geriatric Physical Therapy*, *31*(1).

Nike FuelBand. [Online]. Available: http://www.nike.com/us/en_us/c/nikeplus-fuelband Accessed 28.07.13.

Nike+. [Online]. Available: http://nikeplus.nike.com/plus/ Accessed 28.07.13.

Paradiso, R., Loriga, G., & Taccini, N. (2005). Wealthy, a wearable healthcare system: New frontier on e-textile. *Journal of Telecommunications and Information Technology*, *4*, 105–113.

Shimmer Research. [Online]. Available: http://www.shimmersensing.com/ Accessed 28.07.13.

Sony Ericsson LiveView. [Online]. Available: http://www.sonymobile.com/global-en/products/accessories/liveview/ Accessed 28.07.13.

University of the Third Age (U3A). [Online]. Available: http://www.u3a.org.uk/ Accessed 27.07.13.

Vergados, D., Alevizos, A., Mariolis, A., & Caragiozidis, M. (2008). Intelligent services for assisting independent living of elderly people at home. In *Proceedings of the 1st international conference on pervasive technologies related to assistive environments* (pp. 79:1–79:4).

Villalba, E., & Salvi, D. (2009). Wearable and mobile system to manage remotely heart failure. *Biomedicine, IEEE*, *13*(6), 990–996.

World Health Organization (WHO).(1948). *Preamble to the Constitution of the World Health Organization as adopted by the international health conference*. New York.

Index

Note: Page numbers followed by f indicate figures; t, tables.

I

J

K

Q

R

S

X

Y

Z

Lightning Source UK Ltd.
Milton Keynes UK
UKOW05n0747240914

239093UK00014B/265/P

9 780857 095381